THE RIBBON OF GEOLOGIC TIME SHOWS THE COMPLETE GEOLOGIC TIME SCALE.

THE ESSENTIAL EARTH

THE ESSENTIAL EARTH

Thomas H. Jordan
University of Southern California

John Grotzinger
California Institute of Technology

W. H. FREEMAN AND COMPANY
New York

Publisher: Clancy Marshall
Acquisitions Editor: Valerie Raymond
Developmental Editor: Randi Rossignol
Senior Marketing Manager: Scott Guile
Assistant Editor: Whitney Clench
Media and Supplements Editor: Amy Thorne
Photo Editor: Ted Szczepanski
Photo Researcher: Rae Grant
Cover and Text Designer: Blake Logan
Senior Project Editor: Mary Louise Byrd
Illustrations: Precision Graphics
Illustration Coordinator: Bill Page
Production Coordinator: Julia DeRosa
Composition: Sheridan Sellers, W. H. Freeman and Company, Electronic Publishing Center
Printing and Binding: RR Donnelley

Library of Congress Control Number: 2007940625

ISBN-13: 978-1-4292-0429-3 (paperback); 978-1-4292-1824-5 (hardcover)
ISBN-10: 1-4292-0429-X (paperback); 1-4292-1824-X (hardcover)

©2008 by W. H. Freeman and Company
All rights reserved
Printed in the United States of America

Second printing

W. H. Freeman and Company
41 Madison Avenue, New York, NY 10010
Houndmills, Basingstoke RG21 6XS, England

www.whfreeman.com

We dedicate this book to Frank Press and Ray Siever, pioneering educators in the era of modern geology. This book was possible only because they led the way.

MEET THE AUTHORS

TOM JORDAN is a geophysicist interested in the composition, dynamics, and evolution of the solid Earth. He has conducted research into the nature of deep subduction, the formation of thickened keels beneath the ancient continental cratons, and the question of mantle stratification. He has developed a number of seismological techniques for investigating Earth's interior that bear on geodynamic problems. He has also worked on modeling plate movements, measuring tectonic deformation, quantifying seafloor morphology, and characterizing large earthquakes. He received his Ph.D. in geophysics and applied mathematics at the California Institute of Technology (Caltech) in 1972 and taught at Princeton University and the Scripps Institution of Oceanography before joining the Massachusetts Institute of Technology (MIT) faculty as the Robert R. Shrock Professor of Earth and Planetary Sciences in 1984. He served as the head of MIT's Department of Earth, Atmospheric and Planetary Sciences for the decade 1988–1998. He moved from MIT to the University of Southern California (USC) in 2000, where he is University Professor, W. M. Keck Professor of Earth Sciences, and Director of the Southern California Earthquake Center.

Dr. Jordan received the Macelwane Medal of the American Geophysical Union in 1983, the Woollard Award of the Geological Society of America in 1998, and the Lehmann Medal of the American Geophysical Union in 2005. He is a member of the American Academy of Arts and Sciences, the U.S. National Academy of Sciences, and the American Philosophical Society.

JOHN GROTZINGER is a field geologist interested in the evolution of Earth's surface environments and biosphere. His research addresses the chemical development of the early oceans and atmosphere, the environmental context of early animal evolution, and the geologic factors that regulate sedimentary basins. He has contributed to developing the basic geologic framework of a number of sedimentary basins and orogenic belts in northwestern Canada, northern Siberia, southern Africa, and the western United States. He received a B.S. in geoscience from Hobart College in 1979, an M.S. in geology from the University of Montana in 1981, and a Ph.D. in geology from Virginia Polytechnic Institute and State University in 1985. He spent three years as a research scientist at the Lamont-Doherty Geological Observatory before joining the MIT faculty in 1988. From 1979 to 1990, he was engaged in regional mapping for the Geological Survey of Canada. He currently works as a geologist on the Mars Exploration Rover team, the first mission to conduct ground-based exploration of the bedrock geology of another planet, which has resulted in the discovery of sedimentary rocks formed in aqueous depositional environments.

In 1998, Dr. Grotzinger was named the Waldemar Lindgren Distinguished Scholar at MIT, and in 2000, he became the Robert R. Schrock Professor of Earth and Planetary Sciences. In 2005, he moved from MIT to Caltech, where he is the Fletcher Jones Professor of Geology. He received the Presidential Young Investigator Award of the National Science Foundation in 1990, the Donath Medal of the Geological Society of America in 1992, and the Henno Martin Medal of the Geological Society of Namibia in 2001. He is a member of the American Academy of Arts and Sciences and the U.S. National Academy of Sciences.

BRIEF CONTENTS

CHAPTER 1	WHY WE STUDY EARTH		1
CHAPTER 2	HOW WE STUDY EARTH		31
CHAPTER 3	PLATE TECTONICS: The Unifying Theory of Geology		51
CHAPTER 4	EARTH MATERIALS: Minerals and Rocks		75
CHAPTER 5	IGNEOUS PROCESSES AND VOLCANISM		107
CHAPTER 6	SEDIMENTATION: Rocks Formed by Surface Processes		143
CHAPTER 7	DEFORMATION AND METAMORPHISM		171
CHAPTER 8	TIMING THE GEOLOGIC RECORD		199
CHAPTER 9	HISTORY OF EARTH		215
CHAPTER 10	THE CLIMATE SYSTEM AND GLACIATION		251
CHAPTER 11	THE HYDROLOGIC CYCLE AND GROUNDWATER		281
CHAPTER 12	SHAPING EARTH'S SURFACE: Streams, Coastlines, and Wind		305
CHAPTER 13	EARTHQUAKES AND EARTH STRUCTURE		343
CHAPTER 14	HUMAN IMPACT ON EARTH'S ENVIRONMENT		379

CONTENTS

Meet the Authors — vi
Preface — xiii
Acknowledgments — xviii

CHAPTER 1 WHY WE STUDY EARTH — 1
Greetings from Los Angeles! — 1
Spaceship Earth — 3
Natural Resources — 4
Natural Hazards — 16
The Environment and Global Change — 23
Our Reasons for Studying Earth — 28

CHAPTER 2 HOW WE STUDY EARTH — 31
Measuring Mount Everest — 31
The Scientific Method — 32
Geology as a Science — 33
Earth's Shape and Surface — 36
Peeling the Onion: Discovery of a Layered Earth — 38
Earth as a System of Interacting Components — 42
An Overview of Geologic Time — 46

CHAPTER 3 PLATE TECTONICS: The Unifying Theory of Geology — 51
Geology at Sea — 51
Discovery of Plate Tectonics — 52
Types of Plate Boundaries — 56
Rates and History of Plate Movements — 63
The Grand Reconstruction — 67
Mantle Convection: The Engine of Plate Tectonics — 72

CHAPTER 4 EARTH MATERIALS: Minerals and Rocks — 75
The Diamond Invention — 75
What Are Minerals? — 76
The Structure of Matter — 77
The Formation of Minerals — 80
Rock-Forming Minerals — 83
Physical Properties of Minerals — 87
What Are Rocks? — 93
Igneous Rocks — 94
Sedimentary Rocks — 95
Metamorphic Rocks — 97
The Rock Cycle: Interactions Between the Plate Tectonic and Climate Systems — 98
Concentrations of Valuable Mineral Resources — 100

CHAPTER 5 IGNEOUS PROCESSES AND VOLCANISM — 107
Mount St. Helens: Dangerous but Predictable — 107
How Do Igneous Rocks Differ from One Another? — 109
How Do Magmas Form? — 115
Magmatic Differentiation — 117
Forms of Igneous Intrusions — 119
Magmatic Processes: Mid-Ocean Ridges, Subduction Zones, and Mantle Plumes — 122
Volcanic Processes and Materials — 126
Volcanoes Build Landforms — 133
Hazards Created by Volcanoes — 136

CHAPTER 6 SEDIMENTATION: Rocks Formed by Surface Processes — 143
Black Gold — 143
Weathering and Surface Processes of the Rock Cycle — 145
Sedimentary Basins: The Sinks for Sediments — 151
Sedimentary Environments — 152
Sedimentary Structures — 156

Burial and Diagenesis: From Sediment to Rock	159
Classification of Siliciclastic Sediments and Sedimentary Rocks	160
Classification of Chemical and Biological Sediments and Sedimentary Rocks	163

CHAPTER 7 DEFORMATION AND METAMORPHISM 171

The Inspiration of Marble	171
Mapping Geologic Structure	173
How Rocks Deform	175
Basic Deformation Structures	176
Styles of Continental Deformation	181
Causes of Metamorphism	185
Types of Metamorphism	187
Metamorphic Textures	188
Regional Metamorphism and Metamorphic Grade	192
Plate Tectonics and Metamorphism	195

CHAPTER 8 TIMING THE GEOLOGIC RECORD 199

The Abyss of Time	199
Reconstructing Geologic History from the Stratigraphic Record	201
The Geologic Time Scale: Relative Ages	206
Measuring Absolute Time with Radioactive Clocks	209
The Geologic Time Scale: Absolute Ages	211

CHAPTER 9 HISTORY OF EARTH 215

Geologists on the Moon	215
Origin of the Solar System	216
Early Earth: Formation of a Layered Planet	218
Diversity of the Planets	221
Evolution of the Continents	228
Biological Events in Earth's History	239

CHAPTER 10 THE CLIMATE SYSTEM AND GLACIATION 251

The Ice Hunter	251
Components of the Climate System	252
The Greenhouse Effect	260
Climate Variation	262
Glacial Landscapes	267
Pleistocene Glaciations	272
The Record of Ancient Glaciations	276

CHAPTER 11 THE HYDROLOGIC CYCLE AND GROUNDWATER 281

Drought in the Southwest	281
Flows and Reservoirs	282
Hydrology and Climate	285
Groundwater	290
Human Effects on Groundwater Resources	296
Erosion by Groundwater	297
Water Quality	299
Water Deep in the Crust	301

CHAPTER 12 SHAPING EARTH'S SURFACE: Streams, Coastlines, and Wind 305

Shipwrecks: Consequences of Coastal Geology	305
Earth's Surface Is Shaped by Currents	306
The Flow of Water and Wind	308
Sediment Transport and Deposition	310
Streams: Transport from Mountains to Oceans	313
Coastal Processes: Waves, Tides, and Hurricanes	325
The Shaping of Shorelines	331
Offshore: The Continental Shelf and Beyond	335
The Desert Environment	336

CHAPTER 13 EARTHQUAKES AND EARTH STRUCTURE 343

The Great Alaska Earthquake	343
What Is an Earthquake?	344
Measuring Earthquakes Using Seismic Waves	349
Exploring Earth's Interior Using Seismic Waves	356
Earthquakes and Patterns of Faulting	363
Earthquake Destructiveness	366
Can Earthquakes Be Predicted?	375

CHAPTER 14 HUMAN IMPACT ON EARTH'S ENVIRONMENT 379

The Carbon Crisis	379
The Carbon Economy	380
Fossil-Fuel Resources	382
Alternative Energy Resources	390
The Carbon Cycle	395
Global Climate Change	400
Effects of Climate Change	403
Earth System Engineering and Management	409

Appendix 1: Conversion Factors	AP-1
Appendix 2: Numerical Data Pertaining to Earth	AP-2
Appendix 3: Chemical Reactions	AP-3
Appendix 4: Properties of the Most Common Minerals of Earth's Crust	AP-4
Glossary	GL-1
Index	I-1

PREFACE

A NOTE FROM THE AUTHORS

Our intention in writing *The Essential Earth* is to offer a well-focused introduction to the basics of physical geology that emphasizes the process of science, the societal impact of the field, and topics that inherently grab students' attention. While *The Essential Earth* is an offspring of *Understanding Earth*, it is a wholly re-envisioned book, written specifically for instructors and students who want a scientifically sound text that a nonscientific audience will embrace. We have drawn on our teaching experiences and our involvement in the scientific community to offer what we hope is the best of both worlds, a text that is accessible but does not compromise the science.

FOCUS ON ESSENTIALS

In 14 chapters, *The Essential Earth* covers the fundamental concepts of physical geology and closely related subjects such as climate science. We introduce plate tectonics right up-front in Chapter 3, so that we can use its basic ideas throughout our development of physical geology. We introduce the student to Earth's climate in Chapter 10, before we tackle the various subjects of surface processes in Chapters 11 and 12. To create a textbook that is manageable in depth and breadth, **we have integrated topics that are bound together by common geologic processes.** Thus, igneous rocks and volcanoes are discussed together in Chapter 5, "Igneous Processes and Volcanism."

Similarly, Chapter 7, "Deformation and Metamorphism," examines the basic processes that deform and metamorphose continental crust. The integration of these topics will help students see that deformation usually accompanies metamorphism and that both are intimately associated with plate tectonic processes.

Chapter 9, "History of Earth," weaves together several topics that provide context for an understanding of deep time. These topics include the origin of the solar system, the formation of Earth, and how Earth differs from other planets. The chapter also tells the remarkable story of how continents were formed and how they have changed over geologic time. It ends with a view of biological development across the span of Earth history, showing how organisms and their environment have interacted, sometimes in disastrous ways.

Chapter 10, "The Climate System and Glaciation," examines the geologic record of past climate changes, particularly the ice ages that sent vast glaciers across the northern continents, and how these changes have modified the landscape. Many spectacular features of the landscapes we see today show the student why glaciation is the most important process of the climate system.

Chapter 12, "Shaping Earth's Surface: Streams, Coastlines, and Wind," discusses how water and air currents erode rock and transport sediments. The chapter addresses questions such as, Why do deserts form? How do streams form, and why do they flow to the edges of continents? Why does the coastline have such a distinctive shape? This diverse list of topics has a common thread: moving currents of air and water.

FOCUS ON THE INTERACTIONS OF GEOLOGY AND SOCIETY

Chapter 1, "Why We Study Earth," launches our discussion of geology's role in solving some of the great problems faced by civilization, from developing natural resources and dealing with natural disasters to preserving our environment and adapting to global change.

A large part of Chapter 1 is devoted to society's need for sustainable **natural resources**—energy, water, and raw materials. Chapter 4, "Earth Materials: Minerals and Rocks," expands the discussion of minerals in rocks as valuable natural resources. Chapter 11, "The Hydrologic Cycle and Groundwater," shows how geologic processes determine water supplies. Chapter 14, "Human Impact on Earth's Environment," discusses the formation and depletion of our energy resources and the effects of our use of those resources on global climate change.

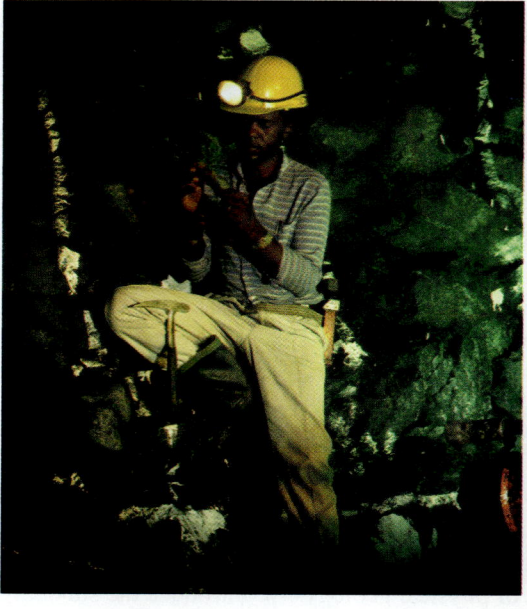

FIGURE 4.29 A geologist examines rock samples in an underground gold mine in Zimbabwe, southern Africa. [Peter Bowater/Photo Researchers.]

Natural hazards illuminate the interaction of geology and society in an exciting and visible way. Chapter 1 introduces the culprits as well as the practical tools Earth science gives us to characterize natural hazards and evaluate their risks to society. In later chapters, we use these hazards to illustrate key geologic processes: volcanic eruptions in Chapter 5, bolide impacts in Chapter 9, severe storms and floods in Chapter 12, and earthquakes in Chapter 13.

Environmental issues are discussed in the context of how human activities have affected, both negatively and positively, the geologic processes that shape our physical environment.

- Chapter 5 describes the hazards created by volcanoes, how people are affected by events such as flank collapse, and how we can predict eruptions.

- Chapter 9 shows how mass extinctions are related to bolide impacts, volcanism, climate warming, and other global environmental crises.

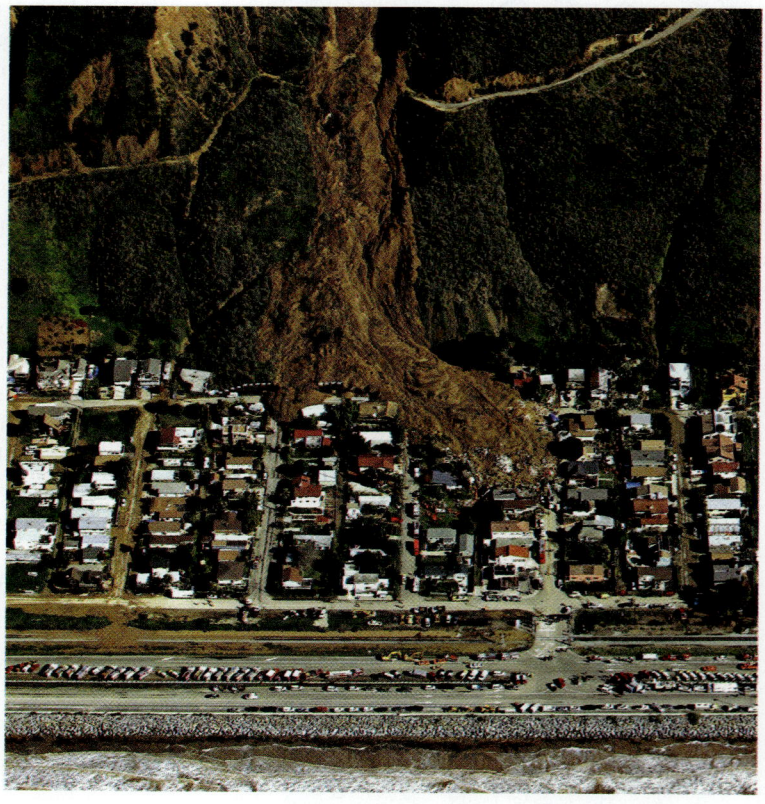

FIGURE 1.31 A massive mudslide buried homes in La Conchita, California, in 2005. [AP Photo/Kevork Djansezian.]

- Chapter 10 describes the greenhouse effect and the effects of human activities on global temperatures.

- Chapter 11 explores the effects of sustained drought on agriculture, and the effects of human activities on our precious and increasingly limited water resources.

- Chapter 12 provides numerous examples of how river flooding, tsunamis, and hurricanes affect human society. The role of human activities in desertification is also discussed.

- Chapter 13 describes damage caused by earthquakes and ways of predicting and lessening the risks of earthquakes.

- Chapter 14 uses Earth's carbon cycle to correct two of society's most pressing issues: the depletion of petroleum as an energy source and the potential for anthropogenic climate change.

FIGURE 14.14 Aerial view of the north entrance to the Yucca Mountain Nuclear Waste Repository being developed at the Nevada Test Site, north of Las Vegas. Yucca Mountain is the ridge to the right of the entrance. [U.S. Department of Energy.]

Chapter opening stories illustrate how humans have affected geology and how geology has affected humans. From the De Beers diamond marketing blitz (Chapter 4) to the beginning of the oil drilling industry (Chapter 6) to drought in the southwestern United States (Chapter 11), these stories touch on economic, social, and political aspects of geology. They also show geologists as explorers who trek into remote jungles (Chapter 2), survey the ocean floor in rough seas (Chapter 3), blast into the sky on space missions (Chapter 9), and study nature in the harshest environments (Chapter 10).

PLATE TECTONICS: The Unifying Theory of Geology

CHAPTER 3

Geology at Sea

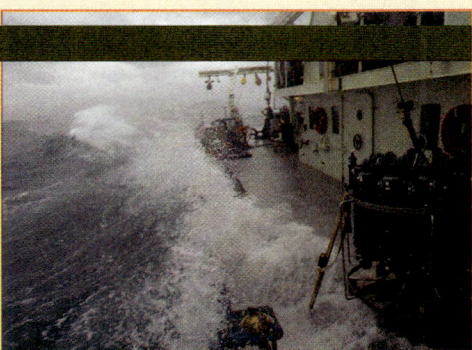

An oceanographic vessel, like most working ships, does not provide posh accommodations. As the lyrics of Jimmy Buffett's song "Landfall" describe it, "you live half your life in an eight-by-five room, just cruisin' to the sound of the big diesel boom." During the other half, you stand watch in the lab, attending to the hull-mounted sonar array and magnetometers hung over the side to map the seafloor as the ship steams along, or out on deck, washed by spray, using thick steel cables to lower instruments into the water or dredge rocks from craggy outcrops several kilometers below the rough sea surface. Sometimes you light the fuse on a kilogram or two of high explosives and toss it off the fantail, making a bang big enough to send sound waves deep into the crust, which then bounce back and tell you something interesting about the properties of the submerged rocks.

Marine geology is a tedious and sometimes dangerous business. Maurice "Doc" Ewing, who was responsible for collecting much of the data that led to the discoveries highlighted in this chapter, was once washed overboard with two mates and his brother John, also a marine geologist, by a freak wave in a storm north of Bermuda. Skillful maneuvering by the ship's captain saved the brothers and one mate from drowning, but Doc Ewing walked with a limp for the rest of his life. Bruce Heezen, another marine geologist important to our story, died aboard a research submarine while exploring the Mid-Atlantic Ridge southwest of Iceland.

Seventy percent of Earth's solid surface is hidden beneath the oceans. The first global oceanographic expedition sailed on the HMS *Challenger* (namesake of the space shuttle) from Portsmouth, England, in 1872. During its 3-year cruise, scientists discovered the symmetrical undersea mountains of the Mid-Atlantic Ridge and the Challenger Deep of the Marianas Trench, the deepest spot in all the world's oceans. Nearly a century of further exploration and many untold adventures aboard ships at sea were required to explain these peculiar features and reveal the secrets of plate tectonics—the grand, unifying theory of geology.

The research vessel *Oceanus*, owned by the National Science Foundation and operated by the Woods Hole Oceanographic Institution, battles the high seas and rough weather of the North Atlantic. [Chris Linder, Woods Hole Oceanographic Institution.]

FOCUS ON WHAT GEOLOGISTS DO AND HOW THEY DO IT

Chapter 2, "How We Study Earth," was written with the hope that all students will come away from this course with an appreciation for **how the scientific process works** and **what geologists do.** This chapter starts with the scientific method and shows how scientists gather and interpret information about our planet. It illustrates how the scientific method has been applied to discover some of Earth's most basic features: its shape and internal layering. Photos of geologists at work help to show students that geology is a "hands-on" science with its own particular style and outlook. They will learn how scientists combine field studies of Earth processes in the natural environments with controlled laboratory experiments and computer simulations to make major discoveries about the Earth system.

FIGURE 2.6 Sensitive instruments like this mass spectrometer can determine the age of a rock by measuring minute amounts of radioactive elements and how that radioactivity has decayed since the rock first formed. [Rosenfeld Images Ltd./Photo Researchers.]

FOCUS ON GEOLOGY AS A VISUAL SCIENCE

There is probably no science more visually compelling than geology. We hope to convey that to students by liberally illustrating the text with carefully developed figures and carefully selected photos of geologic structures, geologic processes, and geologists doing geology.

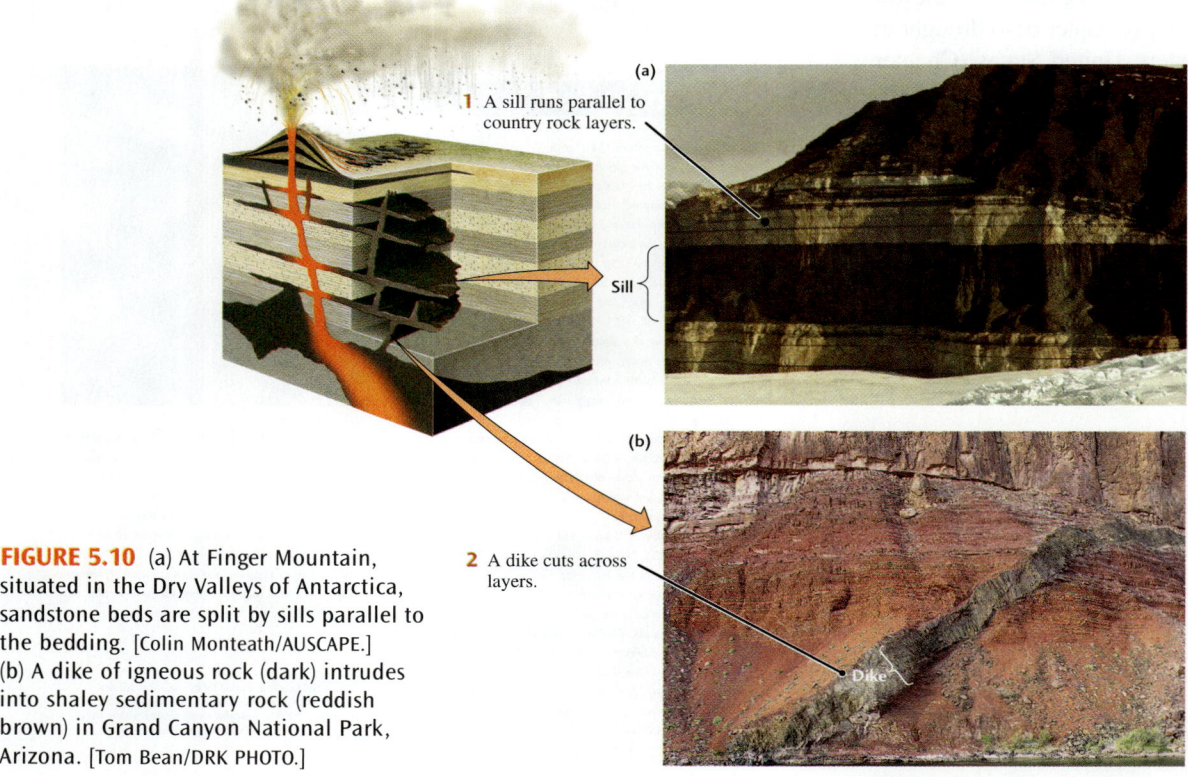

FIGURE 5.10 (a) At Finger Mountain, situated in the Dry Valleys of Antarctica, sandstone beds are split by sills parallel to the bedding. [Colin Monteath/AUSCAPE.]
(b) A dike of igneous rock (dark) intrudes into shaley sedimentary rock (reddish brown) in Grand Canyon National Park, Arizona. [Tom Bean/DRK PHOTO.]

MEDIA AND SUPPLEMENTS

The Essential Earth eBook

http://ebooks.bfwpub.com/essentialearth1e
This online version of *The Essential Earth* combines the text, all existing student media resources, and additional study features, including highlighting, bookmarking, note taking, Google-style searching, in-text definitions of terms, and outlinks to Google. The eBook is available both as part of the GeologyPortal (see below) and as a stand-alone product. Visit http://ebooks.bfwpub.com to preview a sample chapter.

GeologyPortal for *The Essential Earth*

http://portals.bfwpub.com/essentialearth1e
GeologyPortal, W. H. Freeman's nationally hosted learning management solution, is available in a version created specifically to accompany *The Essential Earth*. It incorporates *The Essential Earth* eBook, all student media resources, and powerful learning management capabilities to form a rich, fully integrated online learning environment. The ready-to-use course template is fully customizable. The Portal contains the following features:

- Interactive eBook
- Personalized study plan
- Self-graded, editable quizzes and homework assignments
- Expeditions video clips—exclusive video of Earth in action, filmed by Dr. Jerry Magloughlin of Colorado State University
- Video exercises, based on the Expeditions series, which can be assigned to students and reported to the course gradebook. The exercise questions include multiple-choice, matching, and short-answer questions
- Geology in the News Videos—brief videos from contemporary news highlighting the societal impact of geologic processes
- Animations of key concepts
- Key term flashcards
- Assignments and resources can easily be added via a simple, user-friendly process. For a preview of GeologyPortal, visit http://portals.bfwpub.com/essentialearth1e and click on "Preview GeologyPortal as a Student."

Companion Web Site

The Companion Web Site, at www.whfreeman.com/essentialearth, contains quizzes and key term flashcards to help students prepare for exams. For instructors, the Web site contains the same JPEGs and PowerPoint presentations available on the Instructor's Resource CD-ROM (see below).

Computerized Test Bank CD-ROM

The computerized test bank CD-ROM (with Windows and Mac versions on one disk) offers hundreds of multiple-choice questions. The CD format makes it easy to select, add, edit, and resequence questions, as well as to deliver and grade assignments and exams.

Instructor's Resource CD-ROM

To help instructors create their own Web sites and orchestrate dynamic lectures, this CD-ROM offers all art from the text in PowerPoint and JPEG formats, along with animations of key concepts.

Expeditions in Geology DVD, Volumes 1 and 2
Available to adopters, these two DVDs contain 25 brief videos that explore geological phenomena. These 3- to 7-minute videos are filmed and narrated by Dr. Jerry Magloughlin of Colorado State University.

Online Course Materials (WebCT, Blackboard)
As a service for adopters, we will provide content files in the appropriate online course format, including instructor and student resources for this text.

ACKNOWLEDGMENTS

We have been very fortunate to collaborate with some exceptionally talented people at W. H. Freeman. Since our first involvement with the third edition of *Understanding Earth*, which began in 2001, we have worked closely with Randi Rossignol as our developmental editor and Valerie Raymond as our acquisitions editor. We really appreciate their encouragement and sustained efforts in producing this new textbook. We also wish to thank our valiant senior project editor, Mary Louise Byrd; our outstanding copy editor, Norma Sims Roche; and our resourceful photo editors, Ted Szczepanski and Rae Grant, as well as other capable members of Freeman's staff. In particular, with deadlines looming, Sheridan Sellers has shouldered her heavy responsibilities for book composition with considerable grace and cheer, and Julia DeRosa has been our extremely competent production coordinator.

It is a challenge to both geology instructors and authors of geology textbooks to make decisions about what topics to teach, and at what level, and at the same time to inspire interest and enthusiasm in our students. To meet this challenge, we have called on the advice of many colleagues to guide us as we wrote and rewrote the chapters. We have relied on this consensus of views to shape *The Essential Earth*. Our sincerest thanks go to our colleagues named here.

Stephen Altaner, *University of Illinois at Urbana-Champaign*
Cathy Baker, *Arkansas Tech University*
Solweig Balzer, *University of Alberta*
E. Erik Bender, *Orange Coast College*
Larry Benninger, *University of North Carolina*
J Bret Bennington, *Hofstra University*
Barbara L. Brande, *University of Montevallo*
Susan H. Butts, *Southern Connecticut State University*
Wang-Ping Chen, *University of Illinois at Urbana-Champaign*
Kelly Dilliard, *University of South Dakota*
Grenville Draper, *Florida International University*
Aley El-Shazly, *Marshall University*
Stewart S. Farrar, *Eastern Kentucky University*
P. Jay Fleisher, *SUNY College at Oneonta*
Tracy Furutani, *North Seattle Community College*
Doug Haywick, *University of South Alabama*
Christi Hill, *Grossmont College*
Tami J. Jovanelly, *Berry College*
Nazrul I. Khandaker, *York College of CUNY*
Brian E. Lock, *University of Louisiana at Lafayette*
Ryan Mathur, *Juniata College*
Katherine Milla, *Florida A&M University*
Otto H. Muller, *Alfred University*
Alfred H. Pekarek, *St. Cloud State University*
Michael S. Petronis, *New Mexico Highlands University*
Emma C. Rainforth, *Ramapo College of New Jersey*
Brady P. Rhodes, *California State University, Fullerton*
Joseph M. Smoak, *University of South Florida*
Leif Tapanila, *Idaho State University*
Stephen Taylor, *Western Oregon University*
Paul Umhoefer, *Northern Arizona University*
Marilyn Velinsky Rands, *Lawrence Tech University*
Gary S. Zumwalt, *Louisiana Tech University*

WHY WE STUDY EARTH

CHAPTER 1

Greetings from Los Angeles!

Our names are Tom Jordan and John Grotzinger, and we are the authors of this textbook. We're geologists who have studied Earth from top to bottom and from the inside out. Throughout our professional lives—for seven decades between us—we have traveled around the world in the search for answers to fascinating scientific questions. We've enjoyed many adventures along the way, on lush islands of the South Pacific, in the wilds of Africa, and across the bleak tundra of the Canadian Arctic.

We now reside and teach in metropolitan Los Angeles, where civilization confronts the natural world on an astounding scale. In Southern California, we can see evidence of nature's bounty all around us, and we sometimes feel its wrath. LA was built on a fertile coastal plain underlain by rich reservoirs of life-sustaining water and energy-producing oil. The mighty geologic forces that formed those natural resources are also pushing rocks upward, creating a mountainous breeding ground for natural disasters. In the ranges that rise above us, we can see the scars of earthquakes, wildfires, landslides, and floods.

We view this world from a geological perspective, continually amazed by what we observe: an active planet with a turbulent, glowing-hot interior; an ever-changing surface of oceans and continents; a thin layer between Earth and sky teeming with life; and a human society altering this environment on a global scale. We have written this textbook because we believe our future as a civilization depends on how well we all understand Earth and its challenges.

We want to give you, the student, new eyes to look at the world around you. We also want to describe the scientific methods and tools that you, the next generation, will need to face the challenges. ■

Geologic forces have pushed the San Gabriel Mountains to 3000 m above Los Angeles. Rocks have been brought down by earthquakes, landslides, and floods to form a fertile coastal plain that contains rich natural resources. [Richard Price/Getty Images.]

CHAPTER OUTLINE

Spaceship Earth 3

Natural Resources 4

Natural Hazards 16

The Environment and Global Change 23

Our Reasons for Studying Earth 28

In this opening chapter, we will show you why Earth science is so essential to so many urgent human needs, from developing natural resources and dealing with natural disasters to preserving our environment and adapting to global change. We will also introduce you to the geological quest for fundamental knowledge about the natural world.

The term **geology** (from the Greek words for "earth" and "knowledge") was coined by scientific philosophers more than 200 years ago to describe the study of rock formations and the puzzling evidence from fossils. Through careful observations and reasoning, their successors developed the theories of biological evolution, continental drift, and plate tectonics—major topics of this textbook. Today, *geology* identifies the branch of Earth science that studies all aspects of the planet: its history, its composition and internal structure, and its surface features.

Geologists use many methods and tools to improve our scientific understanding of Earth (**Figure 1.1**). They study rock formations in remote regions with rucksacks on their backs, hammers and maps in their hands. They build ultraprecise laboratory instruments to investigate the physics and chemistry of Earth materials. They download

(a)

(b)

(c)

(d)

FIGURE 1.1 (a) Geologists examine rocks inside Spider Cave at Carlsbad Caverns, New Mexico. (b) A geochemist readies a rock sample for analysis by a mass spectrometer. (c) Geophysicists deploy instruments to measure the underground activity of a volcano. (d) An astronaut checks out instrumentation for monitoring Earth's surface. [(a) AP Photo/Val Hildreth-Werker. (b) John McLean/Photo Researchers. (c) Hawaiian Volcano Observatory/USGS. (d) StockTrek/SuperStock.]

huge datasets from satellites and networks of ground-based sensors, process them on supercomputers, and work hand-in-hand with other Earth scientists to understand our planet as a whole.

Fields of Earth science closely related to geology include *meteorology,* the study of the atmosphere; *oceanography,* the study of the oceans; and *ecology,* which concerns the abundance and distribution of life. *Geophysics, geochemistry,* and *geobiology* are subfields of geology that apply the methods of physics, chemistry, and biology to geologic problems.

Geologists seek answers to many basic questions. Of what material is the planet composed? Why are there continents and oceans? How did the Himalayas, Alps, and Rocky Mountains rise to their great heights? Why are some regions subject to earthquakes and volcanic eruptions while others are not? How did Earth's surface environment, and the life it contains, evolve over billions of years? What changes are likely in the future?

We think you will find the answers to such questions fascinating. Welcome to the science of geology!

SPACESHIP EARTH

Spacecraft launched in the 1960s changed our view of Earth. For the first time, we could look from the outside at our small blue planet spinning through the solar system (**Figure 1.2**). We began to imagine Earth itself as a space-

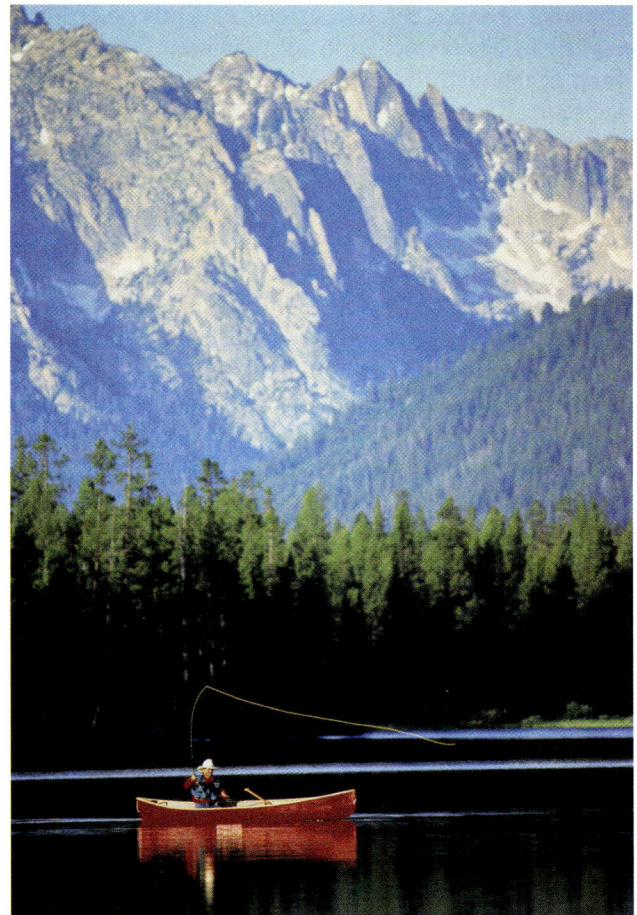

FIGURE 1.3 Our environment is the uniquely habitable region near Earth's surface that humans share with all other organisms. [Gary Brettnacher/Getty Images.]

ship, stocked with limited provisions and furnished with delicate life-support systems. We became more aware that the fast-paced development of modern civilization might eventually deplete those resources and stress our life-support systems to the breaking point.

Half a century into the space age, we have gained a clearer view of our planet. We have been especially successful in understanding our **environment**—the uniquely habitable region near Earth's surface that humans share with all other organisms (**Figure 1.3**). In thinking about how we interact with our surroundings on Spaceship Earth, we sometimes distinguish the *built* environment of civilization—what humans have constructed from steel and concrete—from the *natural* environment of the world around us.

Both land-based and space-based instruments now track changes in the atmosphere, the oceans, and the land surface of Earth. We have become much better equipped to assess the effects of human activities on the natural environment and anticipate future changes.

FIGURE 1.2 North and South America at night. [Image and data processing by NOAA's National Geophysical Data Center, Earth Observation Group (http://www.ngdc.noaa.gov/dmsp).]

What changes can be expected? At this point in the voyage of Spaceship Earth, our environment is not in good health, and its condition appears to be worsening. Substantial increases in atmospheric concentrations of carbon dioxide (CO_2) and other "greenhouse" gases, primarily from the burning of fossil fuels, are warming the climate and melting the polar ice caps. The loss of natural habitat, primarily from the expansion of agriculture and urbanization, is resulting in the extinction of many plant and animal species. What midcourse corrections are available to steer us in a more favorable direction?

In 1963, the architect Buckminster Fuller wrote *An Operating Manual for Spaceship Earth,* in which he noted the need for more knowledge about how Earth works as a mechanical system: "You know that you're either going to have to keep the machine in good order or it's going to be in trouble and fail to function. We have not been seeing our Spaceship Earth as an integrally-designed machine [that] must be comprehended and serviced in total . . . no instruction book came with it."

Figuring out how to operate Spaceship Earth requires a scientific understanding of how the contraption works. We hope that some of the pages you read here will be added to your own personal manual for operating Spaceship Earth.

The space-age view of our planet as a big, complex machine—the *Earth system*—has stimulated a new way of doing Earth science, which we will explain in Chapter 2. It has also become a potent metaphor for people concerned about **sustainable development.** That concept was defined in a famous 1987 United Nations report as "development that meets the needs of the present without compromising the ability of future generations to meet their own needs." It offers an appealing if utopian vision: a promise that the current generation will carefully manage its interactions with Earth to ensure a hospitable environment for future generations.

Sustainability involves many economic and political issues about which many nations do not agree, so forging a global strategy that moves civilization toward this goal will not be easy. People must learn how to work together to accomplish a number of objectives:

1. to manage our natural resources in a way that can continue to provide society with adequate energy, water, and raw materials,

2. to prepare society for inevitable natural disasters such as earthquakes, hurricanes, floods, and volcanic eruptions,

3. to repair the environmental damage we have already done, and

4. to maintain a habitable environment on Spaceship Earth over the long haul.

> *Sustainable development meets the needs of the present without compromising the ability of future generations to meet their own needs.*

These four objectives are interrelated, and they involve political and economic issues that extend far beyond Earth science. But reaching them will depend on a scientific understanding of the natural world and on the new technology that science will help us create.

In this chapter, we sketch a broad picture of the practical issues addressed by Earth science. Later chapters will color in the details by describing the geologic processes that govern the natural environment, produce natural resources, and cause natural disasters.

NATURAL RESOURCES

Human populations and their effects are multiplying at phenomenal rates. Between 1930 and 2000—in the span of just one human lifetime—the global population rose threefold, from 2 billion to 6 billion, and is now increasing by 78 million people per year. A glowing lattice of civilization is spreading rapidly across Earth's surface (see Figure 1.2). Within 20 years, the number of people will probably exceed 8 billion.

This global human society is consuming vast amounts of **natural resources.** This general term refers to the energy, water, and raw materials that are available from the natural environment. The demand for natural resources is skyrocketing as civilization expands and people around the world strive to improve the quality of their lives. Geology provides the scientific knowledge needed to find, exploit, and conserve natural resources.

> *Human effects on our environment are growing with the human population, which now exceeds 6 billion.*

Resources and Reserves

The supply of any material mined from the ground is finite. Its availability depends on its distribution in accessible deposits, as well as on how much we are willing to pay to get it out of the ground. Geologists use two measures of supply. **Reserves** are supplies that have already been discovered and can be exploited economically and legally at the present time. In contrast, *resources* constitute the entire amount of a given material, including the amount that may become available for use in the future. Resources include reserves, plus known but currently unrecoverable supplies, plus undiscovered supplies that geologists think may eventually be found (**Figure 1.4**).

In many cases, resources that are too poor in quality or quantity to be worth exploiting now, or that are too difficult to retrieve, become profitable when new technology is developed or prices rise. A recent example is the production of oil and gas from large reservoirs in the Gulf of Mexico at water depths greater than 300 m. In

FIGURE 1.4 Resources include reserves, plus known but currently unrecoverable deposits, plus undiscovered deposits that geologists think may eventually be found.

the decade from 1993 to 2003, high-tech offshore platforms were able to increase U.S. deep-water oil production from 100,000 barrels per day to almost 1 million (**Figure 1.5**).

Geologists are experts at discovering new resources. It should be kept in mind, however, that the assessment of resources is much less certain than the assessment of reserves. Any figure cited as representing the resources of a particular material is only an educated guess as to how much will be available in the future.

> *Reserves are known supplies of natural resources that could be exploited economically under current conditions.*

Energy Resources

Energy is required do work, so it is fundamental to all aspects of human activity. A crisis in the supply of energy can bring a modern society to a halt. Wars have been fought over access to supplies of fuel; economic recession and destructive currency inflation have resulted from fluctuations in the price of oil. It is no surprise that the fossil-fuel industry is the world's biggest business!

Basic Resource Types Before the industrial revolution of the mid-nineteenth century, most of the energy used in the United States came from the burning of wood (**Figure 1.6**). A wood fire, in chemical terms, is the combustion of *biomass:* organic matter consisting of carbon and hydrogen compounds. The ultimate source of the energy is the sunlight green plants use to grow. In this capacity, the biomass acts as a short-term reservoir for storing solar energy. It is a **renewable resource** because the environment is constantly producing new biomass; for example, a forest chopped down for wood can be regrown and harvested again. Humans have used a variety of renewable energy sources to power mills and other machinery for

FIGURE 1.5 New technologies used aboard offshore platforms in the Gulf of Mexico can recover oil and gas from rock reservoirs below very deep waters. [Larry Lee Photography/CORBIS.]

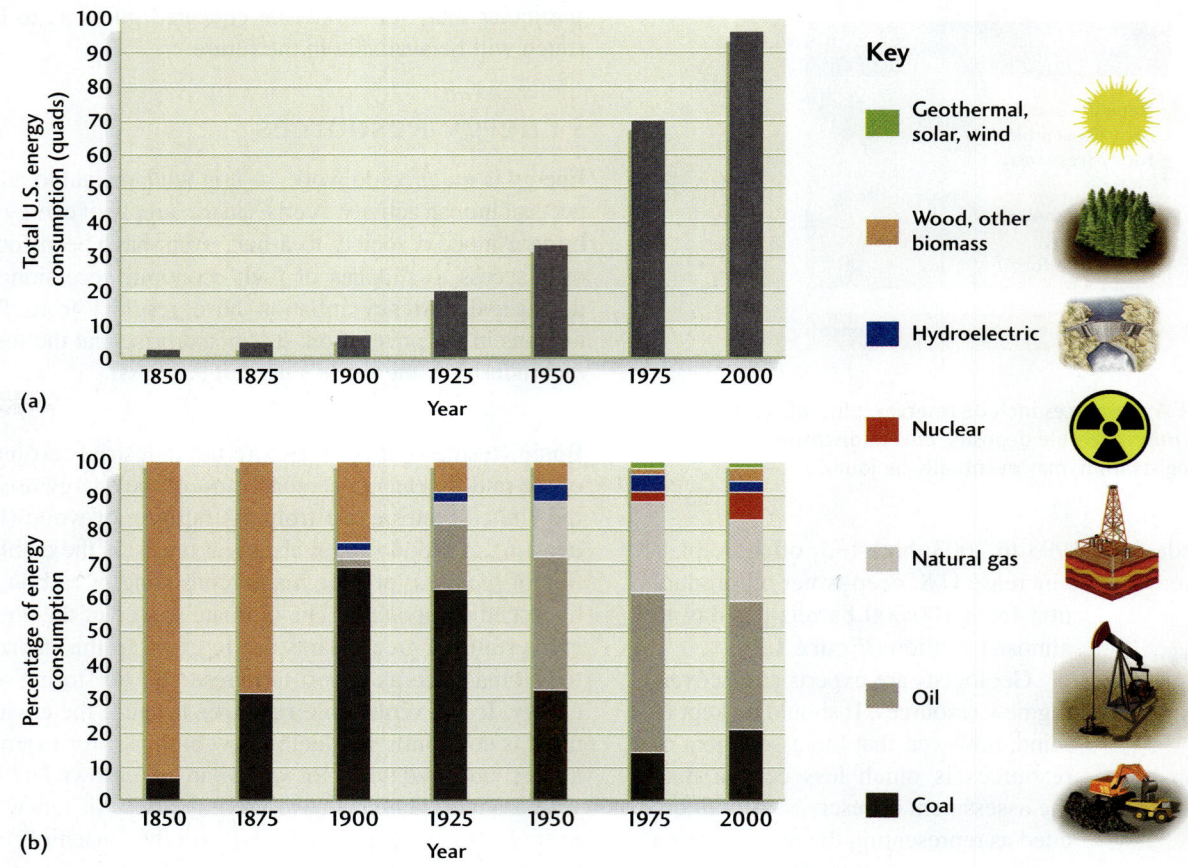

FIGURE 1.6 (a) Total energy consumption in the United States from 1850 to 2000 in quads (1 quad = 10^{15} Btu). (b) Percentages of various types of energy consumed in the United States from 1850 to 2000. [U.S. Energy Information Agency.]

thousands of years, including moving air, falling water, and the work of oxen, horses, and elephants.

Some of the biomass from plants that grew millions of years ago was buried beneath Earth's surface and transformed into the combustible rock called *coal* (**Figure 1.7**). When we burn coal, we are using "fossilized" energy stored by plants from ancient sunlight. Our other major fuels, crude oil (petroleum) and natural gas, were also created by the burial and heating of dead organic matter. These **fossil fuels** are considered to be **nonrenewable resources** because geologic processes produce them much more slowly than we are using them up. Our reserves of oil, natural gas, and coal will be exhausted long before Earth can replenish them.

Rise of the Fossil-Fuel Economy Industrialization increased the demand for energy beyond what the traditional renewable sources could supply. By the late eighteenth century, James Watt and others had developed coal-fired steam engines that could do the work of hundreds of horses. Steam technology lowered the price of energy dramatically, in part because it made coal mining possible on a very large scale—in Britain, from the coalfields of England and Wales; in continental Europe, from the coal basins of western Germany and bordering countries; and in North America, from the Appalachian coalfields of Pennsylvania and West Virginia. The availability of cheap energy sparked the industrial revolution. By the end of the nineteenth century, coal accounted for more than 60 percent of the U.S. energy supply (see Figure 1.6).

The first oil well was drilled in Pennsylvania by Colonel Edwin L. Drake in 1859. The idea that petroleum could be profitably mined like coal provoked skeptics to call the project "Drake's Folly" (**Figure 1.8**). They were wrong, of course; by the early twentieth century, oil and natural gas were beginning to displace coal as the fuels of choice. Not only did they burn more cleanly, producing no ash, but they could be transported by pipeline as well as by rail and ship. Moreover, gasoline and diesel fuels refined from crude oil were suitable for burning in the newly invented internal combustion engine. By 1950, the use of energy from oil and natural gas exceeded that from coal, and these fossil fuels taken together supplied almost 90 percent of U.S. energy needs.

FIGURE 1.7 Coal is mined at Peabody Energy's North Antelope Rochelle coal mine, near Gillette, Wyoming. [Matthew Staver/The New York Times/Redux.]

FIGURE 1.8 Edwin L. Drake (*right*) in front of the oil well that initiated the "age of petroleum." This photo was taken by John Mather in 1866. Titusville, Pennsylvania. [Bettmann/CORBIS.]

Nuclear reactors that could produce electricity from radioactive elements such as uranium were developed after World War II (**Figure 1.9**). The early expectation that nuclear fuels would provide a large, low-cost, environmentally safe source of energy has not been realized, however. Safety concerns, the inability to dispose of nuclear wastes, and the escalating costs of stringent safety and security measures have slowed the construction of nuclear power plants. Nuclear power supplies a substantial fraction of the electrical energy used by some countries, such as France (76 percent) and Sweden (52 percent), but this proportion is much smaller in the United States (21 percent). Overall, nuclear power accounts for less than a tenth of the total U.S. energy budget.

Today the engine of civilization runs primarily on fossil fuels: coal, oil, and natural gas (methane, CH_4). Taken together, they account for 86 percent of the energy consumed by the United States and the world at large. The use of renewable energy resources, especially of biofuels such as ethanol (made from sugarcane and corn), is increasing, but the total U.S. energy production from all renewable resources was just over 6 percent in 2005.

> *A natural resource is renewable if Earth can replenish it as rapidly as humans consume it.*

Global Energy Use Energy use is often measured in units appropriate to the fuel; for example, barrels of oil, cubic feet of natural gas, and tons of coal. But the comparisons are easier if we use a standard unit of energy such as the British thermal unit (Btu). One Btu is the amount of energy needed to raise the temperature of 1 pound of water by 1°F (1054 joules). In 2004, the United States used 100 quadrillion Btu, or *quads,* of energy per year (1 quad = 10^{15} Btu), compared with a

FIGURE 1.9 Japan's Kashiwazaki-Kariwa facility is the world's largest nuclear power plant, with seven reactors and a total generating capacity exceeding 8200 megawatts. It was damaged by a powerful earthquake (magnitude 6.8) that struck the region on July 16, 2007. The plant was shut down and required extensive repairs. [STR/AFP/Getty Images.]

FIGURE 1.10 Actual and projected energy consumption, in quads, 1990–2020, by world regional groupings. [U.S. Energy Information Agency.]

global total of about 440 quads (**Figure 1.10**). Thus, the United States, with 5 percent of the world's population, consumes about four and a half times more energy per person than the global average.

There are promising signs that new technologies and increasing energy conservation are beginning to lessen U.S. energy appetites. The total annual U.S. consumption of energy of all types actually went down in 2006, the first drop in modern times. However, this small reduction (–1 percent) was more than offset by the rapid growth in energy use by the world's two most populous countries, China (+8.4 percent) and India (+5.4 percent).

Overall, worldwide energy consumption has risen during the last 6 years by about 2.7 percent per year. In 2007, China's total energy consumption exceeded that of the United States for the first time. Still, China's average individual (or *per capita*) energy use remains almost eight times less, due to its much larger population. As it and other developing economies strive to improve their standards of living, global energy use per capita is bound to rise, accelerating overall energy consumption.

Petroleum Exploration and Production Geologists are refining methods for finding new oil and gas resources in remote regions of the continents and on the continental margins beneath deep water (**Figure 1.11**). If discoveries in these regions significantly expand our petroleum resources, oil may stay relatively cheap for a longer time. The U.S. Geological Survey estimates that

> *Energy use in the developing world is increasing rapidly, but the average consumption per person remains higher in the United States than in any other large nation.*

(a)

(b)

FIGURE 1.11 (a) The *Geco Topaz*, a vessel operated by WesternGeco Inc., conducting a three-dimensional seismic survey in the North Sea. The bubbles behind the ship are compressed-air explosions that send out sound waves; the reflections from the rocks below are recorded by the ship to produce an image of the subsurface structure. (b) Depiction of a three-dimensional seismic survey, showing acquisition system and a subsurface "cube" of seismic data. The colors show the layers of sediments beneath the seafloor, some trapping oil and natural gas. [(a) Courtesy of Oil & Gas UK. (b) Courtesy of Satoil, Veritas, and BP.]

FIGURE 1.12 Casing crew working on an oil rig drill floor. [Peter Bowater/Photo Researchers.]

the world's total oil resources are at least 2 trillion barrels, equivalent to almost 12,000 quads of energy, and it has drawn the optimistic conclusion that any peak in petroleum production is decades away. The rate at which new petroleum resources are being discovered is not keeping up with rising oil production, however, leading some pessimists to estimate that global production will begin to decline in the next few years. We will revisit this debate in Chapter 14.

More petroleum will be available for future needs if reserves can be more efficiently extracted from deep rock formations. The subterranean reservoirs that hold oil and natural gas are complex geologic systems. Geologists can map the reservoir rocks in three dimensions using various techniques, such as seismic imaging (see Figure 1.11). The three-dimensional pictures they obtain show them where the bulk of the oil and gas is located and allow them to predict how it will flow out of holes drilled into the reservoir.

Drilling holes deep into Earth's rocky crust has become a very sophisticated and expensive business (**Figure 1.12**). The drilling from a single oil platform in the deep waters of the Gulf of Mexico (see Figure 1.5) can cost over $100 million. Petroleum engineers use three-dimensional models to steer drill bits on swooping paths into the richest parts of a reservoir. To coax oil out of stubborn formations, they inject water and carbon dioxide down strategically positioned drill holes to push the oil into areas where it can be more efficiently pumped through other drill holes. These methods have increased the fraction of oil that can be extracted from known oil fields, increasing our reserves.

> *Geologists are improving methods for finding new oil and gas resources and for extracting known resources more efficiently.*

Our Energy Future Despite these efforts to locate more reserves and improve production, our major petroleum reservoirs will become depleted sometime in the next few decades. Oil prices will rise to uneconomic levels, and alternative energy resources will have to take up more and more of the demand. How quickly will this transition to a post-petroleum economy occur? Which alternative sources of energy will replace oil? The predictions and choices, which are fiercely debated, will depend rather heavily on progress in Earth science and technology.

The financial costs of replacing oil (and eventually natural gas) with alternative energy sources must be balanced with the environmental costs. Nuclear power is an attractive option because it generates no greenhouse gases and because enough uranium resources are available as nuclear fuels to produce an estimated 240,000 quads of energy, more than twice the reserves of all fossil fuels combined. However, switching wholesale to nuclear power would create other environmental problems—people don't want nuclear power plants or nuclear waste dumps in their backyards—and would heighten concerns about the proliferation of nuclear weapons.

Developing solar power (**Figure 1.13**) and other renewable energy sources—such as wind, water, and geothermal power, as well as biofuels—will certainly

FIGURE 1.13 Solar cells convert sunlight, a renewable resource, into electrical energy at this utility in a remote village in Nepal. [Ned Gillette/CORBIS.]

help, but current projections indicate that the total energy production from these resources will fall short of our needs for many decades to come, unless there are unanticipated technological breakthroughs.

Coal, a plentiful and cheap alternative to oil and natural gas, is expected to fill the gap. Major deposits of coal are found on all continents of the Northern Hemisphere; the total resource amounts to over 3 trillion metric tons, which is enough to generate 67,000 quads of energy. But coal is a notoriously dirty fuel, polluting the atmosphere with gases and particles that cause smog and acid rain. Moreover, for a given amount of energy, it produces 25 percent more carbon dioxide than oil and 70 percent more than natural gas. Other "unconventional" fossil fuels, such as the tar sands of Alberta and the oil shales of Colorado, are also abundant but polluting.

New technologies could help us to reduce coal pollution. If the carbon dioxide emitted from coal-fired power plants could be economically captured and stored underground, the world's abundant coal resources would become much more attractive as a replacement for petroleum. For example, "clean coal" energy could be used to generate hydrogen fuel for transportation. Turning carbon capture and storage into a feasible technology will be a great geological challenge.

> *The depletion of petroleum reserves forces us to balance the availability, financial costs, and environmental costs of alternative energy sources. Coal is likely to reemerge as the world's primary energy resource.*

Minerals and Other Raw Materials

Coal and oil are examples of raw materials we extract from Earth's crust. Many other raw materials are needed to meet modern society's demands, particularly *minerals,* the crystalline components of rocks. Gem minerals such as diamonds and rubies are prized for their beauty, but their economic value is tiny compared with that of the more mundane minerals and other raw materials that are mined in huge quantities. Industrial mining provides stone for buildings and roads, phosphates for fertilizers, cement for construction, clays for ceramics, sand for silicon chips and fiber-optic cables, and the metals used everywhere in everyday life.

> *In addition to energy resources, human activities, especially construction, require raw materials such as stone, minerals, and metals.*

Rising Economic Demand In 2000, the U.S. consumption of raw materials, exclusive of fuels and food, was about 3.6 Gt (gigatons, or billion metric tons). This use, almost 13,000 kg per capita, was double the amount in 1960 (**Figure 1.14**). Three-fourths was bulk material for construction, such as crushed stone, sand, and gravel. The remainder comprised industrial minerals, such as sand for making glass (10 percent), metals (5 percent), nonrenewable organics, such as asphalt and lubricants (4 percent), and agricultural, fishery, and forestry products (6 percent).

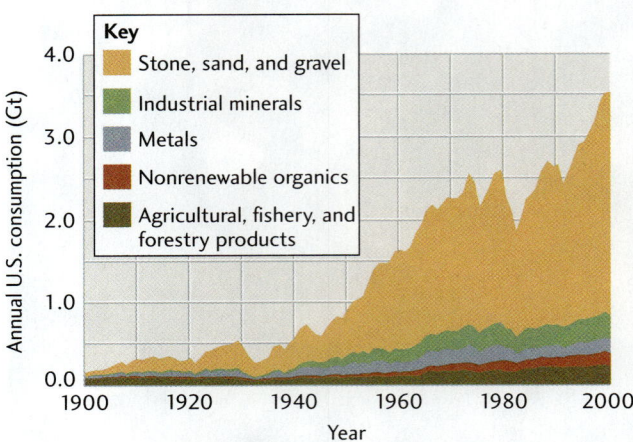

FIGURE 1.14 Annual U.S. consumption of non-fuel, non-food raw materials from 1900 to 2000, by weight in billions of metric tons (Gt).

U.S. consumption of raw materials, about one-third of the global total, is increasing at 2.8 percent per year, more than twice the rate of U.S. population growth. Global demand is rising even faster, due to the rapid expansion of the Asian economies. Worldwide, the average person is using more raw materials each year.

Growing consumption of raw materials has led to a boom in the mining industry, particularly in the mining of metals. After declining in the 1990s, the prices of important metals such as iron, copper, zinc, and nickel doubled or tripled between 2001 and 2006 (**Figure 1.15**), pushed upward by the strong demand for industrial products in developing countries such as China and India. In the case of uranium minerals, which are refined to make fuels for nuclear reactors (and nuclear bombs), the price increased by a factor of 6.

The need for metals can be accommodated in part by *recycling* the metal content of discarded goods and by substituting low-cost ceramics, composites, and plastics. For example, over one-third of the total U.S. aluminum supply now comes from recycling. However, the stocks of copper, zinc, and other essential metals, even if they are recycled, may not meet the needs of the global population over the long haul. According to one recent study, the amount of copper that would be required to supply the world with the amount now used per capita by developed nations equals all of the copper in known reserves, plus all of the copper currently in use as electrical wiring and

> *Despite recycling and substitution of lower-cost materials, the demand for useful metals is increasing rapidly as developing countries become more industrialized.*

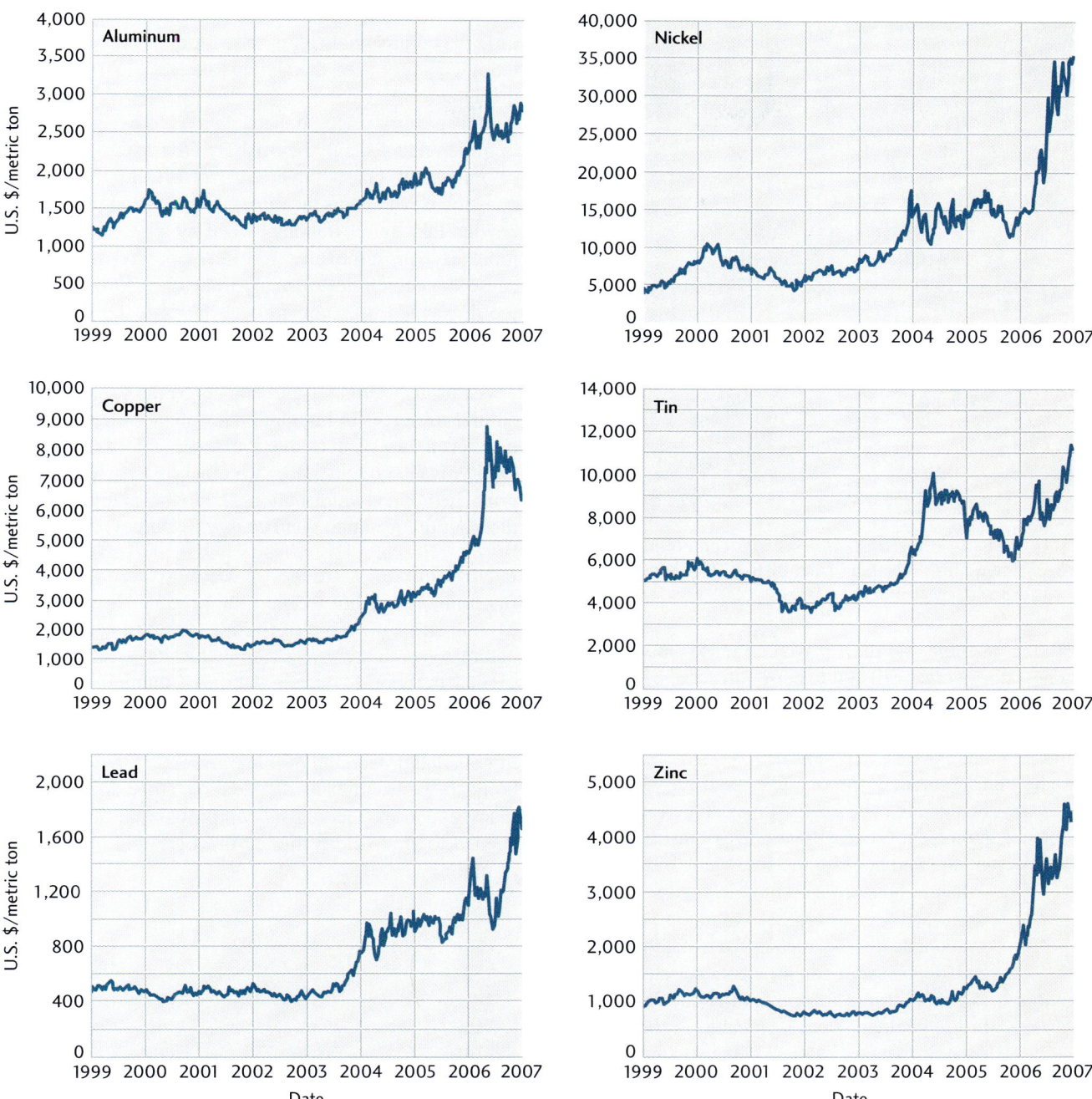

FIGURE 1.15 The prices of six important metals have risen substantially during the last several years, fueled by increasing global demand. All prices are in U.S. dollars per metric ton. [London Metal Exchange.]

in computers, medicines, and a wide array of manufactured goods.

Mineral Exploration Rising demand is driving an increasingly systematic search for new mineral resources. Geology provides us with knowledge about how mineral deposits are formed and where to look for more of them (**Figure 1.16**). It also guides us in exploiting these natural resources more efficiently and in minimizing the damage caused by mining to the environment.

Useful chemical elements, such as metals, are widely distributed in many common rocks and minerals. Theoretically, with enough money and energy, we could extract both abundant and rare elements from many types of rocks. A more practical concern, however, is the identification of new reserves, those mineral deposits that are

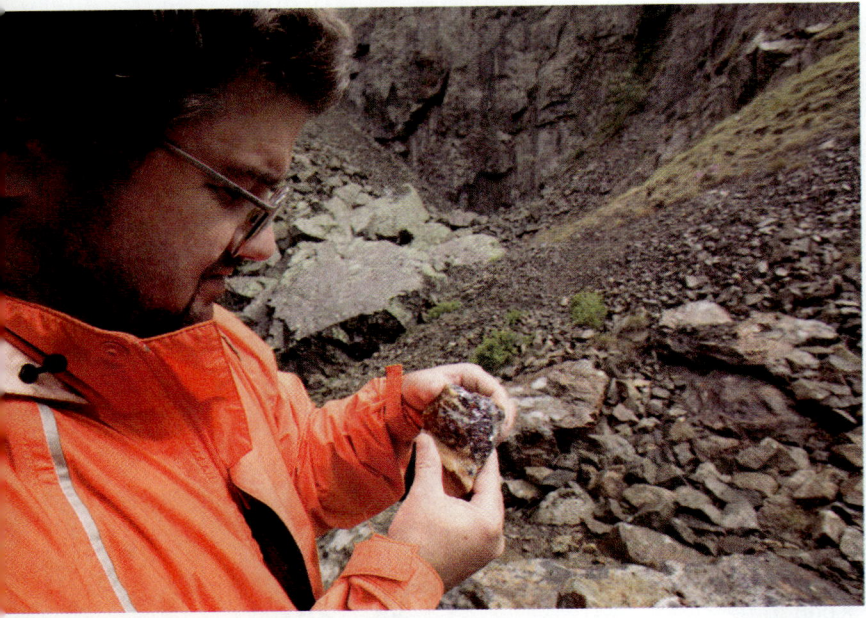

FIGURE 1.16 A geologist examines a rock containing lead at Greenside Mines, Helvellyn, Lake District, England. [Ashley Cooper/Alamy.]

profitable to mine and purify. We can reasonably expect that new discoveries will add to current reserves, but at an uncertain rate. When the richest deposits have been mined out, we will be forced to rely on lower-grade deposits that could be more expensive to recover.

The high-grade deposits of minerals from which we can extract metals are called *ores*. Ores can be found in a variety of geological environments, in young rocks and old rocks, on land and beneath the sea. Many of the most valuable ores were formed in regions of volcanic activity, where hot circulating groundwater leached the metals from nearby rocks and deposited them in a concentrated form. For this reason, information about volcanism in the geologic past is valuable in the search for new metal resources. We will examine this connection in Chapter 4.

Undiscovered mineral resources may lie in regions that are difficult to explore, such as the ocean floor. How important the deep sea will become as a source of minerals is an open question. The answer depends in part on the development of efficient marine technology for deep-sea exploration and mining, as well as on the resolution of legal issues regarding ownership of deep-sea deposits.

In 1982, the United Nations adopted an agreement known as the Convention on the Law of the Sea, which set up a legal and regulatory system for the development of deep-sea resources; the vote was 130 in favor, 4 opposed. The United States was one of the four nations opposed because it was concerned that the agreement might unfairly limit the commercialization of seafloor resources. A quarter of a century later, the United States still has not signed the agreement.

It is generally agreed that a nation has exclusive rights to mineral deposits in the offshore area within 200 nautical miles of its coast—its *exclusive economic zone*. Still in question is who owns the mineral deposits on the seafloor beyond this zone. Are they owned by all nations in common? By the discoverer and developer? By both? Because the economies of many mineral-exporting nations are threatened by competition from new sources on the seafloor, those nations would like to limit such development and obtain a share of the profits.

Crafting acceptable agreements on the exploitation of natural resources in international waters such as the Arctic Ocean, as well as on the continent of Antarctica (which is also regulated by a global treaty), will require good geological information about the abundance and distribution of mineral and energy resources.

Environmental Damage by Mining Despite its benefits, mining has been called the original dirty industry. Mining strips off more of Earth's surface each year than natural erosion by rivers. Annually, its waste products exceed the world's accumulation of municipal garbage (**Figure 1.17**). There has been progress in developing clean and environmentally benign mining operations, but mining and processing of minerals remain major contributors to environmental degradation and poor health in many countries. Geologists can provide the information needed to guide clean mining practices and the disposal of mining wastes.

> Most metals are extracted from high-grade mineral deposits called ores. Many of the most valuable ores were deposited by hot groundwater circulating in volcanically active regions.

FIGURE 1.17 Mine dumps near Johannesburg, South Africa, an impressive example of the wastes produced by mining. [George Steinmetz.]

Some people argue that searching for more ore deposits to mine is not the solution to shortages of raw materials because the extraction, processing, and waste disposal connected with exploiting those materials are so damaging to the environment and human health that they cannot be sustained. They propose a new "materials economy" based on recycling, greater efficiency in the use of raw materials, and substitution of advanced, environmentally benign materials for minerals. One example is the use of fiber-optic glass cables made from sand instead of copper wire.

Mining has considerable negative consequences: it produces a tremendous amount of waste as well as environmental degradation and health problems.

In any case, society will have to find new mineral resources, and we will have to weigh the short-term economic gains of mining against its long-term environmental and health costs.

Water Resources

Water is vital to all life on Earth. Even the hardiest desert plants and animals require water, and humans cannot survive more than a few days without it. However, the amount of water consumed by modern society far exceeds what humans need for simple physical survival. A human can make do with a daily ration of about 2 liters of water, but today in the United States, the average individual uses about 250 liters per day. Industrial and agricultural uses increase the U.S. per capita total to 6000 liters per day. This per capita consumption is two to four times greater than in western Europe, where people pay up to three times as much for their water.

Fresh water is consumed in immense quantities by agriculture, industry, and sewerage systems. On average, U.S. agriculture takes about 35 percent of the water withdrawn from the nation's water supplies, and industry takes about 53 percent, primarily for electricity generation in power plants fired by fossil fuels. The remaining 12 percent accounts for all domestic and other commercial consumption, including sewerage.

Per capita water use in the western United States, mostly for agricultural irrigation, is 10 times greater than that in the eastern states, and the prices of water are lower, even though the arid western states receive only one-fourth of the average national rainfall. In California, for instance, 81 percent of the water is used for irrigation (**Figure 1.18**), 16 percent for municipalities, and 3 percent for industry. A 20 percent reduction in irrigation use would double the amount of water available for use by cities and industries. Such conversions may become necessary if water shortages accompany continuing population growth in the western states.

Per capita water use in the United States, primarily for agriculture, industry, and sewerage, is much higher than in other countries.

FIGURE 1.18 Irrigation in California's Imperial Valley, a natural desert. [David McNew/Getty Images.]

Water Shortages Shortages due to increased demand for water can be compounded by *droughts,* dry periods of months or years that damage crops and cause significant reductions in water supplies. Droughts can occur in all climates, but arid regions are especially vulnerable. Lacking replenishment from rainfall, rivers may shrink and dry up, reservoirs may evaporate, and the soil may dry up and turn to dust, killing vegetation and other life.

The southwestern United States is experiencing its worst drought in more than a hundred years of recorded history. In the lower Colorado River basin, water reservoirs such as Lake Powell and Lake Mead have dropped to less than 60 percent of their capacity (**Figure 1.19**). An extreme drought also grips the southern states.

Our climate history can give us perspectives on the severity of droughts. For example, during the 400-year period from 1500 to 1900, the Southwest was drier, on average, than it has been during the last century. Moreover, the geologic record shows droughts that were more severe and of longer duration than the present drought has been (at least so far). By exploring the past, geologists and

FIGURE 1.19 Lake Powell, in Utah, showing "bathtub rings" from recent drought. [Christopher J. Morris/Redux.]

climate scientists are trying to predict the future. Are the recent droughts just short-term fluctuations in climate, or do they signal a return to an extended dry period? How will global climate change affect rainfall in the Southwest?

Despite its drought problems, the United States is relatively well endowed with water resources. Other countries are less fortunate. A severe drought from 1960 to 1980 destroyed farming and grazing along the southern border of the Sahara Desert, in a region called the Sahel, causing a famine that killed a million people.

According to the United Nations Environment Programme, more than one-third of the world's population is without a safe water supply, and one-fourth will experience chronic water shortages during the next decade. Within two decades, dozens of countries may suffer from water shortages, running up against a "water barrier" to further development.

These problems could be exacerbated by global climate change, which could bring more droughts to already dry regions. Some climate scientists think that global warming is permanently expanding the Sahara Desert southward into the Sahel and may further dry out the already parched southwestern United States.

As the demand on limited water supplies increases, the geological study of water resources, called *hydrology* (from *hydros,* the Greek word for "water"), is becoming an increasingly important field of Earth science.

Groundwater Resources The world's water supply, about 1.46 billion cubic kilometers, is truly enormous. But most of that water, about 96 percent, is salt water in the oceans, and another 3 percent is locked up in glaciers and the polar ice caps. Only 1 percent is fresh water, the hydrologic resource most needed for agriculture and civilization. A small fraction of this fresh water occupies lakes and rivers; most of it is **groundwater** contained in rocks beneath the land surface.

Water is continuously on the move, cycling among its natural reservoirs (**Figure 1.20**). The movement of water from the oceans into the atmosphere by evaporation and its subsequent precipitation as rain or snow on land convert salt water into fresh water. Precipitation recharges groundwater reservoirs, and the runoff flows through rivers and streams back into the oceans. Geologists work with meteorologists, oceanographers, and environmental engineers to investigate this *hydrologic cycle,* which we will discuss further in Chapter 11.

> Most of Earth's fresh water is groundwater, contained in rocks beneath the land surface.

Large parts of North America rely on **aquifers**— reservoirs of groundwater contained in porous rock formations—for almost all their water needs. In the natural course of the hydrologic cycle, the groundwater depleted by evaporation and runoff is recharged by precipitation. In this sense, aquifers are a renewable resource. However, if rapid pumping of water out of the ground depletes the

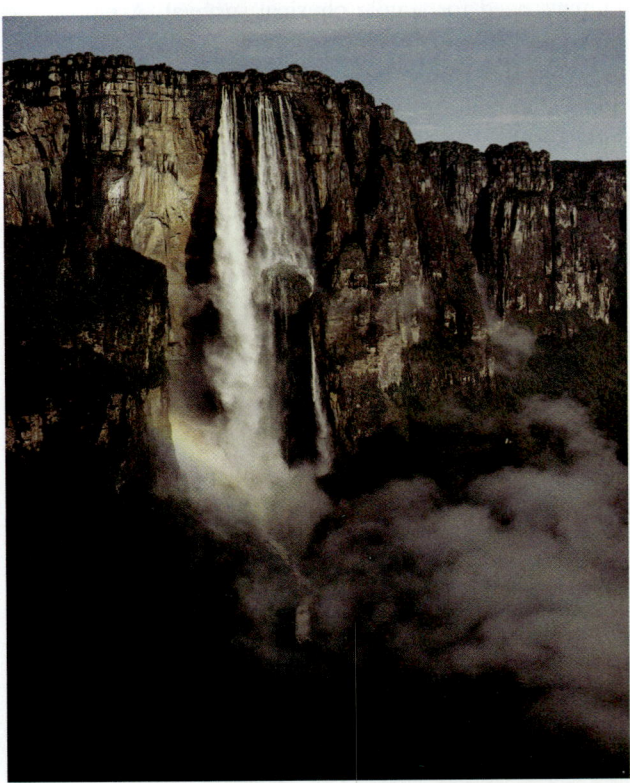

FIGURE 1.20 Angel Falls, Venezuela, is the highest waterfall on Earth. The falls plunge 914 m from a flat-topped mountain composed of 1.7-billion-year-old sandstones. [Michael K. Nichols/National Geographic/Getty Images.]

aquifer, precipitation may take hundreds or even thousands of years to replenish it. At our current high rates of water use, many aquifers are effectively a nonrenewable resource.

To illustrate this problem, we consider the huge Ogallala aquifer, which consists of porous formations of sand and gravel that lie beneath the southern Great Plains (**Figure 1.21**). For more than 100 years, the Ogallala aquifer has supplied fresh groundwater to the towns, ranches, and farms of region, now home to more than a million people. Pumping of water from the aquifer, primarily for irrigation, has been so extensive—about 6 billion cubic meters of water per year from 170,000 wells—that recharge from rainfall cannot keep up. Pressure in the wells has declined steadily, and the water table has dropped by 30 m or more.

Because water use is outpacing recharge, the Ogallala aquifer will decline as a resource during the early decades of the twenty-first century. If the water it produces cannot be replaced, about 5.1 million acres of irrigated land in western Texas and eastern New Mexico will dry up—and so will 12 percent of the country's supply of cotton, corn, sorghum, and wheat and a significant fraction of the feedlots for the nation's cattle.

Aquifers in the northern Great Plains and elsewhere in North America are also being rapidly depleted. Geologists will be helping communities assess and deal with these problems.

Improving Water Resources The traditional ways of increasing water supplies, such as building dams and reservoirs and drilling wells, have become extremely costly because most of the best (and therefore cheapest) sites have already been used. Furthermore, the building of more dams to hold larger reservoirs carries environmental costs, such as the flooding of inhabited areas, detrimental changes in river flows above and below the dams, and the disturbance of fish and other wildlife habitats. Factoring in these costs has led to delays in dam projects and rejection of proposals for new dams.

Informed decision making about the conservation and management of water resources requires knowledge of how water moves through the hydrologic cycle and how this flow responds to natural changes and human modifications. At what rates can we pump water from aquifers without depleting them? What will be the effects of climate change on water supplies?

Geological studies of rivers and aquifers allow us to improve the quality of our water resources as well as their quantity. Water supplies can be contaminated by a variety of human activities—sewage, industrial wastes, and mining wastes are particular concerns—but also through natural processes involving water-rock interactions. Geochemical techniques have been used, for example, to investigate elevated levels of arsenic in groundwater, which pose a serious health hazard in many parts of the world. The problem is particularly acute in Bangladesh, where groundwater provides 97 percent of the drinking water supply. Geologists are helping to guide the placement of new wells that draw water with acceptable concentrations of arsenic.

> *If groundwater is pumped from an aquifer more rapidly than it can be recharged by precipitation, the aquifer is effectively a nonrenewable resource.*

Soils

Soils are an intricate combination of organic material and weathered rock, found in great variety on almost all land surfaces. They constitute an essential and immensely valuable natural resource (**Figure 1.22**). Soils are the primary reservoir of nutrients for agriculture and the ecological systems that produce renewable natural resources. They are also essential for waste disposal and water filtration. Clays and other soil materials are used for construction and in manufacturing activities.

FIGURE 1.21 A map of the Ogallala aquifer, which supplies the southern Great Plains with much of its fresh water. [U.S. Geological Survey.]

FIGURE 1.22 Furrows of green beans growing in the rich soil of Monterey County, California. [CORBIS.]

Soils are a huge reservoir of carbon, containing twice as much as the atmosphere and three times more than all of the world's vegetation. They recycle the organic carbon from dead plants and animals and release it into the atmosphere. They also have a major influence on our water supplies. The quality of the fresh water most people use is determined largely by the soils it passes through.

Soil management is thus crucial to sustaining and improving the human environment, and issues related to soil quality now figure prominently in many policy decisions. Industrial-scale farming has worsened the environmental problems related to soil erosion and the accumulation of toxic elements and salt in soils, as well as downstream contamination of aquatic systems from agricultural runoff. In the United States, new agricultural practices are being adopted to counter these detrimental effects, but much has yet to be learned.

Soils are an important natural resource that must be carefully managed.

Good soil management will require increasing investments in the geological study of soils, including research on the fundamental physical, chemical, and biological processes involved in soil development.

NATURAL HAZARDS

Earth is an active, constantly changing planet. Sunlight falling on Earth's surface energizes swirling currents in the atmosphere and oceans, and heat coming from deep inside Earth melts rock, pushes up mountains, and moves continents. This activity is essential for our survival, creating natural resources and sustaining our environment, but it can also be dangerous. The Earth system produces an interesting array of hazardous phenomena, ranging from severe storms, landslides, and flooding to earthquakes and volcanic eruptions. Material from the solar system—asteroids, comets, and chunks of broken-up planets—has blasted Earth's surface with the energy of thousands of nuclear bombs.

Events produced by natural processes that have the potential to kill people and damage our built environment are called **natural hazards.** Earth science gives us practical tools to characterize natural hazards and evaluate their risks to society. How frequently will a major U.S. city get hit by a hurricane? Is the earthquake hazard for Hilo, Hawaii, as high as that for San Francisco, California? How should towns along the Mississippi River prepare for floods? What poses a greater hazard for the Seattle-Tacoma area, an eruption of Mount Rainier—the active volcano that towers over the metropolis (see Figure 5.33)—or an earthquake caused by the geologic faults that lie beneath it?

Answering questions such as these is complicated by the need to consider the sizes, locations, and frequencies of hazardous events. While bigger events are likely to cause more damage than smaller events, they do not occur as often. The very biggest events are very rare indeed; geologists call them *extreme events*.

About 65 million years ago, an asteroid the size of Manhattan smacked into Mexico's Yucatán Peninsula, blowing a huge amount of dust and smoke into the atmosphere (**Figure 1.23**). The environmental changes caused by this extreme event killed off the dinosaurs and many

FIGURE 1.23 An artist's rendition of an asteroid impact on Mexico's Yucatán Peninsula 65 million years ago. Such extreme events are very rare but can have major geological and biological consequences. This impact wiped out the dinosaurs and accelerated the evolution of mammals. [From "The Day the World Burned" by D. A. Kring and D. D. Durda, *Scientific American*, December 2003. Art by Chris Butler.]

other species, allowing mammals (and eventually humans) to evolve. Geologists have found evidence of similar bad-news collisions at other times in Earth history. However, because the span of Earth history is so incredibly long—billions of years—the chances of an asteroid striking next year, or even during the next century, turn out to be extremely small. More common natural hazards, such as hurricanes and earthquakes, pose greater threats to society in the short term.

A Rogue's Gallery of Natural Hazards

Let's take a brief tour, introducing ourselves to the most important natural hazards and some of their effects on society. Later chapters will describe the natural phenomena that produce these hazards, as well as their geologic consequences, in more detail.

Hurricanes The greatest storms on Earth are **hurricanes,** swirling masses of dense clouds hundreds of kilometers across that suck their energy from the warm surface of tropical seas. In the Northern Hemisphere, they form during late summer when sea surface temperatures are highest; the resulting winds circulate counterclockwise around a calm, central "eye," where the atmospheric pressures are low (Figure 1.24). Their name originates

FIGURE 1.24 Hurricane Katrina on August 28, 2005, a few hours before it struck New Orleans. It was one of the most powerful hurricanes on record, with maximum sustained winds of up to 280 km/hour (175 miles/hour) and gusts up to 360 km/hour (225 miles/hour). [NASA/Jeff Schmaltz, MODIS Land Rapid Response Team.]

FIGURE 1.25 Disaster in Chittagong, Bangladesh, caused by a cyclone on February 5, 1991. [Pablo Bartholomew/Liaison Agency/Getty Images.]

from Huracan, a god of storms to the Mayan people of Central America. In the western Pacific and the China Sea, hurricanes are known as *typhoons,* from a Cantonese word meaning "great wind." In Australia, Bangladesh, Pakistan, and India, they are called *cyclones.*

Whatever you call them, these intense tropical storms can wreak havoc with high winds and torrential rains as they move from the ocean onto land. Low pressures in and around the storm's eye raise water levels, sometimes by many meters, and the resulting "storm surge" can flood coastal areas, particularly when it coincides with normal high tides. The storm surge is the most deadly of a hurricane's hazards.

> *Hurricanes are intense storms that draw their energy from warm tropical waters.*

In 2005, when Hurricane Katrina struck Louisiana, more than 1800 people lost their lives, primarily by drowning in the storm surge. A catastrophic cyclone struck the coastal lowlands of Bangladesh in 1970, drowning as many as 500,000 people—perhaps the deadliest natural disaster of modern times. Another cyclone hit the same region in 1991, drowning at least 140,000 (Figure 1.25). The 1991 storm was more intense, but its death toll was lower due to better disaster preparations; 2 million people were evacuated.

Hurricanes pose tremendous economic risks to states along the Gulf Coast and Eastern Seaboard. In 2004, nine hurricanes made landfall in those areas, causing more than $42 billion in damage. In 2005, seven hurricanes caused over $100 billion in damages. Katrina alone accounted for more than $80 billion, making it the most costly disaster in U.S. history.

The tropical storms that grow into hurricanes can be tracked by satellites, and weather conditions inside the storms can be probed by airplanes. By feeding many types of data into computer models, meteorologists can do a pretty good job of predicting the storm's track and changes in its intensity up to several days in advance. The National Hurricane Center accurately predicted that Katrina would hit New Orleans as a severe hurricane 3 days before it actually did.

Floods When water flows into a region faster than it can flow out, the result is a **flood.** Floods can result from inundation by the sea (e.g., during storm surges), but the most frequent cause is water flowing into a river system more rapidly than the system can discharge it. Some rivers overflow their banks almost every year when the snows melt or rains arrive; others flood at irregular intervals. The floods deposit sediments containing minerals and organic nutrients on the floodplains of the river, making them fertile lands for agriculture (see Chapter 12). Because rivers also supply abundant water and ready-made transportation, floodplains have attracted human settlement since the beginning of civilization.

Small floods are common and usually cause little damage, but the larger episodes that happen every few decades or so can be quite destructive. In 1993, the Mississippi River and its tributaries went on a rampage with a flood that resulted in 487 deaths and more than $15 billion in property damage (**Figure 1.26**). At St. Louis, Missouri, the Mississippi stayed above flood stage for 144 of the 183 days between April and September. In an unexpected secondary effect, the floodwaters leached agricultural chemicals from farmlands and deposited them in the flooded areas, causing widespread pollution.

Today, most large cities are protected by artificial levees and dams, which can help control the flooding, though they cannot entirely eliminate the risk. The flooding of New Orleans by the storm surge from Hurricane Katrina was primarily due to levee failures.

Figuring out how to protect society from floods presents some knotty problems in geological engineering. A river hemmed in by levees can no longer erode its banks and widen its channel to accommodate additional water during periods of high flow, and its floodplain no longer receives deposits of sediment. In the case of New Orleans, the floodplain has sunk below the level of the Mississippi River, making future flooding more likely.

In a dramatic step after the 1993 Mississippi flood, the citizens of Valmeyer, Illinois, voted to move the entire

Rivers flood when water from rain or snowmelt flows into the river system more rapidly than it can be discharged from the system.

FIGURE 1.26 Satellite photo of the great flood of the Mississippi River, 1993. [NASA images created by Jesse Allen, Earth Observatory, from data courtesy of the Landsat Project Science Office.]

town to high ground several miles away. The new site was chosen with the help of a team of geologists from the Illinois Geological Survey. Yet some people who have lived on floodplains all their lives want to stay there and are prepared to live with the risk. The costs of protecting some floodplains are prohibitive, and these places will continue to pose public policy problems.

The prediction of river floods and their heights has become much more reliable as more rainfall and river measurements have become available. Geologists can now forecast the rise and fall of river levels as much as several months in advance, and they can issue reliable flood warnings days in advance.

Earthquakes Shaking episodes caused by the breaking of rocks along geologic faults are called **earthquakes.** *Faults* are surfaces of weakness in Earth's crust where the rocks on one side slip past those on the other side (**Figure 1.27**). The rupture starts at some point on the fault and propagates quickly (at more than 6000 miles per hour), like a crack through glass. The vibrations generated by this cracking travel away from the fault and, if the crack gets big enough, these *seismic waves* can cause strong shaking tens or even hundreds of kilometers away.

Earthquakes can be incredibly destructive to the built environment, ripping apart urban infrastructure and causing buildings to collapse. In cities, most casualties during

FIGURE 1.27 This scarp was formed by slip (left side downward) along a dipping fault surface during the 1954 Fairview Peak earthquake in Nevada. [Karl V. Steinbrugge Collection, Earthquake Engineering Research Center.]

earthquakes are caused by falling buildings and their contents. The death tolls can be especially high in densely populated areas of developing countries, where buildings are often constructed from bricks and mortar without steel reinforcement. More than 75,000 people were killed by a strong earthquake in the Kashmir region between Pakistan and India on October 8, 2005, including many children in poorly built schools (**Figure 1.28**). An earthquake that struck the city of Tangshan, China, in 1976 killed more than 240,000 people.

Secondary effects of earthquakes, such as landslides and fires, can also take their toll. The city of San Francisco was largely destroyed by the fires that followed the great earthquake of April 18, 1906; nearly 3000 of its inhabitants died. A great earthquake in China's Kansu Province in 1920 triggered an extensive landslide that covered a region larger than 100 km^2 and resulted in roughly 200,000 deaths.

When large earthquakes lift up the seafloor, they can generate water waves that propagate across the ocean and increase in size when they pile up along the shore. These **tsunamis** (Japanese for "harbor wave") can inundate and devastate coastal regions. On December 26, 2004, an undersea fault broke west of the Indonesian island of Sumatra, lifting up the seafloor and sending a tsunami across the Indian Ocean. The giant waves drowned more than 300,000 people living on coastlines from Thailand to Africa.

Earthquakes can be understood in terms of the basic geological machinery of plate tectonics, which moves continents and shapes the face of the planet (see Chapter 3). We can predict where and how frequently earthquakes will occur and how big they will be, but no one has yet figured out how to predict *when* large earthquakes will occur.

Even so, geologists can reduce earthquake damage by identifying dangerous faults and helping engineers to

> *Earthquakes are generated by the breaking of rocks along surfaces of weakness in Earth's crust, called faults. The resulting shaking can knock down buildings and trigger many secondary hazards, including fires, landslides, and tsunamis.*

FIGURE 1.28 Searching for victims in a school collapse caused by an earthquake in 2005 in Kashmir. [AP Photo/ Aamir Qureshi, pool.]

design buildings, dams, bridges, and other structures to withstand earthquake shaking. Sensitive instruments can pinpoint the location of the disturbance within seconds of a fault rupture, providing timely information to firefighters and medical personnel who must respond to a disaster.

Volcanic Eruptions Like earthquakes, **volcanic eruptions** are a result of plate tectonics. These events deposit molten rock from Earth's interior onto its surface and spew gases into its atmosphere. Like many other geologic phenomena, volcanoes are a mixed blessing, providing society with mineral resources, fertile soils, and thermal energy, but also posing threats to the communities around them.

The Pacific Ocean is rimmed by a "Ring of Fire" (see Figure 3.6) that contains many active volcanoes, including the majestic cones of Mount Rainier in the Pacific Northwest and Mount Fuji in Japan. These mountains can remain dormant for hundreds of years and then erupt in violent spasms, sometimes blowing their tops into the stratosphere and covering huge areas with hot rocks and ash.

Other types of volcanoes, like those found on the island of Hawaii in the middle of the Pacific, erupt more slowly and steadily, releasing hot, fluid lavas that fill up their craters and flood downhill in glowing rivers that bury everything in their path (Figure 1.29). Over hundreds of thousands of years, these flows can build up huge piles of lava. Measured from the seafloor, the Hawaiian volcanoes are the world's tallest mountains!

Like earthquakes, volcanic eruptions can trigger other hazards. The 1883 explosion of Krakatau, a volcano in Indonesia, generated a tsunami that reached 40 m in height and drowned 36,000 people on nearby coastlines. In 1985, the eruption of Nevado del Ruiz in the Colombian Andes melted ice and snow on its flanks, which rushed down the mountain in a huge mudflow that buried the town of Armero 50 km away, killing more than 25,000 people (Figure 1.30).

> Volcanic eruptions deposit molten rocks from Earth's interior onto its surface and spew gases into its atmosphere.

Geologists can identify dangerous volcanoes and characterize their hazards by studying the deposits laid down in earlier eruptions. Such investigations indicate that Mount Rainier, which looms over the heavily populated cities of Seattle and Tacoma, Washington (see Figure 5.33), probably poses the greatest volcanic risk in the United States. Hundreds of thousands of people live in areas where mudflows have swept down from the volcano over the past 6000 years.

In many cases, volcanic eruptions can be predicted. Sensitive instruments can detect activity associated with the upward movement of molten rocks, such as earthquakes, swelling of the volcano, and gas emissions. People at risk can be evacuated if the authorities are organized and prepared. Scientists monitoring Mount St. Helens, for example, were able to warn people of its eruption in 1980, as described in Chapter 5.

Landslides Downhill movements of rock, mud, and soil are called **landslides.** They occur when the friction that keeps the material from slipping downslope is overcome by the downward force of gravity, often due to water saturation. High seasonal rainfall in 2005 triggered many

FIGURE 1.29 A partly buried school bus in Kalapana, Hawaii. The village was buried by a lava flow from the Kilauea volcano. [Roger Ressmeyer/CORBIS.]

FIGURE 1.30 Armero, Colombia, submerged by mudflows, after an eruption of the long-dormant Nevado del Ruiz volcano, in 1985. [STF/ASP/Getty Images.]

Natural Hazards

landslides in Southern California, including the deadly La Conchita slide that killed 10 people on the night of January 10 (**Figure 1.31**).

Earthquakes, hurricanes, and other natural hazards can also trigger landslides. In 1998, one of the most catastrophic storms of the twentieth century, Hurricane Mitch, dropped torrential rains on Central America; at least 9000 people were killed in the ensuing floods and landslides. One of the hardest-hit places was near the Nicaragua-Honduras border, where a series of landslides and mudflows buried at least 1500 people. Dozens of villages were simply obliterated, engulfed by a sea of mud. The flanks of a crater on the Casita volcano collapsed and started a series of slides and flows that were described as a moving wall of mud more than 7 m high. Those in the direct path of the landslides could not escape, and many were buried alive as they tried to outrun the fast-moving mud. An immense landslide (over 80 million cubic meters) triggered by a 1970 earthquake in Peru destroyed the mountain towns of Yungay and Ranrahirca, killing more than 66,000 people (**Figure 1.32**).

Geologists can map the scars and deposits from previous landslides and issue timely warnings about future earth movements. They can also identify regions that are susceptible to landslides due to steep terrain and weak rock types. Geologic maps showing where landslides are most likely to occur can help land-use planners and developers avoid this type of hazard.

A landslide is a downhill movement of rocks, mud, and soil that is often due to the weakening of a slope by water saturation.

Bolides Chunks of rock and other debris from outer space, called **bolides,** may explode as fireballs in Earth's

FIGURE 1.31 A massive mudslide buried homes in La Conchita, California, in 2005. [AP Photo/Kevork Djansezian.]

FIGURE 1.32 In 1970, an earthquake-induced landslide on Mount Huascarán, Peru, buried the towns of Yungay and Ranrahirca. The landslide traveled 17 km at a speed of up to 280 km/hour and is estimated to have consisted of up to 100 million cubic meters of water, mud, and rocks. The death toll from the earthquake and landslide was 66,700 persons. (*left*) Towns of Yungay and Ranrahirca before the landslide; (*right*) aftermath of the landslide. [Both photos by Lloyd Cluff/CORBIS.]

FIGURE 1.33 Tree fall caused by the Tunguska explosion in 1908, photographed by the Kulik expedition of 1927.

atmosphere or strike its surface. Early in the history of the solar system, before about 3.9 billion years ago, large bolide impacts were much more common than they are today (most of the craters on the Moon date from this period). Collisions with planets have since swept the solar system fairly clean of debris.

Nevertheless, some 40,000 tons of extraterrestrial material continue to fall on Earth each year, mostly as dust and unnoticed small objects. Smaller pieces of debris heat up and vaporize in Earth's atmosphere before they reach the ground, but a few larger pieces make it through. In 1908, a bolide about 20 m in diameter exploded 5 to 10 km above the land surface near Tunguska, Siberia, flattening the forests across an area of 2000 km^2 (**Figure 1.33**). Collisions of this size, which release as much energy as 10 megatons of TNT (the equivalent of a large nuclear blast), probably take place every 500 to 1000 years. Larger chunks of matter 1 to 2 km in diameter impact Earth every few million years or so, releasing as much as 10,000 megatons of energy.

Because an event this size would pose a global threat to civilization, NASA and other space agencies have assigned telescopes to search the solar system for orbiting objects that might someday slam into Earth, and a systematic inventory of near-Earth objects is being conducted. Plans have even been formulated to use spacecraft to deflect such objects away from an Earth collision.

Large bolide impacts on Earth are infrequent, but space agencies are monitoring objects that might someday collide with Earth.

Reducing Risks from Natural Hazards

In discussing how Earth science deals with potential disasters, we need to differentiate between *hazard* and *risk*. The hazard depends on the intensity of the phenomenon, such as the wind speed during a hurricane or the intensity of shaking during an earthquake, and how often it occurs. In other words, hazard describes the chances that nature might throw something big at us.

Risk, on the other hand, describes the damage that is likely to be caused when a disaster actually strikes, usually measured in lost lives and dollars. Unlike hazard, risk depends on the population and number of buildings in a region, as well as on how easily the structures might be damaged. The risk to society from natural hazards is increasing, primarily because our exposure to those hazards is going up through the process of *urbanization,* or city building.

Consider earthquakes as an example. The coastal regions of both California and Alaska experience lots of earthquakes and therefore have comparable earthquake hazards. However, the population of California now exceeds 36 million, while fewer than 1 million people live in Alaska, so the risk that somebody will get hurt or something will be damaged by an earthquake is much greater in California.

The risk goes up as the hazard increases. The earthquake risk in Los Angeles is far greater than in Miami because destructive earthquakes are rare in southern Florida. The hurricane risk is reversed, since destructive hurricanes are common in Miami, but not in Los Angeles.

The risk associated with a natural hazard estimates how much damage might be done, which depends on a region's population.

It is an interesting fact, though an unfortunate one, that the most severe natural disasters—earthquakes, hurricanes, and volcanic eruptions—tend to be concentrated in coastal regions with warm climates, where otherwise favorable environments support booming populations and intense economic development. Consequently, the exposure of society to natural hazards is growing most rapidly where the hazards are highest.

Natural disasters cannot be prevented, but their catastrophic effects can be reduced by using Earth science in formulating public policies.

- Land-use policy can discourage urbanization in regions of high hazard, such as in the likely path of lava flows down active volcanoes.

- Regulations can be enacted to require builders and building owners to reinforce structures that could be

FIGURE 1.34 Workers of a rescue operation stand before a building completely demolished in the September 1985 earthquake in Mexico City, Mexico. More than 10,000 people were killed in Mexico City by this earthquake. [Owen Franken/CORBIS.]

damaged by high winds from hurricanes or intense ground shaking from earthquakes.

- Cities can prepare for anticipated disasters by stockpiling supplies and training people how to respond to emergencies (Figure 1.34).

Characterizing natural hazards and developing techniques for predicting them are important jobs for Earth scientists. History can be a guide: a volcano that has erupted in the past is likely to erupt again. However, the largest disasters are rare enough that the last big one may have happened before anyone was recording history. Earth scientists must therefore extend the historical record by searching for clues to prehistoric extreme events. For example, they can map unusual layers of sand along coastlines that mark old storm surges and tsunamis.

Enlightened public policies informed by Earth science and improved prediction methods can reduce the risk of many natural hazards.

THE ENVIRONMENT AND GLOBAL CHANGE

Life is everywhere on Earth's surface, from its equator to its poles, in its atmosphere, throughout its oceans, and even kilometers down into the rocky outer layers of its solid body. Life has been found in steaming volcanic vents at the bottom of the ocean, in lakes beneath the Antarctic ice, on the frozen peaks of the highest mountains, and in the deepest mines. All of the living organisms everywhere, taken together, form the **biosphere,** *the part of Earth that is alive* (Figure 1.35).

Geologists use methods from both biology and geology to understand how the biosphere interacts with the physical environment. Life constitutes only a small fraction of the environment in terms of mass or volume, but

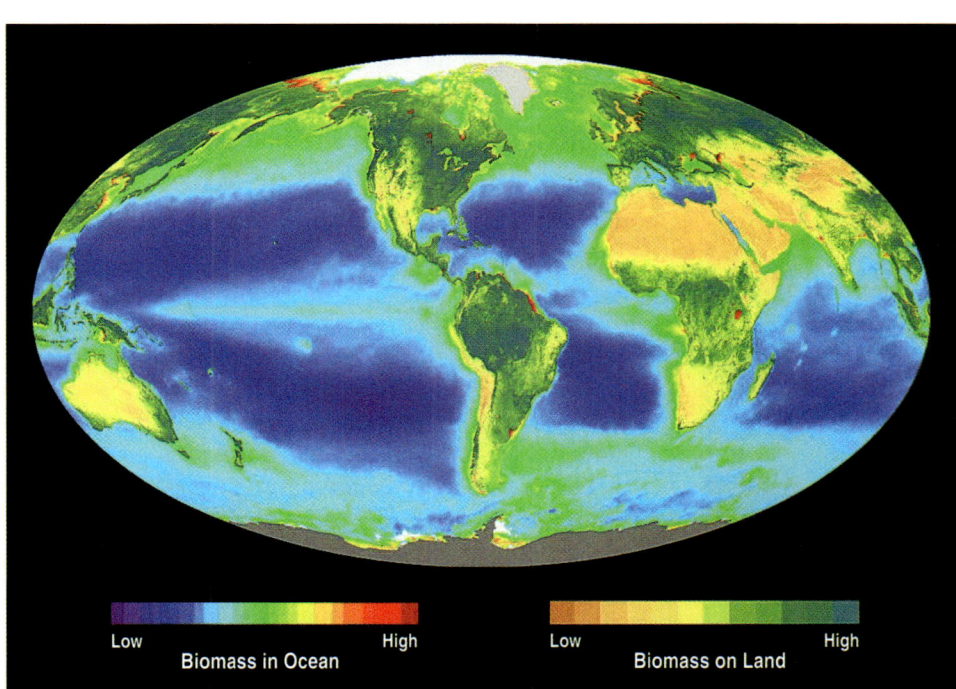

FIGURE 1.35 Earth's biosphere, represented by the global distribution of plant material in the oceans and on land, as mapped by NASA's SeaWiFS satellite. [NASA/Goddard Space Flight Center.]

it plays a very active role in maintaining hospitable conditions for future generations of life. When plants grow, they produce the oxygen that we animals breathe as well as the food we eat. Microorganisms—life we can see only under a microscope—are particularly important in sustaining the environment. Some types of microorganisms feed on rocks, mobilizing the mineral nutrients essential for plant growth. Others recycle dead organic material and return its components to the environment.

You will notice that people often discuss the environment from a local point of view, as when you see a report on "the environmental pollution of New York City" or read an "environmental impact statement" written to win approval for a shopping mall. We try to "save the environment" by recycling our trash and cleaning up toxic chemicals dumped into our soils and rivers. Geological knowledge helps us understand how local environments depend on natural resources and how we might adapt human activities to maintain local environmental quality.

More and more of our environmental concerns are global in scale, however. To address such concerns, we must consider the environment in its most general sense: the collection of all habitats of all living species—*the environment of the biosphere as a whole.*

The Human Factor

Humans have altered the environment by deforestation, agriculture, and other types of land uses since the first civilizations arose about 10,000 years ago (**Figure 1.36**). But our effects in earlier times were usually restricted to local or regional habitats. Energy production on an industrial scale now makes it possible for humans to compete with geologic processes in modifying the environment on a global scale. We are transforming the environment in fundamental ways, as illustrated by some startling observations:

- Dams and reservoirs built by humans now trap about 30 percent of the sediments transported down the world's rivers.

- In most developed countries, construction workers move more tons of soil and rock each year than do all natural processes of erosion combined.

- Within 50 years after the invention of the artificial coolant freon, enough of it had leaked out of refrigerators and air conditioners and floated into the upper atmosphere to damage Earth's protective ozone layer.

- Humans have converted about one-third of the world's forested area to other land uses, primarily agriculture, in the last half-century.

- Since the industrial revolution began in the early nineteenth century, deforestation and the burning of fossil fuels have increased the concentration of carbon dioxide in the atmosphere by almost 40 percent. The atmospheric concentration of carbon dioxide is rising at the unprecedented rate of 4 percent per decade and is likely to cause significant global climate change during the lifetimes of most people now in college.

> Modern industrialized society is capable of changing the environment on an unprecedented global scale.

The expression **global change** entered the world's vocabulary when it became clear that emissions from fossil-fuel burning and other human activities were beginning to alter the chemistry of the atmosphere. People became concerned about changes such as

- mass die-offs of plants and animals due to acid rain

- increased exposure to the Sun's ultraviolet rays due to ozone depletion in the upper atmosphere

- global warming due to increased carbon dioxide and other greenhouse gases in the atmosphere

These changes are all **anthropogenic,** or human-generated (from *anthropos,* the Greek word for "man"). However, they differ considerably in scope and magnitude. Acid rain is primarily a regional problem, although if left unabated, it could have global consequences. Ozone depletion in the upper atmosphere is clearly a global problem, but its cause is mainly chemicals of industrial origin, which are now strictly regulated through an international treaty—a major environmental success. In contrast, there appears to be little agreement over what we should do about global warming and related aspects of human-induced climate change.

FIGURE 1.36 In Indonesia, deforestation by slash and burn. [Charles O'Rear/CORBIS.]

We will discuss the problems of acid rain, stratospheric ozone depletion, and climate change to illustrate the problems of anthropogenic global change and some of their possible solutions.

Acid Rain

In many industrialized areas of the world, the air is severely polluted with sulfur-containing gases. These gases are emitted primarily from the smokestacks of power plants that burn coal containing large amounts of sulfur and from smelters where ores are melted to produce iron and other metals. Coals mined in the eastern and midwestern regions of the United States contain more of these pollutants than coals mined in the western states. Although volcanoes and coastal marshes also add sulfur gases to the atmosphere, more than 90 percent of the sulfur emissions in eastern North America are of anthropogenic origin.

Sulfur gases in the atmosphere react with oxygen and water to form sulfuric acid. Small amounts of sulfuric acid can turn harmless rainwater into **acid rain.** Although it is much too weak to sting human skin, acid rain causes widespread damage to delicate organisms and solid rock.

By acidifying sensitive lakes and streams, acid rain has caused massive kills of fish in many lakes in Canada, the northeastern United States, and Scandinavia. A survey of more than 1000 lakes and thousands of miles of streams in the northeastern United States showed that acid rain has affected 75 percent of the lakes and 50 percent of the streams. In some acidified lakes and streams, fish species such as the brook trout have been completely eradicated. The salmon habitat of eastern Canada has been greatly affected. Acid rain also damages mountain forests, particularly those at high elevations. Acid moisture in the air reduces visibility.

Acid rain causes noticeable damage to fabrics, paints, metals, and rocks, and it rapidly weathers stone monuments and outdoor sculptures (Figure 1.37). In Canada alone, acid rain causes about $1 billion in damage every year to buildings and monuments.

Most of the emissions of sulfur gases in North America come from coal-fired electric utility plants in the midwestern and eastern United States, and most of the acid rain in North America occurs in these regions and in eastern Canada. About 50 percent of the acid rain in Canada originates in the United States—a source of friction between the two countries. The problem of acid rain is equally critical in Europe and Asia, where scientists have also noted damage to lakes and forests. The damage is particularly heavy in China, in eastern European countries, and in Russia.

Concerned scientists have long recommended the restriction of sulfur emissions from power plants and smelters. For a long time, however, political resistance prevented effective control. Finally, in 1990, after many years of wrangling, the U.S. Congress passed major amendments to the 1970 Clean Air Act. The legislation required coal-burning power plants to reduce their annual sulfur emissions by 10 million tons by the year 2000. Deregulation of U.S. railroads lowered the costs of transporting cleaner western coal to midwestern and eastern power plants, thus making it easier to meet the new emissions standards. The

Sulfur-containing gases, emitted primarily from coal-burning power plants, react with oxygen and water in the atmosphere to produce acid rain.

FIGURE 1.37 A monument before and after deterioration caused by acid rain.
[Westfalisches Amt fur Denkmalpflege.]

Bush administration's efforts to relax the 1990 standards, however, have reopened this contentious debate.

Stratospheric Ozone Depletion

Ozone (O_3^+) is a reactive form of oxygen gas. Near Earth's surface, ozone is a major constituent of smog. It undermines health, damages crops, and corrodes materials. Low-lying ozone forms when sunlight interacts with chemical wastes from industrial processes and automobile exhausts.

Ozone also exists in Earth's upper atmosphere, or *stratosphere*, where it is concentrated in a layer 25 to 30 km above the surface. There, solar radiation transforms ordinary oxygen gas into ozone, forming a protective layer that shields Earth by absorbing cell-damaging ultraviolet (UV) radiation from the Sun. Skin cancer, cataracts, impaired immune systems, and reduced crop yields are attributable to excessive UV exposure.

In 1995, the Nobel Prize in chemistry was awarded to Mario Molina, Sherwood Rowland, and Paul Crutzen for the hypotheses they had advanced more than 20 years earlier: that the protective ozone layer can be depleted by chemical reactions involving anthropogenic compounds. One class of compounds, chlorofluorocarbons (CFCs)—which were used as refrigerants, spray-can propellants, and cleaning solvents—raised special concerns. These compounds are stable and harmless—except when they migrate to the stratosphere. Molina and Rowland proposed that these chemicals undergo reactions that destroy ozone in the stratosphere, thinning the protective ozone layer. Molina and Rowland's hypothesis was confirmed when a large hole in the ozone layer was discovered over Antarctica in 1985 (**Figure 1.38**). Subsequently, **stratospheric ozone depletion** was found to be a global phenomenon. Elevated surface levels of UV radiation have been observed to correspond to low stratospheric ozone levels since the early 1990s.

In the 1980s, when scientists were trying to convince government and industry officials that the ozone layer was being depleted due to CFCs, a senior government official remarked that the solution was for people to wear hats, sunscreen, and dark glasses. Fortunately, political wisdom prevailed. In 1989, a group of nations entered into a global treaty to protect the ozone layer. This treaty, called the Montreal Protocol, now includes 140 countries. The Montreal Protocol required that CFC production be phased out by 1996 and set up a fund, paid for by developed nations, to help developing nations switch to ozone-safe chemicals. The manufacturing of safer alternatives has led to a successful phase-out of CFCs, and long-term projections indicate that the depletion of the ozone layer will diminish over the next several decades. The Montreal Protocol has become a model for how scientists, industrial leaders, and government officials can work together to head off an environmental disaster.

The ozone layer in the stratosphere, which shields the biosphere from harmful ultraviolet radiation, has been damaged by chlorofluorocarbons. Use of these chemicals has now been curtailed.

FIGURE 1.38 In September 2001, the area of the Antarctic ozone hole extended over 50 million square kilometers, larger than the combined area of the United States, Canada, and Mexico. [NASA.]

Global Warming

Carbon dioxide (CO_2) is a **greenhouse gas;** it helps to trap solar heat in the atmosphere, much as the frosted windows in a greenhouse trap solar heat. Fossil-fuel burning, deforestation, and agriculture since the beginning of the industrial era have caused a significant rise in the concentrations of carbon dioxide and other greenhouse gases, such as methane, in the atmosphere.

Recognizing the potential problems that this trend poses for global climate, the United Nations established an Intergovernmental Panel on Climate Change (IPCC) in 1988 to assess the risk of anthropogenic climate change, its potential effects, and possible solutions to those problems. The IPCC provides a continuing forum for hundreds of scientists, economists, and policy experts to work together to understand these issues.

In major assessment reports, published in 2001 and 2007, the IPCC drew the following conclusions:

- Since the beginning of the twentieth century, the average temperature of Earth's surface has risen, on average, by about 0.6°C (**Figure 1.39**).

- Much of this "twentieth-century warming" has been caused by anthropogenic increases in greenhouse gases.

- Anthropogenic warming is already causing changes in many natural systems, including reductions in Arctic sea ice and mountain glaciers, warming of lakes and rivers, earlier timing of spring events such as leaf unfolding and bird migrations, and longer growing seasons for plants.

- Levels of greenhouse gases will continue to increase throughout the twenty-first century, primarily because of human activities.

- The rate of increase will depend on a number of socioeconomic factors that will govern the rate of greenhouse gas emissions, including active steps by society to limit those emissions.

- The increase in greenhouse gases will cause significant global warming during the twenty-first century. Projections of the amount of warming are highly uncertain because of doubt about the amount of future greenhouse gas emissions and an incomplete knowledge of how the climate system works. The range of likely temperature increases accepted by most experts is 1° to 6°C (see Figure 1.39).

According to the IPCC, global warming during the twenty-first century is likely to be accompanied by substantial global and regional changes. The warming will probably be greater over land than over the ocean, and regional climates will probably become more variable. Other likely effects include increasing rainfall in the tropics, decreasing rainfall and increasing droughts in temperate zones such as the United States, stronger hurricanes, greater flooding, more heat waves, melting of polar ice, and a rising sea level.

Some of the effects of global warming can already be seen and felt (Figure 1.40). The 10 warmest years ever recorded have all occurred since 1989. In 1998 and 2005, the two warmest years since accurate records began in 1880, the global mean temperature was 0.7°C above the 1961–1990 annual average. The number of very strong hurricanes has almost doubled over the past three decades, and increases in tropical storm intensity have been observed around the world. The amount of sea ice in the Arctic Ocean is decreasing, and the melting of ice in Antarctica and Greenland is accelerating rapidly. Sea level is expected to rise by half a meter or more during the twenty-first century, creating serious problems for low-lying countries such as Bangladesh, as well as the Eastern Seaboard and Gulf Coast of the United States, where flooding during storm surges could become much worse.

> *Anthropogenic global warming is caused primarily by rising concentrations of greenhouse gases from the burning of fossil fuels, deforestation, and agriculture.*

> *Anthropogenic global warming is causing polar ice to melt and storm intensities to increase. It is likely to cause the sea level to rise and species to go extinct.*

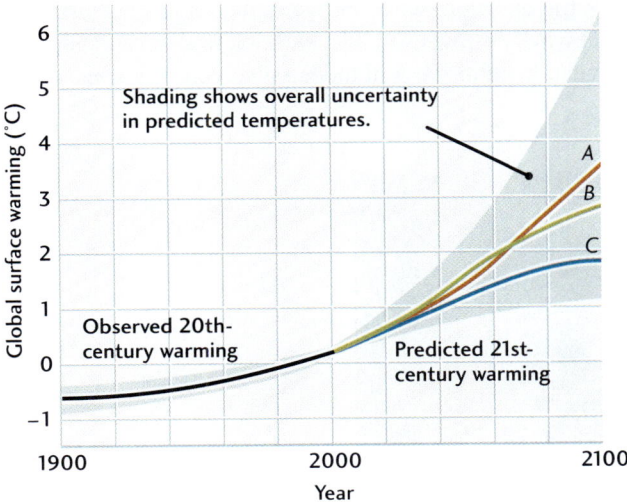

FIGURE 1.39 Global warming, past and future. The black line shows the average surface temperatures observed during the twentieth century (heavily smoothed). Colored lines show three possible scenarios for the twenty-first century. (A) Slow economic growth and continued reliance on fossil fuels. (B) Rapid economic growth and conversion to non–fossil fuels at a moderate rate. (C) Rapid economic growth and rapid conversion to non–fossil fuels. Not shown is a scenario with rapid economic growth and continued reliance on fossil fuels, which would result in global warming in excess of 4°C. The shaded region represents the overall uncertainties in the predicted temperatures from lack of knowledge of the climate system, as well as uncertainties in economic growth and energy technologies. [IPCC, *Climate Change 2007: The Physical Scientific Basis.* Cambridge University Press, 2007.]

FIGURE 1.40 Global warming is reducing Arctic Sea ice, adversely affecting the habitat of Arctic animals such as polar bears. [Thomas and Pat Leeson/Photo Researchers.]

Because anthropogenic warming will be rapid, many plant and animal species will have difficulty adjusting or migrating. Those that cannot cope with rapid warming may become extinct. Global warming is already being blamed for a variety of adverse ecological effects, such as the disruption of Arctic ecosystems as ice lines move northward and permafrost begins to melt, and the spread of malaria as more of the world experiences a tropical climate.

The potentially dire consequences of anthropogenic global change are motivating politicians to work together in ways they never have before, as we all try to avoid the "tragedy of the commons"—the spoiling of our commonly held environmental resources. Neighboring nations are enacting mutually beneficial regulations to address regional environmental problems, and new multinational treaties are being formulated in attempts to manage anthropogenic effects on the global environment. Earth science gives us the knowledge to make rational choices about global environmental management.

OUR REASONS FOR STUDYING EARTH

As you have seen in this chapter, Earth science is a very practical business. It provides our society with the knowledge to manage our natural resources; strategies for characterizing natural hazards and reducing risk; and the ability to understand the environment and predict global change. The authors of this textbook have worked as geologists on many aspects of these practical problems, and we have been rewarded for our efforts (**Figure 1.41**).

Yet our satisfaction as scientists comes from a deeper, more abstract motivation: to understand the world around us, how it works, and how it has evolved. Although we use the tools of physics, chemistry, biology, and mathematics, our inspiration comes from the complexity and beauty of the Earth system, this great spaceship on which we sail.

When we descend into the deep gold mines of South Africa, we are aware of the riches being pulled from the excavations, but we also image the primeval conditions that deposited the gold in those rocks billions of years ago. The stories of those rocks are tales we try to unravel.

When we hike up the slopes of an active volcano, we recognize the danger it poses to the communities below, but we also realize that the molten rock and gases blowing out its top come from many kilometers below Earth's surface. To us, it is a window through which we can look into Earth's deep interior.

When we drive or fly across a landscape, we always ask the question, what are we seeing, and how did it get that way? Follow us in this textbook to sharpen your geological insight. We will indeed give you new eyes!

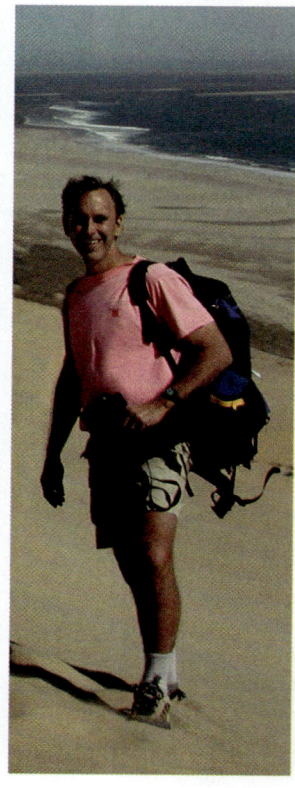

FIGURE 1.41 The authors at work. (*left*) Tom Jordan at a depth of 2800 m in the Mponeng gold mine, South Africa. (*right*) John Grotzinger on Namibia's Skeleton Coast.

SUMMARY

What is geology? Geology is the science that deals with Earth—its history, its composition and internal structure, and its surface features. Geology provides the scientific knowledge needed to develop natural resources, deal with natural disasters, preserve the environment, and adapt to global change. Geophysics, geochemistry, and geobiology are subfields of geology that apply the methods of physics, chemistry, and biology to geologic problems.

What is the goal of sustainable development? Sustainable development is the idea that human activities can use natural resources in a way that satisfies the present needs of society without compromising the ability of future generations to meet their own needs. This goal will require managing natural resources to ensure that adequate energy, water, and raw materials are available; preparing society for inevitable natural disasters; repairing environmental damage; and maintaining a habitable environment over the long term.

How do we categorize our natural resources? Reserves are the known supplies of natural resources that can be exploited economically under current conditions. The most important resources available from the natural environment are energy, minerals and other mined raw materials, water, and soils. Natural resources can be classified as renewable and nonrenewable resources, depending on whether they are replenished at rates comparable to the rates at which we are consuming them. Fossil fuels are examples of nonrenewable energy resources; solar power and biofuels are examples of renewable energy resources.

How important are fossil fuels to our energy economy? Fossil fuels, which include oil, natural gas, and coal, dominate our present energy use. Taken together, these resources account for 86 percent of the energy consumed by the United States and the world at large. Despite efforts to locate more reserves and improve production, the major reservoirs of oil and natural gas will become depleted sometime in the next few decades. Renewable energy sources, which include biofuels and solar, wind, water, and geothermal power, are likely to fall short of replacing oil unless there are unanticipated technological breakthroughs. Many observers expect that coal, which is geologically plentiful and economically cheap to develop, will fill the gap.

What resources come from mining? Mining provides the minerals and other raw materials that are used in huge quantities by our industrialized society. Demand for these materials, especially metals, is increasing due to the rapid expansion of the Asian economies. Ores, the high-grade deposits of minerals from which metals are extracted, can be found in a variety of geological environments. Many of the most valuable ores were formed in regions of volcanic activity, where hot circulating groundwater leached metals from rocks and deposited them in a concentrated form. Geologists are exploring for new mineral resources in remote areas, including polar regions and the ocean floor. In exploiting mineral resources, society must balance the economic gains of mining against its considerable environmental and health costs.

Where does society get its water resources? About 1 percent of Earth's water is fresh water. Only a small fraction of this fresh water occupies lakes and rivers; most is groundwater contained in rocks beneath the land surface. Many regions depend on reservoirs of groundwater in porous rock formations, called aquifers, for their water needs. Pumping water out of aquifers more rapidly than precipitation recharges them can deplete them. In these situations, groundwater becomes a nonrenewable resource.

Why are soils classified as a natural resource? Soils are the primary reservoir of nutrients for agriculture and the ecological systems that produce renewable natural resources. They serve as raw materials for construction and manufacturing. They contain twice as much carbon as the atmosphere and three times more than all of the world's vegetation. They recycle the organic carbon from dead plants and animals and release it into the atmosphere. The quality of the fresh water most people use is determined largely by the soils it passes through.

What are the principal natural hazards? Natural hazards are events produced by natural processes that have the potential to kill people and damage our built environment. Hurricanes (also called typhoons and cyclones) are great storms that suck energy into the atmosphere from the warm surface of tropical seas. Floods can result from inundation by the sea, but the most frequent cause is river flooding, which occurs when water from rainfall and snowmelt flows into a river system faster than it can flow out. Earthquakes are shaking episodes caused by the breaking of rocks along surfaces of weakness in Earth's crust, called faults. Volcanic eruptions deposit molten rock from Earth's interior on its surface. Landslides are downhill movements of rock, mud, and soil that occur when the friction that keeps the material from slipping is overcome by the downward force of gravity, often due to saturation by water. Bolides are chunks of rock and other debris from outer space that explode as fireballs in Earth's atmosphere or impact its surface.

What is the difference between hazard and risk? The hazard to society from a natural phenomenon depends on the intensity of the phenomenon and how often it occurs. Risk describes the damage that is likely to be caused when a disaster actually strikes; it is usually measured in lost lives and dollars. The risk depends on the population and structures exposed to the hazard, as well as how easily the structures might be damaged, whereas the hazard is independent of these factors. The risk to society from natural hazards is increasing, primarily because our exposure to natural hazards is increasing through the process of urbanization.

In what ways are humans affecting the environment? The environment is the region near Earth's surface inhabited by living organisms (which, taken together, constitute the biosphere). The effects of human activities on the environment are growing with the human population, which now exceeds 6 billion people. Humans have always affected their environment on a local scale, but the industrialization of society has harnessed enough energy to compete with geologic processes in causing changes on a global scale. Examples of anthropogenic global change include acid rain, stratospheric ozone depletion by chlorofluorocarbons, and global warming due to the emission of greenhouse gases.

How might human activities change global climate? Fossil-fuel burning and deforestation release carbon dioxide, a greenhouse gas. The rising concentration of carbon dioxide and other greenhouse gases in the atmosphere is likely to cause significant global climate change during the twenty-first century. The nature of that climate change remains uncertain, but scientists estimate that the average global temperature will increase by 1° to 6°C before 2100. This warming is likely to have a number of adverse environmental effects, such as melting of polar ice, a rising sea level, the intensification of storms, prolonged droughts, and extinctions of some plant and animal species.

KEY TERMS AND CONCEPTS >>

acid rain (p. 25)
anthropogenic (p. 24)
aquifer (p. 14)
biosphere (p. 23)
bolide (p. 21)
earthquake (p. 18)
environment (p. 3)
flood (p. 18)
fossil fuels (p. 6)
geology (p. 2)
global change (p. 24)
greenhouse gas (p. 26)
groundwater (p. 14)
hurricane (p. 17)
landslide (p. 20)
natural hazards (p. 16)
natural resources (p. 4)
nonrenewable resource (p. 6)
renewable resource (p. 5)
reserves (p. 4)
risk (p. 22)
soils (p. 15)
stratospheric ozone depletion (p. 26)
sustainable development (p. 4)
tsunami (p. 19)
volcanic eruption (p. 20)

EXERCISES >>

1. What are the main goals of sustainable development?

2. Approximately what percentage of the U.S. energy supply comes from renewable resources?

3. How do geologists contribute to the expansion of our oil reserves?

4. Why have the worldwide prices of many useful metals, such as iron, copper, zinc, and nickel, increased so rapidly in the last few years?

5. What human activity consumes the most water resources in the western United States?

6. Give an example of a situation in which water has become a nonrenewable resource.

7. Why are soils classified as a valuable natural resource? Are they a renewable or nonrenewable resource?

8. The earthquake hazards for the cities of Anchorage, Alaska, and San Francisco, California, are both high. Which place has the higher earthquake risk? Why?

9. Which natural hazard poses the greatest risk to your community?

10. A major bolide impact could destroy civilization, yet the risks from hurricanes and earthquakes are considered to be much greater than the risk from bolides. Why is this the case?

11. In what ways can geology help people living near volcanoes prepare for potentially disastrous eruptions?

12. What factors have made anthropogenic effects on the environment a global problem, rather just a regional one? What are some of those effects?

13. What causes acid rain, and what are its effects?

14. What is causing the depletion of ozone in the stratosphere, and what are its effects?

15. How does the burning of fossil fuels cause global warming?

16. In one sentence or two, why do you think it is important to study Earth?

HOW WE STUDY EARTH

CHAPTER 2

Measuring Mount Everest

On February 6, 1800, Colonel William Lambton of the 33rd Foot Regiment received orders to begin the Great Trigonometrical Survey of India, the most ambitious scientific project of the nineteenth century. Over the next several decades, intrepid British explorers led by Lambton and his successor, George Everest, hauled bulky telescopes and heavy surveying equipment through the jungles of the Indian subcontinent, triangulating the positions of reference monuments established at high points in the terrain, from which they could accurately establish Earth's size and shape. Along the way, in 1852, they discovered that an obscure Himalayan peak, known on their maps only as "Peak XV," was the highest mountain on Earth. They promptly named it Mount Everest, in honor of their former boss.

On February 11, 2000, almost exactly 200 years after Lambton commenced his exploration, NASA launched another great survey, the Shuttle Radar Topography Mission (SRTM). The space shuttle *Endeavour* carried two large radar antennas into low Earth orbit, one in the cargo bay and the second mounted on a mast that could extend outward to 60 m. Working together like a pair of eyes, these antennas mapped the height of the land surface below the shuttle on a very dense geographic grid, rendering the terrain in unprecedented three-dimensional detail. Remarkably, the height of Mount Everest as confirmed by the SRTM (8850 m, or 29,035 ft) turned out to be only 10 m more than the original 1852 estimate.

Though the accuracy of the Great Trigonometrical Survey was impressive, collecting data was a slow process. It took the British over 70 years to measure the positions of 2700 stations across the Indian subcontinent, an average of about one position every 3 months. In comparison, the SRTM collected about 3000 position measurements *each second*. In just 11 days, the SRTM mapped 2.6 billion points covering 80 percent of Earth's land surface, including many remote areas that had not been previously surveyed. And, unlike the British, the shuttle crew did not have to contend with malaria or tigers!

Artist's depiction of the space shuttle *Endeavour* during the Shuttle Radar Topography Mission. The SRTM mapped 80 percent of Earth's land surface using a dual-antenna radar. [NASA.]

CHAPTER OUTLINE

The Scientific Method 32

Geology as a Science 33

Earth's Shape and Surface 36

Peeling the Onion: Discovery of a Layered Earth 38

Earth as a System of Interacting Components 42

An Overview of Geologic Time 46

This chapter paints a broad picture of how we study Earth. It starts with the scientific method, the observational approach to the physical universe on which all scientific inquiry is based. Throughout this textbook, you will see the scientific method in action as we describe how Earth scientists gather and interpret information about our planet. In this chapter, we will illustrate how the scientific method has been applied to discover some of Earth's basic features—its shape and its internal layering. We will also introduce you to a geologist's view of time.

You may start to think about time differently as you begin to comprehend the immense span of geologic history. Earth and the other planets in our solar system formed about 4.5 billion years ago. More than 3 billion years ago, living cells developed on Earth's surface, and life has been evolving ever since. Yet our human origins date back only a few million years—a mere few hundredths of a percent of Earth's existence. The scales that measure individual lives in decades and mark off periods of human history in hundreds or thousands of years are inadequate to study Earth history.

To explain features that are millions or even billions of years old, we look at what is happening on Earth today. We study our complex natural world as an *Earth system* involving many interrelated components. Some of these components, such as the atmosphere and oceans, are clearly visible above Earth's solid surface; others lie hidden deep within its interior. By observing the ways these components interact, scientists have built up an understanding of how the Earth system has changed through geologic time.

THE SCIENTIFIC METHOD

Scientists believe that physical events have physical explanations, even if they may be beyond our present capacity to understand them. The **scientific method,** on which all scientists rely, is the general procedure for discovering how the physical universe works through systematic observations and experiments. Using the scientific method to make new discoveries and to confirm old ones is the process of *scientific research* (**Figure 2.1**).

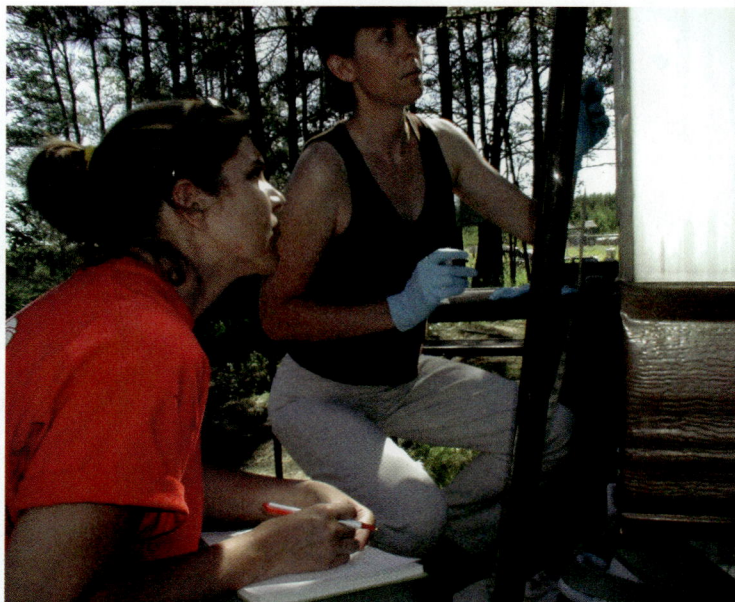

FIGURE 2.1 Research is the process of discovery and confirmation through observation of the natural world. These geologists are researching soil samples near a lake in Minnesota. [U.S. Geological Survey.]

When scientists propose a **hypothesis**—a tentative explanation based on observational data and experiments—they present it to the community of scientists for criticism and repeated testing. A hypothesis that is confirmed by other scientists gains credibility, particularly if it explains new data or predicts the outcome of new experiments.

Here are four interesting scientific hypotheses we have already encountered in the first chapter of this textbook:

- Earth is billions of years old.
- Coal is a rock formed primarily from dead plants.
- Earthquakes are caused by the breaking of rocks along geologic faults.
- The burning of fossil fuels is causing global warming.

The first hypothesis agrees with the ages of thousands of ancient rocks as derived from precise laboratory measurements, and the next two have also been confirmed by many independent observations. The fourth hypothesis has been more controversial, though so many new data support it that most scientists now accept it as true (see Chapter 14). Identifying other examples of hypotheses from Chapter 1 is a good exercise for you to try.

A coherent set of hypotheses that explains some aspect of nature constitutes a *theory*. Good theories are supported by substantial bodies of data and have survived repeated challenges. They usually obey *physical laws,* general principles about how the universe works that can be applied in every situation, such as Newton's law of gravity.

Some hypotheses and theories have been so extensively tested that all scientists accept them as true, at least to a good approximation. For instance, the theory that Earth is almost spherical, which follows from Newton's law of gravity, is supported by so much experience and direct evidence (ask any astronaut) that we take it to be a fact. The longer a theory holds up to all scientific challenges, the more confidently it is held.

Yet theories can never be considered completely proven. The essence of science is that no explanation, no matter how believable or appealing, is closed to questioning. If convincing new evidence indicates that a theory is wrong, scientists will discard it or modify it to account for the data. A theory, like a hypothesis, should always be testable; any proposal about the universe that can't be evaluated by observing nature should not be called a scientific theory!

The scientific method is a procedure for discovering how the universe works through systematic observations and experiments.

For scientists engaged in research, the most interesting hypotheses are often the most controversial, rather than the most widely accepted. The hypothesis that fossil-fuel burning causes global warming has been widely debated. Because the long-term predictions of this hypothesis are so important, many Earth scientists are now involved in testing it.

Knowledge based on many hypotheses and theories can be used to create a *scientific model*—a representation of how a natural process operates or how a natural system behaves. Scientists combine related ideas in a model to test the consistency of their knowledge and to make predictions. Like a good hypothesis or theory, a good model makes predictions that agree with observations.

A scientific model is often formulated as a computer program that simulates the behavior of a natural system through numerical calculations. The forecast of rain or shine you may see on TV tonight comes from a computer model of the weather. A computer can be programmed to simulate geologic phenomena that are too big to replicate in a laboratory or that operate over periods of time that are too long for humans to observe. For example, weather models have been extended to predict climate changes decades in the future.

To encourage discussion of their ideas, scientists share them and the data on which they are based. They present their findings at professional meetings (Figure 2.2), publish them in professional journals, and explain them in informal conversations with colleagues. Most of the great concepts of science, whether they emerge as a flash of insight or in the course of painstaking analysis, result from untold numbers of such interactions. Albert Einstein put it this way: "In science . . . the work of the individual is so bound up with that of his scientific predecessors and contemporaries that it appears almost as an impersonal product of his generation."

FIGURE 2.2 Presentations at scientific meetings, such as this one at a German university, are part of the scientific process. [Technische Universität Berlin/ Sabine Böck.]

Because such free intellectual exchange can be subject to abuses, a code of ethics has evolved among scientists. Scientists must acknowledge the contributions of all others on whose work they have drawn. They must not falsify data, use the work of others without recognizing them, or be otherwise deceitful in their work. They must also accept responsibility for training the next generation of researchers and teachers. These principles are supported by the basic values of scientific cooperation, which a president of the National Academy of Sciences, Bruce Alberts, has aptly described as "honesty, generosity, a respect for evidence, openness to all ideas and opinions."

GEOLOGY AS A SCIENCE

In the popular media, scientists are often portrayed as people who do experiments wearing white coats. That stereotype is not inappropriate: many scientific problems are best investigated in the laboratory. What forces keep atoms together? How do chemicals react? Can viruses cause cancer? The phenomena that scientists observe to answer such

FIGURE 2.3 Marine scientists Craig Marquette and Will Ostrom, from the Woods Hole Oceanographic Institution, deploy a mooring to measure temperatures from the Research Vessel *Oceanus* during a gale off Cape Hatteras. [Chris Linder, Woods Hole Oceanographic Institution.]

questions are sufficiently small and happen quickly enough to be studied in the controlled environment of the laboratory.

The major issues of Earth science, however, involve processes that operate on much larger scales. Controlled laboratory measurements yield critical data for testing hypotheses and theories—the ages and properties of rocks, for instance—but they are usually insufficient to solve major geologic problems. Almost all of the great discoveries described in this textbook were made by observing Earth processes in their uncontrolled, natural environment.

For this reason, geology is an outdoor science with its own particular style and outlook. Geologists "go into the field" to observe nature directly. They learn how mountains were formed by climbing up steep slopes and examining the exposed rocks. They discover how ocean basins have evolved by sailing rough seas to study the ocean floor (**Figure 2.3**).

Geology is also a high-tech science. In the opening story of this chapter, we saw how instruments aboard spacecraft can amass huge quantities of information about Earth's surface. This type of *remote sensing* allows us to map the continents, chart motions of the atmosphere and oceans, and monitor how our environment is changing. Geologists also deploy sensitive instruments on land and on the seafloor to collect data on earthquakes, volcanic eruptions, and other activity within the solid Earth (**Figure 2.4**).

A very special aspect of geology is its ability to probe Earth's long history by reading what has been "written in stone." The **geologic record** is the information preserved in the rocks that have been formed at various times throughout a history more than 4 billion years long. Geologists decipher the geologic record by combining information from many kinds of work: examination of rocks in the field (**Figure 2.5**); careful mapping of their positions relative to older and younger rock formations; collection of representative samples; and determination of their ages using sensitive laboratory instruments (**Figure 2.6**).

> *The information preserved in rocks—the geologic record—can be deciphered to understand Earth's long history.*

FIGURE 2.4 Geologist Peter Gray welding one of the five Global Positioning System stations placed on the flanks of Mount St. Helens. The stations will monitor the changing shape of the land surface as molten rock moves upward within the volcano. [Lyn Topinka/USGS.]

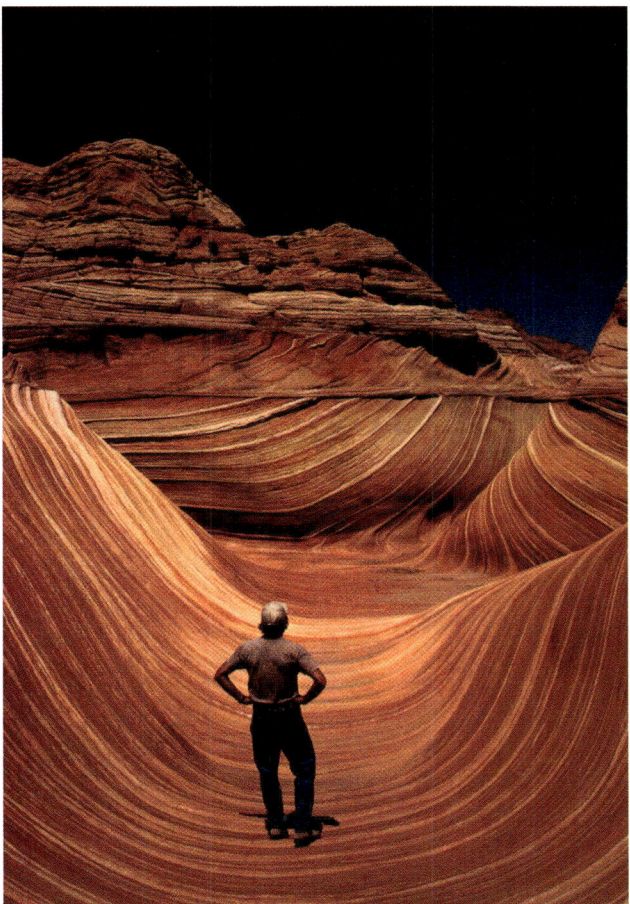

FIGURE 2.5 The geologic record preserves evidence of Earth's long history. These multicolored layers of sandstone at Colorado National Monument were deposited more than 200 million years ago when this part of the western United States was a vast Sahara-like desert. They were subsequently overlain by other rocks, welded by pressure into sandstone, uplifted by mountain-building events, and eroded by wind and water into today's stunning landforms. [Lonely Planet Images/Mark Newman.]

FIGURE 2.6 Sensitive instruments like this mass spectrometer can determine the age of a rock by measuring minute amounts of radioactive elements and how that radioactivity has decayed since the rock first formed. [Rosenfeld Images Ltd./Photo Researchers.]

In *Annals of the Former World,* a compendium of colorful stories about geologists, the popular writer John McPhee offers his view of how geologists bring field and laboratory observations together to visualize the big picture:

They look at mud and see mountains, in mountains oceans, in oceans mountains to be. They go up to some rock and figure out a story, another rock, another story, and as the stories compile through time they connect— and long case histories are constructed and written from interpreted patterns of clues. This is detective work on a scale unimaginable to most detectives, with the notable exception of Sherlock Holmes.

The geologic record tells us that, for the most part, the processes we see in action on Earth today have worked in much the same way throughout the geologic past. This important concept is known as the **principle of uniformitarianism.** It was stated as a scientific hypothesis in the late eighteenth century by a Scottish physician and geologist, James Hutton. In 1830, the British geologist Charles Lyell summarized the concept in a memorable line: "The present is the key to the past."

The principle of uniformitarianism does not mean that all geologic phenomena proceed at the same gradual pace. Some of the most important geologic processes happen as sudden events. A large meteorite that impacts Earth can gouge out a vast crater in a matter of seconds. A volcano can blow its top, and a fault can rupture the ground in an earthquake, almost as quickly. Other processes occur much more slowly. Millions of years are required for continents to drift apart, for mountains to be raised and eroded, and for river systems to deposit thick layers of sediments. Geologic processes take place over a tremendous range of scales in both space and time (**Figure 2.7**).

Nor does the principle of uniformitarianism mean that we have to observe a geologic event to know that it is important in the current Earth system. Humans have never witnessed a large meteorite impact in recorded history, but we know these impacts have occurred many times in the geologic past and will certainly happen again. The same can be said for the vast volcanic outpourings that have covered areas bigger than Texas with lava and poisoned the global atmosphere with volcanic gases. The long history of Earth is punctuated by many such extreme, though infrequent, events that result in rapid changes in the Earth system. Geology is the study of extreme events as well as gradual change.

From Hutton's day onward, geologists have observed nature at work and used the principle of uniformitarianism to interpret features

> *The principle of uniformitarianism states that the geologic processes we see in action on Earth today have worked in much the same way throughout the geologic past.*

Over millions of years, layers of sediments built up over the oldest rocks. The most recent layer—the top—is about 250 million years old.

The rocks at the bottom of the Grand Canyon are 1.7–2.0 billion years old.

About 50,000 years ago, the explosive impact of a meteorite (perhaps weighing 300,000 tons) created this 1.2-km-wide crater in just a few seconds.

FIGURE 2.7 Geologic phenomena can stretch over thousands of centuries or can occur with dazzling speed. (*left*) The Grand Canyon, Arizona. [John Wang/PhotoDisc/Getty Images.] (*right*) Meteor Crater, Arizona. [John Sanford/Photo Researchers.]

found in rock formations. This approach has been very successful. However, Hutton's principle is too confining for geologic science as it is now practiced. Modern geology must deal with the entire range of Earth's history, which began more than 4.5 billion years ago. As we will see in Chapter 9, the violent processes that shaped Earth's early history were distinctly different from those that operate today.

EARTH'S SHAPE AND SURFACE

In 1492, Columbus set a westward course for India because he believed in a theory that had been favored by Greek philosophers: *we live on a sphere.* His math was poor, however, so he badly underestimated Earth's circumference. Instead of a shortcut, he took the long way around, finding a New World instead of the Spice Islands! Had Columbus properly understood the ancient Greeks, he might not have made this fortuitous mistake, because they had accurately measured Earth's size more than 17 centuries earlier.

The credit goes to Eratosthenes, a Greek librarian who lived in Alexandria, Egypt. Sometime around 250 B.C., a traveler told him about a very interesting observation. At noon on the first day of summer (June 21), a deep well in the city of Syene, about 800 km south of Alexandria, was completely lit up by sunlight because the Sun was directly overhead. Acting on a hunch, Eratosthenes did an experiment. He set up a vertical pole in his own city, and at high noon on the summer solstice, the pole cast a shadow.

Because the Sun cast a shadow in Alexandria but was directly overhead at Syene, Eratosthenes concluded that the land surface must be curved. He hypothesized that

FIGURE 2.8 How Eratosthenes measured Earth's circumference. At noon on the summer solstice (June 21), a vertical well in the southern Egyptian town of Syene (near modern Aswan) was filled with light, indicating that the Sun was directly overhead, whereas a vertical pole in his hometown of Alexandria cast a shadow. From the length of the shadow, he found that Syene and Alexandria were separated by about 1/50 of Earth's circumference, about 7° out of 360°. Since the measured distance between the two towns was approximately 800 km, he calculated a circumference close to its modern value of 40,000 km.

Earth had a perfectly spherical shape (the Greeks admired geometrical perfection), and using the distance between the two cities and the size of the pole's shadow, he then deduced a circumference close to its modern value of 40,000 km (**Figure 2.8**). His experiment thus led to a simple model: *Earth is a sphere with a radius of about 6370 km.*

The model of a spherical Earth developed by the ancient Greeks proved to be a good one because it correctly predicted other types of measurements, such as the distance at which a ship's tall mast would disappear over the horizon. This example makes clear why well-designed experiments and good measurements are central to the scientific method: they can give us new information about the natural world.

Much more precise measurements, such as those made by the Great Trigonometrical Survey of India, have shown that Earth is not a perfect sphere. Because of its daily rotation, it bulges out slightly at the equator and is slightly squashed at the poles.

In addition, Earth's smooth curvature is broken by mountains and valleys and other ups and downs. This **topography** is measured with respect to *sea level,* a smooth surface that conforms closely to the squashed spherical shape of the rotating Earth. The largest features are continents, which have typical elevations of 0 to 1 km above sea level, and ocean basins, which have typical depths of 4 to 5 km below sea level.

Many features of geological significance stand out in Earth's topography (**Figure 2.9**), such as continental mountain belts and deep-sea trenches. The elevation of Earth's surface varies by nearly 20 km from the highest point (Mount Everest, in the Himalayas, at 8850 m above sea level) to the lowest point (Challenger Deep, in the Marianas Trench in the Pacific Ocean, at 11,030 m below sea level). Although the Himalayas may loom large to us, their elevation is a small fraction of Earth's radius, only about one part in a thousand, which is why the globe looks like a smooth sphere when seen from outer space.

The topography of Earth's surface is created by great forces acting in Earth's deep interior. To begin to understand these forces, we must look inside the planet, which is layered like an onion.

> *Earth is almost spherical, but it bulges out slightly at the equator and is slightly squashed at the poles, due to its rotation. Its topography varies by about 20 km from the highest point above sea level to the lowest point below sea level.*

FIGURE 2.9 Earth's topography is measured with respect to sea level. Elevations vary from almost 9 km above sea level to about 11 km below sea level. In this picture, the elevation scale is greatly exaggerated.

PEELING THE ONION: Discovery of a Layered Earth

Ancient thinkers such as Eratosthenes divided the universe into two parts, the Heavens above and Hades below. The sky was transparent and full of light, and they could directly observe its stars and track its wandering planets. But Earth's interior was dark and closed to human view. In some places, the ground quaked and erupted hot lava. Surely something terrible was going on down there!

About a century ago, geologists began to peer downward into Earth's interior, not with waves of light (which can't penetrate rock), but with waves produced by earthquakes. An earthquake occurs when geologic forces cause brittle rocks to fracture, sending out vibrations like the cracking of ice on a river. These **seismic waves** (from the Greek word for "earthquake," *seismos*) illuminate Earth's interior and can be recorded on *seismographs,* sensitive instruments that allow geologists to make pictures of Earth's inner workings, much as doctors use ultrasound and CAT scans to image the inside of your body.

When the first networks of seismographs were installed around the world at the end of the nineteenth century, geologists began to discover that Earth's interior was divided into concentric layers of different compositions, separated by sharp, nearly spherical boundaries (Figure 2.10).

> Seismic waves, which are produced by earthquakes, can be used to create images of Earth's interior.

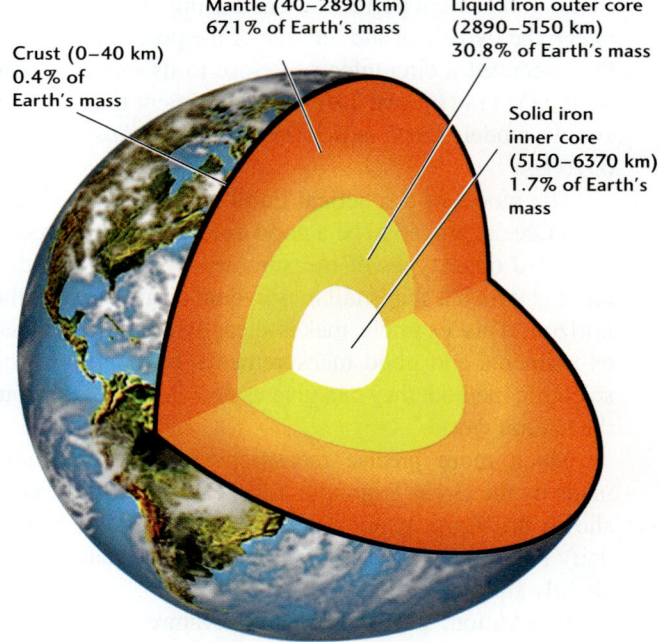

FIGURE 2.10 Earth's major layers, showing their volume and mass as a percentage of Earth's total volume and mass.

Earth's Density

Evidence for Earth's layering was first proposed by the German physicist Emil Wiechert at the end of the nineteenth century, before much seismic data had become available. He wanted to understand why our planet is so heavy, or more precisely, so *dense*. The density of a substance is easy to calculate: just measure its mass on a scale and divide by its volume. A typical rock, such as the granite used for tombstones, has a density of about 2.7 g/cm^3.

Estimating the density of an entire planet is a little harder, but not much. Eratosthenes had shown how to measure Earth's volume in 250 B.C., and sometime around 1680, the great English scientist Isaac Newton figured out how to calculate its mass from the force of gravity that pulls objects to its surface. The details, which involved careful laboratory experiments to calibrate Newton's law of gravity, were worked out by another Englishman, Henry Cavendish. In 1798, he calculated Earth's average density to be about 5.5 g/cm^3, twice as dense as tombstone granite.

Wiechert was puzzled. He knew that a planet made entirely of common rocks, which are composed mostly of *silicates* (minerals containing silicon and oxygen), could not have such a high density. Some iron-rich rocks brought to the surface by volcanoes have densities as high as 3.5 g/cm^3, but no ordinary rock—even when subjected to great pressures—approached Cavendish's value.

> Earth's average density is higher than would be expected if it were made entirely of common rocks, implying that it has a dense core.

The Mantle and Core

In thinking about what lay beneath his feet, Wiechert turned outward to the solar system and, in particular, to meteorites, which are pieces of the solar system that have fallen to Earth. He knew that some meteorites are made of a mixture of two heavy metals, iron and nickel, and thus have densities as high as 8 g/cm^3 (**Figure 2.11**). He also knew that these elements are relatively abundant throughout our solar system. So, in 1896, he stated a grand hypothesis: sometime in Earth's past, most of the iron and nickel in its interior had dropped inward to its center under the force of gravity. This created a dense **core,** which was surrounded by a shell of silicate-rich rocks that he called the **mantle** (using the German word for "coat").

With this hypothesis, he could come up with a two-layer Earth model that agreed with Cavendish's value for Earth's average density. He could also explain the existence of iron-nickel meteorites: they were chunks of the core from an Earthlike planet (or planets) that had broken apart, most likely by collision with other planets.

Wiechert got busy testing his hypothesis using seismic waves recorded by seismographs located around the globe (he designed one himself). The first results showed a shadowy inner mass that he took to be the core, but he had problems identifying some of the seismic waves. These waves come in two basic types: *compressional waves,* which expand and compress the material as they travel through a solid, liquid, or gas; and *shear waves,* which move the material from side to side (shearing motion). Shear waves can propagate only through solids, which resist shearing, and not through fluids such as air and water, which have no resistance to this type of motion.

FIGURE 2.11 Two common types of meteorites. (a) This stony meteorite, which is similar in composition to Earth's silicate mantle, has a density of about 3 g/cm^3. (b) This iron-nickel meteorite, which is similar in composition to Earth's core, has a density of about 8 g/cm^3. [John Grotzinger/Ramón Rivera-Moret/Harvard Mineralogical Museum.]

In 1906, a British seismologist, Robert Oldham, was able to sort out the paths traveled by these two types of seismic waves and show that shear waves did not propagate through the core. The core, at least in its outer part, is liquid! This turns out to be not too surprising. Iron melts at a lower temperature than silicates, which is why metallurgists can use containers made of ceramics (which are silicate materials) to hold molten iron. Earth's deep interior is hot enough to melt iron and nickel, but not silicate rock. Beno Gutenberg, one of Wiechert's students, confirmed Oldham's observations and, in 1914, determined that the depth to the *core-mantle boundary* is just shy of 2900 km (see Figure 2.10).

> Observations of seismic waves indicate that Earth's outer core is liquid while its mantle is solid.

The Crust

Five years earlier, a Croatian scientist had detected another boundary at the relatively shallow depth of 40 km beneath the European continent. This boundary, named the *Mohorovičić discontinuity* (Moho for short) after its discoverer, separates a **crust** composed of low-density silicates, which are rich in aluminum and potassium, from higher-density mantle silicates, which contain more magnesium and iron.

Like the core-mantle boundary, the Moho is a global feature. However, it was found to be substantially shallower beneath the oceans than beneath the continents. On average, the thickness of oceanic crust is only about 7 km, compared with almost 40 km for continental crust. Moreover, rocks in the oceanic crust contain more iron, and are therefore denser than continental rocks. Because the continental crust is thicker but less dense than the oceanic crust, the continents ride higher by floating like buoyant rafts on the denser mantle (**Figure 2.12**), much as icebergs float on the ocean. Continental buoyancy explains the most striking feature of Earth's surface topography: why the elevations shown in Figure 2.9 fall into two main groups, 0 to 1 km above sea level for much of the land surface and 4 to 5 km below sea level for much of the deep sea.

> The continental crust is less dense than either the oceanic crust or the mantle, and so rides higher on the mantle than the oceanic crust.

Shear waves travel well through the mantle and crust, so we know that both are solid rock. How can continents float on solid rock? Rock can be solid and strong over the short term (seconds to years) but weak over the long term (thousands to millions of years). The mantle below a depth of about 100 km has little strength, and over very long periods, it flows as it adjusts to support the weight of continents and mountains.

The Inner Core

Because the mantle is solid and the outer part of the core is liquid, the core-mantle boundary reflects seismic waves just as a mirror reflects light waves. In 1936, Danish seismologist Inge Lehmann discovered another sharp spherical boundary at the much greater depth of 5150 km, indicating a central mass with a higher density than the liquid core. Studies following her pioneering research showed that the inner core can transmit both shear waves and compressional waves. The **inner core** is therefore a solid metallic sphere suspended within the liquid **outer core**—a "planet within a planet." The radius of the inner core is 1220 km, about two-thirds the size of the Moon (see Figure 2.10).

Geologists were puzzled by the existence of this "frozen" inner core. From other considerations, they knew that temperatures inside Earth should increase with depth. According to the best current estimates, Earth's temperature rises from about 3500°C at the core-mantle boundary to almost 5000°C at its center. If the inner core is hotter, how could it be solid while the outer core is molten? The mystery was eventually solved by laboratory experiments on iron-nickel alloys, which showed that the "freezing" was due to higher pressures, rather than lower temperatures, at Earth's center.

> Earth's inner core is solid despite its high temperature because of the extremely high pressures at Earth's center.

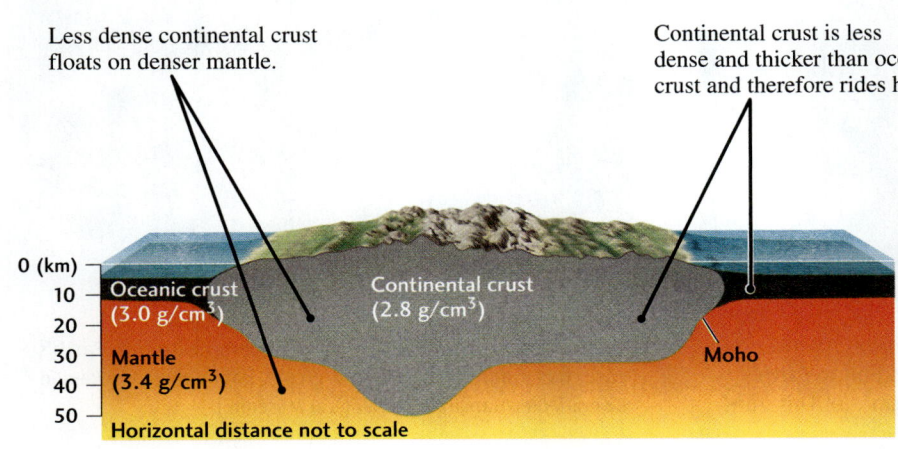

FIGURE 2.12 Crustal rocks are denser than mantle rocks, allowing the crust to float on the mantle. Continental crust is thicker and has a lower density than oceanic crust, which causes it to ride higher, explaining the elevation difference between continents and the deep sea floor.

Chemical Composition of Earth's Major Layers

By the mid-twentieth century, geologists had discovered all of Earth's major layers—crust, mantle, outer core, and inner core—plus a number of more subtle features in its interior. They found, for example, that the mantle itself is layered into an *upper mantle* and a *lower mantle,* separated by a *transition zone* where the rock density increases in a series of steps. These density steps are not caused by changes in the rock's chemical composition, but rather by changes in its compactness due to the increasing pressure with depth. The two largest density jumps in the transition zone are located at depths of about 400 km and 650 km, but they are smaller than the density increases across the Moho and core-mantle boundary, which are due to changes in chemical composition (see Figure 2.10).

Geologists were also able to show that Earth's outer core cannot be made of a pure iron-nickel alloy, because the densities of these metals are higher than the observed density of the outer core. About 10 percent of the outer core's mass must be made of lighter elements, such as oxygen and sulfur. On the other hand, the density of the solid inner core is slightly higher than that of the outer core and is consistent with a nearly pure iron-nickel alloy.

By bringing together many lines of evidence, geologists have put together a model of the composition of Earth and its various layers. In addition to the seismic data, that evidence includes the compositions of crustal and mantle rocks as well as the compositions of meteorites, thought to be samples of the cosmic material from which planets like Earth were originally made.

Only 8 elements, out of more than 100, make up 99 percent of Earth's mass (**Figure 2.13**). In fact, about 90 percent of Earth consists of only four elements: iron, oxygen, silicon, and magnesium. The first two are the most abundant elements, each accounting for nearly a

> *The boundaries between the inner core, outer core, mantle, and crust mark changes in chemical composition.*

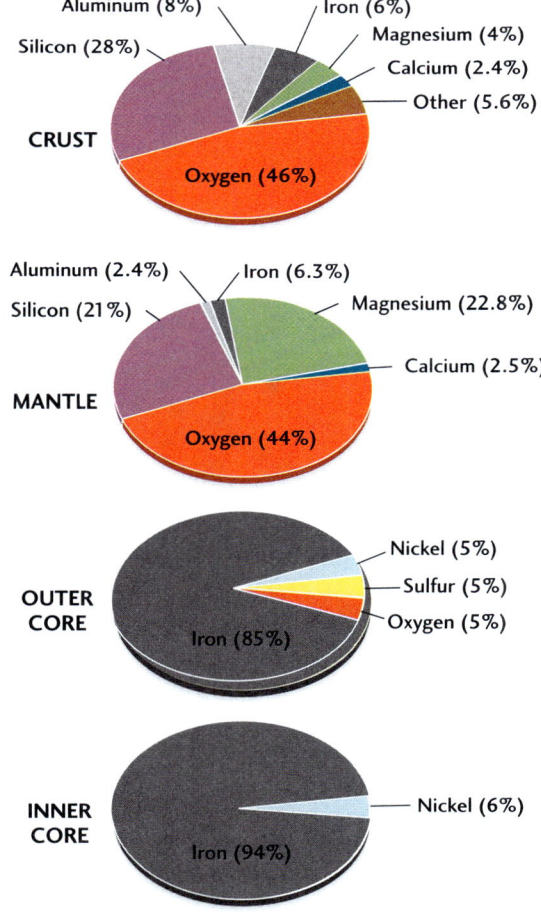

FIGURE 2.13 Jumps in density between Earth's major layers are caused by differences in their chemical composition.

> Ninety percent of Earth's mass is composed of only four elements: iron, oxygen, silicon, and magnesium.

third of the planet's overall mass, but they are distributed very differently. Iron, the densest common element, is concentrated in the core, whereas oxygen, the lightest common element, is concentrated in the crust and mantle. The crust contains more silica (silicon plus oxygen) than the mantle. These relationships show that the different compositions of Earth's layers are primarily the work of gravity. As you can see in Figure 2.13, the crustal rocks on which we stand are almost 50 percent oxygen.

EARTH AS A SYSTEM OF INTERACTING COMPONENTS

Earth is a restless planet, continually changing through geologic activity such as earthquakes, volcanoes, and glaciation. This activity is powered by two heat engines: one internal, the other external (**Figure 2.14**). Earth's internal engine is powered by the heat energy trapped in its deep interior during its violent origin and released inside the planet by radioactivity. The internal heat drives the motions in the mantle and core and supplies the energy that melts rock, moves continents, and lifts up mountains. Earth's external engine is driven by solar energy—heat supplied to Earth's surface by the Sun. Heat from the Sun energizes the atmosphere and oceans and is responsible for our climate and weather. Rain, wind, and ice erode mountains and shape the landscape, and the shape of the landscape, in turn, changes the climate.

All the parts of our planet and all their interactions, taken together, constitute the **Earth system.** Although Earth scientists have long thought in terms of natural systems, it was not until the late twentieth century that they had the tools to investigate how the Earth system actually works. Networks of instruments and Earth-orbiting satellites now collect information about the Earth system on a global scale, and computers are now powerful enough to calculate the mass and energy transfers within the system.

> Earth's activity is powered by two energy sources: solar energy and the heat energy of its deep interior.

The major components of the Earth system can be represented as a set of domains, or "spheres" (**Figure 2.15**). In our discussion of the environment in Chapter 1, we have already encountered some of them: the atmosphere, the hydrosphere (all the liquid water in the oceans, lakes, and rivers), the cryosphere (all the ice in the polar caps and mountain glaciers), and the biosphere. We will define the others shortly.

We will talk about the Earth system throughout this textbook. Let's get started by looking at some of its basic features. The Earth system is an open system that exchanges energy and mass with its surroundings. Radiant energy from the Sun energizes the weathering and erosion of Earth's surface, as well as the growth of plants, which feed almost all living things. Our climate is con-

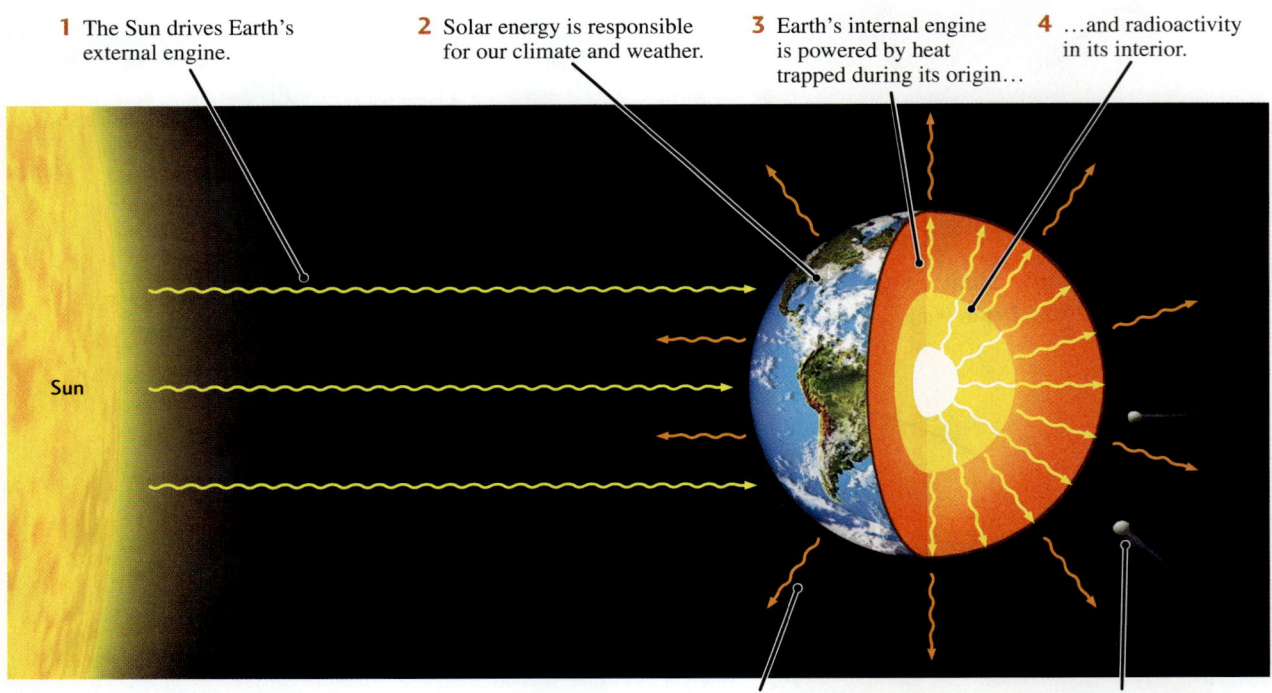

FIGURE 2.14 The Earth system is an open system that exchanges energy and mass with its surroundings.

1 The Sun drives Earth's external engine.
2 Solar energy is responsible for our climate and weather.
3 Earth's internal engine is powered by heat trapped during its origin…
4 …and radioactivity in its interior.
5 Heat radiating from Earth balances solar input and heat from interior.
6 Meteors move mass from the cosmos to Earth.

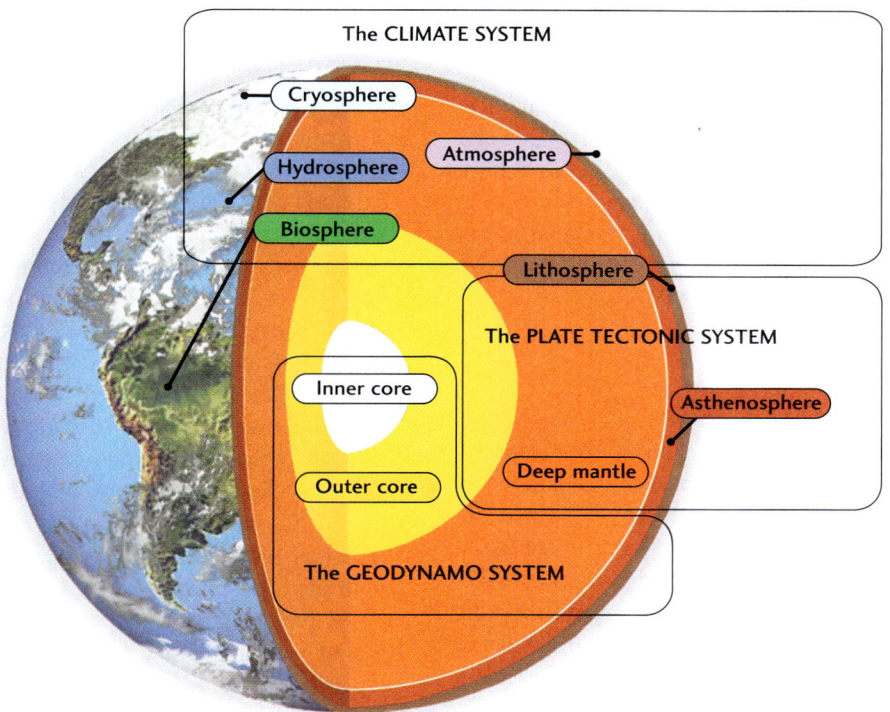

FIGURE 2.15 The Earth system encompasses all parts of our planet and their interactions. This diagram illustrates the basic components of the Earth system, how they are energized, and which among them interact to form the climate system, the plate tectonic system, and the geodynamo system.

trolled by the balance between the solar energy coming into the Earth system and the energy Earth radiates back into space. These days, the exchange of mass between Earth and space is relatively small—only about 40,000 tons of meteorites, equivalent to a cube 24 m on a side, fall to Earth each year. That mass transfer was much greater during the early life of the solar system, however.

Although we think of Earth as a single system, it is a challenge to study the whole thing all at once. Instead, we will focus our attention on those parts of the Earth system (subsystems) we are trying to understand. For instance, in our discussion of global climate change, we will primarily consider interactions among the atmosphere, hydrosphere, cryosphere, and biosphere that are driven by solar energy. Our coverage of how the continents are deformed to make mountains will focus on interactions between the lithosphere and asthenosphere that are driven by Earth's internal energy. Specialized subsystems that produce specific types of behaviors, such as climate change or mountain building, are called **geosystems.** The Earth system can be thought of as the collection of many open, interacting (and often overlapping) geosystems.

In this chapter, we will introduce three important geosystems that operate on a global scale: the climate system, the plate tectonic system, and the geodynamo. Later in this textbook, we will have occasion to discuss a number of smaller geosystems, such as volcanoes that erupt hot lava (Chapter 5), hydrologic systems that give us our drinking water (Chapter 11), and petroleum reservoirs that produce oil and gas (Chapter 14).

The Climate System

Weather is the term we use to describe the temperature, precipitation, cloud cover, and winds observed at a particular location and time on Earth's surface. We all know how variable the weather can be—hot and rainy one day, cool and dry the next—depending on the movements of storm systems, warm and cold fronts, and other atmospheric disturbances. Because the atmosphere is so complex, even the best forecasters have a hard time predicting the weather more than 4 or 5 days in advance. However, we can guess in rough terms what our weather will be much further into the future, because weather is governed primarily by the changes in solar energy input on seasonal and daily cycles: summers are hot, winters cold; days are warmer, nights cooler. The **climate** produced by these weather cycles can be described by averaging temperature and other variables over many years of observation. A complete description of climate also includes measures of how variable the weather has been, such as the highest and lowest temperatures ever recorded on a given day of the year.

The **climate system** includes all the Earth system components that determine climate on a global scale and how climate changes with time. In other words, the climate system describes not only the behavior of the atmosphere, but also how climate is influenced by the hydrosphere, cryosphere, biosphere, and lithosphere (see Figure 2.15).

When the Sun warms Earth's surface, some of the heat is trapped by water vapor, carbon dioxide, and other gases in the atmosphere, much as heat is trapped by frosted glass in a greenhouse. This *greenhouse effect* explains why Earth has a pleasant climate that makes life possible. If its atmosphere contained no greenhouse gases, Earth's surface would be frozen solid! Therefore, greenhouse gases, particularly carbon dioxide, play an essential role in regulating climate. As we will learn in later chapters, the concentration of carbon dioxide in the atmosphere is a balance between the amount spewed out of Earth's interior in volcanic eruptions and the amount withdrawn during the weathering of silicate rocks. In this way, the behavior of the atmosphere is regulated by interactions with the lithosphere.

A particularly urgent problem is to understand the global warming that is being caused by anthropogenic emissions of carbon dioxide and other greenhouse gases. Part of the public debate about global warming centers on the accuracy of predictions based on computer models. Skeptics argue that even the most sophisticated computer models are unreliable because they lack many features of the real Earth system. In Chapter 10, we will discuss some aspects of how the climate system works, and in Chapter 14, the practical problems of climate change caused by human activities.

> Solar radiation energizes the climate system, which involves the atmosphere, hydrosphere, cryosphere, biosphere, and lithosphere.

The Plate Tectonic System

Earth's most dramatic geologic events—volcanic eruptions and earthquakes, for example—also result from interactions within the Earth system. These phenomena are driven by Earth's internal heat, which is transferred upward through the circulation of material in Earth's solid mantle.

1 Convection moves hot water from the bottom to the top…

2 …where it cools, moves laterally, sinks,…

3 …warms, and rises again.

4 Hot matter from the mantle rises,…

5 …causing plates to form and diverge.

6 Where plates converge, a cooled plate is dragged under the neighboring plate,…

7 …sinks, warms, and rises again.

FIGURE 2.16 Convection carries heat upward through the motion of matter.

We have seen that Earth is zoned by chemistry: its crust, mantle, and core are chemically distinct layers. Earth is also zoned by *strength,* a property that measures how much an Earth material can resist being deformed. Material strength depends on both chemical composition (bricks are strong, soap bars are weak) and temperature (cold wax is strong, hot wax is weak).

In some ways, the outer part of the solid Earth behaves like a ball of hot wax. Cooling of the surface forms a strong outer shell, or **lithosphere** (from the Greek *lithos,* meaning "stone"), which encases a hot, weak **asthenosphere** (from the Greek *asthenes,* meaning "weak"). The lithosphere includes the crust and the top part of the mantle down to an average depth of about 100 km. The asthenosphere is the portion of mantle, perhaps 300 km thick, immediately below the lithosphere. When subjected to force, the lithosphere tends to behave like a nearly rigid and brittle shell, whereas the underlying asthenosphere flows like a moldable, or ductile, solid.

According to the remarkable theory of **plate tectonics,** the lithosphere is not a continuous shell; it is broken into about a dozen large plates that move over Earth's surface at rates of a few centimeters per year. Each plate is a rigid unit that rides on the asthenosphere, which is also in motion. The lithosphere that forms a plate may be just a few kilometers thick in volcanically active areas and as much as 200 km thick beneath the older, colder parts of the continents. The discovery of plate tectonics in the 1960s led to the first unified theory that explained the worldwide distribution of earthquakes and volcanoes, continental drift, mountain building, and many other geologic phenomena. Chapter 3 is devoted to a description of plate tectonics.

Why do the plates move across Earth's surface instead of locking up into a completely rigid shell? The forces that push and pull the plates come from the heat engine in Earth's deep interior. Driven by Earth's internal heat, hot mantle material rises at boundaries where plates form and separate. The lithosphere cools and becomes more rigid as it moves away, eventually sinking into the mantle under the pull of gravity at boundaries where plates converge. This general process, in which hotter material rises and cooler material sinks, is called *convection* (**Figure 2.16**).

> Earth's internal heat energizes the plate tectonic system, which involves the lithosphere, asthenosphere, and deep mantle.

> The lithosphere is composed of rigid plates that ride on a ductile asthenosphere. The plates are pushed and pulled by convection as hot mantle material rises from Earth's interior and cooler material sinks back into the interior.

The Geodynamo System

The third global geosystem involves interactions that produce a **magnetic field** deep inside Earth, in its fluid outer core. This magnetic field reaches far into outer space, causing compass needles to point north and shielding the biosphere from the Sun's harmful radiation. When rocks form, they become slightly magnetized by this magnetic field, so geologists can study how the field behaved in the past and use it to help them decipher the geologic record.

Earth rotates about an axis that goes through its north and south poles. Earth's internal magnetic field behaves as if a powerful bar magnet were located at Earth's center and inclined about 11° from this rotational axis. The magnetic force points into Earth at the north magnetic pole and outward from Earth at the south magnetic pole (**Figure 2.17**). At any place on Earth (except near the

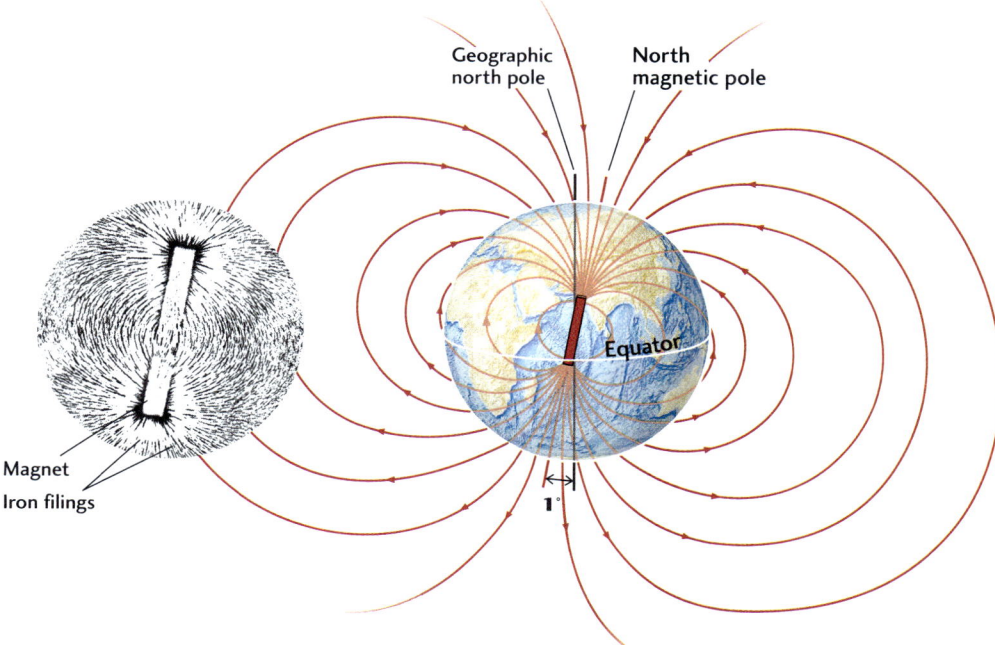

FIGURE 2.17 (*left*) The magnetic field of a bar magnet is revealed by the alignment of iron filings on paper. (*right*) Earth's magnetic field is much like the field that would be produced if a giant bar magnet were placed at Earth's center and slightly inclined (11°) from the axis of rotation. The lines of force produced by Earth's magnetic field are shown. At any place on Earth, a compass needle will point to the north magnetic pole because it orients to the local line of force.

magnetic poles), a compass needle that is free to swing under the influence of the magnetic field will rotate into a position parallel to the local line of force, approximately in the north-south direction.

Although a permanent magnet at Earth's center could explain the dipolar ("two-pole") nature of the observed magnetic field, this hypothesis can be easily rejected. Laboratory experiments demonstrate that the field of a permanent magnet is destroyed when the magnet is heated above about 500°C. We know that the temperatures in Earth's deep interior are much higher than that—thousands of degrees at its center—so, unless the magnetism were constantly regenerated, it could not be maintained.

Scientists theorize that heat flowing out of Earth's core causes convection that generates and maintains the magnetic field. Why is a magnetic field created by convection in the outer core, but not by convection in the mantle? First, the outer core is made primarily of iron, which is a very good electrical conductor, whereas the silicate rocks of the mantle are poor electrical conductors. Second, the convective motions are a million times more rapid in the liquid outer core than in the solid mantle. These rapid motions stir up electric currents in the iron to create a **geodynamo,** which produces the magnetic field.

Convection of liquid iron in Earth's outer core creates a geodynamo, which generates Earth's magnetic field.

For some 400 years, scientists have known that a compass needle points north because of Earth's magnetic field. Imagine how stunned they were when they found geologic evidence that the magnetic field can completely reverse direction—that is, it can flip its north magnetic pole and its south magnetic pole. Over about half of geologic time, a compass needle would have pointed south!

These *magnetic reversals* occur at irregular intervals ranging from tens of thousands to millions of years. The processes that cause them are not well understood, but computer models of the geodynamo show sporadic reversals occurring in the absence of any external factors—that is, purely through interactions within Earth's core. As we will see in the next chapter, magnetic reversals, which leave their imprint on the geologic record, have helped geologists figure out the movements of the tectonic plates.

AN OVERVIEW OF GEOLOGIC TIME

So far, we have discussed Earth's size and shape, its internal layering and composition, and the operation of its three major geosystems. How did Earth get its layered structure in the first place? How have the global geosystems evolved through geologic time? To begin to answer these questions, we present a brief overview of geologic time from the birth of the planet to the present. Later chapters will fill in the details.

Comprehending the immensity of geologic time is a challenge. John McPhee notes that geologists look into the "deep time" of Earth's early history (measured in billions of years), just as astronomers look into the "deep space" of the outer universe (measured in billions of light-years). **Figure 2.18** presents geologic time as a ribbon marked with some major events and transitions.

Origin of Earth and Its Global Geosystems

Using evidence from meteorites, geologists have been able to show that Earth and the other planets formed about 4.56 billion years ago through the rapid condensation of a dust cloud that circulated around the young Sun. This violent process, which involved the aggregation and collision of progressively larger clumps of matter, will be described in more detail in Chapter 9. Within just 100 million years (a relatively short period of time, geologically speaking), the Moon had formed and Earth's core had separated from its mantle. Exactly what happened during the next several hundred million years is hard to figure out. Very little of the rock record survived intense bombardment by the large meteorites that were smashing into Earth. This early period of Earth's history is appropriately called the geologic "dark ages."

The oldest rocks now found on Earth's surface are about 4 billion years old. Rocks as ancient as 3.8 billion years show evidence of erosion by water, indicating the existence of a hydrosphere and the operation of a climate system not too different from that of the present. Rocks only slightly younger, 3.5 billion years old, record a magnetic field about as strong as the one we see today, showing that the geodynamo was operating by that time. By 2.5 billion years ago, enough low-density crust had collected at Earth's surface to form large continental masses. The geologic processes that then modified those continents were very similar to those we see operating in the plate tectonic system today.

Earth formed about 4.56 billion years ago. Rocks about 4 billion years old have been found at Earth's surface.

The Evolution of Life

Life also began very early in Earth's history, as we can tell from the study of **fossils,** traces of organisms preserved in the geologic record. Fossils of primitive bacteria have been found in rocks dated at 3.5 billion years ago. A key event was the evolution of organisms that release oxygen into the atmosphere and oceans. The buildup of oxygen in the atmosphere was under way by 2.5 billion years ago. Atmo-

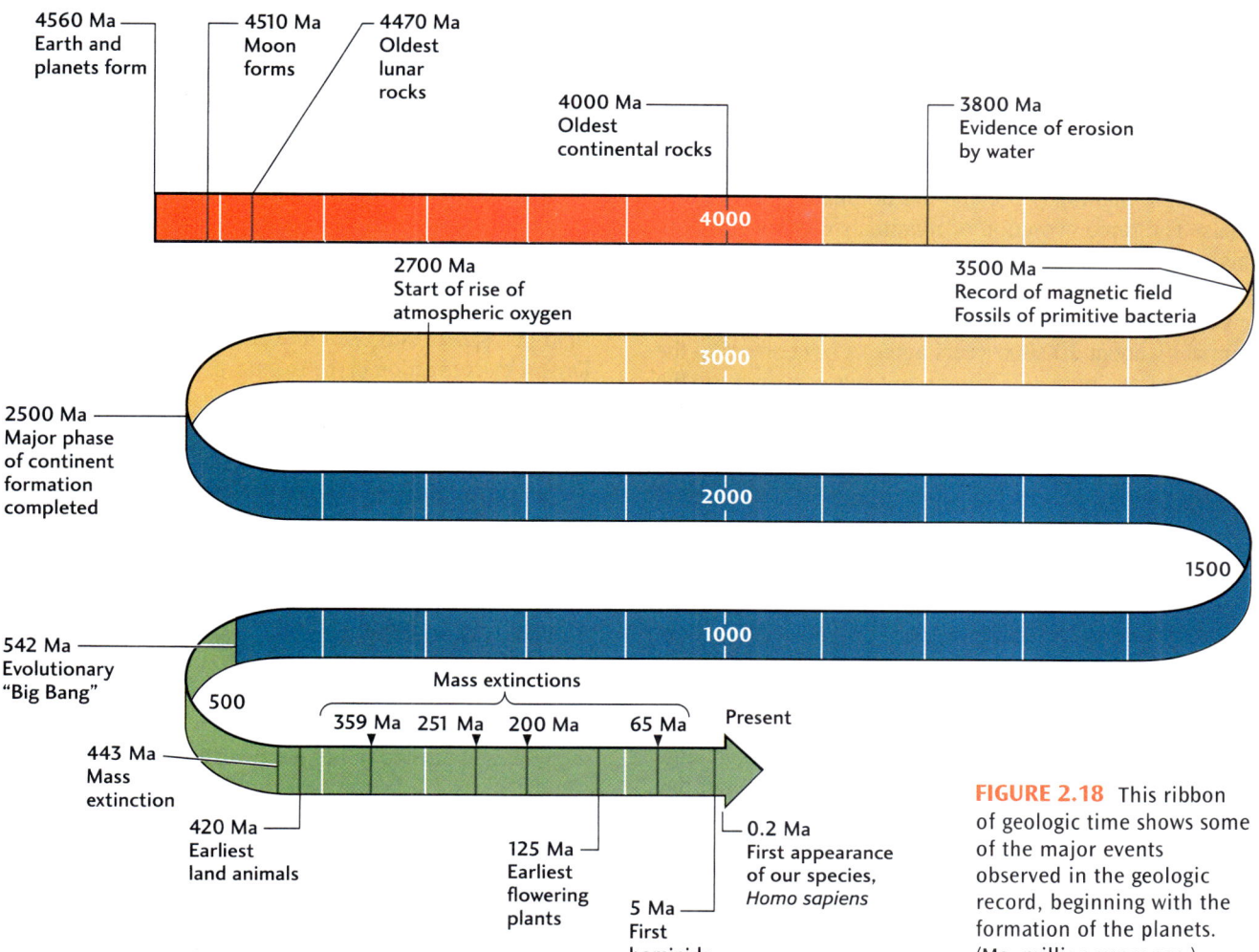

FIGURE 2.18 This ribbon of geologic time shows some of the major events observed in the geologic record, beginning with the formation of the planets. (Ma, million years ago.)

spheric oxygen concentrations probably rose to modern levels in a series of steps over a period as long as 2 billion years.

Life on early Earth was primitive, consisting mostly of small, single-celled organisms that floated near the surface of the oceans or lived on the seafloor. Between 1 billion and 2 billion years ago, more complex life-forms such as algae and seaweed evolved. The first animals appeared about 600 million years ago, evolving in a series of waves. In a period starting 542 million years ago and probably lasting less than 10 million years, eight entirely new branches of the animal kingdom were established, including ancestors to nearly all animals inhabiting Earth today. It was during this evolutionary explosion, sometimes called biology's Big Bang, that animals with shells first left their shelly fossils.

Although biological evolution is often viewed as a very slow process, it is punctuated by brief periods of rapid change. Spectacular examples are *mass extinctions,* during which many types of animals and plants suddenly disappeared from the geologic record. Five of these huge turnovers are marked on the time ribbon in Figure 2.18. The last was caused by a large meteorite impact 65 million years ago. The bolide, about 10 km in diameter, caused the extinction of half of Earth's species, including all the dinosaurs (see Figure 1.23).

The causes of the other mass extinctions are still being debated. In addition to bolide impacts, scientists have proposed other types of extreme events, such as rapid climate changes brought on by glaciations and massive eruptions of volcanic material. The evidence is often ambiguous or inconsistent. For example, the largest extinction event of all time took place about 250 million years ago, wiping out nearly 95 percent of all species. A bolide impact has been proposed by some investigators, but the geologic record shows that ice sheets expanded and seawater chemistry changed at this time, consistent with a major climate crisis. At the same time, an enormous volcanic eruption covered an area in Siberia almost half the size of the United States with 2 million to 3 million cubic kilometers of lava. This mass extinction has been dubbed

> *Primitive fossils indicate that life on Earth began before 3.5 billion years ago.*

"Murder on the Orient Express," because there are so many suspects!

Mass extinctions reduce the number of species competing for space in the biosphere. By "thinning out the crowd," these extreme events can promote the evolution of new species. After the demise of the dinosaurs 65 million years ago, mammals became the dominant class of animals. The rapid evolution of mammals into species with bigger brains and more dexterity led to the emergence of humanlike species (*hominids*) around 5 million years ago and our own species, *Homo sapiens* (Latin for "knowing human"), about 200,000 years ago. As newcomers to the biosphere, we're just beginning to leave our mark on the geologic record. Indeed, our short history as a species can be appreciated by noting that it spans less than a line's width on the ribbon of geologic time in Figure 2.18.

SUMMARY >>

How do geologists study Earth? Geologists, like other scientists, use the scientific method. They develop and test hypotheses, which are tentative explanations for natural phenomena based on observations and experiments. They share their data and test one another's hypotheses. A coherent set of hypotheses that have survived repeated challenges constitutes a theory. Confidence grows in those theories that withstand repeated tests and are able to predict the results of new experiments.

What is Earth's shape? Earth's overall shape is a sphere with an average radius of 6370 km that bulges slightly at the equator and is slightly squashed at the poles due to the planet's rotation. Its topography varies by about 20 km from the highest point to the lowest point. Elevations fall into two main groups: 0 to 1 km above sea level for much of the land surface and 4 to 5 km below sea level for much of the deep sea.

What are Earth's major layers? Earth's interior is divided into concentric layers of different compositions, separated by sharp, nearly spherical boundaries. The outer layer is the crust, which varies in thickness from about 40 km in the case of continental crust to about 8 km for oceanic crust. Below the crust is the mantle, a thick shell of denser silicate rock. The core, which is composed primarily of iron and nickel, is divided into two layers: a liquid outer core and a solid inner core. Jumps in density between these layers are caused by differences in their chemical composition.

How do we study Earth as a system of interacting components? When we try to understand a complex system such as the Earth system, we find that it is often easier to focus on subsystems (which we call geosystems). This textbook discusses three major global geosystems: the climate system, which involves interactions among the atmosphere, hydrosphere, cryosphere, biosphere, and land surface; the plate tectonic system, which involves interactions among Earth's solid components (lithosphere, asthenosphere, and deep mantle); and the geodynamo system, which involves interactions within Earth's core. The climate system is driven by heat from the Sun, whereas the plate tectonic and geodynamo systems are driven by Earth's internal heat.

What are the basic elements of plate tectonics? The rigid lithosphere is broken into about a dozen large plates. Driven by convection in the mantle, these plates move over Earth's surface at rates of a few centimeters per year. Each plate acts as a rigid unit, riding on the asthenosphere, which also is in motion. Hot mantle material rises at boundaries where plates form and separate, cooling and becoming more rigid as it moves away. Eventually, most of it sinks back into the mantle at boundaries where plates converge.

What are some major events in Earth's history? Earth formed 4.56 billion years ago. Rocks as old as 4 billion years have survived in Earth's crust. Liquid water existed on Earth's surface by 3.8 billion years ago. Rocks about 3.5 billion years old show evidence of a magnetic field, and the earliest evidence of life has been found in rocks of the same age. By 2.5 billion years ago, the oxygen content of the atmosphere was rising because of oxygen production by early life, and large continental masses had formed. Animals appeared suddenly about 600 million years ago, diversifying rapidly in a great evolutionary explosion. The subsequent evolution of life was marked by a series of mass extinctions, the last big one caused by a large meteorite impact 65 million years ago, which killed off the dinosaurs. Our species, *Homo sapiens,* first appeared about 200,000 years ago.

KEY TERMS AND CONCEPTS >>

asthenosphere (p. 45)	lithosphere (p. 45)
climate (p. 44)	magnetic field (p. 45)
climate system (p. 44)	mantle (p. 39)
core (p. 39)	outer core (p. 40)
crust (p. 40)	plate tectonic system (p. 45)
Earth system (p. 42)	
fossil (p. 46)	principle of uniformitarianism (p. 35)
geodynamo (p. 46)	
geologic record (p. 34)	
geosystem (p. 43)	scientific method (p. 32)
hypothesis (p. 32)	seismic wave (p. 38)
inner core (p. 40)	topography (p. 37)

EXERCISES

1. Illustrate the differences between a hypothesis, a theory, and a model with some examples drawn from this chapter.

2. How does science differ from religion as a way to understand the world?

3. Give an example of how the model of Earth's spherical shape developed by Eratosthenes could be experimentally tested.

4. Give two reasons why Earth's shape is not a perfect sphere.

5. If you made a model of Earth that was 10 cm in radius, how high would Mount Everest rise above sea level?

6. It is thought that a large meteorite impact 65 million years ago caused the extinction of three-quarters of Earth's species, including all the dinosaurs. Does this event disprove the principle of uniformitarianism? Explain your answer.

7. Imagine you are a tour guide on a journey from Earth's surface to its center. How would you describe the material that your tour group encounters on the way down? Why is the density of the material always increasing as you go deeper?

8. How does the chemical composition of Earth's crust differ from that of its deeper interior? From that of its core?

9. Explain how Earth's outer core can be a liquid while the deep mantle is a solid.

10. How does viewing Earth as a system of interacting components help us to understand our planet? Give an example of an interaction between two or more geosystems that could affect the geologic record.

11. How do the terms *weather* and *climate* differ? Express the relationship between climate and weather using examples from your experience.

12. Earth's mantle is solid, but it undergoes convection as part of the plate tectonic system. Explain why these statements are not contradictory.

13. In what general ways are the climate system, the plate tectonic system, and the geodynamo system similar? In what ways are they different?

14. Not every planet has a geodynamo. Why not? If Earth did not have a magnetic field, what might be different about our planet?

15. Based on the material presented in this chapter, what can we say about how long ago the three major global geosystems began to operate?

16. If no theory can be proved true, why do almost all geologists believe strongly in Darwin's theory of evolution?

PLATE TECTONICS: The Unifying Theory of Geology

CHAPTER 3

Geology at Sea

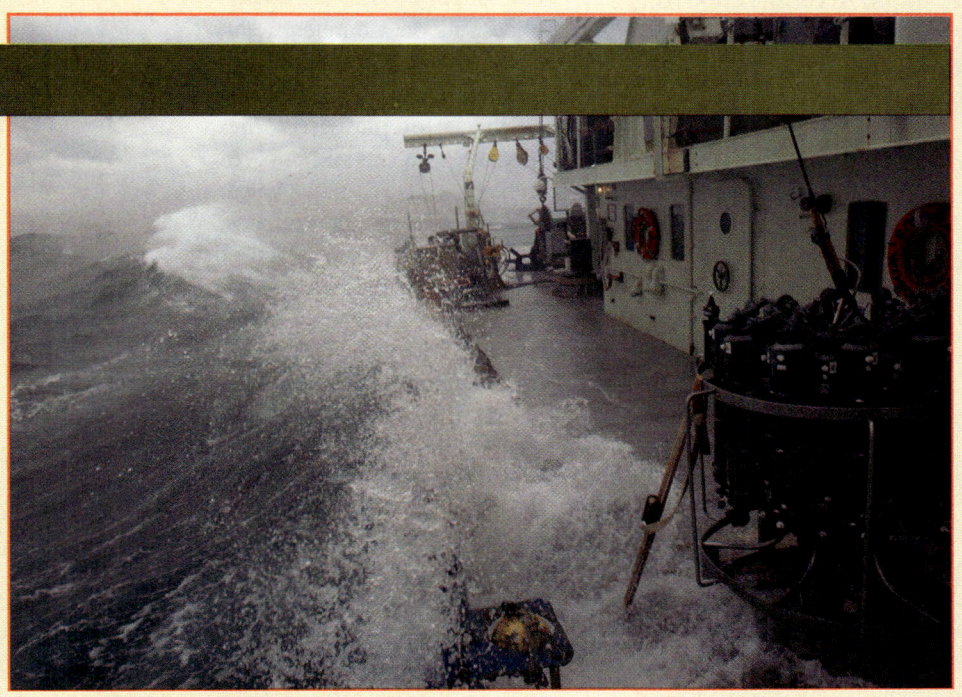

An oceanographic vessel, like most working ships, does not provide posh accommodations. As the lyrics of Jimmy Buffett's song "Landfall" describe it, "you live half your life in an eight-by-five room, just cruisin' to the sound of the big diesel boom." During the other half, you stand watch in the lab, attending to the hull-mounted sonar array and magnetometers hung over the side to map the seafloor as the ship steams along, or out on deck, washed by spray, using thick steel cables to lower instruments into the water or dredge rocks from craggy outcrops several kilometers below the rough sea surface. Sometimes you light the fuse on a kilogram or two of high explosives and toss it off the fantail, making a bang big enough to send sound waves deep into the crust, which then bounce back and tell you something interesting about the properties of the submerged rocks.

Marine geology is a tedious and sometimes dangerous business. Maurice "Doc" Ewing, who was responsible for collecting much of the data that led to the discoveries highlighted in this chapter, was once washed overboard with two mates and his brother John, also a marine geologist, by a freak wave in a storm north of Bermuda. Skillful maneuvering by the ship's captain saved the brothers and one mate from drowning, but Doc Ewing walked with a limp for the rest of his life. Bruce Heezen, another marine geologist important to our story, died aboard a research submarine while exploring the Mid-Atlantic Ridge southwest of Iceland.

Seventy percent of Earth's solid surface is hidden beneath the oceans. The first global oceanographic expedition sailed on the HMS *Challenger* (namesake of the space shuttle) from Portsmouth, England, in 1872. During its 3-year cruise, scientists discovered the symmetrical undersea mountains of the Mid-Atlantic Ridge and the Challenger Deep of the Marianas Trench, the deepest spot in all the world's oceans. Nearly a century of further exploration and many untold adventures aboard ships at sea were required to explain these peculiar features and reveal the secrets of plate tectonics—the grand, unifying theory of geology. ■

The research vessel *Oceanus*, owned by the National Science Foundation and operated by the Woods Hole Oceanographic Institution, battles the high seas and rough weather of the North Atlantic. [Chris Linder, Woods Hole Oceanographic Institution.]

CHAPTER OUTLINE

Discovery of Plate Tectonics 52

Types of Plate Boundaries 56

Rates and History of Plate Movements 63

The Grand Reconstruction 67

Mantle Convection: The Engine of Plate Tectonics 72

The lithosphere—Earth's strong, rigid outer shell of rock—is broken into about a dozen plates, which slide past, converge with, or separate from one another as they move over the weaker, ductile asthenosphere. Plates are created where they separate and recycled where they converge in a continuous process of creation and destruction. Continents, embedded in the lithosphere, drift along with the moving plates. The theory of **plate tectonics** describes the movements of plates and the forces acting on them. It also explains volcanoes, earthquakes, and the distribution of mountain chains, rock assemblages, and structures on the seafloor—all of which result from events at plate boundaries. Plate tectonics provides a conceptual framework for a large part of this textbook and, indeed, for much of geology.

This chapter lays out the theory of plate tectonics, describes plate movements today and in the geologic past, and examines how the forces that drive these movements arise from the mantle convection system.

DISCOVERY OF PLATE TECTONICS

In the 1960s, a great revolution in thinking shook the world of geology. For almost 200 years, geologists had been developing various theories of *tectonics* (from the Greek *tekton,* meaning "builder")—the general term used to describe mountain building, volcanism, and other processes that construct geologic features on Earth's surface. It was not until the discovery of plate tectonics, however, that a single theory could satisfactorily explain the whole range of geologic processes. Physics had a comparable revolution at the beginning of the twentieth century, when the theory of relativity revised the physical laws that govern space, time, mass, and motion. Biology had a comparable revolution in the middle of the twentieth century, when the discovery of DNA allowed biologists to explain how organisms transmit the information that controls their growth and functioning from generation to generation.

In Chapter 2, we discussed how the scientific method guides the work of geologists. In the context of the scientific method, plate tectonics is a confirmed theory whose strength lies in its simplicity, its generality, and its consistency with many types of observations.

The basic ideas of plate tectonics were put together as a unified theory of geology about 50 years ago. The scientific synthesis that led to plate tectonics, however, really began much earlier in the twentieth century, with the recognition of evidence for continental drift.

Continental Drift

Such changes in the superficial parts of the globe seemed to me unlikely to happen if the earth were solid to the center. I therefore imagined that the internal parts might be a fluid more dense, and of greater specific gravity than any of the solids we are acquainted with, which therefore might swim in or upon that fluid. Thus the surface of the earth would be a shell, capable of being broken and disordered by the violent movements of the fluid on which it rested.

(Benjamin Franklin, 1782, in a letter to French geologist Abbé J. L. Giraud-Soulavie)

FIGURE 3.1 The jigsaw-puzzle fit of the continents bordering the Atlantic Ocean formed the basis of Alfred Wegener's theory of continental drift. In his book *The Origin of Continents and Oceans,* Wegener cited as additional evidence the similarity of geologic features on opposite sides of the Atlantic. The matchup of ancient crystalline rocks in adjacent regions of South America and Africa and of North America and Europe is shown here. [Geographic fit from data of E. C. Bullard; geologic data from P. M. Hurley.]

The concept of **continental drift**—large-scale movements of continents—has been around for a long time. In the late sixteenth century and in the seventeenth century, European scientists noticed the jigsaw-puzzle fit of the coasts on both sides of the Atlantic Ocean, as if the Americas, Europe, and Africa had once been part of a single continent and had subsequently drifted apart. By the close of the nineteenth century, the Austrian geologist Eduard Suess had put together some of the pieces of the puzzle. He postulated that the present-day southern continents had once formed a single giant continent called *Gondwana* (or *Gondwanaland*). In 1915, Alfred Wegener, a German meteorologist who was recovering from wounds suffered in World War I, wrote a book on the breakup and drift of continents, in which he laid out the remarkable similarity of geologic features on opposite sides of the Atlantic (Figure 3.1). In the years that followed, Wegener postulated a supercontinent, which he called **Pangaea** (Greek for "all lands"), that broke up into the continents as we know them today.

Although Wegener was correct in asserting that the continents had drifted apart, his hypotheses about how fast they were moving and what forces were pushing them across Earth's surface turned out to be wrong, which reduced his credibility among other scientists. After about a decade of spirited debate, physicists convinced geologists that Earth's outer layers were too rigid for continental drift to occur, and Wegener's ideas were rejected by all but a few geologists.

Wegener and other advocates of the drift hypothesis pointed not only to the geographic matching of geologic features, but also to similarities in rock ages and trends in geologic structures on opposite sides of the Atlantic. They also offered arguments, accepted now as good evidence of drift, based on fossil and climate data. Identical 300-million-year-old fossils of the reptile *Mesosaurus*, for example, are found only in Africa and South America, suggesting that the two continents were joined at that time (Figure 3.2). The animals and plants on different continents showed similarities in their evolution until the postulated breakup time. After that, they followed different evolutionary paths because of the isolation and changing environments of the separating continents. In addition, rocks deposited by glaciers that existed 300 million years ago are now distributed across South America, Africa, India, and Australia. If the southern continents had once been part of Gondwana near the South Pole, a single continental glacier could account for all of these glacial deposits.

The jigsaw-puzzle fit of the continents on both sides of the Atlantic Ocean, as well as rock ages, fossil evidence, and glacial deposits, indicate that those continents were once part of the supercontinent Pangaea.

FIGURE 3.2 Fossils of the freshwater reptile *Mesosaurus*, 300 million years old, are found in South America and Africa and nowhere else in the world. If *Mesosaurus* could swim across the South Atlantic Ocean, it should have been able to cross other oceans and should have spread more widely. The observation that it did not suggests that South America and Africa must have been joined 300 million years ago. [After A. Hallam, "Continental Drift and the Fossil Record," *Scientific American* (November 1972): 57–66.]

Seafloor Spreading

The geologic evidence did not convince the skeptics, who maintained that continental drift was physically impossible. No one had yet come up with a plausible driving force that could have split Pangaea and moved the continents apart. Wegener, for example, thought the continents floated like boats across the solid oceanic crust, dragged along by the tidal forces of the Sun and Moon. His hypothesis was quickly rejected, however, because it could be shown that tidal forces are much too weak to move continents.

The breakthrough came when scientists realized that convection in Earth's mantle (discussed in Chapter 2) could push and pull continents apart, creating new oceanic crust through the process of **seafloor spreading.** In 1928, the British geologist Arthur Holmes proposed that convection currents "dragged the two halves of the original continent apart, with consequent mountain building in the front where the currents are descending, and the ocean floor development on the site of the gap, where the currents are ascending." Many still argued, however, that Earth's crust and mantle are rigid and immobile, and Holmes conceded that "purely speculative ideas of this kind, specially invented to match the requirements, can have no scientific value until they acquire support from independent evidence."

That evidence emerged from exploration of the seafloor after World War II. Doc Ewing, the marine geologist

FIGURE 3.3 This photo, taken in the summer of 1947, shows Maurice "Doc" Ewing (*center*) beaming as he looks at a piece of young basalt dredged from the depths of the Atlantic Ocean by the research vessel *Atlantis I*. On the near left is Frank Press, who initiated the series of geology textbooks that includes this one. [Photo courtesy of Lamont-Doherty Earth Observatory, The Earth Institute at Columbia University.]

mentioned at the opening of this chapter, showed that the seafloor of the Atlantic Ocean is made of young basalt, not old granite, as had been previously thought (**Figure 3.3**). Moreover, the mapping of the undersea Mid-Atlantic Ridge led to the discovery of a deep, cracklike valley, or *rift*, running down its center (**Figure 3.4**). Two of the geologists who mapped this feature were Bruce Heezen and Marie Tharp, assistants to Doc Ewing at Columbia University (**Figure 3.5**). "I thought it might be a rift valley," Tharp said years later. Heezen dismissed the idea as "girl talk," but they soon found that almost all earthquakes in the Atlantic Ocean occurred near the rift, confirming Tharp's hunch (see Figure 3.4). Because most

FIGURE 3.4 The North Atlantic seafloor, showing the cracklike rift valley running down the center of the Mid-Atlantic Ridge and the locations of associated earthquakes (black dots).

FIGURE 3.5 Marie Tharp and Bruce Heezen inspect a map of the seafloor. Their discovery of active rifts on the mid-ocean ridges provided important evidence for seafloor spreading. [The Earth Institute at Columbia University.]

a period in the early 1960s, some physicists and geologists, such as Heezen, actually believed in this idea of an expanding Earth. Other geologists recognized that the seafloor was indeed being recycled. They believed this was happening in regions of intense volcanic and earthquake activity around the margins of the Pacific Ocean basin, known collectively as the Ring of Fire (**Figure 3.6**). The details of the process, however, remained unclear.

In 1965, the Canadian geologist J. Tuzo Wilson first described tectonics around the globe in terms of rigid plates moving over Earth's surface. He characterized three basic types of boundaries where plates move apart, come together, or slide past each other. Soon after, other scientists showed that almost all contemporary tectonic deformation—the process by which rocks are folded, faulted, sheared, or compressed by tectonic forces—is concentrated at these boundaries. They measured the rates and directions of crustal movements and demonstrated that these movements are mathematically consistent with a system of rigid plates moving over the planet's spherical surface. The basic elements of the plate tectonic theory were established by the end of 1968. By 1970, the evidence for plate tectonics had become so persuasive that almost all Earth scientists embraced the theory. Textbooks were revised, and specialists began to consider the implications of the new concept for their own fields.

earthquakes are generated by faulting, their results indicated that the rift was a tectonically active feature. Other mid-ocean ridges with similar rifts and earthquake activity were found in the Pacific and Indian oceans.

In the early 1960s, Harry Hess of Princeton University and Robert Dietz of the Scripps Institution of Oceanography proposed that Earth's crust separates along the rifts in mid-ocean ridges and that new crust forms by the upwelling of hot molten rock into these cracks. The new seafloor—actually the surface of newly created lithosphere—spreads laterally away from the rift and is replaced by even newer crust in a continuing process of plate creation.

> *Seafloor spreading occurs as plates separate along a rift and hot molten rock wells up to create new crust, which moves away from the mid-ocean ridge as more crust is continually created.*

The Great Synthesis: 1963–1968

The seafloor spreading hypothesis put forward by Hess and Dietz in 1962 explained how the continents could drift apart through the creation of new lithosphere at mid-ocean ridges. Could the seafloor and its underlying lithosphere be destroyed by recycling back into Earth's interior? If not, Earth's surface area would have to increase over time. For

FIGURE 3.6 The Pacific Ring of Fire, with its active volcanoes (large red circles) and frequent earthquakes (small black dots), marks plate boundaries where oceanic lithosphere is being recycled.

<< 56 CHAPTER 3 Plate Tectonics: The Unifying Theory of Geology

TYPES OF PLATE BOUNDARIES

According to the theory of plate tectonics, the rigid lithosphere is not a continuous shell, but is broken into a mosaic of 13 large, rigid plates that move over Earth's surface (**Figure 3.7**). Each plate travels as a distinct unit, riding on the asthenosphere, which is also in motion. The largest is the Pacific Plate, which comprises much (though not all) of the Pacific Ocean basin. Some of the plates are named after the continents they include, but in no case is a plate identical with a continent. The North American Plate, for instance, extends from the Pacific coast of North America to the middle of the Atlantic Ocean, where it meets the Eurasian and African plates.

FIGURE 3.7 Earth's surface is a mosaic of 13 major plates of rigid lithosphere, as well as a number of smaller plates, that move slowly over the ductile asthenosphere. Only one of the smaller plates, the Juan de Fuca Plate, off the west coast of North America, is labeled on this map. The arrows show the relative motion of two plates at a point on their boundary. The numbers next to the arrows give the relative plate speeds in millimeters per year. [Plate boundaries by Peter Bird, UCLA.]

DIVERGENT BOUNDARIES

(a) Oceanic Plate Separation

Rifting and spreading along a narrow zone have created the Mid-Atlantic Ridge, a mid-ocean mountain chain where volcanoes and earthquakes are concentrated.

- Mid-Atlantic Ridge
- North American Plate
- Eurasian Plate

(b) Continental Plate Separation

In East Africa, rifting and spreading have created parallel valleys in a zone with volcanoes and earthquakes.

- Great Rift Valley
- African Plate
- Somali Subplate

CONVERGENT BOUNDARIES

(c) Ocean–Ocean Convergence

When two oceanic plates converge, they form a deep-sea trench and a volcanic island arc.

- Mariana Islands
- Marianas Trench

FIGURE 3.8 Interactions at plate boundaries depend on the direction of relative plate movement and the type of crust involved.

The lithosphere is broken into 13 large plates, plus a number of smaller ones, all in constant motion.

In addition to the major plates, there are a number of smaller ones. An example is the Juan de Fuca Plate, a small piece of oceanic lithosphere trapped between the giant Pacific and North American plates just offshore of the northwestern United States. Others are continental fragments, such as the small Anatolian Plate, which includes much of Turkey.

To see plate tectonics in action, go to a plate boundary. Depending on which boundary you visit, you may find earthquakes, volcanoes, rising mountains, long, narrow rifts, folding, or faulting. Many geologic features develop through the interactions of plates at their boundaries.

There are three basic types of plate boundaries (**Figure 3.8**), all defined by the direction of movement of the plates relative to each other:

(d) Ocean–Continent Convergence

When oceanic crust meets continental crust, the oceanic plate is subducted, and a volcanic belt of mountains is formed at the continental plate margin.

Andes · Peru-Chile Trench · Nazca Plate · South American Plate

(e) Continent–Continent Convergence

When two continental plates collide, the crust crumples and thickens, creating high mountains and a wide plateau.

Himalaya · Tibetan Plateau · Indian-Australian Plate · Eurasian Plate · Main thrust fault

TRANSFORM-FAULT BOUNDARIES

(f) Mid-Ocean Ridge Transform Fault

Spreading centers are offset by mid-ocean ridge transform faults, where the two oceanic plates slide horizontally past each other.

Eurasian Plate · North American Plate

(g) Continental Transform Fault

The San Andreas fault in California, where the Pacific Plate slides past the North American Plate, is an example of a transform fault that offsets continental crust.

Pacific Plate · North American Plate

- At **divergent boundaries,** plates move apart and new lithosphere is created (plate area increases).

- At **convergent boundaries,** plates come together and one plate is recycled into the mantle (plate area decreases).

- At **transform faults,** plates slide horizontally past each other (plate area does not change).

Besides these three basic types, there are "oblique" boundaries that combine divergence or convergence with some amount of transform faulting. Moreover, what actually goes on at a plate boundary depends on the type of lithosphere involved, because continental and oceanic lithosphere behave differently. The continental crust is made of rocks that are both lighter and weaker than either the oceanic crust or the mantle beneath the crust

There are three basic types of plate boundaries: divergent boundaries, where plates are moving apart; convergent boundaries, where plates are moving together; and transform faults, where plates are sliding past each other.

(see Figure 2.12). Later chapters will examine these differences in more detail; for now, you need to keep in mind only two of their consequences:

1. Because continental crust is lighter, it is not as easily recycled as oceanic crust.

2. Because continental crust is weaker, plate boundaries that involve continental crust tend to be more spread out and more complicated than oceanic plate boundaries.

Divergent Boundaries

Divergent boundaries within ocean basins are narrow rifts that approximate the idealization of plate tectonics. Divergent boundaries within continents are usually more complicated and distributed over a wider area. This difference is illustrated in Figure 3.8a and b.

Oceanic Plate Separation On the seafloor, the boundary between separating plates is marked by a mid-ocean ridge that exhibits active volcanism, earthquakes, and rifting, all caused by the tensional (stretching) forces of mantle convection that are pulling the two plates apart. The seafloor spreads as hot molten rock, called *magma*, wells up into the rifts to form new oceanic crust. Figure 3.8a shows what happens at one such boundary, the Mid-Atlantic Ridge, where the North American and Eurasian plates are separating. (A more detailed portrait of the Mid-Atlantic Ridge is shown in Figure 3.4.) The island of Iceland exposes a segment of the otherwise submerged Mid-Atlantic Ridge, allowing geologists to view the processes of plate separation and seafloor spreading directly (**Figure 3.9**). The Mid-Atlantic Ridge continues in the Arctic Ocean north of Iceland and connects to a nearly globe-encircling system of mid-ocean ridges that winds through the Indian and Pacific oceans, ending along the western coast of North America. These **spreading centers** have created the millions of square kilometers of oceanic crust that now form the floors of the world's oceans.

At spreading centers on mid-ocean ridges, plates separate and new oceanic crust is formed.

Continental Plate Separation Early stages of plate separation, such as the Great Rift Valley of East Africa (see Figure 3.8b), can be found on some continents. These divergent boundaries are characterized by rift valleys, volcanic activity, and earthquakes distributed over a wider zone than is found at oceanic spreading centers. The Red Sea and the Gulf of California are rifts that are further along in the spreading process (**Figure 3.10**). In these cases, the continents have separated enough for new seafloor to form along the spreading axis, and the ocean has flooded the rift valleys.

FIGURE 3.9 The Mid-Atlantic Ridge, a divergent plate boundary, rises above sea level in Iceland. This cracklike rift valley, filled with newly formed volcanic rocks, indicates that plates are being pulled apart. [Gudmundur E. Sigvaldason, Nordic Volcanological Institute.]

Sometimes continental rifting slows or stops before the continent actually splits apart. The Rhine Valley, along the border of Germany and France in western Europe, is a weakly active continental rift that may be this type of "failed" spreading center. Will the East African Rift continue to open, causing the Somali Subplate to split away from Africa completely and form a new ocean basin, as happened between Africa and the island of Madagascar? Or will the spreading slow and eventually stop, as appears to be happening in western Europe? Geologists don't know the answers.

Convergent Boundaries

Plates cover the globe, so if they separate in one place, they must converge somewhere else, if Earth's surface area is to

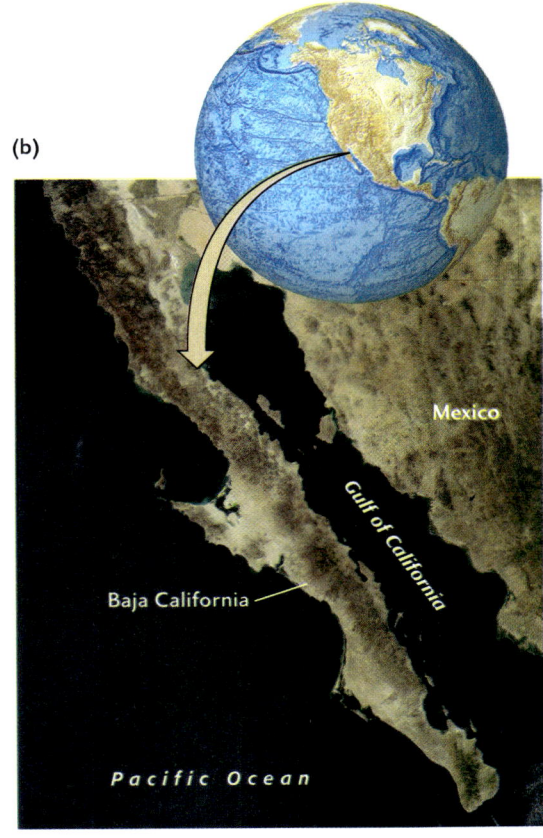

FIGURE 3.10 Continental plate separation. (a) The Red Sea (*lower right*) divides to form the Gulf of Suez on the left and the Gulf of 'Aqaba on the right. The Arabian Peninsula, on the right, splitting away from Africa, on the left, has opened these great rifts, which are now flooded by the sea. The Nile River (*far left*) flows north into the Mediterranean Sea (*top*). [Earth Satellite Corporation.] (b) The Gulf of California, an opening ocean, marks a widening rift between Baja California and the Mexican mainland. [Jeff Schmaltz, MODIS Rapid Response Team, NASA/GSFC.]

> *Where continents split apart, rift valleys, volcanic activity, and earthquakes are distributed over a wider zone than at oceanic spreading centers.*

remain the same. (As far as we can tell, our planet is not expanding!) Thus, where plates come together, they form convergent boundaries. The variety of geologic events resulting from plate convergence makes these boundaries the most complex type observed.

Ocean-Ocean Convergence

If the lithosphere of both plates is oceanic, one plate descends beneath the other in a process known as **subduction** (see Figure 3.8c). The oceanic lithosphere of the subducting plate sinks into the asthenosphere and is eventually recycled by the mantle convection system. This sinking produces a long, narrow deep-sea trench. In the Marianas Trench of the western Pacific, the ocean reaches its greatest depth, about 11 km—deeper than the height of Mount Everest.

As the cold lithospheric slab descends deeper into Earth's interior, the pressure on it increases. Water trapped in the rocks is squeezed out and rises into the asthenosphere above the slab. This fluid causes the mantle to melt, producing a chain of volcanoes, called an **island arc,** behind the trench. The subduction of the Pacific Plate has formed the volcanically active Aleutian Islands west of Alaska as well as the island arcs of the western Pacific. The lithospheric slabs descending into the mantle cause earthquakes as deep as 690 km beneath some island arcs.

Ocean-Continent Convergence

If one plate has a continental edge, it overrides the oceanic plate, because continental crust is lighter and less easily subducted than oceanic crust (see Figure 3.8d). The continental margin crumples, deforming the continental crust and uplifting rocks into a mountain chain roughly parallel to the deep-sea trench. The enormous forces of convergence and subduction produce great earthquakes along the subduction zone. Over time, materials are scraped off the descending

> *Where two oceanic plates converge, one plate is subducted beneath the other and sinks into the asthenosphere.*

Continental crust overrides oceanic crust at a convergent boundary because it is lighter and less easily subducted.

slab and incorporated into the adjacent mountains, leaving geologists with a complex (and often confusing) record of the subduction process. As in the case of ocean-ocean convergence, the water carried downward by the subducting oceanic plate causes mantle material to melt and forms volcanoes in the mountain belt behind the trench.

The western coast of South America, where the South American Plate converges with the oceanic Nazca Plate, is a subduction zone of this type. A great chain of high mountains, the Andes, rises on the continental side of this convergent boundary, and a deep-sea trench lies just off the coast. The volcanoes here are active and deadly. One of them, Nevado del Ruiz in Colombia, killed 25,000 people when it erupted in 1985. Some of the world's greatest earthquakes have been recorded along this boundary.

Another example is the Cascadia subduction zone, where the small Juan de Fuca Plate converges with the North American Plate off the coast of western North America. This convergent boundary gives rise to the dangerous volcanoes of the Cascade Range, such as Mount St. Helens, which had a major eruption in 1980 and a minor one in 2004. There is increasing worry that a great earthquake will occur in the Cascadia subduction zone and cause devastating damage along the coasts of Oregon, Washington, and British Columbia. Such an earthquake could cause a large tsunami like the disastrous one generated by the great Sumatra earthquake of December 26, 2004, which occurred in a subduction zone in the eastern Indian Ocean.

Continent-Continent Convergence Where two continents converge (see Figure 3.8e), the kind of subduction seen at other convergent boundaries cannot occur. The geologic consequences of such a continent-continent collision are impressive. The collision of the Indian and Eurasian plates, both with continents at their leading edges, provides the best example. The Eurasian Plate overrides the Indian Plate, but India and Asia remain afloat on the mantle. The collision creates a double thickness of crust, forming the highest mountain range in the world, the Himalaya, as well as the vast high plateau of Tibet. Severe earthquakes occur in the crumpling crust of this and other continent-continent collision zones. Many episodes of mountain building throughout Earth's history were caused by continent-continent collisions. An example is the Appalachian Mountains that run along the eastern coast of North America. This chain was uplifted when North America, Eurasia, and Africa collided to form the supercontinent Pangaea about 300 million years ago.

Where two continents converge, the collision deforms and thickens the crust, raising high mountains.

Transform-Fault Boundaries

Boundaries where plates slide past each other and lithosphere is neither created nor destroyed are called transform faults (see Figure 3.8f, g). Transform-fault boundaries are typically found along mid-ocean ridges where the continuity of a divergent boundary is broken and the boundary is offset in a steplike pattern.

1 As the Pacific Plate and North American Plate move past each other in opposite directions,…

2 …streambeds crossing the fault have been offset.

FIGURE 3.11 A view northwest along the San Andreas fault in the Carrizo Plain of central California. The San Andreas is a transform fault, forming a portion of the sliding boundary between the Pacific Plate (*left*) and the North American Plate (*right*). Notice how plate movement on the fault has offset the streams flowing across it. [John Shelton.]

The San Andreas fault in California, where the Pacific Plate slides past the North American Plate, is a prime example of a transform fault on land, as shown in **Figure 3.11**. Because the plates have been sliding past each other for millions of years, the rocks facing each other on the two sides of the fault are of different types and ages. Large earthquakes, such as the one that destroyed San Francisco in 1906, can occur on transform-fault boundaries. There is much concern that, within the next several decades, a sudden rupture of the San Andreas fault, or on related faults near Los Angeles and San Francisco, will result in an extremely destructive earthquake.

> Plates slide past each other along transform faults.

Combinations of Plate Boundaries

Each plate is bordered by some combination of divergent, convergent, and transform-fault boundaries. As we can see in Figure 3.7, the Nazca Plate in the Pacific is bounded on three sides by divergence zones, offset in a stepwise pattern by transform faults, and on one side by the Peru-Chile subduction zone, where lithosphere is recycled at a deep-sea trench. The North American Plate is bounded on the east by the Mid-Atlantic Ridge, a divergence zone, and on the west by subduction zones and transform-fault boundaries

RATES AND HISTORY OF PLATE MOVEMENTS

How fast do plates move? Do some plates move faster than others, and if so, why? Is the velocity of plate movements today the same as it was in the geologic past? Geologists have developed ingenious methods to answer these questions and thereby gain a better understanding of plate tectonics. In this section, we will examine three of these methods.

The Seafloor as a Magnetic Tape Recorder

During World War II, extremely sensitive magnetometers were developed to detect submarines by the magnetic fields emanating from their steel hulls. Geologists modified these instruments slightly and towed them behind research ships to measure the local magnetic field created by magnetized rocks beneath the sea. Steaming back and forth across the ocean, seagoing scientists discovered regular patterns in the strength of the local magnetic field that completely surprised them. In many areas, the intensity of the magnetic field alternated between high and low values in long, narrow parallel bands, called **magnetic anomalies,** that were almost perfectly symmetrical with respect to the crest of a mid-ocean ridge (**Figure 3.12**). The detection of these patterns was one of the great discoveries that confirmed the seafloor spreading hypothesis and led to the theory of plate tectonics. It also allowed geologists to measure plate movements far back into geologic time. To understand these advances, we need to look more closely at how rocks become magnetized.

The Rock Record of Magnetic Reversals on Land

Magnetic anomalies are evidence that Earth's magnetic field does not remain constant over time. At present, the north magnetic pole is closely aligned with the geographic north pole (see Figure 2.17), but small changes in the geodynamo can flip the orientation of the north and south magnetic poles by 180°, causing a *magnetic reversal.*

In the early 1960s, geologists discovered that a precise record of this peculiar behavior can be obtained from layered flows of volcanic lava. When iron-rich lavas cool, they become slightly but permanently magnetized in the direction of Earth's magnetic field. This phenomenon is called *thermoremanent magnetization* because the rock "remembers" the magnetization long after the magnetizing field has changed.

In layered lava flows, such as those in a volcanic cone, the rocks at the top represent the most recent layer, while layers deeper in the cone are older. The age of each layer can be determined by various dating methods (described in Chapter 8). Measuring the magnetization of rock samples from each layer then reveals the direction of Earth's magnetic field when that layer cooled. By repeating these measurements at hundreds of places around the world, geologists have worked out the detailed history of magnetic reversals going back in geologic time. The **magnetic time scale** of the past 5 million years is shown in Figure 3.12.

About half of all rocks studied have been found to be magnetized in a direction opposite to that of Earth's present magnetic field. Apparently, the field has flipped frequently over geologic time, so normal fields (same as now) and reversed fields (opposite to now) are equally likely. Major periods when the field is normal or reversed are called *magnetic chrons;* they last about half a million years, on average, although the pattern of reversals becomes highly irregular as we move back in geologic time.

> By measuring the magnetization of volcanic lavas, geologists have determined that Earth's magnetic field has reversed itself many times throughout Earth's history.

Magnetic Anomaly Patterns on the Seafloor
The peculiar banded magnetic patterns found on the seafloor

FIGURE 3.12 Magnetic mapping can measure the rate of seafloor spreading.

(see Figure 3.12) puzzled scientists until 1963, when two Englishmen, F. J. Vine and D. H. Mathews—and, independently, two Canadians, L. Morley and A. Larochelle—made a startling proposal. Based on the new evidence for magnetic reversals that land geologists had collected from lava flows, they reasoned that the bands of high and low magnetic intensity on the seafloor corresponded to bands of rock that were magnetized during ancient episodes of normal and reversed magnetism. That is, when a research ship was above rocks magnetized in the normal direction, it would record a locally stronger field, or a *positive magnetic anomaly*. When it was above rocks magnetized in the reversed direction, it would record a locally weaker field, or a *negative magnetic anomaly*.

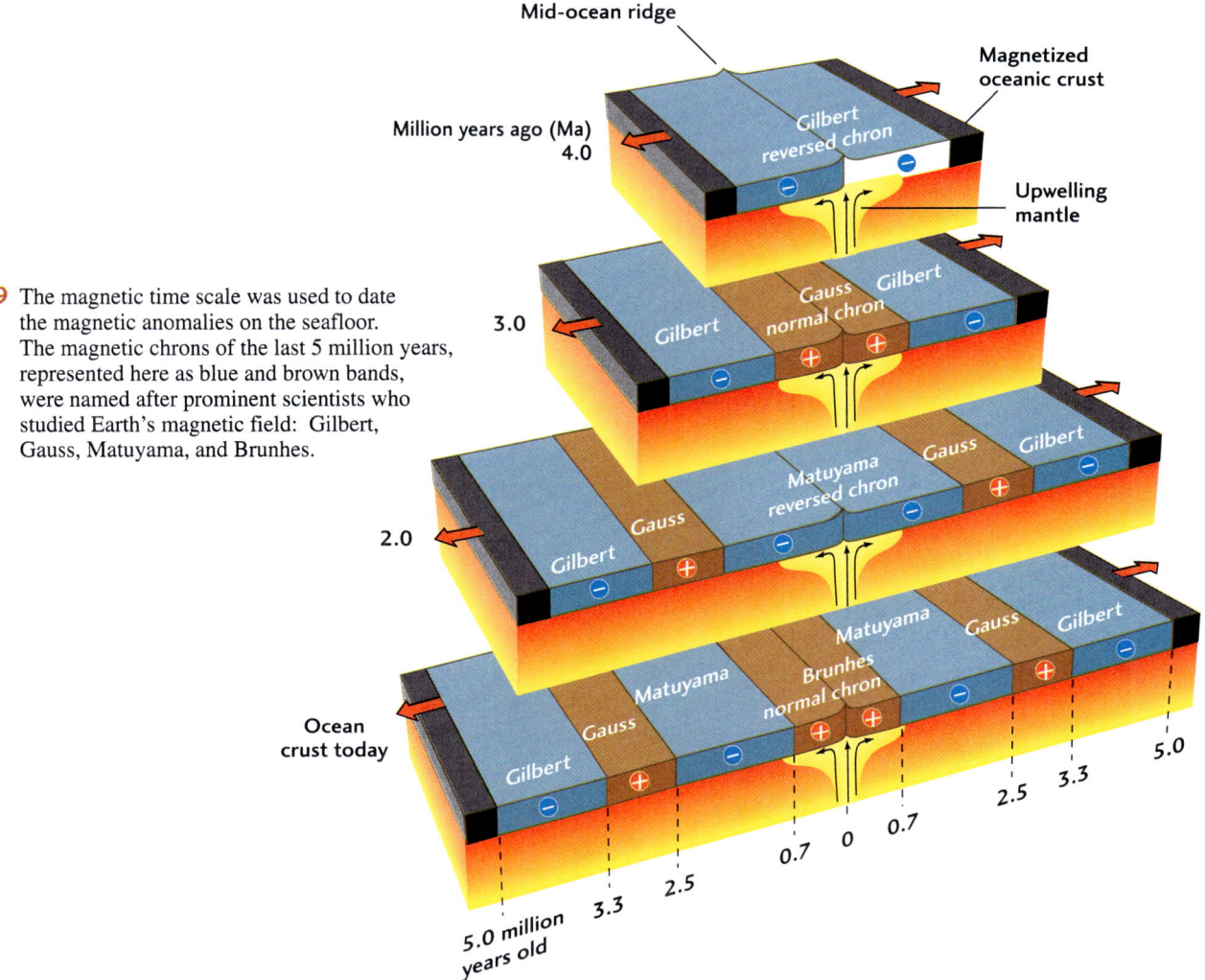

9 The magnetic time scale was used to date the magnetic anomalies on the seafloor. The magnetic chrons of the last 5 million years, represented here as blue and brown bands, were named after prominent scientists who studied Earth's magnetic field: Gilbert, Gauss, Matuyama, and Brunhes.

This idea provided a powerful test of the seafloor spreading hypothesis, which states that new seafloor is created along the rift at the crest of a mid-ocean ridge as the plates move apart (see Figure 3.12). Magma flowing up from Earth's interior cools and solidifies in the rift and becomes magnetized in the direction of Earth's magnetic field at the time. As the seafloor spreads away from the ridge, approximately half of the newly magnetized material moves to one side and half to the other, forming two symmetrical magnetized bands. Newer material fills the rift, continuing the process. In this way, the seafloor acts like a tape recorder that encodes the history of the opening of the oceans.

The seafloor spreading model provides a consistent explanation for the symmetrical patterns of magnetic anomalies found on mid-ocean ridges around the world. Moreover, it gives us a precise tool for measuring rates of seafloor spreading now and in the geologic past. This evidence contributed substantially to the discovery and confirmation of plate tectonics.

Measuring Plate Velocities By using the magnetic time scale that had been worked out from magnetized lavas on land, geologists could assign ages to the bands of magnetized rocks on the seafloor. Moreover, knowing the ages of the seafloor bands and knowing their distance from the ridge, they could calculate how fast the seafloor was spreading. For instance, the magnetic anomaly pattern in Figure 3.12 shows that the boundary between the Gauss normal chron and the Gilbert reversed chron, which was dated from lava flows at 3.3 million years, is located about 30 km away from the Reykjanes Ridge crest, just southwest of Iceland. Thus, seafloor spreading has moved the North American and Eurasian plates apart

Bands of high and low magnetic intensity on the seafloor provided evidence that new seafloor is continually being created at and spreading away from mid-ocean ridges.

by about 60 km in 3.3 million years, giving a spreading rate of 18 km per million years or, equivalently, 18 mm/year.

If you look at Figure 3.7, you will see that the spreading rate at the Mid-Atlantic Ridge south of Iceland is fairly low compared with the rate at many other places. The speed record for spreading can be found on the East Pacific Rise, just south of the equator, where the Pacific and Nazca plates are separating at a rate of about 150 mm/year—an order of magnitude faster than the rate in the North Atlantic. A rough average spreading rate for mid-ocean ridges around the world is 50 mm/year. This is approximately the rate at which your fingernails grow—so, geologically speaking, the plates move very fast indeed!

We can follow the magnetic time scale through many reversals of Earth's magnetic field. The corresponding magnetic bands on the seafloor, which can be thought of as age bands, have been mapped in detail from the ridge crests across the ocean basins over a time span of almost 200 million years.

The power and convenience of using seafloor magnetization to work out the history of ocean basins cannot be overemphasized. Simply by steaming a ship back and forth over the ocean and measuring the magnetic fields of the seafloor rocks, geologists have been able to determine the ages of various regions of the seafloor without directly examining rock samples. In effect, they learned how to "replay the tape."

Although measuring seafloor magnetization is a very effective technique, it is an indirect, or remote, sensing method in that rocks are not recovered from the seafloor and their ages are not directly determined in the laboratory. Direct evidence of seafloor spreading and plate movement was still needed to convince the few remaining skeptics. Deep-sea drilling supplied it.

Deep-Sea Drilling

In 1968, a program of drilling into the seafloor was launched as a joint project of several major oceanographic institutions and the National Science Foundation. Later, many nations joined the effort (**Figure 3.13**). This global experiment aimed to drill through, retrieve, and study seafloor rocks from many places in the world's oceans. Using hollow drills, scientists brought up cores containing sections of seafloor rocks. In some cases, the drilling penetrated thousands of meters below the seafloor surface. This gave geologists an opportunity to work out the history of the ocean basins from direct evidence.

One of the most important facts geologists sought was the age of each rock sample. Small particles falling through the ocean waters—dust from the atmosphere, organic material from marine plants and animals—begin to accumulate as seafloor sediments as soon as new oceanic crust forms. Therefore, the age of the oldest sediments in a core—those immediately on top of the crust—tells the geologist how old the seafloor is at that spot. The ages of sediments can be obtained primarily from the fossil skeletons of tiny single-celled planktonic organisms that live at the surface of the open ocean and sink to the bot-

Direct measurements of rock age obtained by drilling into the seafloor validated the magnetic time scale and confirmed the hypothesis of seafloor spreading.

FIGURE 3.13 The deep-sea drilling vessel *JOIDES Resolution* is 143 m long and carries a drilling derrick 61 m high that is capable of drilling into the seafloor beneath the deepest ocean. Samples from cores of sediment recovered by drilling have confirmed the age of the seafloor deduced from magnetic anomalies. Such samples have also shed new light on the history of ocean basins and ancient climatic conditions. [Courtesy of Ocean Drilling Program/Texas A&M University.]

tom when they die. Geologists found that the ages of the sediments in the cores increased with distance from mid-ocean ridges and that the age of the seafloor at any one place agreed almost perfectly with the age determined from magnetic reversal data. This agreement validated the magnetic time scale and provided strong evidence for seafloor spreading.

GPS Measurements of Plate Movements

In his publications advocating the continental drift hypothesis, Alfred Wegener made a big mistake: he proposed that North America and Europe were drifting apart at a rate of nearly 30 meters per year—a thousand times faster than the Atlantic seafloor is actually spreading! This unbelievably high speed was one of the reasons that many scientists roundly rejected his notions of continental drift. Wegener made his estimate by incorrectly assuming that the continents were joined together as Pangaea as recently as the last ice age (which occurred only about 20,000 years ago).

Wegener imagined that he could measure continental drift in the following way: Two observers, one in Europe and the other in North America, would simultaneously determine their positions relative to the fixed stars. From those positions, they would calculate the distance between their two observation posts at that instant. They would then repeat this distance measurement from the same observation posts sometime later—say, after 1 year. If the continents were drifting apart, then the distance should have increased, and the value of the increase would determine the speed of the drift. Unfortunately, in Wegener's day, the accuracy of astronomical positioning was poor. We now know that the spreading of the Mid-Atlantic Ridge from one year to the next is only a couple of centimeters, a thousand times too small to be observed by the techniques that were then available.

Since the mid-1980s, geologists have used a constellation of 24 Earth-orbiting satellites, called the Global Positioning System (GPS), to make much more accurate positioning measurements, and they can now track the movements of the continents using inexpensive, portable radio receivers not much bigger than this textbook (**Figure 3.14**). This technology is similar to the GPS systems that are now used in automobiles and by hikers, though much more precise.

The Global Positioning System provides geologists with a way to observe plate movements as they are occurring.

The changes in distance between land-based GPS receivers recorded over several years agree in both magnitude and direction with those calculated from magnetic anomalies on the seafloor, indicating that plate movements are remarkably steady over periods

Triangulation using three or more GPS satellites can locate a precise point on Earth.

Displacements of the GPS station locations from one year to the next allow us to measure plate movements.

A GPS station

FIGURE 3.14 Signals from GPS satellites can be used to monitor plate movement. [Photo courtesy Southern California Earthquake Center.]

ranging from just a few years up to millions of years. Geologists are now using GPS to measure plate movements on a regular basis at many locations around the globe.

THE GRAND RECONSTRUCTION

The supercontinent Pangaea was the only major landmass that existed 250 million years ago. One of the great triumphs of modern geology is the reconstruction of the events that led to the assembly of Pangaea and to its later fragmentation into the continents we know today. Let's use what we have learned about plate tectonics to see how this feat was accomplished.

Seafloor Isochrons

The color map in **Figure 3.15** shows the ages of the rocks on the seafloor as determined by magnetic reversal data and fossils from deep-sea drilling. Each colored band represents a span of time corresponding to the age of the crust within that band. The boundaries between bands, called **isochrons,** are contours that connect rocks of equal age. Isochrons tell us the time that has elapsed since the rocks were injected as magma into a spreading zone and, therefore, the amount of spreading that has occurred since they formed. Notice how the seafloor becomes progressively older on both sides of the mid-ocean ridges. For example, the distance from a ridge axis to a 140-million-year isochron (boundary between green and blue bands) indicates the extent of new seafloor created over that time span. The more widely spaced isochrons of the eastern Pacific signify faster spreading rates than those in the Atlantic.

In 1990, after a 20-year search, geologists found the oldest oceanic rocks by drilling into the seafloor of the western Pacific. These rocks turned out to be about 200 million years old, only about 4 percent of Earth's age. This finding indicates how geologically young the seafloor is compared with the continents. Over a period of 100 million to 200 million years in some places and only tens of millions of years in others, oceanic lithosphere forms, spreads, cools, and is recycled into the underlying mantle. In contrast, the oldest known continental rocks are about 4 billion years old.

The seafloor, which is continually recycled into the underlying mantle, is geologically younger than the continental crust.

FIGURE 3.15 This isochron map shows the ages of rocks on the seafloor. The isochrons give the age of the seafloor in millions of years since its creation at mid-ocean ridges. Light gray indicates land; dark gray indicates shallow water over continental shelves. Mid-ocean ridges, along which new seafloor is extruded, coincide with the youngest rocks (red). [*Journal of Geophysical Research* 102 (1997): 3211–3214. Courtesy of R. Dietmar Müller.]

Reconstructing the History of Plate Movements

Earth's plates behave as rigid bodies. That is, the distances between three points on the same rigid plate—say, New York, Miami, and Bermuda on the North American Plate—do not change very much, no matter how far the plate moves. But the distance between, say, New York and Lisbon increases over time because the two cities are on two different plates that are separating along the Mid-Atlantic Ridge. The direction of the movement of one plate in relation to another depends on two geometric principles that govern the behavior of rigid plates on a sphere:

- *Transform-fault boundaries indicate the directions of relative plate movement.* With few exceptions, no overlap, buckling, or separation occurs along typical transform-fault boundaries in the oceans. The two plates merely slide past each other without creating or destroying plate material. Look for a transform-fault boundary if you want to deduce the directions of relative plate movement, because the orientation of the fault is the direction in which one plate slides with respect to the other (see Figure 3.8f, g).

- *Seafloor isochrons reveal the positions of divergent boundaries in earlier times.* Isochrons on the seafloor are roughly parallel and symmetrical to the ridge axis along which they were created (see Figure 3.15). Because each isochron was at the divergent boundary at an earlier time, isochrons that are of the same age but on opposite sides of a mid-ocean ridge can be brought together to show the positions of the plates, and the configuration of the continents embedded in them, as they were in that earlier time.

The Breakup of Pangaea

Using these principles, geologists have reconstructed the opening of the Atlantic Ocean and the breakup of Pangaea (**Figure 3.16**). Figure 3.16e shows the supercontinent Pangaea as it existed 240 million years ago. It began to break apart when North America rifted away from Europe about 200 million years ago (Figure 3.16f). The opening of the North Atlantic was accompanied by the separation of the northern continents (referred to as Laurasia) from the southern continents (Gondwana) and the rifting of Gondwana along what is now the eastern coast of Africa (Figure 3.16g). The breakup of Gondwana separated South America, Africa, India, and Antarctica, creating the South Atlantic and Antarctic oceans and narrowing the Tethys Ocean (Figure 3.16h). The separation of Australia from Antarctica and the ramming of India into Eurasia closed the Tethys Ocean, giving us the world we see today (Figure 3.16i).

The plate movements have not ceased, of course, so the configuration of the continents will continue to evolve. A plausible scenario for the distribution of continents and plate boundaries 50 million years in the future is displayed in Figure 3.16j.

The Assembly of Pangaea

The isochron map in Figure 3.15 tells us that all of the seafloor on Earth's surface today has been created since the breakup of Pangaea. We know from the geologic record in older continental mountain belts, however, that plate tectonics had been operating for billions of years before this breakup. Evidently, seafloor spreading took place just as it does today, and there were previous episodes of continental drift and collision. Subduction into the mantle has destroyed the seafloor created in those earlier times, however, so we must rely on the older evidence preserved on continents to identify and chart the movements of ancient continents (*paleocontinents*).

Old mountain belts, such as the Appalachians of North America and the Urals, which separate Europe from Asia, help us locate ancient collisions of the paleocontinents. In many places, the rocks reveal ancient episodes of rifting and subduction. Rock types and fossils also indicate the distribution of ancient seas, glaciers, lowlands, mountains, and climates. Knowledge of ancient climates enables geologists to locate the latitudes at which continental rocks formed, which in turn helps them to assemble the jigsaw puzzle of paleocontinents. When volcanism or mountain building produces new continental rocks, these rocks also record the direction of Earth's magnetic field, just as oceanic rocks do when they are created by seafloor spreading. Like a compass frozen in time, the fossil magnetism of a continental fragment records its ancient orientation and magnetic latitude.

The left side of Figure 3.16 shows one of the latest efforts to depict the pre-Pangaean configuration of continents. It is truly impressive that modern science can recover the geography of this strange world of hundreds of millions of years ago. The evidence from rock types, fossils, climate, and rock magnetization has allowed scientists to reconstruct an earlier supercontinent, called **Rodinia,** that formed about 1.1 billion years ago and began to break up about 750 million years ago (Figure 3.16a). They have been able to chart its fragments over the subsequent 500 million years as these fragments drifted and reassembled into the supercontinent Pangaea. Geologists continue to sort out the details of this complex jigsaw puzzle, whose individual pieces have changed shape over geologic time.

[*Text continues on page 72.*]

> *Seafloor isochrons have been used to reconstruct the movements of plates since the breakup of the supercontinent Pangaea. Older continental rocks indicate that the same tectonic processes were occurring before Pangaea was formed.*

ASSEMBLY OF PANGAEA

RODINIA (a) Late Proterozoic, 750 Ma

1 The supercontinent of Rodinia formed about 1.1 billion years ago and began to break up about 750 million years ago.

(b) Late Proterozoic, 650 Ma

2 Geologists have used a variety of evidence, including paleomagnetism and information about ancient climates, to reconstruct the pre-Pangaean pattern of continental drift.

(c) Middle Ordovician, 458 Ma

(d) Early Devonian, 390 Ma

PANGAEA (e) Early Triassic, 237 Ma

3 The supercontinent Pangaea was mostly assembled by 237 Ma, surrounded by a superocean called Panthalassa (Greek for "all seas"), the ancestral Pacific Ocean. The Tethys Ocean, between Africa and Eurasia, was the ancestor of the Mediterranean Sea.

FIGURE 3.16 Continental rifting, drifting, and collisions assembled and then disassembled the supercontinent Pangaea. [Paleogeographic maps by Christopher R. Scotese, 2003 PALEOMAP Project (www.scotese.com).]

BREAKUP OF PANGAEA

(f) Early Jurassic, 195 Ma

1 The breakup of Pangaea was signaled by the opening of rifts from which lava poured. Rock assemblages that are relics of this great event can be found today in 200-million-year-old volcanic rocks from Nova Scotia to North Carolina.

(g) Late Jurassic, 152 Ma

2 By about 150 million years ago, Pangaea was in the early stages of breakup. The Atlantic Ocean had partially opened, the Tethys Ocean had contracted, and the northern continents (Laurasia) had all but split away from the southern continents (Gondwana). India, Antarctica, and Australia began to split away from Africa.

(h) Late Cretaceous, Early Tertiary, 66 Ma

3 By 66 million years ago, the South Atlantic had opened and widened. India was well on its way northward toward Asia, and the Tethys Ocean was closing to form the Mediterranean.

THE PRESENT-DAY AND FUTURE WORLD

(i) PRESENT-DAY WORLD

4 The modern world has been produced over the past 65 million years. India collided with Asia, ending its trip across the ocean, and is still pushing northward into Asia. Australia has separated from Antarctica.

(j) 50 million years in the future

MANTLE CONVECTION: The Engine of Plate Tectonics

Everything discussed in this chapter so far might be called descriptive plate tectonics. But a description is not an explanation. We need a more comprehensive theory that explains *why* plates move.

As Arthur Holmes and other early advocates of continental drift realized, mantle convection is the "engine" that drives the large-scale tectonic processes operating on Earth's surface. In Chapter 2, we described the mantle as a moldable, or ductile, solid. Hot mantle material is capable of flowing like a sticky fluid (warm wax or cold syrup, for example). Heat escaping from Earth's deep interior causes this material to undergo convection (circulate upward and downward) at speeds of a few tens of millimeters per year. The details of the mantle convection system have not been completely worked out, but we can gain some insights by trying to answer three important questions.

What Forces Drive Plate Movements?

The lithospheric plates are not just dragged along Earth's surface by mantle convection; they are active participants in the flow. In convergence zones, the gravitational pull exerted by the cold (and thus heavy) slabs of old subducting lithosphere drags the plates downward into the mantle. This hypothesis explains why plates being subducted along a large fraction of their boundaries (e.g., the Pacific, Nazca, Cocos, and Indian plates) move faster, while those that do not have significant fractions of subducting slabs (e.g., the North American, South American, African, Eurasian, and Antarctic plates) move more slowly. In other words, the plates "fall back" into the mantle under their own weight.

There is also evidence that overriding plates, as well as subducting plates, are pulled toward convergent boundaries by gravitational forces. For example, as the Nazca Plate is subducted beneath South America, it causes the convergent boundary at the Peru-Chile Trench to retreat toward the Pacific, "sucking" the South American Plate to the west.

Other complications are evident from the history of plate movements. When the continents came together to form Pangaea, they acted as an insulating blanket, preventing heat from getting out of Earth's mantle (as it otherwise would through the process of seafloor spreading). That heat built up over time, forming hot bulges in the mantle beneath the supercontinent. These bulges raised Pangaea slightly and caused it to rift apart in a kind of "landslide" off the top of the bulges. Gravitational forces continue to drive subsequent seafloor spreading as the plates "slide downhill" off the crest of the Mid-Atlantic Ridge. Earthquakes that sometimes occur in plate interiors provide direct evidence of the compression of plates by these ridge-related gravitational forces.

Oceanic lithosphere slides down and away from mid-ocean ridges and is pulled into subduction zones by its own weight.

How Deep in the Mantle Does Plate Recycling Occur?

For plate tectonics to work, the lithospheric material that descends into the mantle at subduction zones must be recycled through the mantle and eventually come back up as new lithosphere created along the spreading centers of the mid-ocean ridges. How deep into the mantle does this recycling process extend? That is, where is the lower boundary of the mantle convection system?

The deepest that boundary could be is about 2900 km below Earth's outer surface, where a sharp boundary separates the mantle from the core (**Figure 3.17**). As we saw in Chapter 2, the iron-rich liquid below this core-mantle boundary is much denser than the solid rock of the mantle, preventing any significant exchange of material between the two layers. We can thus imagine a system of whole-mantle convection in which the material from the plates circulates all the way through the mantle, down as far as the core-mantle boundary (Figure 3.17a).

However, some scientists think that the mantle is divided into two layers: an upper mantle system in the outer 700 km, where the recycling of lithosphere takes place, and a lower mantle system from a depth of 700 km to the core-mantle boundary, where convection is much more sluggish. According to this hypothesis, called stratified convection, the separation of the two systems is maintained because the upper system consists of lighter rock than the lower system and thus floats on top, in the same way the mantle floats on the core (Figure 3.17b).

To test these two competing hypotheses, scientists have looked for "lithospheric graveyards" below convergent boundaries where plates have been subducted. Old subducted lithosphere is colder than the surrounding mantle and can therefore be "seen" using seismic waves (much as doctors use ultrasound waves to look into your body). Sure enough, scientists have found regions of colder material in the deep mantle under North and South America, eastern Asia, and other sites adjacent to convergent boundaries. These zones occur as extensions of descending lithospheric slabs, and some appear to go down as far as the core-mantle boundary. From this evidence, most scientists have concluded that plate recycling takes place through whole-mantle convection rather than stratified convection.

Mantle Convection: The Engine of Plate Tectonics

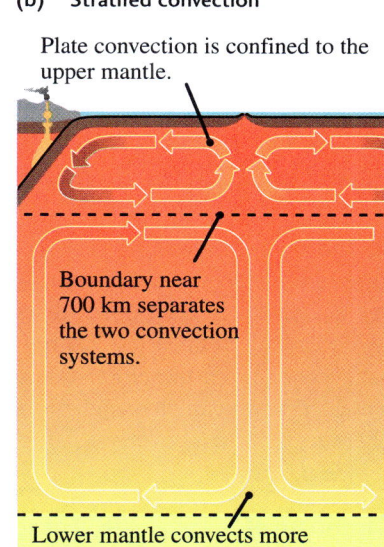

FIGURE 3.17 Two competing hypotheses about the mantle convection system.

(a) Whole-mantle convection — Mantle convection circulates as deep as the core-mantle boundary.

(b) Stratified convection — Plate convection is confined to the upper mantle. Boundary near 700 km separates the two convection systems. Lower mantle convects more sluggishly than upper mantle.

What Is the Nature of Rising Convection Currents?

What about the rising currents of hot mantle material needed to balance subduction? Are there concentrated, sheetlike upwellings directly beneath the mid-ocean ridges? Most scientists who study the problem think not. Instead, they believe that the rising currents are slower and spread out over broader regions. This view is consistent with the idea that seafloor spreading is a rather passive process: pull the plates apart almost anywhere, and you will generate a spreading center.

There is one big exception, however: a type of narrow, jetlike upwelling called a *mantle plume* (Figure 3.18). The best evidence for mantle plumes comes from regions of intense, localized volcanism (called *hot spots*), such as Hawaii, where huge volcanoes form in the middle of a plate, far away from any spreading center. Mantle plumes are thought to be slender cylinders of fast-rising material, less than 100 km across, that come from the deep mantle. Mantle plumes are so intense that they can literally burn holes in the plates and erupt tremendous volumes of lava. Plumes may be responsible for outpourings of lava so massive that they may have changed Earth's climate and

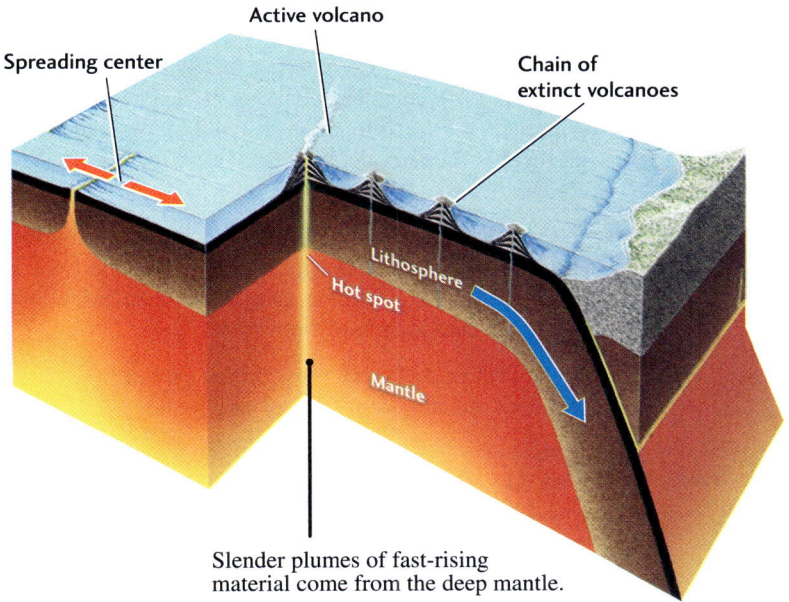

FIGURE 3.18 An illustration of the mantle plume hypothesis.

killed off many life-forms in mass extinction events (see Chapter 9). We will describe mantle plumes in more detail in Chapter 5.

SUMMARY >>

What is the theory of plate tectonics? According to the theory of plate tectonics, the lithosphere is broken into about a dozen rigid, moving plates. Three types of plate boundaries are defined by the direction of the movements of plates in relation to each other: divergent, convergent, and transform-fault boundaries. Earth's surface area does not change over time; therefore, the area of new lithosphere created at divergent boundaries equals the area of lithosphere recycled at convergent boundaries by subduction into the mantle.

What are some of the geologic characteristics of plate boundaries? Many geologic features develop at plate boundaries. Divergent boundaries are typically marked by volcanic activity and earthquakes at the crest of a mid-ocean ridge. Convergent boundaries are marked by deep-sea trenches, earthquakes, mountain building, and volcanoes. Transform faults, along which plates slide past each other, can be recognized by earthquake activity and offsets in geologic features.

How can the age of the seafloor be determined? We can measure the age of the seafloor by using thermoremanent magnetism. Magnetic anomaly patterns mapped on the seafloor can be compared with the magnetic time scale, which was established by measuring the magnetic anomalies of lavas with known ages on land. Seafloor ages have been verified through dating of rock samples obtained by deep-sea drilling. Geologists can now draw isochron maps for most of the world's oceans, which allow them to reconstruct the history of seafloor spreading over the past 200 million years. Using this method and other geologic data, geologists have developed a detailed model of how Pangaea broke apart and the continents drifted into their present configuration.

What is the engine that drives plate tectonics? The plate tectonic system is driven by mantle convection, the energy for which comes from Earth's internal heat. Gravitational forces act on the cooling lithosphere as it slides downhill from spreading centers and sinks into the mantle at subduction zones. Subducted lithosphere extends as deep as the core-mantle boundary, indicating that the whole mantle is involved in the convection system that recycles the plates. Rising convection currents may include mantle plumes, intense upwellings from the deep mantle that cause localized volcanism at hot spots in the middle of a plate.

KEY TERMS AND CONCEPTS >>

continental drift (p. 53)
convergent boundary (p. 59)
divergent boundary (p. 59)
island arc (p. 61)
isochron (p. 68)
magnetic anomaly (p. 63)
magnetic time scale (p. 63)
Pangaea (p. 53)
plate tectonics (p. 52)
Rodinia (p. 69)
seafloor spreading (p. 53)
spreading center (p. 60)
subduction (p. 61)
transform fault (p. 59)

EXERCISES >>

1. Why are there active volcanoes along the Pacific coast in Washington and Oregon but not along the eastern coast of the United States?

2. What mistakes did Wegener make in formulating his theory of continental drift? Do you think the geologists of his era were justified in rejecting his theory?

3. Would you characterize plate tectonics as a hypothesis, a theory, or a fact? Why?

4. How do the differences between continental and oceanic crust affect the way plates interact?

5. From Figure 3.7, trace the boundaries of the South American Plate on a sheet of paper and identify segments that are divergent, convergent, and transform-fault boundaries. Approximately what fraction of the plate area is occupied by the South American continent? Is the fraction of the South American Plate occupied by oceanic crust increasing or decreasing over time? Explain your answer using the principles of plate tectonics.

6. In Figure 3.15, the isochrons are symmetrically distributed in the Atlantic Ocean, but not in the Pacific Ocean. For example, seafloor as much as 180 million years old (in darkest blue) is found in the western Pacific, but not in the eastern Pacific. Why?

7. From the isochron map in Figure 3.15, estimate how long ago the continents of Australia and Antarctica were separated by seafloor spreading. Did this happen before or after South America separated from Africa?

8. Name three mountain belts formed by continent-continent collisions that are occurring now or have occurred in the past.

9. Most active volcanoes are located on or near plate boundaries. Give an example of a volcano that is not on a plate boundary and describe a hypothesis consistent with plate tectonics that can explain it.

EARTH MATERIALS: Minerals and Rocks

CHAPTER 4

The Diamond Invention

The French were bred to die for love
They delight in fighting duels
But I prefer a man who lives
And gives expensive jewels
A kiss on the hand may be quite continental
But diamonds are a girl's best friend
A kiss may be grand but it won't pay the rental
On your humble flat, or help you at the automat
Men grow cold as girls grow old
And we all lose our charms in the end
But square cut or pear shaped
These rocks don't lose their shape
Diamonds are a girl's best friend

"Diamonds Are a Girl's Best Friend" is a song most famously performed by Marilyn Monroe in the 1953 movie *Gentlemen Prefer Blondes*. The song is a testimony to a brilliant marketing campaign, known as the "diamond invention." At its root is De Beers Consolidated Mines, Ltd., whose plan it was to romanticize diamonds by subtly changing the public's perception of the way a man courts—and wins—a woman. The marketing agency recruited movie idols, the symbols of romance for the general public, to advertise diamonds as symbols of enduring love. Magazine stories stressed the large diamonds that celebrities presented to their lovers, and photographs showed prominent women conspicuously flaunting the sparkling stone on their hands.

The diamond invention was a cunning scheme for sustaining the value of diamonds for the few people who produced them. It involved gaining complete control over the production and distribution of all of the world's diamonds. A system was created for allocating the gems to a select number of diamond cutters who all agreed to follow a strict set of rules intended to ensure that the quantity of cut diamonds available never exceeded the public's demand for them.

The most important ingredient in the campaign's success was to instill in the buying public the idea that they must never resell their diamonds. That "Diamonds are Forever" is not true, geologically speaking, but it creates an illusion in the mass mindset that diamonds are "forever" in the sense that they must never be resold. You own them for life, and then pass them on through generations to come. ■

Marilyn Monroe, with diamonds. [LGI Stock/CORBIS.]

CHAPTER OUTLINE

What Are Minerals? 76

The Structure of Matter 77

The Formation of Minerals 80

Rock-Forming Minerals 83

Physical Properties of Minerals 87

What Are Rocks? 93

Igneous Rocks 94

Sedimentary Rocks 95

Metamorphic Rocks 97

The Rock Cycle: Interactions Between the Plate Tectonic and Climate Systems 98

Concentrations of Valuable Mineral Resources 100

The diamond invention has resulted in a staggering stockpile of uncut diamonds in a well-protected vault in London; a billion-dollar cash reserve in European banks; an intelligence network operating out of Antwerp, Tel Aviv, Johannesburg, and London; a global array of advertising agencies, brokers, and distributors; corporate fronts in Africa for concealing massive diamond purchases; and private treaties with nations establishing quotas for annual production. For example, until recently, diamonds surpassed even oil as the way Russia raised most of its cash to underpin its economy, and only through cooperation with De Beers in price fixing.

But the diamond invention goes far beyond being a monopoly whose goal is to fix diamond prices. It is a process for converting tiny crystals of a mineral called diamond, made up of colorless carbon—judged by many to be less beautiful than rubies or sapphires—into universally recognized tokens of power and romance.

The diamond invention is a relatively recent development in the history of social power. Up until the late 1800s, diamonds were genuinely rare gemstones, found only in riverbeds in India and in the tropical rain forests of Brazil. The entire world production amounted to only a few pounds each year. All this changed when "kimberlites" were discovered in South Africa in 1870. Kimberlites are rocks formed by volcanic eruptions containing rocks transported from depths of 100–200 km—all the way from Earth's upper mantle. All diamonds are formed in the upper mantle, where extreme pressures allow carbon to form its distinctive diamond structure.

The kimberlites of South Africa meant that instead of the rare chance of finding diamonds in a riverbed, diamonds now could be gathered by the bucketful. Suddenly the market was swamped by an ever-expanding flood of diamonds. The British capitalists who had financed the operation of the kimberlite mines quickly learned that their investment would be threatened. In fact, diamonds were so plentiful as to be common, and their price was completely dependent on their previous rarity. Produced at their true abundance, diamonds might be even less valuable than semiprecious stones such as tourmaline and garnet.

The investors in these mines had no choice but to merge their interests into a single company that would be powerful enough to control the production of diamonds and perpetuate the illusion of their scarcity. Thus, De Beers Consolidated, a company incorporated in South Africa, was born. While other commodities fluctuate wildly in price due to various global economic influences, diamonds continue to appreciate in value each year. However, the diamond invention is not forever. If De Beers either loses its control of diamond production or fails to continue to convince the general public never to sell their diamonds, the diamond invention could fail.

We begin this chapter with a study of minerals, often using diamond as a model of how minerals form. Minerals are the building blocks of rocks, which are, in turn, the records of geologic history.

To tell Earth's story, geologists often adopt a "Sherlock Holmes" approach: they use current evidence to deduce the tectonic processes and events that occurred in the past at some particular place. The kinds of minerals found in volcanic rocks, for example, provide evidence of eruptions that brought molten rock to Earth's surface, while the minerals in granite reveal that it crystallized deep in the crust under the very high temperatures and pressures produced when two continental plates collide. Understanding the geology of a region also allows us to make informed guesses about where undiscovered deposits of economically important mineral resources might lie.

The focus of this chapter is **mineralogy,** the branch of geology that studies the composition, structure, appearance, stability, occurrence, and associations of minerals.

WHAT ARE MINERALS?

Minerals are the building blocks of rocks. With the proper tools, most rocks can be separated into their constituent minerals. A few rocks, such as limestone, contain only a single mineral (in this case, calcite). Other rocks, such as granite, are made of several different minerals. To identify and classify the many kinds of rocks that compose Earth and understand how they formed, we must know about minerals.

Geologists define a **mineral** as a *naturally occurring, solid crystalline substance, usually inorganic, with a specific chemical composition.* Minerals are homogeneous: they cannot be divided mechanically into smaller components.

Let's examine each part of our definition of a mineral in a little more detail.

FIGURE 4.1 The mineral calcite is found in the shells of many organisms, such as foraminifers. [(*left*) John Grotzinger/Ramón Rivera-Moret/Harvard Mineralogical Museum; (*right*) Andrew Syred/Photo Researchers.]

> *Minerals are the building blocks of rocks. They are solid, usually inorganic, naturally occurring, crystalline structures that have a specific chemical composition.*

Naturally occurring To qualify as a mineral, a substance must be found in nature. The diamonds mined in South Africa are minerals. The synthetic versions produced in industrial laboratories are not minerals, nor are the thousands of laboratory products invented by chemists.

Solid crystalline substance Minerals are solid substances—they are neither liquids nor gases. When we say that a mineral is *crystalline,* we mean that the tiny particles of matter, or atoms, that compose it are arranged in an orderly, repeating, three-dimensional array. Solid materials that have no such orderly arrangement are referred to as *glassy* or *amorphous* (without form) and are not conventionally called minerals. Windowpane glass is amorphous, as are some natural glasses formed during volcanic eruptions. Later in this chapter, we will explore in detail the process by which crystalline materials form.

Usually inorganic Minerals are defined as inorganic substances and so exclude the organic materials that make up plant and animal bodies. Organic matter is composed of organic carbon, the form of carbon found in all organisms, living or dead. Decaying vegetation in a swamp may be geologically transformed into coal, which is also made of organic carbon, but although it is found in naturally occurring deposits, coal is not considered a mineral. Many minerals, however, are secreted by organisms. One such mineral, calcite (**Figure 4.1**), which forms the shells of oysters and many other organisms, contains inorganic carbon. The calcite of these shells, which constitute the bulk of many limestones, fits the definition of a mineral because it is inorganic and crystalline.

With a specific chemical composition The key to understanding the composition of Earth's materials lies in knowing how the chemical elements are organized into minerals. What makes each mineral unique is its chemical composition and the arrangement of its atoms in an internal structure. A mineral's chemical composition either is fixed or varies within defined limits. The mineral quartz, for example, has a fixed ratio of two atoms of oxygen to one atom of silicon. This ratio never varies, even though quartz is found in many different kinds of rocks. Similarly, the chemical elements that make up the mineral olivine—iron, magnesium, oxygen, and silicon—always have a fixed ratio. Although the numbers of iron and magnesium atoms may vary, the sum of those two atoms in relation to the number of silicon atoms always forms a fixed ratio.

THE STRUCTURE OF MATTER

In 1805, John Dalton hypothesized that each of the various chemical elements consists of a different kind of atom, that all atoms of any given element are identical, and that chemical compounds are formed by various combinations of atoms of different elements in definite proportions. By the early twentieth century, physicists, chemists, and mineralogists, building on Dalton's ideas, had come to understand the structure of matter much as we do today. We now know that an **atom** is the smallest unit of an element that retains the physical and chemical properties of that element.

We also know that atoms are the small units of matter that combine in chemical reactions and that atoms themselves are divisible into even smaller units.

The Structure of Atoms

Understanding the structure of atoms allows us to predict how chemical elements will react with one another and form new crystal structures. The structure of an atom is defined by a nucleus, which contains protons and neutrons and which is surrounded by electrons. (For a more detailed review of the structure of atoms, see Appendix 3.)

At the center of every atom is a dense *nucleus* containing virtually all the mass of the atom in two kinds of particles: protons and neutrons (**Figure 4.2**). A *proton* has a positive electrical charge of +1. A *neutron* is electrically neutral—that is, uncharged. Atoms of the same chemical element may have different numbers of neutrons, but the number of protons does not vary. For instance, all carbon atoms have six protons.

Surrounding the nucleus is a cloud of moving particles called *electrons,* each with a mass so small that it is conventionally taken to be zero. Each electron carries a negative electrical charge of −1. The number of protons in the nucleus of any atom is balanced by the same number of electrons in the cloud surrounding the nucleus, so that the atom is electrically neutral. Thus, the nucleus of a carbon atom is surrounded by six electrons (see Figure 4.2).

Atomic Number and Atomic Mass

The number of protons in the nucleus of an atom is called its *atomic number.* Because all atoms of the same element have the same number of protons, they also have the same atomic number. All atoms with six protons, for example, are carbon atoms (atomic number 6). In fact, the atomic number of an element can tell us so much about that element's behavior that the periodic table organizes elements according to their atomic number (see Appendix 3). Elements in the same vertical column of the periodic table, such as carbon and silicon, tend to react similarly.

The *atomic mass* of an element is the sum of the masses of its protons and its neutrons. (Electrons, because they have so little mass, are not included in this sum.) Atoms of the same chemical element always have the same number of protons but may have different numbers of neutrons and therefore different atomic masses. Atoms of the same element with different numbers of neutrons are called **isotopes.** Isotopes of the element carbon, for example, all have six protons but may have six, seven, or eight neutrons, giving atomic masses of 12, 13, and 14.

In nature, the chemical elements exist as mixtures of isotopes, so their atomic masses are never whole numbers. Carbon's atomic mass, for example, is 12.011. It is close to 12 because the isotope carbon-12 is by far the most abundant. The relative abundances of the various isotopes of an element on Earth are determined by processes that enhance the abundances of some isotopes over others. Carbon-12, for example, is favored by some chemical reactions, such as photosynthesis, in which organic carbon compounds are produced from inorganic carbon compounds.

> Atoms are the smallest units of matter. The structure of an atom is defined by a nucleus, which contains protons and neutrons and is surrounded by electrons.

Chemical Reactions

The structure of an atom determines its chemical reactions with other atoms. **Chemical reactions** are interactions of the atoms of two or more chemical elements in certain fixed proportions that produce chemical compounds. For example, when two hydrogen atoms combine with one oxygen atom, they form a new chemical compound, water (H_2O). The properties of a chemical compound may be entirely dif-

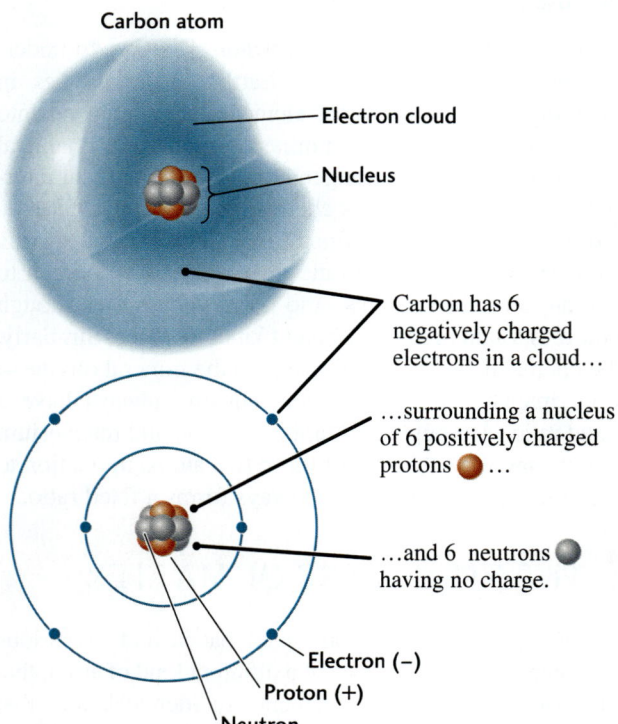

FIGURE 4.2 Electron structure of the carbon atom (carbon-12). The electrons, each with a charge of −1, are represented as a negatively charged cloud surrounding the nucleus, which contains six protons, each with a charge of +1, and six neutrons, each with no charge. The size of the nucleus is greatly exaggerated in these drawings; it is much too small to show at a true scale.

ferent from those of its constituent elements. For example, when an atom of sodium, a metal, combines with an atom of chlorine, a noxious gas, they form the chemical compound sodium chloride, better known as table salt. We represent this compound by the chemical formula NaCl, in which the symbol Na stands for the element sodium and the symbol Cl for the element chlorine. (Every chemical element has been assigned its own symbol, which we use as a kind of shorthand for writing chemical formulas and equations; these symbols are given in the periodic table in Appendix 3.)

Chemical compounds, such as minerals, are formed either by *electron sharing* or by *electron transfer* between the reacting atoms. Carbon and silicon tend to form compounds by electron sharing. Diamond is a compound composed entirely of carbon atoms sharing electrons (**Figure 4.3**).

In chemical reactions, new chemical compounds are formed.

In the reaction between sodium (Na) and chlorine (Cl) atoms to form sodium chloride (NaCl), electrons are transferred. The sodium atom loses one electron, which the chlorine atom gains

FIGURE 4.3 Some atoms, such as the carbon atoms in diamond, share electrons to form covalent bonds.

(**Figure 4.4**). An atom or group of atoms that has an electrical charge, either positive or negative, because of the loss or gain of one or more electrons is called an **ion.** Because the chlorine atom has gained a negatively charged electron, it is now a negatively charged ion, Cl⁻. Likewise, the loss

(a) When sodium (Na) and chlorine (Cl) react, the sodium atom loses one electron.

The chlorine atom gains that electron. In this reaction, an orderly array of ions is formed.

Sodium atom (1 electron in outer shell)   Chlorine atom (7 electrons in outer shell)   Electrical attraction   Cation (+)   Anion (−)

(b) Sodium and chloride ions pack together in a cubic structure.

Each sodium ion (circled in red) is surrounded by six chloride ions (circled in yellow), and vice versa.

Chloride ion   Sodium ion

NaCl crystal   Halite (table salt)

FIGURE 4.4 Some atoms, such as those in sodium chloride, transfer electrons to form ionic bonds.
[John Grotzinger/Ramón Rivera-Moret/Harvard Mineralogical Museum.]

of an electron gives sodium a positive charge, Na$^+$. The compound NaCl itself remains electrically neutral because the positive charge on Na$^+$ is exactly balanced by the negative charge on Cl$^-$. A positively charged ion is a *cation,* and a negatively charged ion is an *anion.*

Chemical Bonds

When a chemical compound is formed either by electron sharing or by electron transfer, the ions or atoms that make up the compound are held together by electrostatic attraction between negatively charged electrons and positively charged protons. These attractions, or **chemical bonds,** between shared electrons or between gained and lost electrons may be strong or weak. Strong bonds keep a substance from decomposing into its elements or into other compounds. They also make minerals hard and keep them from cracking or splitting. Two major types of bonds are found in most rock-forming minerals: ionic bonds and covalent bonds.

- The simplest form of chemical bond is the *ionic bond.* Bonds of this type form by electrostatic attraction between ions of opposite charge, such as Na$^+$ and Cl$^-$ in sodium chloride (see Figure 4.4a), when electrons are transferred. This attraction is of exactly the same nature as the static electricity that can make nylon or silk clothing cling to the body. The strength of an ionic bond decreases greatly as the distance between ions increases, and it increases as the electrical charges of the ions increase. Ionic bonds are the dominant type of chemical bonds in mineral structures: *about 90 percent of all minerals are essentially ionic compounds.*

- Elements that do not readily gain or lose electrons to form ions and instead form compounds by sharing electrons are held together by *covalent bonds.* These bonds are generally stronger than ionic bonds. One mineral with a covalently bonded crystal structure is diamond, which consists of a single element, carbon. Each carbon atom can share four of its own electrons with other carbon atoms and can acquire another four electrons by sharing with other carbon atoms. In diamond, every carbon atom is surrounded by four others arranged in a regular *tetrahedron,* a four-sided pyramidal form, each side a triangle (see Figure 4.3). In this configuration, each carbon atom shares an electron with each of its four neighbors, resulting in a very stable configuration. Figure 4.9 (page 83) shows a network of carbon tetrahedra linked together.

Atoms of metallic elements, which have strong tendencies to lose electrons, pack together as cations, and the freely mobile electrons are shared and dispersed among the ions. This free electron sharing results in a kind of covalent bond that we call a *metallic bond.* It is found in a small number of minerals, among them the metal copper and some sulfides.

The chemical bonds of some minerals are intermediate between pure ionic and pure covalent bonds because some electrons are exchanged and others are shared.

THE FORMATION OF MINERALS

Let's look more closely at the orderly forms that result from the chemical bonds we have just described. In this section, we examine the crystal structures of minerals and the conditions under which minerals form. Later in this chapter, we will see how those crystal structures affect the physical properties of minerals.

Minerals form by the process of **crystallization,** in which the atoms of a gas or liquid come together in the proper chemical proportions and crystalline arrangement to form a solid substance. (Remember that the atoms in a mineral are arranged in an orderly three-dimensional array.) The bonding of carbon atoms in diamond, a covalently bonded mineral, is one example of crystallization. Under the very high pressures and temperatures in Earth's mantle, carbon atoms bond together in tetrahedra, and each tetrahedron attaches to another, building up a regular three-dimensional structure from a great many atoms. As a diamond crystal grows, it extends its tetrahedral structure in all directions, always adding new atoms in the proper geometric arrangement. Diamonds can be artificially synthesized under very high pressures and temperatures that mimic the conditions in Earth's mantle, where natural diamonds form.

The Crystal Structure of Minerals

The sodium and chloride ions that make up sodium chloride, an ionically bonded mineral, also crystallize in an orderly three-dimensional array. In Figure 4.4b, we can see the geometry of their arrangement, with each ion of one kind surrounded by six ions of the other kind in a series of *cubic* structures extending in three directions. We can think of ions as solid spheres, packed together in close-fitting structural units. Figure 4.4b also shows the relative sizes of the ions in NaCl. There are six neighboring ions in NaCl's basic structural unit. The relative sizes of the sodium and chloride ions allow them to fit together in a closely packed arrangement.

Many of the cations of abundant minerals are relatively small, while most anions are large (**Figure 4.5**). This is the case with the most common anion on Earth, oxygen (O^{2-}). Because anions tend to be larger than cations, most of the space of a crystal is occupied by the

> In chemical reactions, new bonds are formed. Some electrons are shared in a covalent bond or transferred from one atom to another to form an ionic bond.

CATIONS	Silicon (Si^{4+})	Aluminum (Al^{3+})	Iron (Fe^{3+})	Magnesium (Mg^{2+})	Iron (Fe^{2+})	Sodium (Na^+)	Calcium (Ca^{2+})	Potassium (K^+)
	0.27	0.53	0.65	0.72	0.73	0.99	1.00	1.38

ANIONS	Oxygen (O^{2-})	Chloride (Cl^-)	Sulfide (S^{2-})
	1.40	1.81	1.84

FIGURE 4.5 Sizes of ions as they are commonly found in rock-forming minerals. Ionic radii are given in 10^{-8} cm. [After L. G. Berry, B. Mason, and R. V. Dietrich, *Mineralogy*. San Francisco: W. H. Freeman, 1983.]

> The crystal structure of a mineral is defined by an orderly three-dimensional arrangement of atoms in specific proportions.

anions, and the cations fit into the spaces between them. As a result, crystal structures are determined largely by how the anions are arranged and how the cations fit between them.

Cations of similar sizes and charges tend to substitute for one another and to form compounds having the same crystal structure but differing in chemical composition. *Cation substitution* is common in minerals that contain the silicate ion (SiO_4^{4-}), such as olivine, which is abundant in many volcanic rocks. Iron (Fe^{2+}) and magnesium (Mg^{2+}) ions are similar in size, and both have two positive charges, so they easily substitute for each other in the structure of olivine. The composition of pure magnesium olivine is Mg_2SiO_4; that of pure iron olivine is Fe_2SiO_4. The composition of olivine containing both iron and magnesium is given by the formula $(Mg,Fe)_2SiO_4$, which simply means that the number of iron and magnesium cations may vary, but their combined total (expressed as a subscript 2) does not vary in relation to each SiO_4^{4-} ion. The proportion of iron to magnesium is determined by the relative abundances of the two elements in the molten material from which the olivine crystallized. In many silicate minerals, aluminum (Al^{3+}) substitutes for silicon (Si^{4+}). Aluminum and silicon ions are so similar in size that aluminum can take the place of silicon in many crystal structures. The difference in charge between aluminum (3+) and silicon (4+) ions is balanced by an increase in the number of other cations, such as sodium (1+).

The Crystallization of Minerals

Crystallization starts with the formation of microscopic single **crystals,** orderly three-dimensional arrays of atoms in which the basic arrangement is repeated in all directions.

The boundaries of crystals are natural flat (*planar*) surfaces called *crystal faces* (**Figure 4.6**). The crystal faces of a mineral are the external expression of the mineral's internal atomic structure. **Figure 4.7** pairs a drawing of a perfect quartz crystal with a photograph of the actual mineral. The six-sided (hexagonal) shape of the quartz crystal corresponds to its hexagonal internal atomic structure.

During crystallization, the initially microscopic crystals grow larger, maintaining their crystal faces as long as

FIGURE 4.6 Crystals of amethyst and quartz, growing on top of epidote crystals (green). The planar surfaces are crystal faces, whose geometries are determined by the underlying arrangement of the atoms that make up the crystals. [John Grotzinger/Ramón Rivera-Moret/Harvard Mineralogical Museum.]

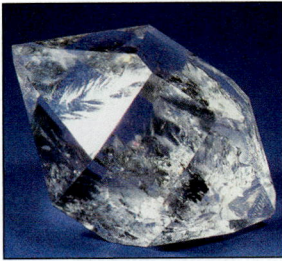

Crystal faces

A perfect quartz crystal   A natural quartz crystal

FIGURE 4.7 A perfect crystal is rare in nature, but no matter how irregular the shapes of the faces may be, the angles are always exactly the same. [Breck P. Kent.]

> As a crystal grows, the basic arrangement of its atoms is repeated in all directions. The crystal faces of a mineral reflect this basic arrangement.

they are free to grow. Large crystals with well-defined faces form when growth is slow and steady and space is adequate to allow growth without interference from other crystals nearby. For this reason, most large mineral crystals form in open spaces in rocks, such as fractures or cavities.

Often, however, the spaces between growing crystals fill in, or crystallization proceeds rapidly. Crystals then grow over one another and coalesce to become a solid mass of crystalline particles, or *grains*. In this case, few or no grains show crystal faces. Large crystals that can be seen with the naked eye are relatively unusual, but many minerals in rocks display crystal faces that can be seen under a microscope.

Unlike crystalline minerals, glassy materials—which solidify from liquids so quickly that they lack any internal atomic order—do not form crystals with planar faces. Instead, they are found as masses with curved, irregular surfaces. The most common glass is volcanic glass.

How Do Minerals Form?

Lowering the temperature of a liquid below its freezing point is one way to start the process of crystallization. In water, for example, 0°C is the temperature below which crystals of ice—a mineral—start to form. Similarly, **magma**—hot, molten liquid rock—crystallizes into solid minerals when it cools. As a magma falls below its melting point, which may be higher than 1000°C, crystals of silicate minerals such as olivine or feldspar begin to form. (Geologists usually refer to melting points of magmas rather than freezing points, because freezing implies cold.)

Crystallization can also occur as liquids evaporate from a solution. A *solution* is a homogeneous mixture of one chemical substance with another, such as salt and water. As the water evaporates from a salt solution, the concentration of salt eventually gets so high that the solution can hold no more salt and is said to be *saturated*. If evaporation continues, the salt starts to *precipitate,* or drop out of solution as crystals. Deposits of table salt, or halite, form under just these conditions when seawater evaporates to the point of saturation in some hot, arid bays or arms of the ocean (**Figure 4.8**).

Diamond and graphite (the material used as the "lead" in pencils) exemplify the dramatic effects that temperature and pressure can have on mineral formation. These two minerals are **polymorphs,** alternative structures formed from a single chemical compound (**Figure 4.9**). They are both formed from carbon but have different crystal structures and very different appearances. From experimentation and geological observation, we know that diamond forms and remains stable at the very high pressures and temperatures found in Earth's mantle. High pressures force the atoms in diamond into a closely packed structure.

> *Crystallization occurs when a liquid is cooled below its melting point or when a liquid precipitates from an evaporating solution.*

> *Polymorphs are alternative mineral structures with the same chemical composition that are formed under different temperatures and pressures.*

FIGURE 4.8 Halite crystals precipitating within a modern hypersaline lagoon, San Salvador Island, Bahamas. Note the cubic shape of the crystals. [John Grotzinger.]

Graphite is formed at lower pressures and temperatures than diamond. Its carbon forms sheets whose atoms are more loosely packed than those in diamond.

Within its sheets, carbon atoms are joined by strong bonds.

Weak bonds connect carbon atoms in alternating sheets.

Natural **diamond** is formed by very high pressures and temperatures in Earth's mantle. Its carbon atoms are closely packed.

All carbon atoms in diamond are closely packed, and all the bonds are very strong.

Graphite

Diamond

FIGURE 4.9 Graphite and diamond are polymorphs, alternative structures formed from a single chemical compound, carbon. [John Grotzinger/Ramón Rivera-Moret/Harvard Mineralogical Museum.]

Diamond therefore has a higher **density** (mass per unit volume, usually expressed in grams per cubic centimeter, g/cm^3), than graphite, which is less closely packed: diamond has a density of 3.5 g/cm^3, while that of graphite is only 2.1 g/cm^3. Graphite forms and is stable at moderate pressures and temperatures, such as those in Earth's crust.

Low temperatures can also produce close packing. Quartz and cristobalite are polymorphs of silica (SiO$_2$). Quartz forms at low temperatures and is relatively dense (2.7 g/cm^3). Cristobalite, which forms at higher temperatures, has a more open structure and is therefore less dense (2.3 g/cm^3).

ROCK-FORMING MINERALS

All minerals on Earth have been grouped into eight classes according to their chemical composition (Table 4.1). Some minerals, such as copper, occur naturally as un-ionized pure elements, and they are classified as *native elements*. Most others are classified by their anions.

Although many thousands of minerals are known, geologists commonly encounter only about 30 of them. These minerals are the building blocks of most crustal rocks and are called *rock-forming minerals*. Their relatively small

TABLE 4.1	Some Chemical Classes of Minerals	
Class	Defining Anions	Example
Native elements	None: no charged ions	Copper metal (Cu)
Oxides	Oxygen ion (O^{2-})	Hematite (Fe$_2$O$_3$)
Hydroxides	Hydroxyl ion (OH$^-$)	Brucite (Mg[OH]$_2$)
Halides	Chloride (Cl$^-$), fluoride (F$^-$), bromide (Br$^-$), iodide (I$^-$)	Halite (NaCl)
Carbonates	Carbonate ion (CO$_3^{2-}$)	Calcite (CaCO$_3$)
Sulfates	Sulfate ion (SO$_4^{2-}$)	Anhydrite (CaSO$_4$)
Silicates	Silicate ion (SiO$_4^{4-}$)	Olivine (Mg$_2$SiO$_4$)
Sulfides	Sulfide ion (S^{2-})	Pyrite (FeS^2)

number corresponds to the small number of elements that are abundant in Earth's crust. As we learned in Chapter 2, 99 percent of the crust is made up of only eight elements.

In the following pages, we consider the five most common classes of rock-forming minerals:

- *Silicates*, the most abundant class of minerals in Earth's crust, are composed of oxygen (O) and silicon (Si)—the two most abundant elements in the crust—mostly in combination with cations of other elements.
- *Carbonates* are minerals composed of carbon and oxygen in the form of the carbonate anion (CO_3^{2-}) in combination with calcium and magnesium. Calcite ($CaCO_3$) is one such mineral.
- *Oxides* are compounds of the oxygen anion (O^{2-}) and metallic cations; an example is the mineral hematite (Fe_2O_3).
- *Sulfides* are compounds of the sulfide anion (S^{2-}) and metallic cations; an example is the mineral sphalerite (ZnS).
- *Sulfates* are compounds of the sulfate anion (SO_4^{2-}) and metallic cations; an example is the mineral anhydrite ($CaSO_4$).

The other chemical classes of minerals, including native elements and halides, are not as common as the rock-forming minerals.

> The most common classes of rock-forming minerals are the silicates, carbonates, oxides, sulfides, and sulfates.

Silicates

The basic building block of all silicate mineral structures is the silicate ion. It is a tetrahedron—a pyramidal structure with four sides—composed of a central silicon ion (Si^{4+}) surrounded by four oxygen ions (O^{2-}), giving the formula SiO_4^{4-} (**Figure 4.10**). Because the silicate ion has a negative charge, it often bonds to cations to form electrically neutral minerals. The silicate ion typically bonds with sodium (Na^+), potassium (K^+), calcium (Ca^{2+}), magnesium (Mg^{2+}), and iron (Fe^{2+}). Alternatively, it can share oxygen ions with other silicon-oxygen tetrahedra. Tetrahedra may be isolated (linked only to cations), or they may be linked to other silicate tetrahedra in rings, single chains, double chains, sheets, or frameworks. Some of these structures are shown in Figure 4.10.

> Silicates are the most abundant class of minerals in Earth's crust. Their basic structure is a tetrahedron, which links to other silica tetrahedra or to cations to form a variety of structures.

Isolated Tetrahedra Isolated tetrahedra are linked by the bonding of each oxygen ion of the tetrahedron to a cation (see Figure 4.10a). The cations, in turn, bond to the oxygen ions of other tetrahedra. The tetrahedra are thus isolated from one another by cations on all sides. Olivine is a rock-forming mineral with this structure.

Single-Chain Structures Single chains form by the sharing of oxygen ions. Two oxygen ions of each tetrahedron bond to adjacent tetrahedra in an open-ended chain (see Figure 4.10b). Single chains are linked to other chains by cations. Minerals of the pyroxene group are single-chain silicate minerals. Enstatite, a pyroxene, is composed of iron or magnesium ions, or both, and is limited to a chain of tetrahedra in which the two cations may substitute for each other, as in olivine. The formula $(Mg,Fe)SiO_3$ represents this structure.

Double-Chain Structures Two single chains may combine to form double chains linked to each other by shared oxygen ions (see Figure 4.10c). Adjacent double chains linked by cations form the structure of minerals in the amphibole group. Hornblende, a member of this group, is an extremely common mineral in both igneous and metamorphic rocks. It has a complex composition that includes calcium (Ca^{2+}), sodium (Na^+), magnesium (Mg^{2+}), iron (Fe^{2+}), and aluminum (Al^{3+}).

Sheet Structures In sheet structures, each tetrahedron shares three of its oxygen ions with adjacent tetrahedra to build stacked sheets of tetrahedra (see Figure 4.10d). Cations may be interlayered with tetrahedral sheets. The micas and clay minerals are the most abundant sheet silicates. Muscovite, $KAl_2(AlSi_3O_{10})(OH)_2$, is one of the most common sheet silicates and is found in many types of rocks. It can be separated into extremely thin, transparent sheets. Kaolinite, $Al_2Si_2O_5(OH)_4$, which also has this structure, is a common clay mineral found in sediments and is the basic raw material for pottery.

Frameworks Three-dimensional frameworks may form as each tetrahedron shares all its oxygen ions with other tetrahedra. Feldspars, the most abundant minerals in Earth's crust, are framework silicates (see Figure 4.10e), as is another of the most common minerals, quartz (SiO_2).

Silicate Compositions Chemically, the simplest silicate is silicon dioxide, also called silica (SiO_2), which is found most often as the mineral quartz. When the silicate tetrahedra of quartz are linked, sharing two oxygen ions for each silicon ion, the total formula adds up to SiO_2.

In other silicate minerals, the basic structural units—rings, chains, sheets, and frameworks—are bonded to cations such as sodium (Na^+), potassium (K^+), calcium (Ca^{2+}), magnesium (Mg^{2+}), and iron (Fe^{2+}). As noted in the discussion of cation substitution, aluminum (Al^{3+}) substitutes for silicon in many silicate minerals.

Carbonates

The mineral calcite (calcium carbonate, $CaCO_3$) is one of the most abundant minerals in Earth's crust and is the

Mineral	Chemical formula	Cleavage planes and number of cleavage directions	Silicate structure	Specimen
(a) Olivine	$(Mg,Fe)_2SiO_4$	1 plane	Isolated tetrahedra	
(b) Pyroxene	$(Mg,Fe)SiO_3$	2 planes at 90°	Single chains	
(c) Amphibole	$Ca_2(Mg,Fe)_5Si_8O_{22}(OH)_2$	2 planes at 60° and 120°	Double chains	
(d) Micas	Muscovite: $KAl_2(AlSi_3O_{10})(OH)_2$ Biotite: $K(Mg,Fe)_3AlSi_3O_{10}(OH)_2$	1 plane	Sheets	
(e) Feldspars	Orthoclase feldspar: $KAlSi_3O_8$ Plagioclase feldspar: $(Ca,Na)AlSi_3O_8$	2 planes at 90°	Three-dimensional frameworks	

FIGURE 4.10 The silicate ion is the basic building block of all silicate mineral structures. [John Grotzinger/Ramón Rivera-Moret/Harvard Mineralogical Museum.]

FIGURE 4.11 Carbonate minerals, such as calcite (calcium carbonate, $CaCO_3$), have a layered structure. (a) Calcite. [John Grotzinger/Ramón Rivera-Moret/Harvard Mineralogical Museum.] (b) Top view of the carbonate ion, composed of a carbon ion surrounded in a triangle by three oxygen ions, with a net charge of –2. (c) View of the alternating layers of calcium and carbonate ions in calcite.

chief constituent of a group of rocks called limestones (Figure 4.11). Its basic building block, the carbonate ion (CO_3^{2-}), consists of a carbon ion surrounded by three oxygen ions in a triangle, as shown in Figure 4.11b. The carbon atom shares electrons with the oxygen atoms. Groups of carbonate ions are arranged in sheets somewhat like those of the sheet silicates, which are bonded together by layers of cations (see Figure 4.11c). The sheets of carbonate ions in calcite are separated by layers of calcium ions. Dolomite ($CaMg[CO_3]_2$), another major mineral of crustal rocks, is made up of the same carbonate sheets separated by alternating layers of calcium ions and magnesium ions.

Oxides

Oxide minerals are compounds in which oxygen is bonded to atoms or cations of other elements, usually metallic cations such as iron (Fe^{2+} or Fe^{3+}). Most oxide minerals are ionically bonded, and their structures vary with the size of the metallic cations. This class of minerals has great economic importance because it includes the ores containing many of the metals, such as chromium and titanium, used in the industrial and technological manufacture of metallic materials and devices. Hematite (Fe_2O_3), shown in Figure 4.12, is a chief ore of iron.

Another of the abundant minerals in this class, spinel, is an oxide of two metals, magnesium and aluminum ($MgAl_2O_4$). Spinel (see Figure 4.12) has a closely packed cubic structure and a high density (3.6 g/cm³), reflecting the conditions of high pressure and temperature under which it forms. Transparent gem-quality spinel resembles ruby and sapphire and is found in the crown jewels of England and Russia.

Sulfides

The chief ores of many valuable minerals—such as copper, zinc, and nickel—are members of the sulfide class. This class includes compounds in which the sulfide ion (S^{2-}) is bonded to metallic cations. In the sulfide ion, a sulfur atom has gained two electrons in its outer shell. Most sulfide

FIGURE 4.12 Oxide minerals. (*left*) hematite; (*right*) spinel. [John Grotzinger/Ramón Rivera-Moret/Harvard Mineralogical Museum.]

FIGURE 4.13 Pyrite, a sulfide mineral, is also known as "fool's gold." [John Grotzinger/Ramón Rivera-Moret/Harvard Mineralogical Museum.]

minerals look like metals, and almost all are opaque. The most common sulfide mineral is pyrite (FeS_2), often called "fool's gold" because of its yellowish metallic appearance (**Figure 4.13**).

Sulfates

The basic building block of all sulfates is the sulfate ion (SO_4^{2-}). It is a tetrahedron made up of a central sulfur atom surrounded by four oxygen ions (O^{2-}). One of the most abundant minerals of this class is gypsum (**Figure 4.14**), the primary component of plaster. Gypsum, a calcium sulfate, forms when seawater evaporates. During evaporation, Ca^{2+} and SO_4^{2-}, two ions that are abundant in seawater, combine and precipitate as layers of sediment, forming calcium sulfate ($CaSO_4 \cdot 2H_2O$). (The dot in this formula signifies that two water molecules are bonded to the calcium and sulfate ions.)

Another calcium sulfate, anhydrite ($CaSO_4$), differs from gypsum in that it contains no water. Its name is derived from the word *anhydrous,* meaning "free from water." Gypsum is stable at the low temperatures and pressures found at Earth's surface, whereas anhydrite is stable at the higher temperatures and pressures where sedimentary rocks are buried.

As we discovered in 2004, sulfate minerals were precipitated from water early in the history of Mars. Much as on Earth, these minerals were precipitated when lakes and shallow seas dried up and formed sedimentary layers. However, many of these sulfate minerals are quite different from the sulfate mineral precipitates commonly found on Earth and include strange iron-bearing sulfates that precipitated from very harsh, acidic waters.

PHYSICAL PROPERTIES OF MINERALS

Geologists use their knowledge of mineral composition and structure to understand the origins of rocks. First they must identify the minerals that make up a rock. To do so, they rely greatly on chemical and physical properties that can be observed relatively easily. In the nineteenth and early twentieth centuries, geologists carried field kits for rough chemical analyses of minerals that would help in their identification. One such test is the origin of the phrase "the acid test." It consists of dropping diluted hydrochloric acid (HCl) on a mineral to see if it fizzes (**Figure 4.15**).

FIGURE 4.14 Gypsum is a sulfate formed when seawater evaporates. [John Grotzinger/Ramón Rivera-Moret/Harvard Mineralogical Museum.]

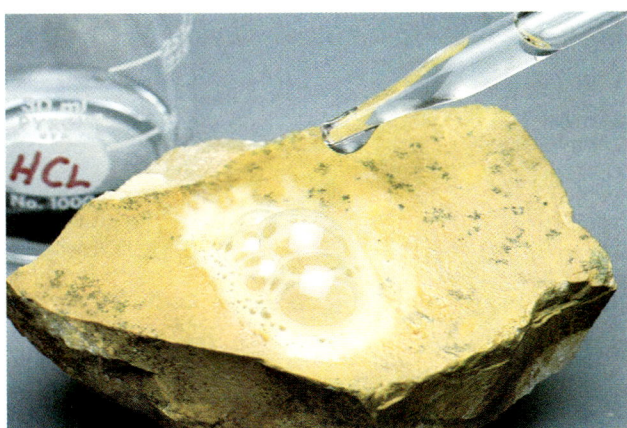

FIGURE 4.15 The acid test. One easy but effective way to identify certain minerals is to drop diluted hydrochloric acid (HCl) on the substance of interest. If it fizzes, indicating the escape of carbon dioxide, the mineral is likely to be calcite. [Chip Clark.]

Geologists identify minerals by measuring key physical properties that include hardness, cleavage, fracture, luster, color, density, and crystal habit.

The fizzing indicates that carbon dioxide (CO_2) is escaping, which means that the mineral is likely to be calcite, a carbonate.

In this section, we review the physical properties of minerals, many of which contribute to their practical and decorative value.

Hardness

Hardness is a measure of the ease with which the surface of a mineral can be scratched. Just as diamond, the hardest mineral known, scratches glass, a quartz crystal, which is harder than feldspar, scratches a feldspar crystal. In 1822, Friedrich Mohs, an Austrian mineralogist, devised a scale (now known as the **Mohs scale of hardness**) based on the ability of one mineral to scratch another. At one extreme is the softest mineral (talc); at the other, the hardest (diamond) (Table 4.2). The Mohs scale is still one of the best practical tools for identifying an unknown mineral. With a knife blade and a few of the minerals on the hardness scale, a field geologist can gauge an unknown mineral's position on the scale. If the unknown mineral is scratched by a piece of quartz but not by the knife, for example, it lies between 5 and 7 on the scale.

Recall that covalent bonds are generally stronger than ionic bonds. The hardness of any mineral depends on the strength of its chemical bonds: the stronger the bonds, the harder the mineral. Hardness varies with crystal structure within the silicate class of minerals. For example, hardness varies from 1 in talc, a sheet silicate, to 8 in topaz, a silicate with isolated tetrahedra. Most silicates fall in the 5 to 7 range on the Mohs scale. Only sheet silicates are relatively soft, with hardnesses between 1 and 3.

Within groups of minerals having similar crystal structures, increasing hardness is related to other factors that also increase bond strength:

TABLE 4.2 Mohs Scale of Hardness

Mineral	Scale Number	Common Objects
Talc	1	
Gypsum	2	Fingernail
Calcite	3	Copper coin
Fluorite	4	
Apatite	5	Knife blade
Orthoclase	6	Window glass
Quartz	7	Steel file
Topaz	8	
Corundum	9	
Diamond	10	

- *Size* The smaller the atoms or ions, the smaller the distance between them and the greater the electrical attraction—and thus the stronger the bond.
- *Charge* The larger the charge of ions, the greater the attraction between them and thus the stronger the bond.
- *Packing of atoms or ions* The closer the packing of atoms or ions, the smaller the distance between them and thus the stronger the bond.

Size is an especially important factor for most metallic oxides and for most sulfides of metals with high atomic numbers—such as gold, silver, copper, and lead. Minerals of these groups are soft, with hardnesses of less than 3, because their metallic cations are so large. Carbonates and sulfates, whose structures are not closely packed, are also soft, with hardnesses of less than 5.

FIGURE 4.16 Cleavage of mica. The diagram shows the cleavage plane in the crystal structure, oriented perpendicular to the plane of the page. Horizontal lines mark the interfaces of silica-oxygen tetrahedral sheets and the sheets of aluminum hydroxide bonding the two tetrahedral sheets into a sandwich. Cleavage takes place between tetrahedral–aluminum hydroxide sandwiches. The photograph shows thin sheets of mica separating along the cleavage planes. [Chip Clark.]

Cleavage

Cleavage is the tendency of a crystal to split along planar surfaces (**Figure 4.16**). The term *cleavage* is also used to describe the geometric pattern produced by such breakage. Cleavage varies inversely with bond strength: high bond strength produces poor cleavage, while low bond strength produces good cleavage. Because of their strength, covalent bonds generally give poor or no cleavage. Ionic bonds are relatively weak, so they give good cleavage. However, bond strength also varies along the different planes within a mineral that is entirely covalently bonded or entirely ionically bonded. For example, all of the bonds in diamond are covalent bonds, which are very strong, but some planes are more weakly bonded than others. Thus, diamond, the hardest mineral of all, can be cleaved along these weaker planes to produce perfect planar surfaces.

If the bonds between some of the planes of atoms or ions in a crystal are relatively weak, the mineral can be made to split along those planes. Muscovite, a mica sheet silicate, splits along smooth, lustrous, flat, parallel surfaces, forming transparent sheets less than a millimeter thick. Mica's excellent cleavage results from the relative weakness of the bonds between the sandwiched layers of cations and tetrahedral silica sheets (see Figure 4.16).

Cleavage is classified according to two primary sets of characteristics: the number of planes and pattern of cleavage, and the quality of surfaces and ease of cleaving.

Number of Planes and Pattern of Cleavage The number of planes and pattern of cleavage are identifying hallmarks of many rock-forming minerals. Muscovite, for example, has only one plane of cleavage, whereas calcite and dolomite crystals have three excellent cleavage planes that give them a rhomboidal shape (**Figure 4.17**).

FIGURE 4.17 Example of rhomboidal cleavage in calcite. [Charles D. Winters/Photo Researchers.]

A crystal's structure determines its cleavage planes and its crystal faces. Crystals have fewer cleavage planes than possible crystal faces. Faces may be formed along any of numerous planes defined by rows of atoms or ions. Cleavage occurs along any of those planes across which the bonding is weak. All crystals of a mineral exhibit its characteristic cleavage planes, whereas only some crystals display particular faces.

Galena (lead sulfide, PbS) and halite (sodium chloride, NaCl) cleave along three planes, forming perfect cubes. Distinctive angles of cleavage help identify two important groups of silicates, the pyroxenes and amphiboles, that otherwise often look alike (**Figure 4.18**). Pyroxenes have a single-chain structure and are bonded so that their cleavage planes are almost at right angles (about 90°) to each other. In cross section, the cleavage pattern of

Pyroxene

Pyroxene is a single-chain silicate mineral. When the weak bonds between chains break, the angle is about 90°.

Amphibole

Amphibole is a double-chain silicate. When the bonds between the double chains break, the angle is about 120°.

FIGURE 4.18 Comparison of cleavage planes in pyroxene and amphibole. These two minerals often look very much alike, but their angles of cleavage differ. These angles are frequently used to identify and classify them.

pyroxene is nearly a square. In contrast, amphiboles, which have a double-chain structure, are bonded so as to give two cleavage planes, at about 60° and 120° to each other. They produce a diamond-shaped cross section.

Quality of Surfaces and Ease of Cleaving A mineral's cleavage is assessed as perfect, good, or fair, according to the quality of surfaces produced and the ease of cleaving. Muscovite can be cleaved easily, producing extremely high quality, smooth surfaces; its cleavage is *perfect*. The single-chain and double-chain silicates (the pyroxenes and amphiboles, respectively) show *good* cleavage. Although these minerals split easily along the cleavage plane, they also break across it, producing cleavage surfaces that are not as smooth as those of mica. *Fair* cleavage is shown by the ring silicate beryl. Beryl's cleavage is irregular, and the mineral breaks relatively easily along directions other than cleavage planes.

Many minerals are so strongly bonded that they lack even fair cleavage. Quartz, a framework silicate, is so strongly bonded in all directions that it breaks only along irregular surfaces. Garnet, an isolated tetrahedral silicate, is also bonded strongly in all directions and so shows no cleavage. This absence of a tendency to cleave is found in most framework silicates and in silicates with isolated tetrahedra.

Fracture

Fracture is the tendency of a crystal to break along irregular surfaces other than cleavage planes. All minerals show fracture, either across cleavage planes or—in such minerals as quartz—with no cleavage in any direction. Fracture is related to how bond strengths are distributed in directions that cut across cleavage planes. Breakage of these bonds results in irregular fractures. Fractures may be *conchoidal,* showing smooth, curved surfaces like those of a thick piece of broken glass. Another common fracture surface with an appearance like split wood is described as *fibrous* or *splintery*. The shapes and appearances of many kinds of irregular fractures depend on the particular structure and composition of the mineral.

Luster

The way the surface of a mineral reflects light gives it a characteristic **luster.** Mineral lusters are described by the terms listed in Table 4.3. Luster is controlled by the kinds of atoms present and their bonding, both of which affect the way light passes through or is reflected by the mineral. Ionically bonded crystals tend to have a glassy, or vitreous, luster, but covalently bonded materials are more variable. Many tend to have an adamantine luster, like that of diamond. Metallic luster is shown by pure metals, such as

TABLE 4.3	Mineral Luster
Luster	Characteristics
Metallic	Strong reflections produced by opaque substances
Vitreous	Bright, as in glass
Resinous	Characteristic of resins, such as amber
Greasy	The appearance of being coated with an oily substance
Pearly	The whitish iridescence of such materials as pearl
Silky	The sheen of fibrous materials such as silk
Adamantine	The brilliant luster of diamond and similar minerals

gold, and by many sulfides, such as galena (lead sulfide, PbS). A pearly luster results from multiple reflections of light from planes beneath the surfaces of translucent minerals, such as the mother-of-pearl inner surfaces of many clamshells, which are made of the mineral aragonite. Luster, although an important criterion for field classification, depends heavily on the visual perception of reflected light. Textbook descriptions fall short of the actual experience of holding the mineral in your hand.

Color

The **color** of a mineral is imparted by light, either transmitted through or reflected by crystals or irregular masses of the mineral or a streak of mineral powder. **Streak** refers to the color of the fine deposit of mineral powder left on an abrasive surface, such as a tile of unglazed porcelain, when a mineral is scraped across it. Such a tile, called a *streak plate* (**Figure 4.19**), is a good diagnostic tool because the uniform small grains of mineral that are present in the powder are revealed on the plate. Hematite (Fe_2O_3), for example, may be black, red, or brown, but this mineral will always leave a trail of reddish brown powder on a streak plate.

Color is a complex and not yet fully understood property of minerals. It is determined both by the kinds of ions found in the pure mineral and by trace impurities.

Ions and Mineral Color The color of pure minerals depends on the presence of certain ions, such as iron or chromium, that strongly absorb portions of the light spectrum. Olivine that contains iron, for example, absorbs all colors except green, which it reflects, so we see this type of olivine as green. We see pure magnesium olivine as white (transparent and colorless).

FIGURE 4.19 Hematite may be black, red, or brown, but it always leaves a reddish brown streak when scraped along a ceramic streak plate. [Breck P. Kent.]

Trace Impurities and Mineral Color All minerals contain impurities. Instruments can now measure even very small quantities of some elements—as little as a billionth of a gram in some cases. Elements that make up much less than 0.1 percent of a mineral are reported as "traces," and many of them are called *trace elements.*

Some trace elements can be used to interpret the origins of the minerals in which they are found. Others, such as the traces of uranium in some granites, contribute to local natural radioactivity. Still others, such as small dispersed flakes of hematite that color a feldspar crystal brownish or reddish, are notable because they give a general color to an otherwise colorless mineral. Many of the gem varieties of minerals, such as emerald (green beryl) and sapphire (blue corundum), get their color from trace elements dissolved in the solid crystal (Figure 4.20).

FIGURE 4.20 Trace elements give gems their colors. Sapphire (*left*) and ruby (*center*) are formed of the same common mineral, corundum (aluminum oxide). Small amounts of impurities produce the intense colors that we value. Ruby, for example, is red because of small amounts of chromium, the same element that gives emerald (*right*) its green color. [John Grotzinger/Ramón Rivera-Moret/Harvard Mineralogical Museum.]

Emerald derives its color from chromium; the sources of sapphire's blue color are iron and titanium.

The color of a mineral may be distinctive, but it is not the most reliable clue to its identity. Some minerals always show the same color; others may have a range of colors. Many minerals show a characteristic color only on freshly broken surfaces or only on weathered surfaces. Some—precious opals, for example—show a stunning display of colors on reflecting surfaces. Others change color slightly with a change in the angle of the light shining on their surfaces.

Density

You can easily feel the difference in weight between a piece of hematite iron ore and a piece of sulfur of the same size by lifting the two pieces. A great many common rock-forming minerals, however, are too similar in density for such a simple test. Scientists therefore need some easy method to measure this property of minerals. A standard measure of density is **specific gravity,** which is the weight of a mineral divided by the weight of an equal volume of pure water at 4°C.

Density depends on the atomic mass of a mineral's ions and how closely they are packed in its crystal structure. Consider the iron oxide magnetite, with a density of 5.2 g/cm^3. This high density results partly from the high atomic mass of iron and partly from the closely packed structure that magnetite has in common with the other members of the spinel group of oxides (see page 87). The density of the iron silicate olivine, at 4.4 g/cm^3, is lower than that of magnetite for two reasons. First, the atomic mass of silicon, one of the elements that make up olivines, is lower than that of iron. Second, iron olivine has a more openly packed structure than minerals of the spinel group. The density of magnesium olivine is even lower, at 3.32 g/cm^3, because magnesium's atomic mass is much lower than that of iron.

Increases in density caused by increases in pressure affect the way minerals transmit light, heat, and seismic waves. Experiments at extremely high pressures have shown that the structure of olivine converts into the denser structure of the spinel group at pressures corresponding to a depth of 400 km. At a greater depth, 670 km, mantle materials are further transformed into silicate minerals with the even more densely packed structure of perovskite ($CaTiO_3$). Because of the huge volume of the lower mantle, perovskite is probably the most abundant mineral in Earth as a whole. Some perovskite minerals have been synthesized to be used as high-temperature semiconductors, which conduct electricity without loss of current and may have great commercial potential. Mineralogists experienced with natural perovskites helped unravel the structure of these newly created materials. Temperature

FIGURE 4.21 Some caves are filled with giant crystals of minerals. These selenite crystals are a gem-quality form of gypsum (calcium sulfate). [Javier Trueba/MSF/Photo Researchers.]

FIGURE 4.22 Asbestos (chrysotile). Fibers are readily combed from the solid mineral. [Runk/Schoenberger/Grant Heilman Photography.]

also affects density: the higher the temperature, the more open and expanded the structure of the mineral, and thus the lower its density.

Crystal Habit

A mineral's **crystal habit** is the shape in which individual crystals or aggregates of crystals grow (Figure 4.21). Crystal habits are often named after common geometric shapes, such as blades, plates, and needles. Some minerals have such a distinctive crystal habit that they are easily recognizable. An example is quartz, with its six-sided column topped by a pyramid-like set of faces (see Figure 4.7). These shapes indicate not only the planes of atoms or ions in the mineral's crystal structure but also the typical speed and direction of crystal growth. Thus, a needlelike crystal is one that grows very quickly in one direction and very slowly in all other directions. In contrast, a plate-shaped crystal (often referred to as *platy*) grows fast in all directions that are perpendicular to its single direction of slow growth. *Fibrous* crystals take shape as multiple long, narrow fibers, essentially aggregates of long needles (Figure 4.22).

TABLE 4.4	Physical Properties of Minerals
Property	Relation to Composition and Crystal Structure
Hardness	Strong chemical bonds result in hard minerals. Covalently bonded minerals are generally harder than ionically bonded minerals.
Cleavage	Cleavage is poor if bonds in crystal structure are strong, good if bonds are weak. Covalent bonds generally give poor or no cleavage; ionic bonds are weaker and so give good cleavage.
Fracture	Related to distribution of bond strengths across irregular surfaces other than cleavage planes.
Luster	Tends to be glassy for ionically bonded crystals, more variable for covalently bonded crystals.
Color	Determined by kinds of atoms or ions and trace impurities. Many ionically bonded crystals are colorless. Iron tends to color strongly.
Streak	Color of fine mineral powder is more characteristic than that of massive mineral because of uniformly small size of grains.
Density	Depends on atomic weight of atoms or ions and their closeness of packing in crystal structure. Iron minerals and metals have high density; covalently bonded minerals have more open packing and so have lower density.
Crystal habit	Depends on planes of atoms or ions in a mineral's crystal structure and the typical speed and direction of crystal growth.

Asbestos is a generic name for a group of silicates with a more or less fibrous habit that allows the crystals to become embedded in the lungs if they are inhaled. Other minerals with deleterious effects include arsenic-containing pyrites, some of which are poisonous when ingested and others that release toxic fumes when heated. Mineral dust diseases affect many miners, who may face large occupational exposures. An example is silicosis, a disease of the lungs caused by inhaling quartz dust.

Table 4.4 summarizes the physical properties of minerals that we discussed in this section.

WHAT ARE ROCKS?

A geologist's primary aim is to understand the properties of rocks and to deduce their geologic origins from those properties. Such deductions further our understanding of our planet, and they also provide important information about fuel reserves. For example, knowing that oil forms in certain kinds of sedimentary rocks that are rich in organic matter allows us to explore for new oil reserves more intelligently. Similarly, our knowledge of the properties of rocks helps us find new reserves of other useful and economically valuable mineral and energy resources, such as natural gas, coal, and metallic ores. Understanding how rocks form also guides us in solving environmental problems. For example, the underground storage of radioactive and other wastes depends on analysis of the rock to be used as a repository: Will this rock be prone to earthquake-triggered landslides? How might it transmit polluted waters in the ground?

A **rock** is a naturally occurring solid aggregate of minerals or, in some cases, nonmineral solid matter. Some rocks, such as white marble, are composed of just one mineral (in this case, calcite). A few rocks are composed of nonmineral matter. These include the noncrystalline, glassy volcanic rocks obsidian and pumice as well as coal, which is made up of compacted plant remains. In an *aggregate,* minerals are joined in such a way that they retain their individual identity (**Figure 4.23**).

> A rock is a naturally occurring solid aggregate of minerals or, in some cases, nonmineral solid matter.

What determines the physical appearance of a rock? Rocks vary in color, in the sizes of their crystals or grains, and in the kinds of minerals that compose them. Along a road cut, for example, we might find a rough white and

FIGURE 4.23 Rocks are naturally occurring aggregates of minerals.
[John Grotzinger/Ramón Rivera-Moret/Harvard Mineralogical Museum.]

FIGURE 4.24 The three families of rocks are formed in different environments by different geologic processes. [Granite and gneiss: John Grotzinger/Ramón Rivera-Moret/Harvard Mineralogical Museum. Sandstone: John Grotzinger/Ramón Rivera-Moret/MIT.]

pink–speckled rock composed of interlocking crystals large enough to be seen with the naked eye. Nearby, we might see a grayish rock containing many large glittering crystals of mica and some grains of quartz and feldspar. Overlying both the white and pink rock and the gray one, we might see horizontal layers of a striped white and mauve rock that appear to be made up of sand grains cemented together. And these rocks might all be overlain by a dark fine-grained rock with tiny white dots in it.

The identity of a rock is determined partly by its mineralogy and partly by its texture. Here, the term *mineralogy* refers to the relative proportions of a rock's constituent minerals. **Texture** describes the sizes and shapes of a rock's mineral crystals or grains and the way they are put together. If the crystals or grains, which are only a few millimeters in diameter in most rocks, are large enough to be seen with the naked eye, the rock is categorized as *coarse-grained*. If they are not large enough to be seen, the rock is categorized as *fine-grained*.

The mineralogy and texture that determine a rock's appearance are themselves determined by the rock's geologic origin—where and how it formed (**Figure 4.24**). The dark rock that caps the sequence of rocks in our road cut, called basalt, was formed by a volcanic eruption. Its mineralogy and texture depend on the chemical composition of rocks that were melted deep within Earth. All rocks formed by the solidification of molten rock are called **igneous rocks.**

The striped white and mauve layers in the road cut are sandstone, formed as sand particles accumulated, perhaps on an ancient beach, and eventually were covered over, buried, and cemented together. All rocks formed as the burial products of layers of sediments (such as sand, mud, or calcium carbonate shells), whether they were laid down on land or under the sea, are called **sedimentary rocks.**

The grayish rock of our road cut, a schist, contains crystals of mica, quartz, and feldspar. It formed deep in Earth's crust as high temperatures and pressures transformed the mineralogy and texture of a buried sedimentary rock. All rocks formed by the transformation of preexisting solid rocks under the influence of high pressure and temperature are called **metamorphic rocks.**

The three types of rocks seen in our road cut represent the three great families of rock: igneous, sedimentary, and metamorphic. Let's take a closer look at each of these families and at the geologic processes that form them.

The physical appearance of a rock is defined primarily by its color, its mineralogy, and the sizes of its crystals or grains.

The mineralogy and texture of rocks are determined by the geologic processes by which they form.

IGNEOUS ROCKS

Igneous rocks (from the Latin *ignis*, meaning "fire") form by crystallization from a magma, a mass of melted rock that originates deep in the crust or upper mantle, where temperatures reach the 700°C or more needed to melt most rocks. When a magma cools slowly in Earth's interior, microscopic crystals start to form. As the magma cools below its melting point, some of these crystals have time to grow to several millimeters in diameter or larger before the whole mass crystallizes as a coarse-grained igneous rock. But when a magma erupts from a volcano onto Earth's surface, it cools

Igneous extrusion (lava)

Extrusive igneous rocks form when magma erupts at the surface, rapidly cooling to fine ash or lava and developing tiny crystals.

The resulting rock, such as the basalt sample here, is fine-grained or has a glassy texture.

Igneous intrusion

Intrusive igneous rocks crystallize when molten rock intrudes into unmelted rock masses in Earth's crust.

Large crystals grow during the slow cooling process, producing coarse-grained rocks such as the granite sample shown here.

FIGURE 4.25 The formation of extrusive and intrusive igneous rocks. [John Grotzinger/Ramón Rivera-Moret/Harvard Mineralogical Museum.]

Igneous rocks are formed by solidification of molten rock.

and solidifies so rapidly that individual crystals have no time to grow gradually. In that case, many tiny crystals form simultaneously, and the result is a fine-grained igneous rock. Geologists distinguish two major types of igneous rocks—intrusive and extrusive—on the basis of the sizes of their crystals.

Intrusive and Extrusive Igneous Rocks

Intrusive igneous rocks crystallize when magma intrudes into unmelted rock masses deep in Earth's crust. Large crystals grow as the magma slowly cools, producing coarse-grained rocks. Intrusive igneous rocks can be recognized by their large, interlocking crystals (**Figure 4.25**). Granite is an intrusive igneous rock.

Extrusive igneous rocks form from magmas that erupt at the surface through volcanoes and cool rapidly. Extrusive igneous rocks, such as basalt, are easily recognized by their glassy or fine-grained texture (see Figure 4.25).

Common Minerals in Igneous Rocks

Most of the minerals of igneous rocks are silicates, partly because silicon is so abundant and partly because many silicate minerals melt at the high temperatures and pressures reached in deeper parts of the crust and in the mantle. The common silicate minerals found in igneous rocks include quartz, feldspar, mica, pyroxene, amphibole, and olivine (Table 4.5).

TABLE 4.5	Some Common Minerals of Igneous, Sedimentary, and Metamorphic Rocks		
	Igneous Rocks	Sedimentary Rocks	Metamorphic Rocks
	Quartz	Quartz	Quartz
	Feldspar	Clay minerals	Feldspar
	Mica	Feldspar	Mica
	Pyroxene	*Calcite	Garnet
	Amphibole	*Dolomite	Pyroxene
	Olivine	*Gypsum	Staurolite
		*Halite	Kyanite

Note: Asterisks indicate nonsilicate minerals.

SEDIMENTARY ROCKS

Sediments, the precursors of sedimentary rocks, are found on Earth's surface as layers of loose particles, such as sand, silt, and the shells of organisms. These particles originate in the processes of weathering and erosion. **Weathering** refers to all of the chemical and physical processes that break up and decay rocks into fragments and dissolved substances of various sizes. These particles are then transported by **erosion,** the set of processes that loosen soil and rock and move them downhill or downstream to the spot

FIGURE 4.26 Weathering breaks down rocks into fragments and dissolved materials that are then carried downhill and downstream by erosion to be deposited as layers of sediment. (*left*) sandstone; (*right*) fossilized coral. [John Grotzinger/Ramón Rivera-Moret/MIT.]

1 Particles of rock created by weathering…
2 …are transported downhill by erosion…
3 …and deposited as layers of sediment on land or in water,…
4 …where they form parallel layers, or bedding.
5 Buried sediments lithify by compaction and cementation.
6 Siliciclastic sediments are made of deposited particles of sand, silt, and gravel.
7 Chemical and biological sediments are precipitated from seawater or made by coral reefs and shells.

where they are deposited as layers of sediment (**Figure 4.26**). Sediments are deposited in two ways:

- *Siliciclastic sediments* are made up of physically deposited particles, such as grains of quartz and feldspar derived from weathered granite. (*Clastic* is derived from the Greek word *klastos,* meaning "broken.") These sediments are laid down by running water, wind, and ice and form layers of sand, silt, and gravel.

- *Chemical and biological sediments* are new chemical substances that form by precipitation when some of a rock's components dissolve during weathering and are carried in river waters to the sea. These sediments include layers of such minerals as halite (sodium chloride) and calcite (calcium carbonate, most often found in the form of reefs and shells).

From Sediment to Solid Rock

Lithification is the process that converts sediments into solid rock. It occurs in one of two ways:

- In *compaction,* particles are squeezed together by the weight of overlying sediment into a mass denser than the original.

- In *cementation,* minerals precipitate around deposited particles and bind them together.

Sediments are compacted and cemented after they are buried under additional layers of sediment. Sandstone forms by the lithification of sand particles, and limestone forms by the lithification of shells and other particles of calcium carbonate.

Sedimentary rocks are the result of surface processes on Earth, including weathering, erosion, and the precipitation of dissolved minerals.

Layers of Sediment

Sediments and sedimentary rocks are characterized by **bedding,** the formation of parallel layers of sediment as particles settle to the bottom of a water body or on a land surface. Because sedimentary rocks are formed by surface processes, they cover much of Earth's land surface and seafloor. In terms of surface area, most rocks found at Earth's surface are sedimentary, but these rocks weather easily, so their volume is small compared with that of the igneous and metamorphic rocks that make up the main volume of the crust.

Common Minerals of Sedimentary Rocks

The common minerals of siliciclastic sediments are silicates, because silicate minerals predominate in the rocks that weather to form sedimentary particles (see Table 4.5). The most abundant silicate minerals in siliciclastic sedimentary rocks are quartz, feldspar, and clay minerals. Clay minerals form by weathering and alteration of preexisting silicate minerals, such as feldspar.

The most abundant minerals of chemically or biologically precipitated sediments are carbonates, such as calcite, the main constituent of limestone. Dolomite, also found in limestone, is a calcium-magnesium carbonate formed by precipitation during lithification. Two other chemical sediments—gypsum and halite—form by precipitation as seawater evaporates.

METAMORPHIC ROCKS

Metamorphic rocks take their name from the Greek words for "change" (*meta*) and "form" (*morphe*). These rocks are produced when high temperatures and pressures deep within Earth cause any kind of preexisting rock—igneous, sedimentary, or other metamorphic rock—to change its mineralogy, texture, or chemical composition while maintaining its solid form. The temperatures of metamorphism are below the melting points of the rocks (about 700°C), but high enough (above 250°C) for the rocks to be changed by recrystallization and chemical reactions.

Regional and Contact Metamorphism

Metamorphism may take place over a widespread area or a limited one (**Figure 4.27**). **Regional metamorphism** occurs where high pressures and temperatures extend over

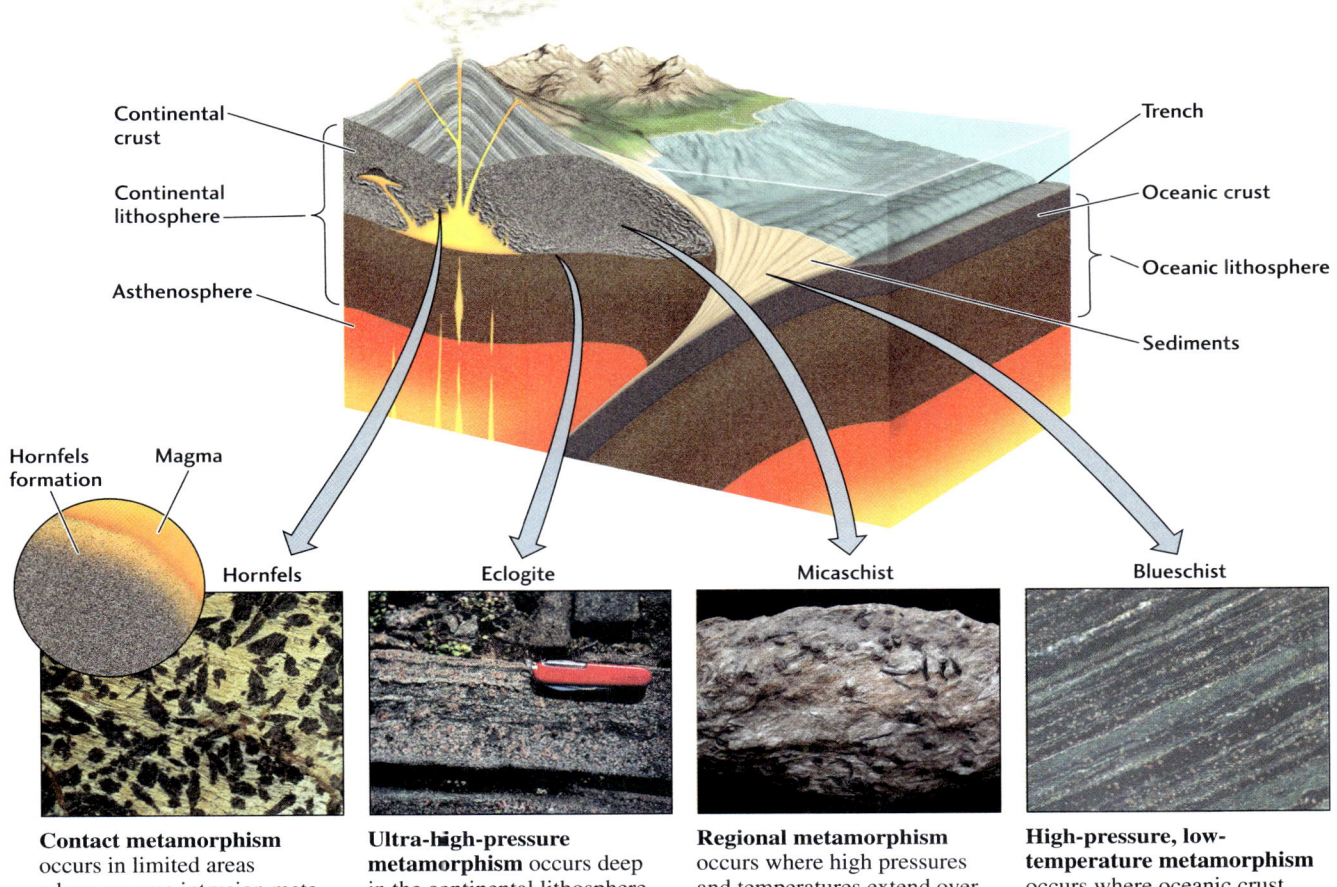

Contact metamorphism occurs in limited areas where magma intrusion metamorphoses neighboring rock by its heat, forming hornfels.

Ultra-high-pressure metamorphism occurs deep in the continental lithosphere and oceanic crust.

Regional metamorphism occurs where high pressures and temperatures extend over large regions.

High-pressure, low-temperature metamorphism occurs where oceanic crust subducts beneath the leading edge of a continental plate.

FIGURE 4.27 Metamorphic rocks form under conditions of high temperatures and pressures. [*left to right:* hornfels: Biophoto Associates/Photo Researchers; eclogite: Julie Baldwin; micaschist: John Grotzinger; blueschist: Mark Cloos.]

large regions, as happens where plates collide. Regional metamorphism accompanies plate collisions that result in mountain building and the folding and breaking of sedimentary layers that were once horizontal. Where high temperatures are restricted to smaller areas, as in the rocks near and in contact with a magmatic intrusion, rocks are transformed by **contact metamorphism.**

Many regionally metamorphosed rocks, such as schists, have characteristic **foliation,** wavy or flat planes produced when the rock was structurally deformed into folds. Granular textures are more typical of most contact metamorphic rocks and of some regional metamorphic rocks formed by very high pressures and temperatures.

> *Metamorphic rocks are produced by modification of preexisting rocks under conditions of high temperature and pressure deep within Earth.*

Common Minerals of Metamorphic Rocks

Silicates are the most abundant minerals of metamorphic rocks because the parent rocks from which they are formed are also rich in silicates (see Table 4.5). Typical minerals of metamorphic rocks are quartz, feldspar, mica, pyroxene, and amphibole—the same kinds of silicates characteristic of igneous rocks. Several other silicates—kyanite, staurolite, and some varieties of garnet—are characteristic of metamorphic rocks alone. These minerals form under conditions of high pressure and temperature in the crust and are not characteristic of igneous rocks. They are therefore good indicators of metamorphism. Calcite is the main mineral of marbles, which are metamorphosed limestones.

THE ROCK CYCLE: Interactions Between the Plate Tectonic and Climate Systems

Earth scientists have known for over 200 years that the three families of rocks—igneous, metamorphic, and sedimentary—all can evolve from one to another. The **rock cycle** explains how each type of rock is converted into one of the other two types.

We also now know that the cycling of rock is the result of interactions between plate tectonics and climate. For example, the melting of subducting lithospheric slabs and the formation of magma result from processes operating within the plate tectonic system. When these molten rocks erupt to the land surface, the newly formed igneous rocks are subject to weathering by the climate system. The same process injects volcanic ash and carbon dioxide gas high into the atmosphere, where they may affect global climate. As global climate changes, perhaps becoming warmer or cooler, the rate of rock weathering changes, which in turn influences the rate at which rock material (sediment) is returned to Earth's interior.

Let's trace one turn of the rock cycle, beginning with the creation of new oceanic lithosphere at a mid-ocean ridge spreading center as two continents drift apart (**Figure 4.28**). The ocean gets wider and wider, until at some point the process reverses itself and the ocean closes. As the ocean basin closes, igneous rocks created at the mid-ocean ridge eventually descend into a subduction zone beneath a continental plate. Sediments that were formed on the continent and transported to its edge may also be dragged down into the subduction zone. Ultimately, the two continents, which were once drifting apart, may now move closer together and collide. As the

> *The rock cycle is a set of processes that can transform each type of rock into the other two types.*

1 The cycle begins with rifting within a continent. Sediments erode from the continental interior and are deposited in rift basins, where they are buried to form sedimentary rocks.

6 Streams transport sediment away from collision zones to oceans, where it is deposited as layers of sand and silt. Layers of sediment are buried and lithify to form sedimentary rock.

FIGURE 4.28 The rock cycle results from the interaction of the plate tectonic and climate systems.

igneous rocks and sediments that descend into the subduction zone go deeper and deeper into Earth's interior, they begin to melt to form a new generation of igneous rocks. The great heat associated with the intrusion of these igneous rocks, coupled with the heat and pressure that comes with being pushed to levels deep within Earth, transforms these igneous rocks—and other surrounding rocks—into metamorphic rocks. When the continents collide, these igneous and metamorphic rocks are uplifted into a high mountain chain as a section of Earth's crust crumples and deforms.

Weathering of these uplifted igneous and sedimentary rocks results in the formation of loose materials that erosion then strips away. Water and wind transport some of these materials across the continents and eventually to the edges of the continents, where they are deposited as sediments. The sediments laid down where the land meets the ocean are buried under successive layers of sediments, where they slowly lithify into sedimentary rock. These oceans, like those mentioned at the beginning of the cycle, may also have formed by spreading along mid-ocean ridges, thus completing the rock cycle.

The particular pathway illustrated here—that of a continent breaking apart, forming a new ocean basin, then closing back up again—is only one variation among the many that may take place in the rock cycle. Any type of rock—igneous, sedimentary, or metamorphic—can be uplifted during a mountain-building event and then weathered and eroded to form new sediments. Some stages may be omitted: as a sedimentary rock is uplifted and eroded, for example, metamorphism and melting are skipped. In some cases, the rock cycle may proceed very slowly. For example, we know that some igneous and metamorphic rocks many kilometers deep in the crust

2 Rifting and spreading continue, and a new ocean basin develops. Magma rises from the asthenosphere at mid-ocean ridges and chills to form basalt, an igneous rock.

3 Subsidence of the continental margin—sinking of Earth's lithosphere—leads to accumulation of sediment and formation of sedimentary rock during burial.

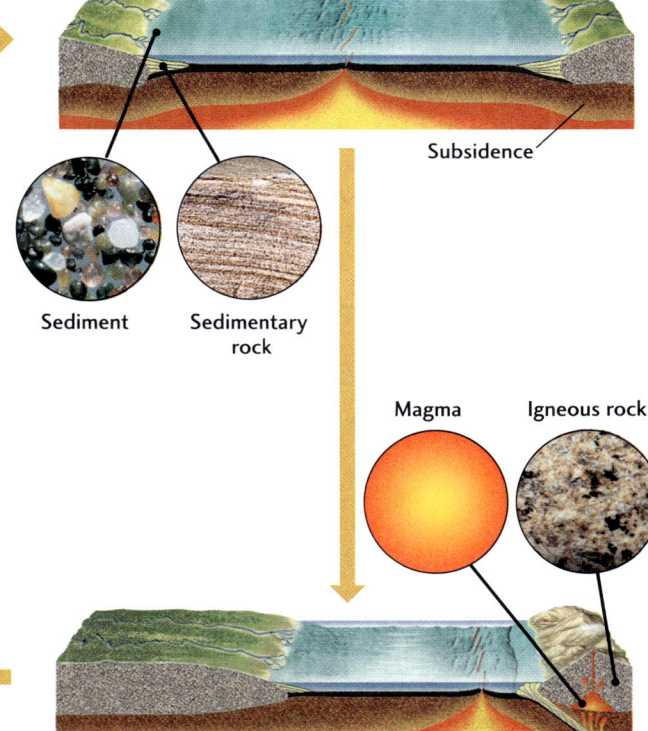

4 Oceanic crust subducts beneath a continent, building a volcanic mountain chain. The subducting plate melts as it descends. Magma rises from the melting plate and mantle and cools to make granitic igneous rocks.

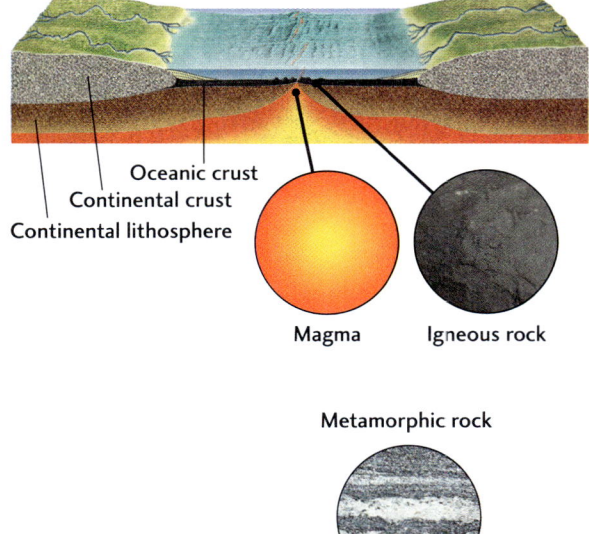

5 Further closing of the ocean basin leads to continental collision, forming high mountain ranges. Where continents collide, rocks are buried deeper or modified by heat and pressure, forming metamorphic rocks. Uplifted mountains force moisture-laden air to rise, cool, and release its moisture as precipitation. Weathering creates loose material—soils and sediment—that erosion strips away.

may be uplifted or exposed to weathering and erosion only after billions of years have passed.

The essence of the rock cycle is that plate tectonic processes drive rocks at the surface deep into Earth, where they may become metamorphosed or even melt, and then return these rocks to the surface. Weathering at the surface decomposes these rocks to produce loose materials that may accumulate as thick sediments, only to be driven down deep into Earth again by plate tectonics.

The rock cycle never ends. It is always operating at different stages in various parts of the world, forming and eroding mountains in one place and laying down and burying sediments in another. The rocks that make up the solid Earth are recycled continuously, but we can see only the surface parts of the cycle. We must deduce the recycling of the deep crust and the mantle from indirect evidence.

CONCENTRATIONS OF VALUABLE MINERAL RESOURCES

The rock cycle turns out to be crucial in creating economically important concentrations of valuable minerals in Earth's crust. Finding these minerals and extracting them is a vital job for Earth scientists, so we now turn our attention to how and where some of these geological prizes are formed.

The chemical elements of Earth's crust are widely distributed in many kinds of minerals, and those minerals are found in a great variety of rocks (see Appendix 4 for a summary of the properties of the most common minerals of Earth's crust). In most places, any given element will be found homogenized with other elements in amounts close to its average concentration in the crust. An ordinary granitic rock, for example, may contain a small percentage of iron, close to the average concentration of iron in Earth's crust.

When an element is present in higher concentrations, it means that the rock underwent some geologic process that concentrated larger quantities of that element than normal. The *concentration factor* of an element in a mineral deposit is the ratio of the element's abundance in the deposit to its average abundance in the crust. High concentrations of elements are found in a limited number of specific geological settings. These settings are of economic interest because the higher the concentration of a resource in a given deposit, the lower the cost to recover it.

Ore Minerals

Ores are rich deposits of minerals from which valuable metals can be recovered profitably. The minerals containing these metals are *ore minerals*. Ore minerals include sulfides (the largest group), oxides, and silicates. The ore minerals in each of these groups are compounds of metallic elements with sulfur, oxygen, and silicon oxide, respectively. The copper ore mineral covelite, for example, is a copper sulfide (CuS). The iron ore mineral hematite (Fe_2O_3) is an iron oxide. The nickel ore mineral garnierite is a nickel silicate, $Ni_3Si_2O_5(OH)_4$. In addition, some metals, such as gold, are found in their native state—that is, uncombined with other elements (**Figure 4.29**).

FIGURE 4.29 (*left*) A geologist examines rock samples in an underground gold mine in Zimbabwe, southern Africa. [Peter Bowater/Photo Researchers.] (*right*) Native gold on a quartz crystal. [Chip Clark.]

FIGURE 4.30 Many ore mineral deposits are found in veins formed by hydrothermal solutions rising from magmatic intrusions. This quartz vein deposit (about 1 cm thick) in Oatman, Arizona, contains gold and silver ores. [Peter Kresan.]

Recall in our discussion of the rock cycle that continental margins where subduction occurs may be associated with melting of oceanic lithosphere to form igneous rocks. Very large ore deposits can be formed in such a tectonic setting when hot water solutions—also known as **hydrothermal solutions**—are formed around bodies of molten rock. This happens when circulating groundwater or seawater comes into contact with a magmatic intrusion, reacts with it, and carries off significant quantities of elements and ions released by the reaction. These elements and their ions then interact with one another to deposit ore minerals, usually as the solution cools.

Veins

Hydrothermal solutions moving through fractured rocks often deposit ore minerals. These fluids flow easily through the fractures and joints of the rocks, cooling rapidly in the process. Quick cooling causes fast precipitation of the ore minerals. The resulting *tabular* (sheetlike) deposits of precipitated minerals in the fractures and joints are called **veins.** Some ore minerals are found in veins; others are found in the rocks surrounding the veins, which are altered when the hot solutions heat and infiltrate those rocks. As the solutions react with surrounding rocks, they may precipitate ore minerals together with quartz, calcite, or other common vein-filling minerals. Vein deposits are a major source of gold (**Figure 4.30**).

Hydrothermal vein deposits are among the most important sources of metallic ores. Typically, metallic ores exist as sulfides, such as iron sulfide (pyrite), lead sulfide (galena), zinc sulfide (sphalerite), and mercury sulfide (cinnabar)—shown in **Figure 4.31**—as well as copper sulfide (covelite and chalcocite). Hydrothermal solutions

FIGURE 4.31 Metal sulfide ores. Sulfides are the most common types of metallic ores. [Chip Clark.]

FIGURE 4.32 Copper ores. Chalcopyrite and chalcocite are copper sulfide ores. Malachite is a carbonate of copper found in association with sulfides of copper. [Chip Clark.]

Chalcopyrite (a copper sulfide)   Malachite (a copper carbonate)   Chalcocite (a copper sulfide)

reach the surface as hot springs and geysers, many of which precipitate metallic ores—including ores of lead, zinc, and mercury—as they cool.

Disseminated Deposits

Mineral deposits that are scattered through volumes of rock much larger than veins are called *disseminated deposits.* In both igneous and sedimentary rocks, minerals are disseminated along abundant cracks and fractures. Among the economically important disseminated deposits are the copper deposits of Chile and the southwestern United States. These deposits develop in geologic provinces with abundant igneous rocks, usually emplaced as large intrusive bodies. In Chile, these intrusive igneous rocks are related to the subduction of oceanic lithosphere beneath the Andes, an event very similar to what was described in our example of the rock cycle. The most common copper mineral in these deposits is chalcopyrite, a copper sulfide (**Figure 4.32**). The copper was deposited when ore-forming minerals were introduced into a great number of tiny fractures in granitic intrusive rocks and in the rocks surrounding the higher parts of the igneous intrusions. Some unknown process associated with the magmatic intrusion or its aftermath broke these rocks into millions of pieces. Hydrothermal solutions penetrated and recemented the rocks by precipitating ore minerals throughout the extensive network of tiny fractures. This widespread dispersal produced a low-grade but very large resource of many millions of tons of ore, which can be mined economically by large-scale methods (**Figure 4.33**).

Disseminated hydrothermal deposits are also present in sedimentary rocks. This is the case in the lead-zinc province of the Upper Mississippi Valley, which extends from southwestern Wisconsin to Kansas and Oklahoma. The ores in this province are not associated with a known magmatic intrusion that could have been a source of hydrothermal fluids, so their origin must be very different. Some geologists speculate that the ores were deposited by groundwater that was driven out of the ancestral Appalachian Mountains when they were much higher. A continental collision between North America and Africa may have created a continental-scale squeegee that pushed fluids from deep within the collision zone all the way into the continental interior of North America. Groundwater may have penetrated hot crustal rocks at great depths and dissolved soluble ore minerals, then moved upward into the overlying sedimentary rocks, where it precipitated the minerals as fillings in cavities. In some cases, it appears that these solutions infiltrated limestone formations and dissolved some carbonates, then replaced the carbonates with equal volumes of new sulfide crystals. The major minerals of the hydrothermal deposits in this province are lead sulfide (galena) and zinc sulfide (sphalerite).

Igneous Deposits

The most important *igneous ore deposits*—deposits of ore in igneous rocks—are found as segregations of ore minerals near the bottoms of intrusions. These deposits form when minerals crystallize from molten magma, settle, and accumulate on the floor of a magma chamber. Most of the chromium and platinum ores in the world, such as the deposits in South Africa and Montana, are found as layered accumulations of minerals that formed in this way (**Figure 4.34**). One of the richest ore bodies ever found, at

FIGURE 4.33 Kennecott Copper Mine, Utah, an open-pit mine. Open-pit mining is typical of the large-scale methods used to exploit widely disseminated ore deposits. [David R. Frazier/The Image Works.]

Sudbury, Ontario, is a large mafic igneous intrusion containing great quantities of layered nickel, copper, and iron sulfides near its base. Geologists believe that these sulfide deposits formed from crystallization of a dense, sulfide-rich liquid that separated from the rest of the cooling magma and sank to the bottom of the chamber before it congealed.

Pegmatites are extremely coarse grained intrusive rocks of granitic composition. As the magma in a large granitic intrusion cools, the last material to crystallize forms pegmatites, in which minerals present in only trace amounts in the parent rock are concentrated. Pegmatites may contain rare mineral deposits rich in such elements as boron, lithium, fluorine, niobium, and uranium and in such gem minerals as tourmaline.

Sedimentary Deposits

Sedimentary mineral deposits include some of the world's most valuable mineral sources. Many economically important minerals segregate as an ordinary result of sedimentary processes. Sedimentary mineral deposits are also important sources of copper, iron, and other metals. These deposits are chemically precipitated in sedimentary environments to which large quantities of metals are transported in solution. Some of the important sedimentary

FIGURE 4.34 Chromite (chromium ore, visible as dark layers) in a layered igneous intrusion in the Bushveldt Complex, South Africa. [Spence Titley.]

FIGURE 4.35 Panning for gold was popularized by "forty-niners" during the California gold rush and is still popular in the San Gabriel River today. [(top) Bo Zaunders/CORBIS; (bottom) David Butow/CORBIS SABA.]

copper ores, such as those of the Permian Kupferschiefer (German for "copper slate") beds of Germany, may have precipitated from hydrothermal solutions rich in metal sulfides that interacted with sediments on the seafloor. The tectonic setting of these deposits may have been something like the mid-ocean ridge described in our example of the rock cycle, except that it developed within a continent. Here, rifting of the continental crust led to development of a deep trough, where sediments and ore minerals were deposited in a very still, narrow sea.

Many rich deposits of gold, diamonds, and other heavy minerals such as magnetite and chromite are found in *placers*, sedimentary ore deposits that have been concentrated by the mechanical sorting action of river currents. These ore deposits originate where uplifted rocks weather to form grains of sediment, which are then sorted by weight when currents of water flow over them. Because heavy minerals settle out of a current more quickly than lighter minerals such as quartz and feldspar, the heavy minerals tend to accumulate on river bottoms and sandbars. Here the current is strong enough to keep the lighter minerals suspended and in transport, but too weak to move the heavier minerals. Similarly, ocean waves preferentially deposit heavy minerals on beaches or shallow offshore bars. A gold panner accomplishes the same thing: the shaking of a water-filled pan allows the lighter minerals to be washed away, leaving the heavier gold in the bottom of the pan (**Figure 4.35**). Some placers can be traced upstream to the location of the original mineral deposit, usually of igneous origin, from which the minerals were eroded. Erosion of the Mother Lode, an extensive gold-bearing vein system lying along the western flanks of the Sierra Nevada, produced the placers that were discovered in 1848 and led to the California gold rush. The placers were found before their source was discovered. Placers also led to the discovery of the Kimberley diamond mines of South Africa two decades later.

> Ore deposits are formed in geological settings where elements accumulate in concentrations that far exceed their normal abundances. Precipitation from hydrothermal solutions, crystallization and settlement in magmas, and selective transport of heavy minerals in streams are a few examples of the processes by which elements can be concentrated.

SUMMARY >>

What is a mineral? Minerals, the building blocks of rocks, are naturally occurring, inorganic solids with specific crystal structures and chemical compositions that either are fixed or vary within a defined range. A mineral is constructed of atoms, the small units of matter that combine in chemical reactions. An atom is composed of a nucleus made up of protons and neutrons and surrounded by electrons. The atomic number of an element is the number of protons in its nucleus, and its atomic mass is the sum of the masses of its protons and neutrons.

How do atoms combine to form the crystal structures of minerals? Chemical elements react with one another to form compounds either by gaining or losing electrons to become ions or by sharing electrons. The ions in some chemical compounds are held together by ionic bonds, which form by electrostatic attraction between positive ions (cations) and negative ions (anions). Atoms that share electrons to form a compound are held together by covalent bonds. When a mineral crystallizes, atoms or ions come together in the proper proportions to form a crystal structure—an orderly three-dimensional array in which the basic arrangement of the atoms is repeated in all directions.

What are the major rock-forming minerals? Silicate minerals, the most abundant minerals in Earth's crust, are crystal structures built of silicate tetrahedra linked in various ways. Tetrahedra may be isolated (olivines) or in single chains (pyroxenes), double chains (amphiboles), sheets (micas), or frameworks (feldspars). Carbonate minerals are made of carbonate ions bonded to calcium or magnesium or both. Oxide minerals are compounds of oxygen and metallic elements. Sulfide and sulfate minerals are composed of sulfur atoms in combination with metallic elements.

What are the physical properties of minerals? A mineral's physical properties, which indicate its composition and structure, include hardness—the ease with which its surface is scratched; cleavage—its ability to split or break along flat surfaces; fracture—the way it breaks along irregular surfaces; luster—the way it reflects light; color—imparted by transmitted or reflected light to crystals or irregular masses or visible as streak (the color of a fine powder); density—mass per unit volume; and crystal habit—the shapes of individual crystals or aggregates.

What determines the properties of the various kinds of rocks that form in and on Earth's surface? Mineralogy (the kinds and proportions of minerals that make up a rock) and texture (the sizes, shapes, and spatial arrangement of its crystals or grains) define a rock. The mineralogy and texture of a rock are determined by the geologic processes by which it formed.

What are the three families of rocks and how do they form? Igneous rocks form by the crystallization of magmas as they cool. Intrusive igneous rocks form in Earth's interior and have large crystals. Extrusive igneous rocks, which form at Earth's surface where lavas erupt from volcanoes, have a glassy or fine-grained texture. Sedimentary rocks form by the lithification of sediments after burial. Sediments are derived from the weathering and erosion of rocks exposed at Earth's surface. Metamorphic rocks form by changes in igneous, sedimentary, or other metamorphic rocks that are subjected to high temperatures and pressures in Earth's interior.

How does the rock cycle explain the transformation of rocks from one type into another? The rock cycle relates geologic processes to the formation of the three types of rocks from one another. We can view these processes by starting at any point in the cycle, such as the creation of new oceanic lithosphere at a mid-ocean ridge as two continents drift apart. The ocean basin gets wider until at some point the process reverses itself. As the basin closes, igneous rocks and sediments are subducted beneath a continental plate. As they descend into Earth's interior, they begin to melt to form a new generation of igneous rocks. The heat associated with the intrusion of these igneous rocks, coupled with the heat and pressure that comes with being pushed to levels deep within Earth, transforms these igneous rocks—and other surrounding rocks—into metamorphic rocks. Ultimately, the two continents may collide, and these igneous and metamorphic rocks may be uplifted into a high mountain chain. The uplifted rocks slowly weather, and their fragments are deposited as sediments. Plate tectonics and climate are the mechanisms by which the rock cycle operates.

What is hydrothermal mineral deposition? Hydrothermal deposits, which are some of the most important ore mineral deposits, are formed when circulating groundwater or seawater reacts with a magmatic intrusion to form a hydrothermal solution. The heated water leaches soluble minerals in its path and transports them to cooler rocks, where they are precipitated in fractures, joints, and voids. The resulting ores may be found in veins or in disseminated deposits.

How do igneous ore deposits form? Igneous ore deposits typically form when minerals crystallize from molten magma, settle, and accumulate on the floor of a magma chamber. They are often found as layered accumulations of minerals.

KEY TERMS AND CONCEPTS >>

atom (p. 77)
bedding (p. 96)
chemical bond (p. 80)
chemical reaction (p. 78)
cleavage (p. 89)
color (p. 90)
contact metamorphism (p. 98)
crystal (p. 81)
crystal habit (p. 92)
crystallization (p. 80)
density (p. 83)
erosion (p. 95)

extrusive igneous rock (p. 95)
foliation (p. 98)
fracture (p. 90)
hardness (p. 88)
hydrothermal solution (p. 101)
igneous rock (p. 94)
intrusive igneous rock (p. 95)
ion (p. 79)
isotope (p. 78)
lithification (p. 96)

luster (p. 90)
magma (p. 82)
metamorphic rock (p. 94)
mineral (p. 76)
mineralogy (p. 76)
Mohs scale of hardness (p. 88)
ore (p. 100)
polymorph (p. 82)
regional metamorphism (p. 97)
rock (p. 93)
rock cycle (p. 98)
sediment (p. 95)
sedimentary rock (p. 94)
specific gravity (p. 91)
streak (p. 90)
texture (p. 94)
vein (p. 101)
weathering (p. 95)

EXERCISES >>

1. Define a mineral.

2. What is the difference between an atom and an ion?

3. Name two types of chemical bonds.

4. List the basic crystal structures of silicate minerals.

5. Name three classes of minerals other than silicates.

6. How would a field geologist measure hardness?

7. What are the differences between extrusive and intrusive igneous rocks?

8. What are the differences between regional and contact metamorphism?

9. What are the differences between siliciclastic and chemical or biological sediments?

10. List three common silicate minerals found in each of the three families of rocks: igneous, sedimentary, and metamorphic.

11. Of the three families of rocks, which form at Earth's surface and which in the interior of the crust?

12. What are the characteristics of an economically important ore deposit?

13. Choose two minerals from Appendix 4 that you think might make good abrasive or grinding stones for sharpening steel, and describe the physical property that causes you to believe they would be suitable for that purpose.

14. Aragonite, with a density of 2.9 g/cm^3, has exactly the same chemical composition as calcite, with a density of 2.7 g/cm^3. Other things being equal, which of these two minerals is more likely to have formed under high pressure?

15. There are at least seven physical properties one can use to identify an unknown mineral. Which ones are most useful in discriminating between minerals that look similar? Describe a strategy that would allow you to prove that an unknown clear calcite crystal is not the same as a known clear crystal of quartz.

16. Using the rock cycle, trace the path from a magma to a granite intrusion to a metamorphic gneiss to a sandstone. Be sure to include the role of tectonics and the specific processes that create the rocks.

17. Where are igneous rocks most likely to be found? How could you be certain that the rocks were igneous and not sedimentary or metamorphic?

18. Back in the late 1800s, gold miners used to pan for gold by placing sediment from rivers in a pan and filtering water through the pan while swirling the pan's contents. The miners wanted to be certain that they had found real gold and not pyrite ("fool's gold"). Why did this method work? What mineral property does the process of panning for gold use? What is another possible method for distinguishing between gold and pyrite?

IGNEOUS PROCESSES AND VOLCANISM

CHAPTER 5

Mount St. Helens: Dangerous but Predictable

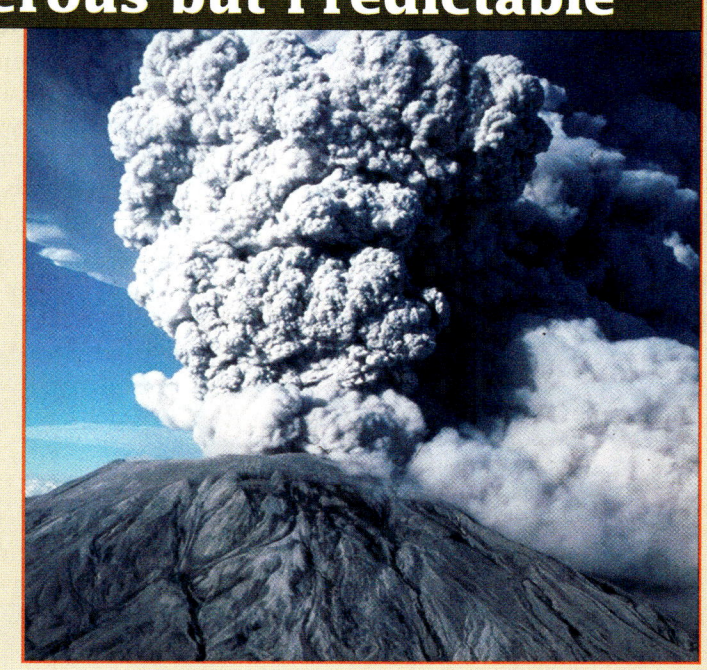

On March 20, 1980, the U.S. Geological Survey issued a formal hazard alert for Mount St. Helens, a volcano in the Cascade Range of the Pacific Northwest. After 123 years of dormancy, a series of small to medium-sized earthquakes under the volcano had signaled the start of a new eruptive phase.

Throughout April, the seismic tremors increased, indicating that magma was moving beneath the summit, and instruments detected an ominous swelling of the northeastern flank of the mountain. The USGS issued a more serious warning, and people were ordered out of the vicinity. On May 18, the main eruption began abruptly. A large earthquake apparently triggered the collapse of the north side of the mountain, causing a massive landslide, the largest ever recorded anywhere. As a huge flow of debris plummeted down the mountain, gas and steam under high pressure were released in a tremendous lateral blast that blew out the northern flank of the mountain

USGS geologist David A. Johnston was monitoring the volcano from his observation post 8 km to the north. He must have seen the advancing blast wave before he radioed his last message: "Vancouver, Vancouver, this is it!" A northward-directed jet of superheated (500°C) ash, gas, and steam roared out of the breach with hurricane force, devastating a zone 20 km outward from the volcano and 30 km wide. A vertical eruption sent an ash plume 25 km into the sky, twice as high as a commercial jet flies. The ash plume drifted to the east and northeast with the prevailing winds, bringing darkness at noon to an area 250 km to the east and depositing a layer of ash as deep as 10 cm over much of Washington State, northern Idaho, and western Montana. The energy of the blast was equivalent to about 25 million tons of TNT. The volcano's summit was destroyed, its elevation was reduced by 400 m, and its northern flank disappeared. In effect, the mountain was hollowed out.

Vancouver, Vancouver, this is it!

Local devastation was spectacular. Within an inner blast zone extending 10 km from the volcano, the thick forest was buried under several meters of volcanic debris. Beyond this zone, out to 20 km, trees were stripped of their branches and blown over like broken matchsticks. As far as 26 km away, the hot blast was so intense that it overturned a truck and melted its plastic parts. Some fishermen were severely burned and survived only by jumping into a river. More than 60 other people were killed by the blast and its effects.

After more than a decade of relative quiescence, the volcano reawoke in September 2004 with a series of minor steam and ash eruptions that continued into 2005. Growth of the central volcanic dome (see Figure 5.27b) suggests that the current phase of eruptive activity may persist for some time into the future. ■

This photo of the July 22, 1980, eruption of Mount St. Helens was taken by Mike Doukas, who was in a U.S. Geological Survey helicopter. Skamania County, Washington. [M. Doukas/U.S. Forest Service.]

CHAPTER OUTLINE

How Do Igneous Rocks Differ from One Another? 109

How Do Magmas Form? 115

Magmatic Differentiation 117

Forms of Igneous Intrusions 119

Magmatic Processes 122

Volcanic Processes and Materials 126

Volcanoes Build Landforms 133

Hazards Created by Volcanoes 136

As we learned in Chapter 4, much of Earth's crust is composed of igneous rock. Understanding the processes that melt and resolidify rocks is a key to understanding how Earth's crust forms.

We also learned that plate tectonic processes create a wide variety of igneous rocks. Specifically, igneous rocks form at spreading centers where plates move apart, along convergent boundaries where one plate descends beneath another, and at hot spots where magmas ascend from great depths in the mantle (Figure 5.1). **In this chapter, we will look in more detail at the igneous processes associated with specific plate tectonic settings. We still have much to learn about the exact mechanisms of melting and solidification: How do igneous rocks differ from one another? Where do igneous rocks form? How do rocks solidify from magma? Where do magmas form? We will give special attention to volcanoes, which allow us to see into Earth's deep interior to understand the igneous processes that have generated Earth's oceanic and continental crust.**

We will examine how magma rises through the crust, emerges onto the surface as lava, and cools into hard volcanic rock. We will see how tectonic processes produce volcanism at plate boundaries. We will also consider the destructive power of volcanoes and the hazards they pose to human society. The vast active volcano that stretches across the wilderness of Yellowstone National Park, for example, expels more energy in the form of heat than all the electric power consumed in the three surrounding states of Wyoming, Idaho, and Montana combined. That energy is not released steadily, however; some of it builds up in hot magma chambers until the volcano blows its top. A cataclysmic eruption of the Yellowstone volcano

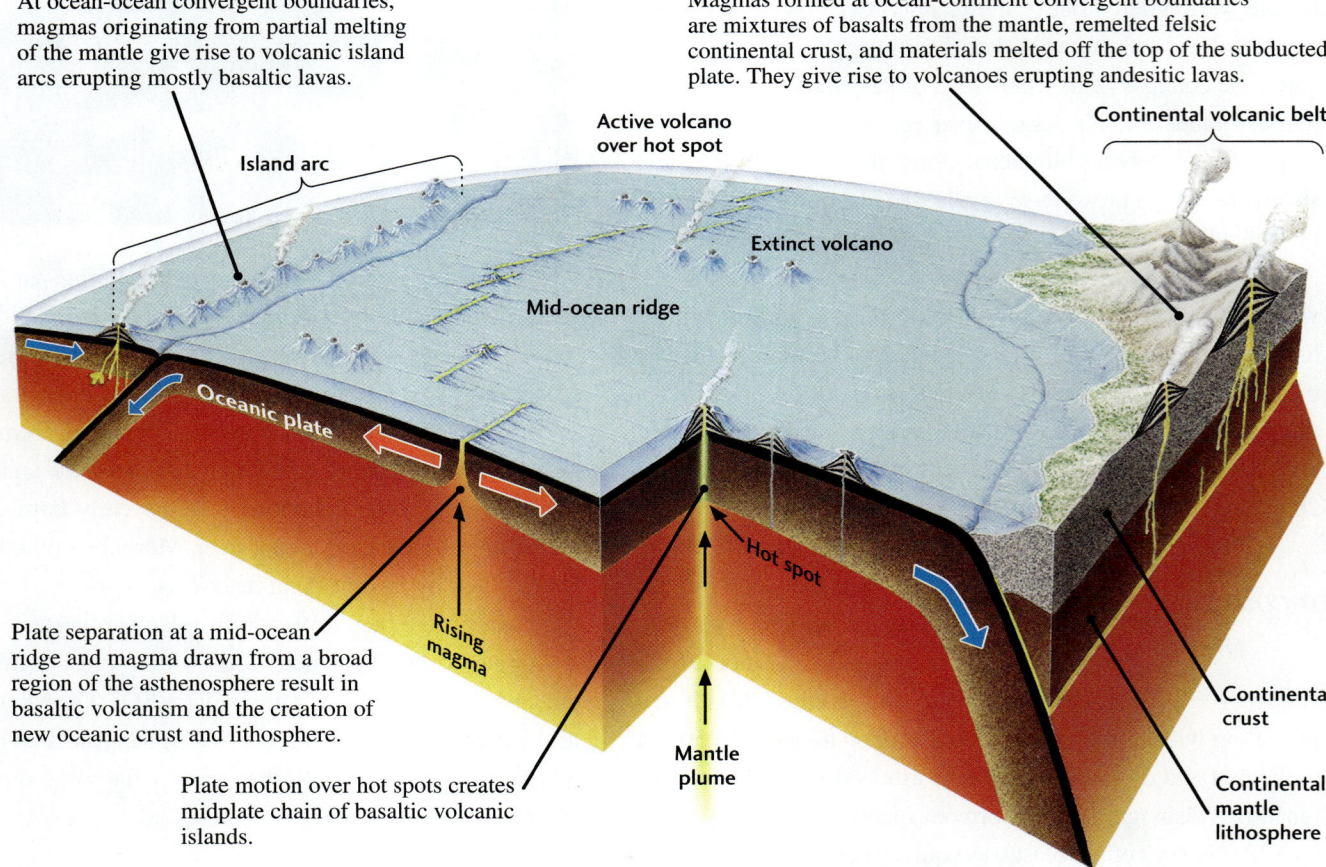

FIGURE 5.1 Plate tectonic processes explain the global pattern of igneous processes.

630,000 years ago ejected 1000 km³ of material into the air, covering regions as far away as Texas and California with a layer of volcanic ash.

The geologic record shows that volcanic explosions nearly as big as the Yellowstone eruption, or even bigger, have occurred in the western United States at least six times during the last 2 million years, so we can be fairly certain that such an eruption will happen again. We can only imagine what it might do to civilization. Hot ash will snuff out all life within 100 km or more, and cooler but choking ash will blanket the ground more than 1000 km away. Dust thrown high into the stratosphere will dim the Sun for several years, dropping temperatures and plunging the Northern Hemisphere into an extended volcanic winter.

HOW DO IGNEOUS ROCKS DIFFER FROM ONE ANOTHER?

Today we classify igneous rock samples in the same way that some geologists did in the late nineteenth century: by their texture and by their mineral and chemical composition.

Texture

Two hundred years ago, the first division of igneous rocks was made on the basis of texture, which largely reflects differences in mineral *crystal size*. Geologists classified rocks as either coarse-grained or fine-grained (see Chapter 4). Crystal size is a simple characteristic that geologists can easily see in the field. A coarse-grained rock such as granite has distinct crystals that are easily visible to the naked eye. In contrast, the crystals of fine-grained rocks such as basalt are too small to be seen, even with a magnifying glass. **Figure 5.2** shows samples of granite and basalt, accompanied by photomicrographs of very thin, transparent slices of each rock. *Photomicrographs,* which are simply photographs taken through a microscope, give us an enlarged view of minerals and their textures. Textural differences were clear to early geologists, but several more clues were needed to unravel the meaning of those differences.

The variation among igneous rocks is due to variations in their texture, which largely reflects crystal size, and variations in their mineral composition.

First Clue: Volcanic Rocks

Early geologists observed volcanic rocks forming from lava during volcanic eruptions. (**Lava** is the term that we apply to magma flowing out onto Earth's surface.) They noted that when lava cooled rapidly, it formed either a fine-grained rock or a glassy one in which no crystals could be distinguished. Where lava cooled more slowly, as in the middle of a thick flow many meters high, somewhat larger crystals were present.

Second Clue: Laboratory Studies of Crystallization

Just over a hundred years ago, experimental scientists began to understand the nature of crystallization. Anyone who has frozen a tray of ice cubes knows that water solidifies to ice in a few hours as its temperature drops below the freezing point. If you have ever attempted to retrieve your ice cubes before they were completely solid, you may have seen thin ice crystals forming at the surface and along the sides of the tray. During crystallization, the water molecules take up fixed positions in the solidifying crystal structure, and they are no longer able to move freely, as they did when the water was liquid. All other liquids, including magmas, crystallize in this way.

The first tiny crystals form a pattern. Other atoms or ions in the crystallizing liquid then attach themselves in such a way that the tiny crystals grow larger. It takes

FIGURE 5.2 Igneous rocks were first classified by texture. Early geologists assessed texture with a small hand-held magnifying glass. Modern geologists have access to high-powered polarizing microscopes, which can produce photomicrographs of thin, transparent rock slices like those shown here. [John Grotzinger/Ramón Rivera-Moret/Harvard Mineralogical Museum. Photomicrographs by Raymond Siever.]

some time for the atoms or ions to "find" their correct places on a growing crystal, so large crystals form only if they have time to grow slowly. If a liquid solidifies very quickly, as a magma does when it erupts onto the cool surface of Earth, the crystals have no time to grow. Instead, a large number of tiny crystals form simultaneously as the liquid cools and solidifies.

Third Clue: Granite—Evidence of Slow Cooling

By studying volcanoes, early geologists determined that finely crystalline textures indicate quick cooling at Earth's surface and that finely crystalline igneous rocks are evidence of former volcanism. But in the absence of direct observation, how could geologists deduce that coarse-grained rocks form by *slow* cooling deep in Earth's interior? Granite—one of the commonest rocks of the continents—turned out to be the crucial clue (**Figure 5.3**). James Hutton, one of geology's founding fathers, saw granite cutting across and disrupting layers of sedimentary rocks as he worked in the field in Scotland. He noticed that the granite had somehow fractured and invaded the sedimentary rocks, as though the granite had been forced into the fractures as a liquid.

As Hutton looked at more and more granites, he began to focus on the sedimentary rocks bordering them. He observed that the minerals of the sedimentary rocks in contact with the granite were different from those found in sedimentary rocks at some distance from the granite. He concluded that the changes in the sedimentary rocks must have resulted from great heat and that the heat must have come from the granite. Hutton also noted that the granite was composed of interlocked crystals (see Figure 5.2). By this time, chemists had established that a slow crystallization process produces this pattern.

With these three lines of evidence, Hutton proposed that granite forms from a hot molten material that solidifies deep within Earth. The evidence was conclusive, because no other explanation could accommodate all the facts. Other geologists, who saw the same characteristics of granites in widely separated places in the world, came to recognize that granite and many similar coarsely crystalline rocks were the products of magma that had crystallized slowly in the interior of Earth.

> The texture of igneous rocks depends on crystal size. Crystal size depends on how fast magma cools—the more slowly it cools, the larger the crystal size in the resulting rock.

Intrusive Igneous Rocks The full significance of an igneous rock's texture is now clear: its texture is linked to the rate, and therefore the place, of cooling. Slow cooling of magma in Earth's interior allows adequate time for

FIGURE 5.3 Granitic intrusions (light-colored) cutting across metamorphosed sedimentary rock. [Tom Bean/DRK PHOTO.]

FIGURE 5.4 Igneous rock types can be identified by texture. [John Grotzinger/Ramón Rivera-Moret/Harvard Mineralogical Museum.]

Extrusive pyroclasts form in violent eruptions from lava thrown high in the air.

Extrusive igneous rocks cool rapidly on Earth's surface and are fine-grained.

Intrusive igneous rocks cool slowly in Earth's interior, allowing large, coarse crystals to form.

Porphyritic crystals start to grow beneath Earth's surface. Some crystals grow large, but the remaining melt cools faster, forming smaller crystals, either because it is erupted to the surface or because it is intruded close to Earth's surface.

Intrusive igneous rocks form by the cooling of magma in Earth's interior. Extrusive igneous rocks form by the cooling of magma erupted onto Earth's surface.

the growth of the interlocking large, coarse crystals that characterize intrusive igneous rocks (**Figure 5.4**). An *intrusive igneous rock* is one that has forced its way into surrounding rock. That surrounding rock is called **country rock.**

Extrusive Igneous Rocks

Rapid cooling at Earth's surface produces the fine-grained texture or glassy appearance of the *extrusive igneous rocks* (see Figure 5.4). These rocks, composed partly or largely of volcanic glass, form when lava or other material erupts from volcanoes. For this reason, they are also known as *volcanic rocks.* They fall into two major categories based on the type of erupted material from which they are formed:

• *Lavas* Volcanic rocks formed from lavas range in appearance from smooth and ropy to sharp, spiky, and jagged, depending on the conditions under which the rocks formed.

• *Pyroclasts* In more violent eruptions, **pyroclasts** form when fragments of lava are thrown high into the air. Pyroclasts range in size from fine particles of volcanic ash to huge boulders. As they fall to the ground and cool, these fragments of volcanic debris may stick together to form rocks.

A **porphyry** is an igneous rock that has a mixed texture in which large crystals "float" in a predominantly fine-grained matrix (see Figure 5.4). The large crystals, called *phenocrysts,* form while the magma is still below Earth's surface. Then, before other crystals can grow, a volcanic eruption brings the magma to the surface, where it cools quickly to a finely crystalline mass. In some cases, porphyries form as intrusive igneous rocks; for example, they may form where magmas cool quickly at very shallow levels in the crust. Porphyry textures are important to geologists because they indicate that different minerals crystallize at different rates.

Later in this chapter, we will examine some special forms of intrusive igneous rocks, and we will look more closely at how extrusive igneous rocks form during volcanism. Now, however, we turn to the second way in which the family of igneous rocks is subdivided.

Extrusive igneous rocks are formed from two kinds of materials, lavas and pyroclasts.

A porphyry is an igneous rock consisting of large crystals dispersed within a finer-grained matrix.

Chemical and Mineral Composition

We have seen that igneous rocks can be subdivided according to their texture. They can also be classified on the basis of their chemical and mineral composition. Volcanic glass, which is formless even under a microscope, is often classified by chemical analysis. One of the earliest classifications of igneous rocks was based on a simple chemical analysis of their silica (SiO_2) content. Silica is abundant in most igneous rocks, as noted in Chapter 4, and accounts for 40 to 70 percent of their total weight. We still refer to rocks rich in silica, such as granite, as *silicic*.

Modern classifications now group igneous rocks according to their relative proportions of silicate minerals (see Table 5.1 and Appendix 5). The silicate minerals—quartz, feldspar (both orthoclase and plagioclase), muscovite and biotite micas, the amphibole and pyroxene groups, and olivine—form a systematic series. *Felsic* minerals are the highest in silica; *mafic* minerals are the lowest in silica. The adjectives *felsic* (from *fel*dspar and *sili*ca) and *mafic* (from *ma*gnesium and *f*er*ric*, from the Latin *ferrum*, "iron") are applied to both the minerals and the rocks that have high contents of these minerals. Mafic minerals crystallize at higher temperatures—that is, earlier in the cooling of a magma—than felsic minerals.

As the mineral and chemical compositions of igneous rocks became known, geologists soon noticed that some extrusive and intrusive rocks were identical in composition and differed only in texture. Basalt, for example, is an extrusive rock formed from lava. Gabbro has exactly the same mineral composition as basalt but forms deep in Earth's crust (see Figure 5.4). Similarly, rhyolite and granite are identical in composition but differ in texture. Thus, extrusive and intrusive rocks form two chemically and mineralogically parallel sets of igneous rocks. Conversely, most chemical and mineral compositions can appear in either extrusive or intrusive rocks. The only exceptions are very highly mafic rocks, which rarely appear as extrusive igneous rocks.

Figure 5.5 is a model that portrays these relationships. The horizontal axis plots silica content as a percentage of a given rock's weight. The per-

> Igneous rocks can be classified by the proportions of silica and silicate minerals they contain.

> Extrusive and intrusive igneous rocks may have the same chemical composition and differ only in texture.

TABLE 5.1 Common Minerals of Igneous Rocks

Compositional Group	Mineral	Chemical Composition	Silicate Structure
FELSIC	Quartz	SiO_2	Frameworks
	Potassium feldspar	$KAlSi_3O_8$	
	Plagioclase feldspar	$NaAlSi_3O_8$; $CaAl_2Si_2O_8$	
	Muscovite (mica)	$KAl_3Si_3O_{10}(OH)_2$	Sheets
MAFIC	Biotite (mica)	K, Mg, Fe, Al } $Si_3O_{10}(OH)_2$	
	Amphibole group	Mg, Fe, Ca, Na } $Si_8O_{22}(OH)_2$	Double chains
	Pyroxene group	Mg, Fe, Ca, Al } SiO_3	Single chains
	Olivine	$(Mg,Fe)_2SiO_4$	Isolated tetrahedra

FIGURE 5.5 Classification model of igneous rocks. The vertical axis shows the minerals contained in a given rock as a percentage of its volume. The horizontal axis includes a scale of silica content by weight. Thus, if you knew by chemical analysis that a coarsely textured rock sample was about 70 percent silica, you could deduce that its composition was about 6 percent amphibole, 3 percent biotite, 5 percent muscovite, 14 percent plagioclase feldspar, 22 percent quartz, and 50 percent orthoclase feldspar. Your rock would be granite. Although rhyolite has the same mineral composition, its fine texture would eliminate it from consideration.

centages given—from high silica content at 70 percent to low silica content at 40 percent—cover the range found in igneous rocks. The vertical axis displays a scale measuring the minerals contained in a given rock as a percentage of its volume. If you know the silica content of a rock sample, you can determine its mineral composition and, from that, the type of rock.

We can use Figure 5.5 to guide our discussion of the intrusive and extrusive igneous rocks. We begin with the felsic rocks, on the extreme left of the model.

Felsic Rocks Felsic rocks are poor in iron and magnesium and rich in minerals that are high in silica. Such minerals include quartz, orthoclase feldspar, and plagioclase feldspar. Orthoclase feldspar is more abundant than plagioclase feldspar. Plagioclase feldspars contain both calcium and sodium. As Figure 5.5 indicates, they are richer in sodium near the felsic end and richer in calcium near the mafic end. Thus, just as mafic minerals crystallize at higher temperatures than felsic minerals, calcium-rich plagioclases crystallize at higher temperatures than sodium-rich plagioclases.

Felsic rocks tend to be light in color. **Granite,** one of the most abundant intrusive igneous rocks, contains about 70 percent silica. Its mineral composition includes abundant quartz and orthoclase feldspar and a smaller amount of plagioclase feldspar. These light-colored felsic minerals give granite its pink or gray color. Granite also contains small amounts of muscovite and biotite micas and amphibole.

Rhyolite is the extrusive equivalent of granite. This light brown to gray rock has the same felsic composition and light coloration as granite, but it is much more fine-grained. Many rhyolites are composed largely or entirely of volcanic glass.

Intermediate Igneous Rocks Midway between the felsic and mafic ends of the series are the **intermediate igneous rocks.** As their name indicates, these rocks are neither as rich in silica as the felsic rocks nor as poor in it as the mafic rocks. We find the intermediate intrusive rocks to the right of granite in Figure 5.5. The first is **granodiorite,** a light-colored rock that looks something like granite. It is also similar to granite in having abundant quartz, but its predominant feldspar is plagioclase, not orthoclase. To its right is **diorite,** which contains still less silica and is dominated by plagioclase feldspar, with little or no quartz. Diorites contain a moderate amount of the mafic minerals biotite, amphibole, and pyroxene. They tend to be darker than granite or granodiorite.

Felsic rocks are rich in minerals with a high silica content. They include granite, an intrusive rock, and rhyolite, an extrusive rock.

Intermediate igneous rocks, which fall between the felsic and mafic rocks in silica content, include intrusive rocks called granodiorite and diorite and extrusive rocks called dacite and andesite.

The volcanic equivalent of granodiorite is **dacite.** To its right in the extrusive series is **andesite,** the volcanic equivalent of diorite. Andesite derives its name from the Andes, the volcanic mountain chain in South America.

Mafic Rocks **Mafic rocks** contain large proportions of pyroxenes and olivines. These minerals are relatively poor in silica but rich in magnesium and iron, from which they get their characteristic dark colors. **Gabbro,** with even less silica than is found in the intermediate igneous rocks, is a coarse-grained, dark gray intrusive igneous rock. Gabbro has an abundance of mafic minerals, especially pyroxene. It contains no quartz and only moderate amounts of calcium-rich plagioclase feldspar.

Basalt, as we have seen, is dark gray to black and is the fine-grained extrusive equivalent of gabbro. Basalt is the most abundant igneous rock of the crust, and it underlies virtually the entire seafloor. In some places on the continents, extensive thick sheets of basalt make up large plateaus. The Columbia River basalts of Washington State and the remarkable formation known as the Giant's Causeway in Northern Ireland are two examples. The Deccan basalts of India and the Siberian basalts of northern Russia are enormous outpourings of basalt that appear to coincide closely with two of the greatest periods of mass extinction in the fossil record.

> Mafic rocks have a low silica content. They include gabbro, an intrusive rock, and basalt, an extrusive rock.

Ultramafic Rocks **Ultramafic rocks** consist primarily of mafic minerals and contain less than 10 percent feldspar. Here, at the far right of Figure 5.5, with a silica content of only about 45 percent, we find **peridotite,** a coarse-grained, dark greenish gray rock made up primarily of olivine with smaller amounts of pyroxene. Peridotites are the dominant rocks in Earth's mantle, and as we will see, they are the source of the basaltic magmas that form rocks at mid-ocean ridges. Ultramafic rocks are rarely found as extrusives. Because they form at such high temperatures, usually through the accumulation of crystals at the bottom of a magma chamber, they are rarely liquid and hence do not form typical lavas.

The names and exact compositions of the various rocks in the felsic-to-mafic series are less important than the trends shown in Table 5.2. There is a strong correlation between a rock's mineralogy and its temperature of crystallization or melting. As Table 5.2 indicates, mafic minerals melt at higher temperatures than felsic minerals. At temperatures below the melting point, minerals crystallize. Therefore, mafic minerals also crystallize at higher temperatures than felsic minerals. We can also see from the table that silica content increases as we move from the mafic group to the felsic group. Increasing silica content results in increasingly complex silicate structures (see Table 5.1), which interfere with a melted rock's ability to flow. As a structure grows more complex, its ability to flow decreases. Thus, **viscosity**—the measure of a liquid's *resistance* to flow—increases as silica content increases. Viscosity is an important factor in the behavior of lavas.

TABLE 5.2 Changes in Some Major Chemical Elements from Felsic to Mafic Rocks

	Felsic	Intermediate		Mafic
Coarse-Grained (intrusive)	Granite	Granodiorite	Diorite	Gabbro
Fine-Grained (extrusive)	Rhyolite	Dacite	Andesite	Basalt

← Silica increasing
← Sodium increasing
← Potassium increasing
Calcium increasing →
Magnesium increasing →
Iron increasing →

← (Viscosity increasing)

(Melting temperature increasing) →

Ultramafic rocks have a very low silica content and include peridotite, an intrusive rock. Extrusive ultramafic rocks are very rare.

It is clear that a rock's minerals are an indication of the conditions under which the rock's parent magma formed and crystallized. To interpret this information accurately, however, we must understand more about igneous processes. We turn to that topic next.

HOW DO MAGMAS FORM?

We know from the way Earth transmits seismic waves that the bulk of the planet is solid for thousands of kilometers down to the core-mantle boundary (see Chapter 2). The evidence of volcanic eruptions, however, tells us that there must also be liquid regions where magmas originate. How do we resolve this apparent contradiction? The answer lies in the processes that melt rocks and create magmas.

Although we do not yet understand the exact mechanisms of melting and solidification within Earth, geologists have learned a great deal from laboratory experiments using high-temperature furnaces. From these experiments, we know that a rock's melting point depends on its composition and on conditions of temperature and pressure (Table 5.3).

Magmas form where the temperature is high enough to melt rock.

TABLE 5.3 Factors Affecting Melting Temperatures

tained. The fraction of rock that has melted at a given temperature is called a *partial melt*. To visualize a partial melt, think of how a chocolate chip cookie would look if you heated it to the point at which the chocolate chips melted while the main part of the cookie stayed solid. The chips represent the partial melt, or magma.

The ratio of solid to partial melt depends on the proportions and melting temperatures of the minerals that make up the original rock. It also depends on the temperature at the depth in the crust or mantle where melting takes place. At the lower end of a rock's melting range, a partial melt might be less than 1 percent of the volume of the original rock. Much of the hot rock would still be solid, but appreciable amounts of liquid would be present as small droplets in the tiny spaces between crystals throughout the mass. In the upper mantle, for example, some basaltic partial melts are produced by only 1 to 2 percent melting of peridotite. However, 15 to 20 percent melting of mantle peridotite to form basaltic magmas is common beneath mid-ocean ridges. At the high end of a rock's melting range, much of the rock would be liquid, with lesser amounts of unmelted crystals in it. An example would be a reservoir of a basaltic magma and crystals just beneath a volcano such as the island of Hawaii.

Geologists used this knowledge of partial melts to determine how different kinds of magma form at different temperatures and in different regions of Earth's interior. As you can imagine, the composition of magma formed from rock in which only the minerals with the lowest melting points have melted may be significantly different from the composition of magma formed from completely melted rock. Thus, basaltic magmas that form in different regions of the mantle may have somewhat different compositions.

Melting is not complete at a given temperature. Instead, rocks undergo partial melting because the minerals within the rocks melt at different temperatures.

Temperature and Melting

A hundred years ago, geologists discovered that a rock does not melt completely at a given temperature. Instead, rocks undergo **partial melting** because the minerals that compose them melt at different temperatures. As temperatures rise, some minerals melt and others remain solid. If the same conditions are maintained at any given temperature, the same mixture of solid and melted rock is main-

Pressure and Melting

To get the whole story on melting, we must consider pressure as well as temperature. Pressure increases with depth as a result of the increased weight of overlying rock. Geologists found that as they melted rocks under various pressures in the laboratory, higher pressures led to higher melting temperatures. Thus, rocks that would melt at a given temperature at Earth's surface would remain solid at the same temperature in Earth's interior. For example, a rock that melts at 1000°C at Earth's surface might have a much higher melting temperature, perhaps 1300°C, deep in the interior, where pressures are many thousands of times greater than those at the surface. It is the effect of pressure that explains why the rocks in most of the crust and mantle do not melt. Rock can melt only when both the temperature and pressure conditions are right.

Just as an increase in pressure can keep rock solid, a decrease in pressure can make rock melt, given a suitably high temperature. Because of convection currents, mantle material rises to Earth's surface at mid-ocean ridges at a more or less constant temperature. As the material rises and the pressure decreases below a critical point, solid rock melts spontaneously, without the introduction of any additional heat. This process, known as **decompression melting,** produces the greatest volume of magma anywhere on Earth. It is the process by which most basalts form on the seafloor.

> Higher pressures result in higher melting temperatures. If pressure is decreased at a given temperature, decompression melting occurs.

Water and Melting

The many experiments on melting temperatures and partial melting paid other dividends as well. One of them was a better understanding of the role of water in rock melting. Geologists working on natural lavas in the field determined that water was present in some magmas. This gave them the idea to add water to their experimental melts back in the laboratory. By adding small but differing amounts of water, they discovered that the compositions of partial melts varied not only with temperature and pressure, but also with the amount of water present.

Consider, for example, the effect of dissolved water on pure albite, a sodium-rich plagioclase feldspar, at the low pressures at Earth's surface. If only a small amount of water is present in the rock, it will remain solid at temperatures just over 1000°C, hundreds of degrees above the boiling point of water. At these temperatures, the water in the albite is present as a vapor (gas). If large amounts of water are present, however, the melting temperature of the albite will decrease, dropping to as low as 800°C. This behavior follows the general rule that dissolving one substance (in this case, water vapor) in another (in this case, albite) lowers the melting temperature of the solution. If you live in a cold climate, you are probably familiar with this principle because you know that salt sprinkled on icy roads lowers the melting temperature of the ice. By the same principle, the melting temperature of albite—and of all the feldspars and other silicate minerals—drops considerably in the presence of large amounts of water. Their melting points decrease in proportion to the amount of water dissolved in the molten silicate.

Water content is a significant factor in determining the melting temperatures of mixtures of sedimentary and other rocks. Sedimentary rocks contain an especially large volume of water in their pore spaces, more than is found in igneous or metamorphic rocks.

> The presence of water in rock lowers its melting temperature.

As we will see later in this chapter, the water in sedimentary rocks plays an important role in melting in Earth's interior.

The Formation of Magma Chambers

Most substances are less dense in their liquid form than in their solid form. The density of a melted rock is lower than the density of a solid rock of the same composition. With this knowledge, geologists reasoned that large bodies of magma could form in the following way. If the less dense melt were given a chance to move, it would move upward—just as oil, which is less dense than water, rises to the surface of a mixture of oil and water. Being liquid, the partial melt could move slowly upward through pores and along the boundaries between crystals of the surrounding rock. As the hot drops of molten rock moved upward, they would mix with other drops, gradually forming larger pools of magma within Earth's solid interior.

The rise of magmas through the mantle and crust may be slow or rapid. Magmas rise at rates of 0.3 m/year to almost 50 m/year, over periods of tens of thousands or even hundreds of thousands of years. As they ascend, magmas may mix with other melts and may also melt portions of the crust. We now know that large pools of molten rock form **magma chambers**—magma-filled cavities in the lithosphere that form as rising pools of melted rock push aside surrounding solid rock. A magma chamber may encompass a volume as large as several cubic kilometers. We cannot yet say exactly how magma chambers form, nor what they look like in three dimensions. We think of them as large, liquid-filled cavities in solid rock, which expand as more of the surrounding rock melts or as magma migrates through cracks and other small openings. Magma chambers contract as they expel magma to the surface in eruptions. We know that magma chambers exist because seismic waves have shown us the depth, size, and general outlines of the magma chambers underlying some active volcanoes.

> A magma chamber is a large cavity in the lithosphere that is filled with molten rock.

Temperature and Depth

Our understanding of igneous processes is based mainly on data from two sources. The first is volcanoes, which give us information about where magmas are located. The second source of data is the record of temperatures measured in deep drill holes and mine shafts. This record shows that the temperature of Earth's interior increases with depth. Using these measurements, scientists have been able to estimate the rate at which temperature rises as depth increases.

The temperatures recorded at a given depth in some locations are much higher than the temperatures recorded at the same depth in other locations. These results indicate that some parts of Earth's mantle and crust are hotter than others. For example, the Great Basin of the western United States is an area where the North American continent is being stretched and thinned, with the result that temperature increases with depth at an exceptionally rapid rate, reaching 1000°C at 40 km, not far below the base of the crust. This temperature is almost high enough to melt basalt. By contrast, in tectonically stable regions, such as the central parts of continents, temperature increases much more slowly, reaching only 500°C at the same depth.

We now know that magmas formed by partial melting can vary in their composition. And we know that temperatures in Earth's interior are sufficient to create magmas. Let's turn now to the question of why there are so many different types of igneous rocks.

MAGMATIC DIFFERENTIATION

The processes we've discussed so far account for the melting of rocks to form magmas. But what accounts for the variety of igneous rocks? Are magmas of different chemical compositions made by the melting of different kinds of rocks? Or do magmatic processes produce a variety of rocks from an originally uniform parent material?

Again, the answers to these questions came from laboratory experiments. Geologists mixed chemical elements in proportions that simulated the compositions of natural igneous rocks, then melted these mixtures. As the melts cooled and solidified, the experimenters observed and recorded the temperatures at which crystals formed as well as the chemical compositions of those crystals. This research gave rise to the theory of **magmatic differentiation,** a process by which rocks of varying composition can arise from a uniform parent magma. Magmatic differentiation occurs because different minerals crystallize at different temperatures. During crystallization, the composition of the magma changes as it is depleted of the chemical elements that form the crystallized minerals.

In a kind of mirror image of partial melting, the last minerals to melt are the first minerals to crystallize from a cooling magma. This initial crystallization withdraws chemical elements from the melt, changing the magma's composition. Continued cooling crystallizes the minerals that melted at the next lower temperature range. Again, the magma's chemical composition changes as various elements are withdrawn. Finally, as the magma solidifies completely, the last minerals to crystallize are the ones that melted first. Thus, the same parent magma, because of its changing chemical composition throughout the crystallization process, can give rise to different igneous rocks.

Because different minerals crystallize at different temperatures, the composition of a magma changes progressively as various elements are withdrawn to form crystallized minerals.

Fractional Crystallization

Fractional crystallization is the process by which the crystals formed in a cooling magma are segregated from the remaining liquid. This segregation happens in several ways, following a sequence commonly described as *Bowen's reaction series* (**Figure 5.6**). In the simplest scenario, crystals formed in a magma chamber settle to the chamber's floor and are thus removed from further reaction with the remaining liquid. The magma then migrates to new locations, forming new chambers. Thus, crystals that form early are segregated from the remaining magma, which continues to crystallize as it cools.

Fractional crystallization is the process by which different minerals in a magma crystallize with progressively decreasing temperatures.

The effects of fractional crystallization can be seen in the Palisades, a line of imposing cliffs that faces the city of New York on the west bank of the Hudson River

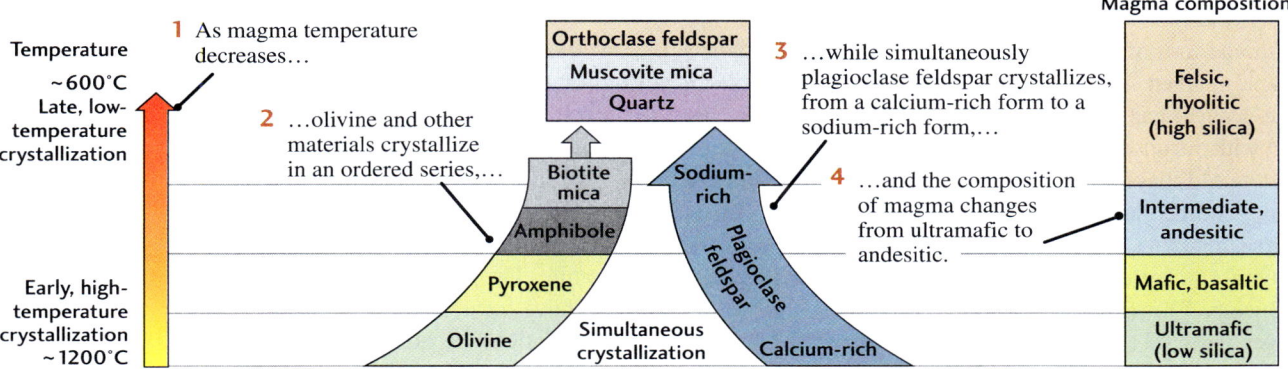

FIGURE 5.6 Bowen's reaction series provides a model of fractional crystallization.

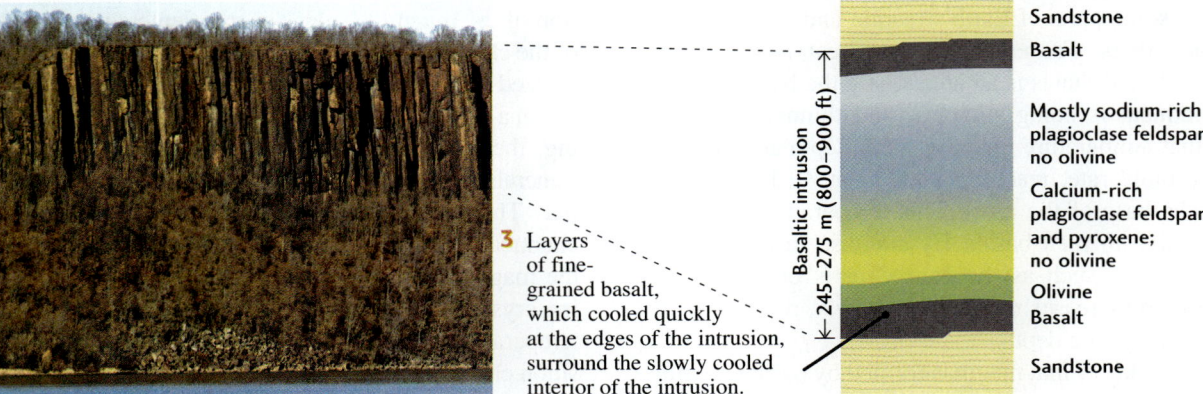

FIGURE 5.7 Fractional crystallization explains the composition of the basaltic intrusion that forms the Palisades. [Zehdreh Allen-Lafayette.]

(Figure 5.7). This igneous formation is about 80 km long and, in places, more than 300 m high. It formed as a magma of basaltic composition intruded into almost horizontal sedimentary rocks. It contains abundant olivine near the bottom, pyroxene and plagioclase feldspar in the middle, and mostly plagioclase feldspar near the top. These variations in mineral composition from bottom to top made the Palisades a perfect site for testing the theory of fractional crystallization. Geologists melted rocks with about the same mineral compositions as those found in the Palisades intrusion and determined that the temperature of the magma had to have been about 1200°C. The parts of the magma within a few meters of the relatively cold sedimentary rocks above and below it cooled quickly. The quick cooling formed a fine-grained basalt and preserved the chemical composition of the original melt. The hot interior cooled more slowly, as evidenced by the slightly larger crystals found in the intrusion's interior.

The theory of fractional crystallization leads us to expect that the first mineral to crystallize from the slowly cooling interior of the Palisades intrusion would have been olivine, as this heavy mineral would sink through the melt to the bottom of the intrusion. It can be found today as a coarse-grained, olivine-rich layer just above the chilled, fine-grained basaltic layer along the bottom zone of contact with the underlying sedimentary rock. Continued cooling would have produced pyroxene crystals, followed almost immediately by calcium-rich plagioclase feldspar. These minerals, too, would have settled out through the magma and accumulated in the lower third of the Palisades intrusion. The abundance of plagioclase feldspar in the upper parts of the intrusion is evidence that the magma continued to change composition until successive layers of settled crystals were topped off by a layer of mostly sodium-rich plagioclase feldspar crystals.

Granite from Basalt: Complexities of Magmatic Differentiation

Studies of the lavas of volcanoes have shown that basaltic magmas are common—far more common than the rhyolitic magmas that correspond in composition to granites. How, then, could the granites have become so abundant in the crust?

The original theory of magmatic differentiation suggested that a basaltic magma would gradually cool and differentiate into a cooler, more silicic melt. The early stages of this differentiation would produce andesitic magma, which might erupt to form andesitic lavas or solidify by slow crystallization to form dioritic intrusives. Intermediate stages would make magmas of granodiorite composition. If this process were carried far enough, its late stages would form rhyolitic lavas and granitic intrusions (see Figure 5.6).

We have since learned that magmatic differentiation is a more complex process than had been originally described. One line of research showed that so much time would be needed for small crystals of olivine to settle through a dense, viscous magma that they might never reach the bottom of a magma chamber. Other researchers demonstrated that many layered intrusions—similar to but much larger than the Palisades—do not show the simple progression of layers predicted by the original theory.

The first sticking point, however, was the source of granite. The great volume of granite found on Earth could not have formed from basaltic magmas by magmatic differentiation, because large quantities of liquid volume would be lost by crystallization during successive stages of differentiation. To produce a given amount of granite, an initial volume of basaltic magma 10 times the size of the granitic intrusion would be required. That observation would imply the crystallization of huge quantities of basalt underlying granitic intrusions. But geologists could not find anything like that amount of basalt. Even where great volumes of basalt are found—at mid-ocean ridges—there is no wholesale conversion into granite through magmatic differentiation.

Most in question is the original idea that all granitic rocks are formed from the differentiation of a single type of magma, a basaltic melt. Instead, geologists now believe that that the melting of varied source rocks of the upper mantle and crust is responsible for much of the observed variation in the composition of magmas:

1. Rocks in the upper mantle might undergo partial melting to produce basaltic magmas.

2. Mixtures of sedimentary rocks and basaltic oceanic rocks, such as those found in subduction zones, might melt to form andesitic magmas.

3. Mixtures of sedimentary, igneous, and metamorphic continental crustal rocks might melt to produce granitic magmas.

Thus, the mechanisms of magmatic differentiation are much more complex than first recognized:

- Magmatic differentiation can be achieved by the partial melting of mantle and crustal rocks over a range of temperatures and water contents.
- Magmas do not cool uniformly; they may exist transiently at a range of temperatures within a magma chamber. Differences in temperature within and among magma chambers may cause the chemical composition of the magma to vary from one region to another.
- A few magmas are *immiscible*—they do not mix with one another, just as oil and water do not mix. When such magmas coexist in one magma chamber, each forms its own crystallization products. Magmas that are *miscible*—that *do* mix—may follow a crystallization path different from that followed by any one magma alone.

FORMS OF IGNEOUS INTRUSIONS

As noted earlier, we cannot directly observe the shapes of the igneous intrusions formed when magmas intrude into the crust. We can deduce their shapes and distributions only by observing them today where intrusive rocks have been uplifted and exposed by erosion, millions of years after the magma intruded and cooled.

We do have indirect evidence of current magmatic activity. Seismic waves, for example, show us the general outlines of the magma chambers that underlie some active volcanoes. But they cannot reveal the detailed shapes or sizes of intrusions arising from those magma chambers. In some nonvolcanic but tectonically active regions, such as an area near the Salton Sea in southern California, measurements of temperatures in deep drill holes reveal a crust much hotter than normal, which may be evidence of an intrusion nearby.

But in the end, most of what we know about igneous intrusions is based on the work of field geologists who have examined and compared a wide variety of outcrops and have reconstructed their history. In the following pages, we consider some of these forms: plutons, sills and dikes, and veins. **Figure 5.8** illustrates a variety of extrusive and intrusive structures.

The most basic forms of igneous intrusions are plutons, sills and dikes, and veins.

FIGURE 5.8 The basic forms of extrusive and intrusive igneous structures.

Dikes cut across layers of country rock…

…but sills run parallel to them.

Batholiths are the largest forms of plutons, covering at least 100 km².

Rising magma wedges open and fractures overlying country rock.

The overlying rock may bow up.

The magma melts surrounding rock,…

…which mixes with the magma and changes its composition.

The magma also breaks off blocks of overlying rock— xenoliths— that sink into the magma.

FIGURE 5.9 Magmas make their way into country rock in three basic ways: by invading cracks and wedging open overlying rock, by breaking off pieces of rock, and by melting country rock. Pieces of broken-off country rock, called xenoliths, can become completely dissolved in the magma. If many xenoliths are dissolved and the country rock differs in composition from the magma, the composition of the magma will change.

Plutons

Plutons are large igneous bodies formed deep in Earth's crust. They range in size from a cubic kilometer to hundreds of cubic kilometers. We can study these large bodies when uplift and erosion uncover them or when mines or drill holes cut into them. Plutons are highly variable, not only in size but also in shape and in their relationship to the country rock.

This wide variation is due in part to the different ways in which magma makes space for itself as it rises through the crust. Most plutons intrude at depths greater than 8 to 10 km. At these depths, few holes or openings exist because the pressure of the overlying rock would close them. But the upwelling magma overcomes even that great pressure.

Magma rising through the crust makes space for itself in three ways (**Figure 5.9**) that may be referred to collectively as *magmatic stoping:*

1. *Wedging open the overlying rock.* As the rising magma lifts the great weight of the overlying rock, it fractures that rock, penetrates the cracks, wedges them open, and so flows into the rock. Overlying rocks may bow upward during this process.

2. *Breaking off large blocks of rock.* Magma can push its way upward by breaking off blocks of the invaded crust. These blocks, known as *xenoliths,* sink into the magma, melt, and blend into the liquid, in some places changing the composition of the magma.

3. *Melting surrounding rock.* Magma also makes its way by melting country rock.

Most plutons show sharp zones of contact with country rock and other evidence of the intrusion of a liquid magma into solid rock. Other plutons grade into country rock and contain structures vaguely resembling those of sedimentary rocks. The features of these plutons suggest that they formed by partial or complete melting of preexisting sedimentary rocks.

Batholiths, the largest plutons, are great irregular masses of coarse-grained igneous rock that by definition cover at least 100 km^2 (see Figure 5.8). Batholiths are thick, horizontal, sheetlike or lobe-shaped bodies extending from a funnel-shaped central region. Their bottoms may extend 10 to 15 km deep, and a few are estimated to go even deeper. The coarse grains of batholiths result from slow cooling at great depths. Other, smaller plutons are called **stocks.** Both batholiths and stocks cut across the layers of the country rock that they intrude.

Sills and Dikes

Sills and dikes are similar to plutons in many ways, but they are smaller and have a different relationship to the layering of the country rock. A **sill** is a sheetlike body formed by the injection of magma between parallel layers of preexisting bedded rock (**Figure 5.10**). The boundaries of sills lie parallel to those layers, whether or not the layers are horizontal. Sills range in thickness from a single centimeter to hundreds of meters, and they can extend over considerable areas. Figure 5.10a shows a large sill at Finger Mountain, Antarctica. The 300-m-thick Palisades intrusion (see Figure 5.7) is another large sill.

Sills may superficially resemble layers of lava flows and pyroclastic deposits, but they differ from those layers in four ways:

Plutons are large igneous bodies formed deep within Earth's crust. They create space for themselves by wedging open, melting, and breaking off pieces of country rock.

FIGURE 5.10 (a) At Finger Mountain, situated in the Dry Valleys of Antarctica, sandstone beds are split by sills parallel to the bedding. [Colin Monteath/AUSCAPE.] (b) A dike of igneous rock (dark) intrudes into shaley sedimentary rock (reddish brown) in Grand Canyon National Park, Arizona. [Tom Bean/DRK PHOTO.]

1. They lack the ropy, blocky, and vesicle-filled structures that characterize many volcanic rocks.

2. They are more coarse-grained than volcanics because they have cooled more slowly.

3. Rocks above and below sills show the effects of heating: their color may have been changed, or their mineral composition may have been altered by contact metamorphism.

4. Many lava flows overlie weathered older flows or soils formed between successive flows; sills do not.

Dikes are the major route of magma transport in the crust. They are like sills in being sheetlike igneous bodies, but dikes cut across the layers in bedded country rock (see Figure 5.10b). Dikes sometimes form by forcing open existing fractures, but more often they create channels through new cracks opened by the pressure of rising magma. Some individual dikes can be followed in the field for tens of kilometers. Their thicknesses vary from many meters to a few centimeters. In some dikes, xenoliths provide evidence of disruption of the surrounding rock during the intrusion process. Dikes rarely exist alone; more typically, swarms of hundreds or thousands of dikes are found in a region that has been deformed by a large igneous intrusion.

The texture of dikes and sills varies. Many are coarse-grained, with an appearance typical of intrusive rocks. Many others are fine-grained and look much more like volcanic rocks. Because we know that texture corresponds to the rate of cooling, we can conclude that the fine-grained dikes and sills invaded country rock nearer Earth's surface, where the country rock was cold compared with the intrusions. Their fine texture is the result of fast cooling. The coarse-grained ones formed at depths of many kilometers and invaded warmer rocks whose temperatures were much closer to their own.

Sills are small intrusive bodies formed by the injection of magma between layers of bedded country rock. Dikes are small intrusive bodies that cut across layers of country rock.

Veins

Veins are deposits of minerals found within a rock fracture that are foreign to the country rock. Irregular pencil- or sheet-shaped veins branch off from the tops and sides of many intrusive bodies. They may be a few millimeters to several meters across, and they tend to be tens of meters to kilometers long or wide. The formation of veins is described in more detail in Chapter 4. The well-known Mother Lode of the Gold Rush of 1849 in California is a vein of gold-bearing quartz crystals. Veins of extremely coarse grained granite cutting across a much finer grained country rock are called **pegmatites** (**Figure 5.11**). Pegmatites crystallize from a water-rich magma in the late stages of solidification.

FIGURE 5.11 A granite pegmatite vein. Note the large crystals. [John Grotzinger/Ramón Rivera-Moret/Harvard Mineralogical Museum.]

MAGMATIC PROCESSES: Mid-Ocean Ridges, Subduction Zones, and Mantle Plumes

Geologists have observed that the facts and theories of igneous rock formation fit nicely into a framework based on plate tectonic theory. Batholiths, for example, are found in the cores of many mountain ranges formed by the con-

FIGURE 5.12 Idealized section of an ophiolite suite. The combination of deep-sea sediments, submarine pillow lavas, sheeted basaltic dikes, and mafic igneous intrusions indicates a deep-sea origin. Ophiolite suites are fragments of ocean lithosphere emplaced on a continent as a result of plate collisions. Peridotite—a dominant rock in the mantle—undergoes decompression melting to form gabbro, which is then erupted to form volcanic pillow lavas (see Figure 5.13). [Photos courtesy of John Grotzinger. Thin sections courtesy of T. L. Grove.]

Magma forms most abundantly at two types of plate boundaries: mid-ocean ridges and subduction zones.

vergence of two plates. This observation implies a connection between pluton formation and the mountain-building process and between both of them and plate tectonics. The geometry of plate motions is the link we need to tie tectonic activity to magmatic processes (see Figure 5.1).

Magma forms most abundantly at two types of plate boundaries: mid-ocean ridges, where two plates diverge and the seafloor spreads, and subduction zones, where one plate dives beneath another. Mantle plumes, though not associated with plate boundaries, are also the result of partial melting and form near the core-mantle boundary deep within Earth's interior.

Spreading Centers as Magma Factories

Most igneous rocks are formed by the globally extensive network of mid-ocean ridges. Each year, approximately 3 km^3 of basaltic lava erupts along the mid-ocean ridges in the process of seafloor spreading. This is a truly enormous volume. In comparison, all the active volcanoes along convergent plate boundaries (about 400) generate volcanic rock at a rate of less than 1 km^3/year. Enough magma has erupted in seafloor spreading over the past 200 million years to create all of the present-day seafloor, which covers nearly two-thirds of Earth's surface. Throughout the mid-ocean ridge network, decompression melting of the mantle creates magmas that well up along rising convection currents and are then erupted onto the seafloor.

Ophiolite suites (**Figure 5.12**) are pieces of oceanic crust that are found on land. These rock assemblages consist of deep-sea sediments, submarine basaltic lavas, and mafic igneous intrusions. They were created by seafloor spreading and then raised above sea level and thrust onto a continent in a later episode of plate collision. Ophiolite suites, along with information gleaned from ocean drilling and sound-wave profiling, tell us what may be happening in sub-seafloor magma chambers.

How does seafloor spreading work? We can think of a spreading center as a huge factory that processes mantle material to produce oceanic crust. **Figure 5.13** is a

Seafloor spreading at mid-ocean ridges generates tremendous volumes of basaltic magma. Peridotite rising from the mantle undergoes decompression, leading to partial melting and the formation of basaltic magmas. These magmas then crystallize to form oceanic crust.

FIGURE 5.13 Decompression melting creates magma at seafloor spreading centers.

highly simplified representation of a seafloor spreading center. The raw material fed into this factory is peridotite that comes from the asthenosphere. The mineral composition of the average peridotite in the mantle is chiefly olivine, with smaller amounts of pyroxene and garnet. Temperatures in the asthenosphere are hot enough to melt a small fraction of this peridotite (less than 1 percent), but not hot enough to generate substantial volumes of magma..

Decompression melting is the process that generates great volumes of magma from peridotite. Recall from earlier in this chapter that the melting temperature of a mineral depends on the pressure at which it melts as well as its composition: decreasing the pressure generally lowers a mineral's melting temperature. Consequently, if a mineral is near its melting point and the pressure on it is lowered while its temperature is kept constant, the mineral will melt. As the plates pull apart, the partially molten peridotite is sucked inward and upward toward the spreading center; the decrease in pressure as the peridotite rises melts a large fraction of the rock (up to 15 percent).

The peridotites subjected to this process do not melt evenly; the garnet and pyroxene minerals they contain melt at a lower temperature than the olivine. For this reason, the magma generated by decompression melting is not peridotitic in composition; rather, it is enriched in silica and iron and has the same composition as basalt (see Figure 5.13). This basaltic melt forms a magma chamber below the mid-ocean ridge crest, from which it separates into three layers:

1. Some magma rises through the narrow cracks that open where the plates are separating and erupts into the ocean, forming the basaltic pillow lavas that cover the seafloor (see Figure 5.13).

2. Some magma freezes in the cracks as vertical, sheeted dikes of gabbro.

3. The remaining magma solidifies as massive intrusions of gabbro as the underlying magma chamber is pulled apart by seafloor spreading.

These igneous units—pillow lavas, sheeted dikes, and massive gabbros—are the basic layers of the crust that geologists have found throughout the world's oceans. Seafloor spreading results in another layer beneath this oceanic crust: the residual peridotite from which the basaltic magma was originally derived. Above the pillow lavas, a thin blanket of deep-sea sediments begins to cover the newly formed oceanic crust. As the seafloor spreads, these layers of sediments, lavas, dikes, and gabbros are transported away from the mid-ocean ridge where this characteristic sequence of rocks is assembled—almost like a production line.

Subduction Zones as Magma Factories

Other kinds of magmas underlie regions in which volcanoes are highly concentrated, such as the Andes mountain range of South America and the Aleutian Islands of Alaska. Both of these regions lie over subduction zones, where one plate sinks under another. Subduction zones are also major magma factories (Figure 5.14). They generate magmas of varying composition, depending on how much and what kinds of materials are subducted. Where subduction takes place beneath a continent, the resulting volcanoes and volcanic rocks form a mountainous volcanic belt on land, such as the Andes, which mark the subduction of the oceanic Nazca Plate beneath continental South America. Similarly, subduction of the small Juan de Fuca Plate beneath western North America has generated the Cascade Range, with its active volcanoes, in northern California, Oregon, and Washington. Where one oceanic plate is subducted under another, a deep-sea trench and a volcanic island arc are formed.

> Subduction zones generate magmas of varying composition.

When an oceanic plate collides with and dives beneath another oceanic plate, several processes are set in motion that cause **fluid-induced melting.** The fluid involved is primarily water, which, as we have seen, lowers the melting temperature of rock. By the time oceanic lithosphere is subducted at a convergent plate boundary, a lot of water has been incorporated into its outer layers. Some of the seawater that circulates through the crust near a spreading center reacts with basalt to form new minerals with water bound into their structures. In addition, as the lithosphere ages and moves across the ocean basin, sediments containing water are deposited on its surface. Some of these sediments get scraped off at the deep-sea trench where the plate subducts, but much of this water-laden material is carried downward into the subduction zone.

As the lithosphere moves downward, it is subjected to increasing pressure. Water is squeezed out of the minerals in the outer layers of the descending crust and rises buoyantly into the mantle wedge above the crust. At depths greater than about 5 km, the temperature increases, and the remaining water is released by metamorphic chemical reactions. All of this water then causes melting of the descending basalt-rich oceanic crust and the overlying peridotite-rich mantle wedge. Most of the resulting mafic magma accumulates at the base of the crust of the overriding plate, and some of it intrudes into the

> Subduction of water-rich oceanic lithosphere leads to fluid-induced melting of the subducting lithosphere.

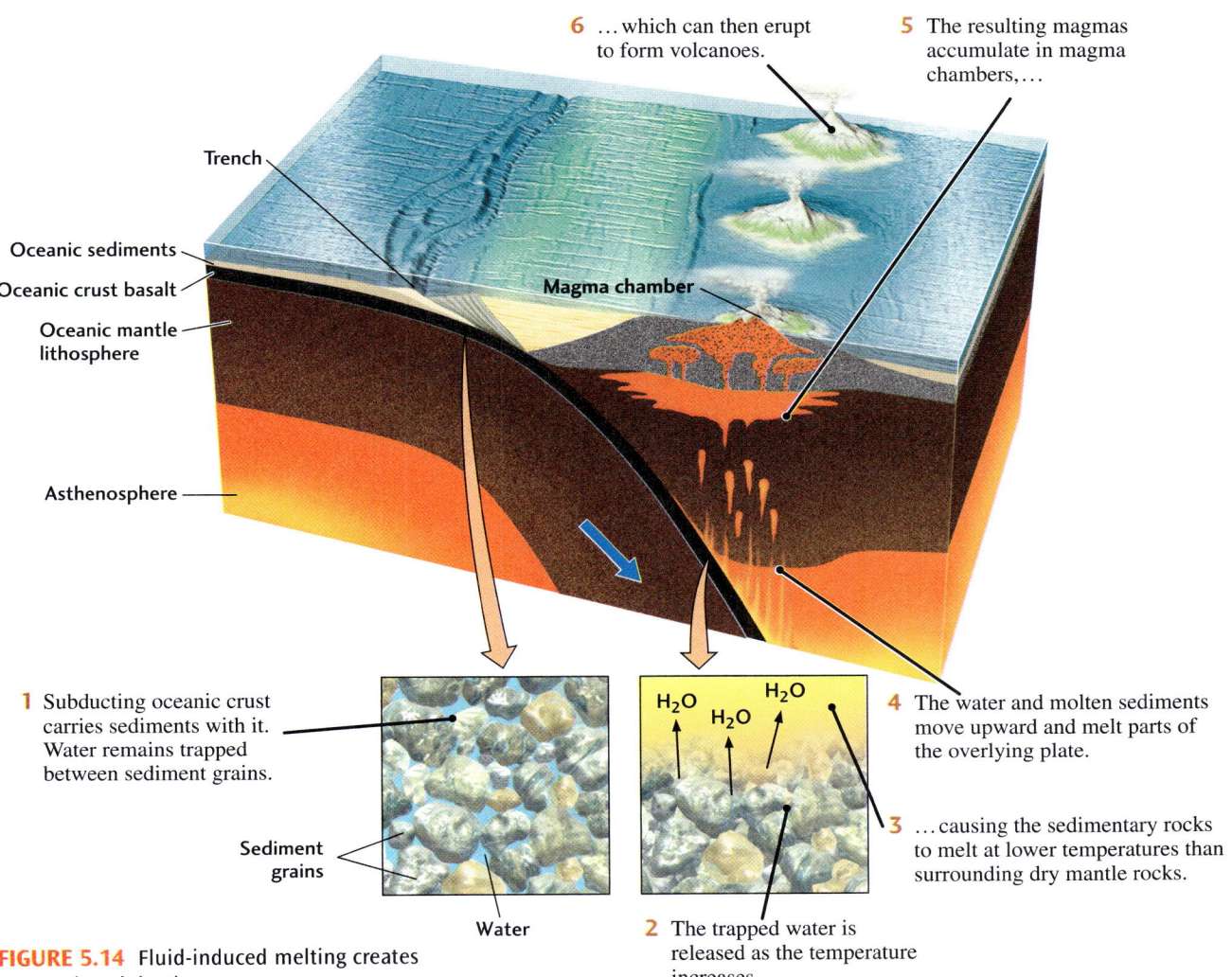

FIGURE 5.14 Fluid-induced melting creates magma in subduction zones.

crust to form magma chambers, which results in the formation of a volcanic island arc. When an oceanic plate is subducted beneath a continental plate, a similar process gives rise to a belt of volcanoes.

The magmas produced by this type of fluid-induced melting are essentially basaltic, although their composition is more variable than that of mid-ocean ridge basalts. The composition of the magmas is further altered during their residence in the crust. Within the magma chambers, the process of fractional crystallization increases the magma's silica content, producing eruptions of andesitic lavas. Where the overlying plate is continental, the heat from the magmas can melt the felsic rocks in the crust, forming magmas with an even higher silica content, such as dacitic and rhyolitic magmas (see Table 5.2). The contribution of lithospheric fluids to these magmas is inferred because trace elements known to be present in oceanic crust and sediments are found in the magmas.

Mantle Plumes as Magma Factories

We learned earlier that the magmas at mid-ocean ridges are formed as a result of decompression melting. This process is also important in the formation of **mantle plumes**. Mantle plumes, like spreading centers, are sites of decompression, but they differ from spreading centers by forming within lithospheric plates rather than along the margins of plates. Basalts similar to those produced at mid-ocean ridges are found in thick accumulations over some parts of continents distant from plate boundaries. In the states of Washington, Oregon, and Idaho, the Columbia and Snake rivers flow over a large area covered by this kind of basalt, which solidified from lavas that flowed out millions of

> Mantle plumes produce hot spots where basaltic magmas form by decompression melting. The movement of plates over these hot spots may produce a trail of volcanoes.

FIGURE 5.15 Plate motion generates a trail of progressively older volcanoes. (a) The Hawaiian Island chain and its extension into the northwestern Pacific reveal the northwestward movement of the Pacific Plate. (b) The Yellowstone volcanic track marks the movement of the North American Plate over a hot spot during the past 16 million years. [Wheeling Jesuit University/NASA Classroom of the Future.]

years ago. Large quantities of basalt are also erupted in isolated volcanic islands far from plate boundaries, such as the Hawaiian Islands. In such places, slender, pencil-like plumes of hot mantle rise from deep within Earth, perhaps as deep as the core-mantle boundary.

Mantle plumes that reach Earth's surface and form **hot spots** are responsible for these outpourings of basalt. The basalt is produced by decompression melting of the mantle. When the peridotites brought up toward the surface in a mantle plume reach low pressures at shallow depths, they begin to melt, producing basaltic magma. The magma eventually penetrates the lithosphere and erupts at the surface. The current position of the plate over the hot spot is marked by an active volcano, which becomes inactive as plate movement pushes it away from the hot spot. Plate motion thus generates a trail of extinct, progressively older volcanoes. As shown in **Figure 5.15**, the Hawaiian Islands fit this pattern well, revealing a rate of movement of the Pacific Plate over the Hawaiian hot spot of about 100 mm/year. The Yellowstone volcano is the product of a hot spot beneath a continental plate.

VOLCANIC PROCESSES AND MATERIALS

Volcanoes and volcanic rocks deserve special attention because they provide geologists with portals into the igneous processes that operate deep within Earth, which might not otherwise be visible. The natural processes that give rise to volcanoes and volcanic rocks are known collectively as *volcanism*.

A **volcano** consists of a magma chamber, to which magma is transported from deep within Earth; a plumbing system that transports the magma to Earth's surface; one or more vents at the surface of Earth; and often a hill or mountain constructed from the accumulation of lava and pyroclasts erupted from the vents. Volcanoes can be viewed as factories that process raw materials (magmas) and transport the end products (lavas and pyroclasts) to the surface.

As we have seen, the magmas that feed volcanoes vary in their chemical composition. Some volcanoes are fed by multiple magma chambers with different compositions, leading to volcanoes made up of stacked deposits of different compositions. The shape of a volcano is controlled both by magma composition and by the eruption process itself. For example, basaltic magmas tend to produce volcanoes with broad, gently sloping forms, whereas andesitic and rhyolitic magmas tend to form volcanoes that have narrower, more conical forms. Volcanoes produced by andesitic and rhyolitic magmas are also more likely to explode, which can dramatically modify their shapes.

> *Volcanoes are formed by the transport of magmas from deep within Earth to magma chambers closer to the surface. Eruptions then transport the magmas to Earth's surface.*

Volcanic Processes

In most cases, the magmas that rise to Earth's surface through volcanoes have been generated by partial melting in the asthenosphere, but in some cases they result from partial melting of the lithosphere. The starting composition of the magma has an important effect on the final composition of the lava that is erupted at Earth's surface. However, chemical processing in magma chambers also has an important effect on lava composition.

Transport of Magmas to Magma Chambers A volcano can be thought of as a plumbing system through which magma is transported to the surface (**Figure 5.16**). Magmas rising buoyantly through the lithosphere pool together in a magma chamber, usually at shallow depths in the crust. This chamber periodically empties to the surface through a pipelike central vent in repeated cycles of eruptions. Lava can also erupt from other vents on the flanks of a volcano or from vertical cracks (fissures).

Chemical Modification of Magmas Volcanoes can also be thought of as chemical factories that process

5 …and gases are injected into the atmosphere.

4 Lavas accumulate on the surface to form a volcano,…

3 Lavas erupt from the magma chamber through a central vent and side vents.

2 …rises through the lithosphere to form a crustal magma chamber.

1 Magma, which originates in the asthenosphere,…

FIGURE 5.16 Volcanoes transport magma from Earth's interior to its surface, where rocks are formed and gases are injected into the atmosphere.

FIGURE 5.17 The active volcanoes of the world with vents on land or above the ocean surface are represented by red dots. Black lines represent plate boundaries. Not shown on this map are the numerous volcanoes of the mid-ocean ridge system below the water's surface.

magmas to produce lavas. As we have seen, a magma gains chemical components as it melts the surrounding rocks while rising through the lithosphere. It loses other components when crystals settle out along the way or sink to the bottom of shallow magma chambers. And its gaseous constituents escape into the atmosphere or ocean as it erupts at the surface. These changes in composition result in changes in the physical properties of the lava, such as its viscosity, which in turn create the variation in the types of volcanoes that we see on Earth.

Eruption of Lavas and Formation of Volcanic Rocks

Volcanic rocks form when magmas erupt at Earth's surface as lavas. As we will learn below, lavas of different composition produce different landforms. The differences depend on the chemical composition, gas content, and temperature of lavas. The higher the silica content and the lower the temperature, for example, the more viscous the lava is, and the more slowly it moves. The more gas a lava contains, the more violent its eruption is likely to be.

Figure 5.17 shows the locations of the world's active volcanoes that occur on land or above the ocean's surface. About 80 percent are found at convergent plate boundaries, 15 percent at divergent plate boundaries, and the remaining few within plate interiors. There are many more active volcanoes than shown on this map, however. Most of the lava erupted on Earth's surface comes from vents beneath the sea, located on the spreading centers of the mid-ocean ridges.

Types of Lava

Erupted lavas usually solidify into one of three major types of igneous rock: basalts, andesites, and rhyolites.

> Lavas can be classified into three major types: basaltic, andesitic, and rhyolitic.

Basaltic Lavas Basaltic lavas have a mafic composition (high in iron, magnesium, and calcium) and the lowest silica content of the three lava types. Basaltic magmas are the most common magma type. They are produced along mid-ocean ridges and at hot spots, as well as in continental rift valleys.

When cool, *basaltic lavas* are black or dark gray, but at their high eruption temperatures (1000° to 1200°C), they glow in reds and yellows (Figure 5.18). Because their temperatures are high and their silica content low, basaltic lavas are extremely fluid and can flow downhill fast and far. Lava streams flowing as fast as 100 km/hour have been observed, although velocities of a few kilometers per hour are more common. In 1938, two daring Russian volcanologists measured temperatures and collected gas samples while floating down a river of molten basalt on a raft of colder solidified lava. The surface temperature of the raft was 300°C, and the river temperature was 870°C. Lava streams have been observed to travel more than 50 km from their sources.

> Basaltic lavas, which are the most common type, have the lowest silica content of the three types of lava.

FIGURE 5.18 A central vent eruption from Kilauea, a shield volcano on the island of Hawaii, produces a river of hot, fast-flowing basaltic lava. [J. D. Griggs/USGS.]

Basaltic eruptions are rarely explosive. Basaltic lavas erupt when hot, fluid magmas fill up the volcano's plumbing system and overflow, sending lava down the flanks of the volcano in great streams that can engulf everything in their path. Basaltic lava flows take on different forms depending on how they cool. On land, they solidify as pahoehoe (pronounced pa-ho'-ee-ho'u-ee) or aa (ah-ah) (**Figure 5.19**).

Pahoehoe (the word is Hawaiian for "ropy") forms when a highly fluid lava spreads in sheets and a thin, glassy, elastic skin congeals on its surface as it cools. As the molten liquid continues to flow below the surface, the skin is dragged and twisted into coiled folds that resemble rope.

"*Aa*" is what the unwary exclaim after venturing barefoot onto lava that looks like clumps of moist, freshly plowed earth. Aa forms when lava loses its gases and consequently flows more slowly than pahoehoe, allowing a thick skin to form. As the flow continues to move, the thick skin breaks into rough, jagged blocks. The blocks pile up in a steep front of angular boulders that advances like a tractor tread. Aa is

> Basaltic lava flows take on different forms depending on how they cool.

FIGURE 5.19 Two forms of basaltic lava, ropy pahoehoe (*left*) and jagged blocks of aa (*right*), produced by Mauna Loa on the island of Hawaii. [John Grotzinger.]

truly treacherous to cross. A good pair of boots may last about a week on it, and the traveler can count on cut knees and elbows.

A single downhill basaltic flow commonly has the features of pahoehoe near its source, where the lava is still fluid and hot, and of aa farther downstream, where the flow's surface—having been exposed to cool air longer—has developed a thicker outer skin.

Basaltic magma that cools under water forms *pillow lavas*—piles of ellipsoidal, pillowlike blocks of basalt about a meter wide (see Figure 5.12). Pillow lavas are an important indicator that a region on dry land was once under water. Scuba-diving geologists have actually observed pillow lavas forming on the ocean floor off Hawaii. Tongues of molten basaltic lava develop a tough, plastic skin on contact with the cold ocean water. Because the lava inside the skin cools more slowly, the pillow's interior develops a crystalline texture, whereas the quickly chilled skin solidifies to a crystal-less glass.

Andesitic Lavas Andesitic magmas have an intermediate silica content and are more viscous than basaltic lavas. They are erupted mainly in the volcanic belts above subduction zones. The name comes from a prime example—the Andes of South America.

> *Andesitic lavas have intermediate silica content and are more viscous than basaltic lavas.*

The temperatures of *andesitic lavas* are lower than those of basalts, and their silica content is higher, so they flow more slowly and lump up in sticky masses. If one of these sticky masses plugs up the central vent of the volcano, gases can build up beneath the plug and eventually blow off the top of the volcano. The explosive eruption of Mount St. Helens in 1980 (**Figure 5.20**) is a famous example.

Some of the most destructive volcanic eruptions in history have been *phreatic,* or steam, explosions, which occur when hot, gas-charged magma encounters groundwater or seawater, generating vast quantities of superheated steam (**Figure 5.21**). The island of Krakatau, an andesitic volcano in Indonesia, was destroyed by a phreatic explosion in 1883. This legendary eruption was heard thousands of kilometers away, and it generated a tsunami that killed more than 40,000 people.

Rhyolitic Lavas Rhyolitic lavas have a felsic composition (high in sodium and potassium) with a silica content

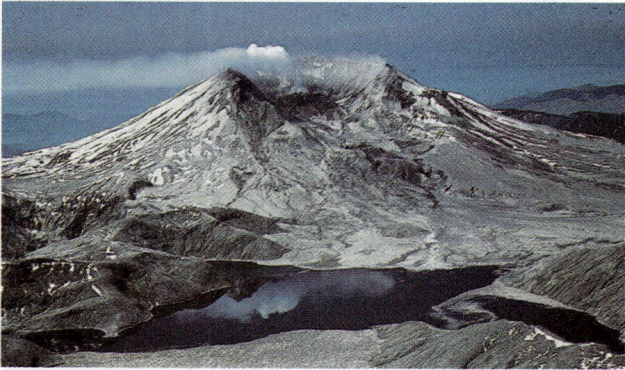

FIGURE 5.20 Mount St. Helens, an andesitic volcano in southwestern Washington State, before, during, and after its cataclysmic eruption in May 1980, which ejected about 1 km³ of pyroclastic material. The collapsed northern flank can be seen in the bottom photo. [*Before:* Emil Muench/Photo Researchers. *Erupting:* U.S. Geological Survey. *After:* David Weintraub/Photo Researchers.]

FIGURE 5.21 Phreatic explosion from Nisino-sima, a new volcano that rose above the sea in 1973 after a submarine eruption in the Pacific Ocean about 900 km south of Tokyo. [Japan Meteorological Agency.]

Rhyolitic lavas have the highest silica content of the three lava types and therefore are the most viscous and potentially explosive. These lavas are produced where melting of continental crust occurs.

greater than 68 percent. Rhyolitic magmas are produced in zones where heat from the mantle has melted large volumes of continental crust. Today, the Yellowstone volcano is producing huge amounts of rhyolitic magmas that are building up in shallow chambers.

Rhyolite is light in color, often a pretty pink. It has a lower melting point than andesite, becoming liquid at temperatures of only 600° to 800°C. Because rhyolitic lavas are richer in silica than any other lava type, they are the most viscous lava type. A rhyolitic flow typically moves more than 10 times more slowly than a basaltic flow, and it tends to pile up in thick, bulbous deposits. Gases are easily trapped beneath rhyolitic lavas, and large rhyolitic volcanoes such as Yellowstone produce the most explosive of all volcanic eruptions.

Textures of Volcanic Rocks

The textures of volcanic rocks, like the surfaces of solidified lava flows, reflect the conditions under which they solidified. Coarse-grained textures with visible crystals can result if lavas cool slowly. Lavas that cool quickly tend to have fine-grained textures. If they are silica-rich, rapidly cooled lavas can form **obsidian,** a glassy volcanic rock that is solid and dense. Chipped or fragmented obsidian produces very sharp edges, and Native Americans and many other hunting groups have used it for arrowheads and a variety of cutting tools.

Volcanic rock often contains little bubbles, created as gases are released during an eruption. Magma is typically charged with gas, like soda in an unopened bottle. When magma rises, the pressure on it decreases, just as the pressure on the soda drops when the bottle cap is removed. And just as the carbon dioxide in the soda forms bubbles when the pressure is released, the water vapor and other dissolved gases escaping from lava as it erupts create gas cavities, or *vesicles* (**Figure 5.22**). **Pumice** is an extremely vesicular volcanic rock, usually rhyolitic in composition. Some pumice has so many vesicles that it is light enough to float on water.

Pyroclastic Deposits

Water and gases in magmas can have even more dramatic effects. Before a magma erupts, the confining pressure of the overlying rock keeps these volatiles from escaping. When the magma rises close to the surface and the pressure drops, the volatiles may be released with explosive force, shattering the lava and any overlying solidified rock into pyroclasts of various sizes, shapes, and textures (**Figure 5.23**).

FIGURE 5.22 Vesicular basalt sample. [John Grotzinger.]

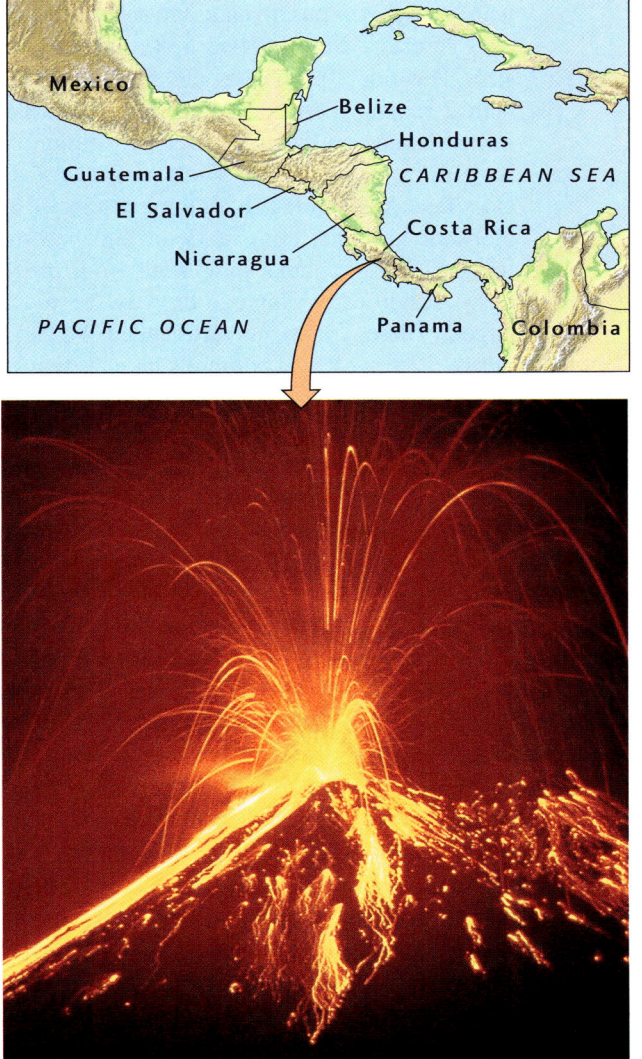

FIGURE 5.23 An explosive eruption at Arenal volcano, Costa Rica. [Gregory G. Dimijian/Photo Researchers.]

FIGURE 5.24 A volcanologist examines a volcanic bomb ejected from Asama volcano, Japan. [Science Source/Photo Researchers.]

FIGURE 5.25 Volcanic breccia. [John Grotzinger.]

Pyroclastic deposits include fragments thrown from volcanoes during explosive eruptions (pyroclasts) and flows of hot gases mixed with dust and ash (pyroclastic flows).

Such explosive eruptions are particularly likely with gas-rich, viscous rhyolitic and andesitic lavas.

Pyroclasts are classified according to their size. The finest pyroclasts are fragments less than 2 mm in diameter, usually of glass, that form when escaping gases force a fine spray of magma from a volcano. These fragments are classified as **volcanic ash.** Fragments ejected as blobs of lava that cool in flight and become rounded, or as chunks torn loose from previously solidified volcanic rock, can be much larger. These fragments are called volcanic **bombs** (Figure 5.24). Volcanic bombs as large as houses have been thrown more than 10 km by explosive eruptions. Volcanic ash fine enough to stay aloft can be carried great distances. Within 2 weeks of the 1991 eruption of Mount Pinatubo in the Philippines, its volcanic ash was traced all the way around the world by Earth-orbiting satellites.

Sooner or later, these pyroclasts fall to Earth, building the largest deposits near their source. As they cool, the hot, sticky fragments become welded together (lithified). The rocks created from smaller fragments are called **tuffs;** those formed from larger fragments are called **breccias** (Figure 5.25).

Pyroclastic flows, which are particularly spectacular and often deadly, occur when hot ash, dust, and gases are ejected in a glowing cloud that rolls downhill at high speeds. The solid particles are buoyed up by the hot gases, so there is little frictional resistance to their movement (Figure 5.26).

In 1902, with very little warning, a pyroclastic flow with an internal temperature of 800°C exploded from the side of Mont Pelée, on the Caribbean island of Martinique. The avalanche of choking hot gas and glowing

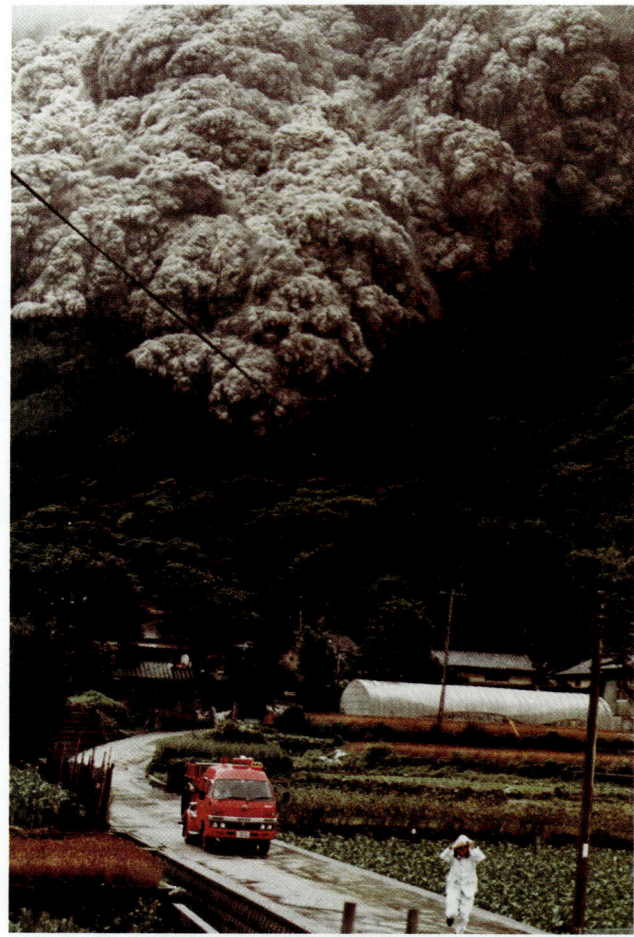

FIGURE 5.26 This pyroclastic flow plunged down the slopes of Mount Unzen, in Japan, in June 1991. Note the fireman and fire engine in the foreground, trying to outrun the hot ash cloud descending on them. Three scientists who were studying this volcano died when they were engulfed by a similar flow. [AP/Wide World Photos.]

volcanic ash plunged down the slopes at a hurricane speed of 160 km/hour. In 1 minute and with hardly a sound, the searing emulsion of gas, ash, and dust enveloped the town of St. Pierre and killed 29,000 people.

VOLCANOES BUILD LANDFORMS

The shape a volcano takes varies with the properties of the magma, especially its chemical composition and gas content, the type of material (lava versus pyroclasts) erupted, and the environmental conditions under which it erupts—on land or under the sea. Volcanic landforms also depend on the rate at which lava is produced and the plumbing system that gets it to the surface (**Figure 5.27**).

> *The shape of a volcano depends on the composition, amount, and type of material erupted as well as on where it erupts.*

Central Eruptions

Central eruptions discharge lava or pyroclasts from a *central vent*. The vent consists of an opening atop a pipelike feeder channel rising from the magma chamber. The magma ascends through this channel to erupt at Earth's surface. Central eruptions create the most familiar of all volcanic features: the volcanic mountain, shaped like a cone.

Shield Volcanoes A lava cone is built by successive flows of lava from a central vent. If the lava is basaltic, it flows easily and spreads widely. If flows are copious and frequent, they create a broad, shield-shaped volcano 2 or more kilometers high and many tens of kilometers in circumference with relatively gentle slopes. Mauna Loa, on the island of Hawaii, is the classic example of a **shield volcano** (Figure 5.27a). Although it rises only 4 km above sea level, it is actually the world's tallest mountain: measured from its base on the seafloor, Mauna Loa is 10 km high, higher than Mount Everest! It has a base diameter of 120 km, covering three times the area of Rhode Island. It grew to this enormous size by the accumulation of thousands of lava flows, each only a few meters thick, over a period of about a million years. The island of Hawaii actually consists of the overlapping tops of active shield volcanoes emerging from the ocean.

Volcanic Domes In contrast to basaltic lavas, andesitic and rhyolitic lavas are so viscous that they can barely flow. They often produce a *volcanic dome,* a rounded, steep-sided mass of rock. Domes look as though lava had been squeezed out of a vent like toothpaste, with very little lateral spreading. Domes often plug vents, trapping gases beneath them. Pressure increases until an explosion occurs, blasting the dome into fragments. Mount St. Helens produced such an explosion when it erupted in 1980 (Figure 5.27b).

Cinder-Cone Volcanoes When volcanic vents discharge pyroclasts, these solid fragments can build up to create *cinder cones*. The profile of a cinder cone is determined by the maximum angle at which the fragments remain stable rather than sliding downhill. The larger fragments, which fall near the vent, form very steep but stable slopes. Finer particles are carried farther from the vent and form gentle slopes at the base of the cone. The classic concave-shaped volcanic cone with its summit vent develops in this way (Figure 5.27c).

Stratovolcanoes When a volcano emits lava as well as pyroclasts, alternating lava flows and beds of pyroclasts build a concave-shaped composite volcano, or **stratovolcano.** Stratovolcanoes are common above subduction zones. Famous examples are Mount Fuji in Japan (Figure 5.27d), Mount Vesuvius and Mount Etna in Italy, and Mount Rainier in Washington State. Mount St. Helens used to have a near-perfect stratovolcano shape until its eruption in 1980 destroyed its northern flank (see Figure 5.20).

Craters A bowl-shaped pit, or **crater,** is found at the summit of most volcanoes, centered on the vent. During an eruption, the upwelling lava overflows the crater walls. When the eruption ceases, the lava that remains in the crater often sinks back into the vent and solidifies. When the next eruption occurs, that material may be blasted out of the crater. The crater later becomes partly filled by rock fragments that fall back into it. Because a crater's walls are steep, they may cave in or become eroded over time. In this way, a crater can grow to several times the diameter of the vent and hundreds of meters deep. The crater of Mount Etna in Sicily, for example, is currently 300 m in diameter (Figure 5.27e).

Calderas When great volumes of magma are discharged rapidly, a large magma chamber cannot support its roof. In such cases, the overlying volcanic structure can collapse catastrophically, leaving a large, steep-walled, basin-shaped depression much larger than the crater, called a **caldera** (Figure 5.27f). Calderas can be impressive features, ranging in size from a few kilometers to 50 km or more in diameter. The Yellowstone volcano, which has an area greater than Rhode Island, is the largest active caldera in the United States.

After some hundreds of thousands of years, enough fresh magma may reenter the collapsed magma chamber to reinflate it, forcing the caldera floor to dome upward again and creating a *resurgent caldera*. The cycle of eruption, collapse, and resurgence may occur repeatedly over

<< 134   CHAPTER 5 Igneous Processes and Volcanism

(a) **Shield volcanoes** are built up by the accumulation of thousands of thin basaltic flows that spread as gently sloping sheets. Each layer in the diagram represents many hundreds of thin flows. Magma can erupt on the flanks of a volcano as well as from the central vent.

(b) **Volcanic domes** are bulbous masses of felsic lava, which are so viscous that instead of flowing, they pile up over the vent. The photo shows a growing dome within the crater of Mount St. Helens after its 1980 eruption.

(c) **Cinder-cone volcanoes** are formed when pyroclasts are deposited as layers that dip away from the crater at the summit. The vent beneath the crater is filled with fragmental debris. The photo is of Cerro Negro, shown erupting in 1968, a cinder cone built on an older terrain of lava flows.

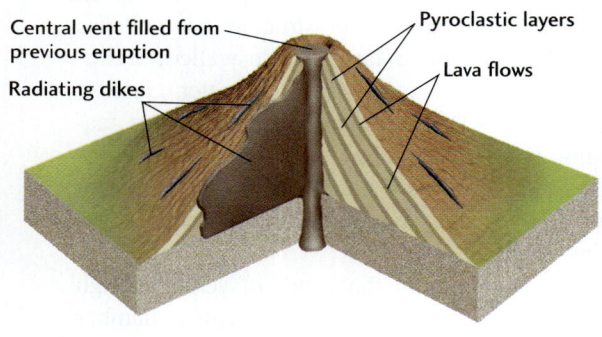

(d) **Stratovolcanoes** are built from alternating layers of pyroclasts and lava flows. Lava that has solidified in fissures forms riblike dikes that strengthen the cone.

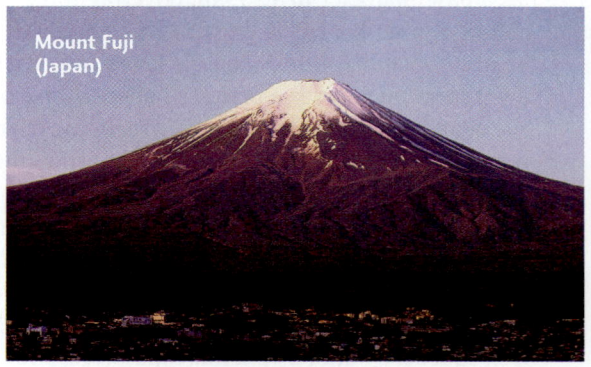

FIGURE 5.27 The eruptive styles of volcanoes and the landforms they create are determined by the composition of magma [(a) U.S. Geological Survey; (b) Lyn Topinka/USGS Cascades Volcano Observatory; (c) Mark Hurd Aerial Surveys; (d) CORBIS; (e) Fabrizio Villa/AP/Wide World Photos; (f) Greg Vaughn/Tom Stack & Associates.]

(e) **Craters** are found at the summits of most volcanoes. After an eruption, lava often sinks back into the vent and solidifies, to be blasted out by a later pyroclastic eruption.

Mt. Etna (Sicily, Italy)

(f) **Calderas** result when a violent eruption empties a volcano's magma chamber, which then cannot support the overlying rock. It collapses, leaving a large, steep-walled basin.

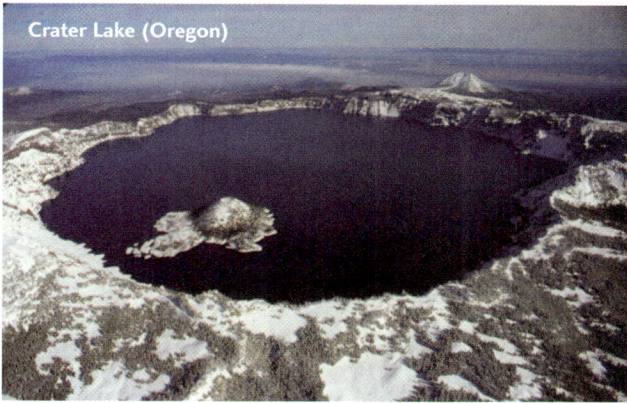

Crater Lake (Oregon)

geologic time. Three times over the last 2 million years, Yellowstone Caldera has erupted catastrophically, in each instance ejecting hundreds or thousands of times more material than the 1980 Mount St. Helens eruption and depositing ash over much of what is now the western United States. Other resurgent calderas are Valles Caldera in New Mexico and Long Valley Caldera in California, which last erupted about 1.2 million and 760,000 years ago, respectively.

Fissure Eruptions

The largest volcanic eruptions do not come from a central vent, but through large, nearly vertical cracks in Earth's surface, sometimes tens of kilometers long (Figure 5.28). Such *fissure eruptions* are the main style of volcanism along mid-ocean ridges, where new oceanic crust is formed. A moderate-sized fissure eruption occurred in 1783 on a segment of the Mid-Atlantic Ridge that comes ashore in

FIGURE 5.28 In a fissure eruption, highly fluid basaltic lava flows rapidly away from fissures and forms widespread layers, rather than building up into a volcanic mountain. [After R. S. Fiske/USGS.] These volcanic cones lie along the Laki fissure in Iceland, which opened in 1783 and erupted the largest flow of lava on land in recorded history. [Tony Waltham.]

FIGURE 5.29 The Columbia River basalts. Successive flows of flood basalts piled up to build this immense plateau, which covers a large area of Washington and Oregon. [Dave Schiefelbein.]

Iceland. One-fifth of Iceland's population perished from starvation as a result. A fissure 32 km long opened and spewed out some 12 km^3 of basalt, enough to cover Manhattan to a height about halfway up the Empire State Building. Fissure eruptions continue on Iceland, although on a smaller scale.

Flood Basalts Highly fluid basaltic lavas erupting from fissures on continental plates can spread out in thin sheets over flat terrain. Successive flows often pile up into immense basaltic lava plateaus, called *flood basalts,* rather than forming a shield volcano as they do when the eruption is confined to a central vent. A huge eruption of flood basalts about 16 million years ago buried 160,000 km^2 of preexisting topography in Washington, Oregon, and Idaho (**Figure 5.29**). Individual flows were more than 100 m thick, and some were so fluid that they traveled more than 500 km from their source. An entirely new landscape with new river valleys has since evolved atop the lava that buried the old surface. Plateaus made by flood basalts are found on every continent.

Ash-Flow Deposits Eruptions of pyroclasts on continents have produced extensive sheets of hard volcanic tuffs called *ash-flow deposits.* A succession of forests in Yellowstone National Park have been buried under such ash flows. Some of the largest pyroclastic deposits on the planet are the ash flows erupted in the mid-Cenozoic era, 45 million to 30 million years ago, through fissures in what is now the Basin and Range province of the western United States. The amount of material released during this pyroclastic flare-up was a staggering 500,000 km^3—enough to cover the entire state of Nevada with a layer of rock nearly 2 km thick! As far as we know, humans have never witnessed one of these spectacular events.

HAZARDS CREATED BY VOLCANOES

Volcanic eruptions have a prominent place in human history and mythology. Ancient philosophers were awed by volcanoes and their fearsome eruptions of molten rock. The myth of the lost continent of Atlantis may have its source in the explosion of Thera (also known as Santorini), a volcanic island in the Aegean Sea. The eruption, which has been dated at 1623 B.C., formed a caldera 7 km by 10 km in diameter, visible today as a lagoon as much as 500 m deep with two small active volcanoes in the center. The eruption and the tsunami that followed it destroyed dozens of coastal settlements over a large part of the eastern Mediterranean. Some scientists have attributed the mysterious demise of the Minoan civilization to this ancient catastrophe.

Of Earth's 500 to 600 active volcanoes, at least one in six is known to have claimed human lives. In the past 500 years alone, more than 250,000 people have been killed by volcanic eruptions. Statistics on these fatalities and their causes are shown in **Figure 5.30**.

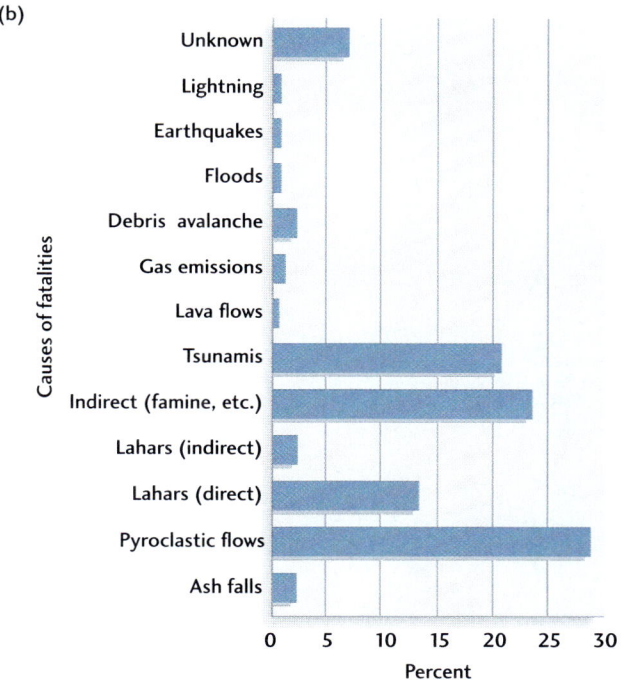

FIGURE 5.30 (a) Cumulative statistics on fatalities caused by volcanoes since A.D. 1500. The seven eruptions that dominate the record, each of which claimed 10,000 or more victims, are named. These eruptions account for two-thirds of the total deaths. (b) Causes of volcano fatalities since A.D. 1500. [After T. Simkin, L. Siebert, and R. Blong. *Science* 291 (2001): 255.]

Volcanoes can kill people and damage property in many ways, some of which are listed in Figure 5.30 and depicted in **Figure 5.31**. We have already mentioned some of these hazards, including pyroclastic flows and tsunamis. Several additional volcanic hazards are of special concern.

Lahars

Among the most dangerous volcanic events are the torrential mudflows of wet volcanic debris called **lahars.** They can occur when a pyroclastic flow meets a river or a snowbank, when the wall of a water-filled crater breaks, when a lava flow melts glacial ice, or when heavy rainfall transforms new ash deposits into mud. One extensive layer of volcanic debris in the Sierra Nevada of California contains 8000 km^3 of material of lahar origin, enough to cover all of Delaware with a deposit more than a kilometer thick. Lahars have been known to carry huge boulders for tens of kilometers. When Nevado del Ruiz in the Colombian Andes erupted in 1985, lahars triggered by the melting of glacial ice near the summit plunged down the slopes and buried the town of Armero 50 km away, killing more than 25,000 people.

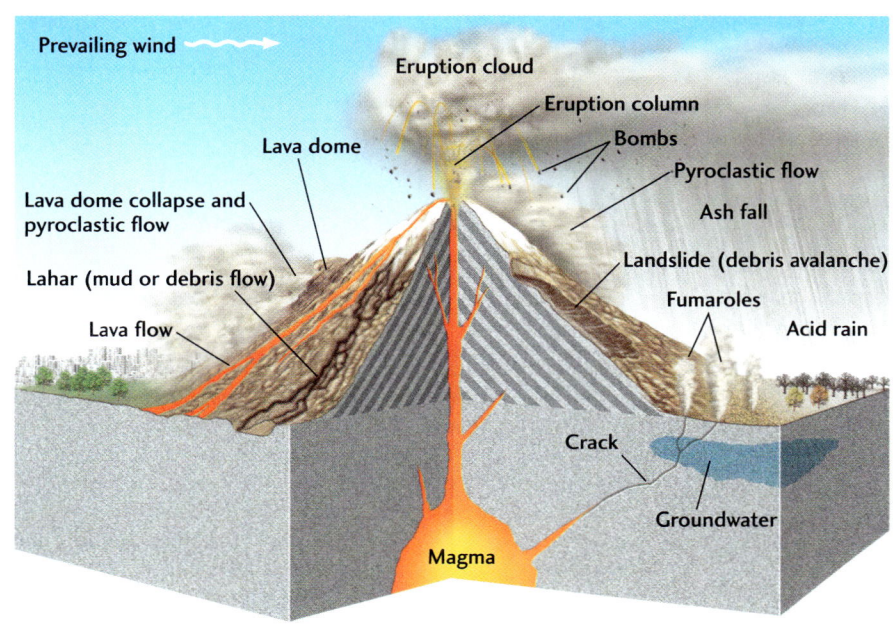

FIGURE 5.31 Some of the volcanic hazards that can kill people and destroy property. [B. Meyers et al./USGS.]

Flank Collapse

A volcano is constructed from thousands of deposits of lava or pyroclasts or both, which is not the best way to build a stable structure. A volcano's sides may become too steep and break or slip off. In recent years, volcanologists have discovered many prehistoric examples of catastrophic structural failures in which a big piece of a volcano broke off, perhaps because of an earthquake, and slid downhill in a massive, destructive landslide. On a worldwide basis, such *flank collapses* occur at an average rate of about four times per century. The collapse of one side of Mount St. Helens was the most damaging part of its 1980 eruption (see Figure 5.20).

Surveys of the seafloor off the Hawaiian Islands have revealed many giant landslides on the underwater flanks of the Hawaiian Ridge. When they occurred, these massive movements of earth probably triggered huge tsunamis. In fact, coral-bearing marine sediments have been found some 300 m above sea level on one of the Hawaiian Islands. These sediments were probably deposited by a giant tsunami that was caused by a prehistoric flank collapse.

The southern flank of Kilauea, on the island of Hawaii, is advancing toward the sea at a rate of 250 mm/year, which is relatively fast, geologically speaking. This advance became even more worrisome when it suddenly accelerated by a factor of several hundred on November 8, 2000, probably as a result of heavy rainfall a few days earlier. A network of motion sensors detected an ominous surge in velocity of about 50 mm/day. The surge lasted for 36 hours, after which the normal motion was reestablished. Someday, maybe thousands of years from now but perhaps sooner, the southern flank of the volcano is likely to break off and slide into the ocean. This catastrophic event would trigger a tsunami that could prove disastrous for Hawaii, California, and other Pacific coastal areas.

Caldera Collapse

Though infrequent, collapses of large calderas are one of the most destructive natural phenomena on Earth. Monitoring the activity of calderas is very important because of their long-term potential for widespread destruction. Fortunately, no catastrophic collapses have occurred in North America during recorded history, but geologists are concerned about an increase in small earthquakes in Yellowstone and Long Valley calderas as well as other indications of activity in their underlying magma chambers. For example, carbon dioxide leaking into the soil from magma in the crust has been killing trees since 1992 on Mammoth Mountain, a volcano on the boundary of Long Valley Caldera. Other indications of resurgence in this caldera have been the uplifting of the center of the caldera by more than half a meter in the past 20 years and the occurrence of nearly

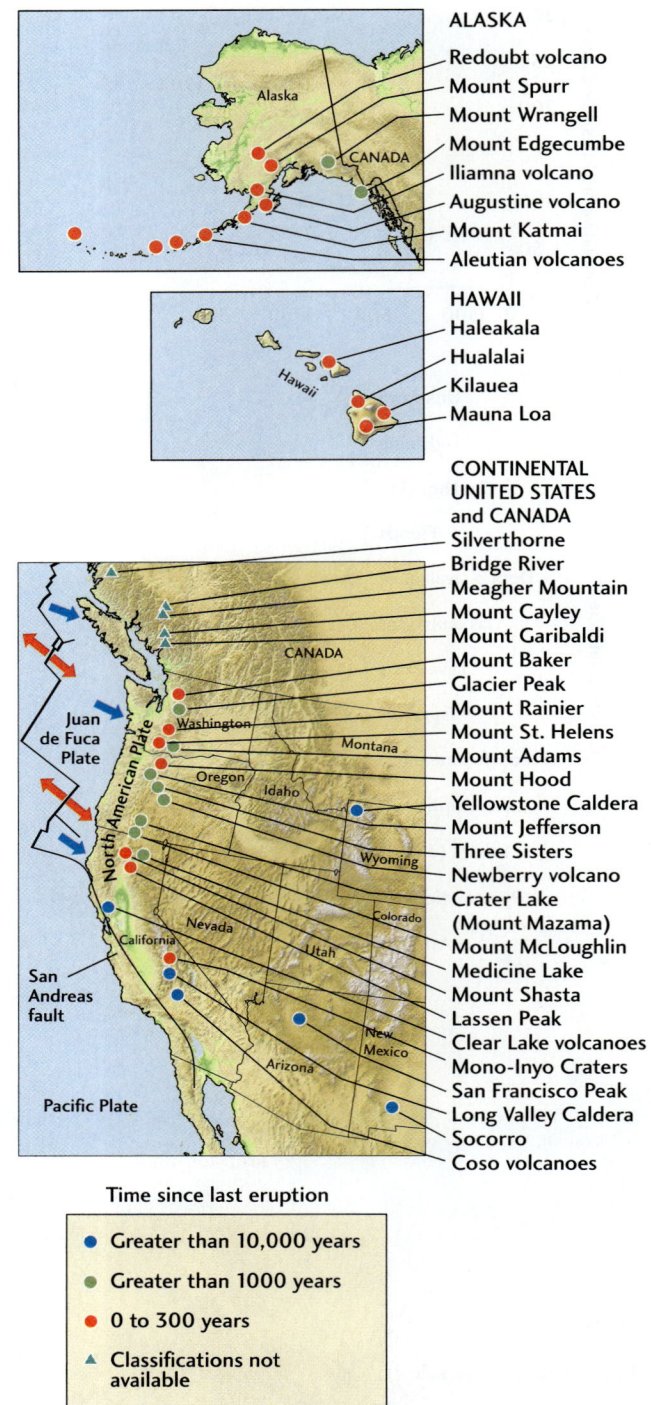

FIGURE 5.32 Locations of potentially hazardous volcanoes in the United States and Canada. Volcanoes within each U.S. group are color-coded by time since their last eruption; those that have erupted most recently are thought to present the greatest cause for concern. (These classifications are subject to revision as studies progress and are not available for Canadian volcanoes.) Note the relationship between the volcanoes extending from northern California to British Columbia and the convergent boundary between the North American Plate and the Juan de Fuca Plate. [After R. A. Bailey, P. R. Beauchemin, F. P. Kapinos, and D. W. Klick/USGS.]

continuous swarms of small earthquakes (more than a thousand in a single day in 1997).

Eruption Clouds

With the growth in air travel, a volcanic hazard that is attracting increased attention is the clouds of volcanic ash lofted into air traffic lanes by erupting volcanoes. Over a period of 25 years, more than 60 commercial jet passenger planes have been damaged by such clouds. One Boeing 747 temporarily lost all four engines when ash from an erupting volcano in Alaska was sucked into the engines and caused them to flame out. Fortunately, the pilot was able to make an emergency landing. Warnings of eruption clouds near air traffic lanes are now being issued by several countries.

Reducing the Risks of Volcanic Hazards

Volcanic eruptions cannot be prevented, but their catastrophic effects can be significantly reduced by a combination of science and enlightened public policy. Volcanology has progressed to the point that we can identify the world's dangerous volcanoes and characterize their potential hazards by studying deposits laid down in earlier eruptions. Some potentially dangerous volcanoes in the United States and Canada are identified in Figure 5.32. Assessments of their hazards can be used to guide zoning regulations to restrict land use—the most effective measure to reduce property losses and casualties.

Such studies indicate that Mount Rainier, because of its proximity to the heavily populated cities of Seattle and Tacoma, probably poses the greatest volcanic risk in the United States (Figure 5.33). Some 150,000 people live in areas where the geologic record shows evidence of floods and lahars that have swept down from the volcano over the past 6000 years. An eruption could kill thousands of people and cripple the economy of the Pacific Northwest.

Predicting Eruptions

These potential risks raise an important question: Can volcanic eruptions be predicted? In many cases, the answer is yes. Instrumented monitoring can detect signals such as earthquakes, swelling of the volcano, and gas emissions that warn of impending eruptions. People at risk can be evacuated if the authorities are organized and prepared. Scientists monitoring Mount St. Helens were able to warn people before its eruption in 1980, as described at the opening of this chapter. Government infrastructure was in place to evaluate the warnings and to enforce evacuation orders, so very few people were killed.

Another successful warning was issued a few days before the cataclysmic eruption of Mount Pinatubo in the Philippines on June 15, 1991. A quarter of a million peo-

FIGURE 5.33 Mount Ranier, seen from Tacoma, Washington. [Alamy.]

ple were evacuated, including some 16,000 residents of the nearby U.S. Clark Air Force Base (which was heavily damaged by the eruption and has since been permanently abandoned). Tens of thousands of lives were saved from the lahars that destroyed everything in their paths. Casualties were limited to the few who disregarded the evacuation order. In 1994, 30,000 residents of Rabaul, Papua New Guinea, were successfully evacuated by land and sea hours before the two volcanoes on either side of the town erupted, destroying or damaging most of it. Many owe their lives to the government, which conducted evacuation drills, and to scientists at the local volcano observatory, who issued a warning when their seismographs recorded the ground tremors that signaled magma moving toward the surface.

Improving our ability to predict eruptions is important because there are about 100 high-risk volcanoes in the world and some 50 volcanic eruptions per year.

Controlling Eruptions

Can we go further by actually controlling volcanic eruptions? Not likely, because large volcanoes release energy on a scale that dwarfs our capabilities for control. However, under special circumstances and on a small scale, the damage can be reduced. Perhaps the most successful attempt to manage volcanic activity was made on the Icelandic island of Heimaey in January 1973. By spraying advancing lava with seawater, residents cooled and slowed the flow, preventing the lava from blocking the port entrance and saving some homes from destruction.

In the years ahead, the best policy for protecting the public will be the establishment of more warning and evacuation systems and more rigorous restriction of settlements in potentially dangerous locations.

SUMMARY >>

How are igneous rocks classified? All igneous rocks can be divided into two broad textural classes: (1) coarse-grained rocks, which are intrusive and therefore cooled slowly; and (2) fine-grained rocks, which are extrusive and cooled rapidly. Within each of these broad categories, igneous rocks are classified as felsic, intermediate, or mafic on the basis of their silica content, or mineralogically on the basis of their proportions of lighter-colored, felsic minerals and darker, mafic minerals.

How and where do magmas form? Magmas form at places in the lower crust and mantle where temperatures are high enough for at least partial melting of rock. Because the minerals within a rock melt at different temperatures, the composition of magmas varies with melting temperature. Pressure raises the melting temperature of rock, and water lowers it. Because melted rock is less dense than solid rock, magma rises through the surrounding rock, and drops of magma come together to form magma chambers.

How does magmatic differentiation account for the variety of igneous rocks? Because different minerals crystallize at different temperatures, the composition of magma changes as it cools as various elements are withdrawn to form crystallized minerals. When a melt undergoes such fractional crystallization, the last rocks to form may be more felsic than the earlier, more mafic rocks. Fractional crystallization does not adequately explain the abundance of granite in Earth's crust, however, nor does differentiation of basaltic magma explain the observed compositions and abundances of igneous rocks. Different kinds of igneous rocks may be produced by variations in the compositions of magmas caused by the melting of different mixtures of sedimentary and other rocks and by the mixing of magmas.

What are the forms of igneous intrusions? Large intrusive igneous bodies are called plutons. The largest plutons are batholiths, which are thick tabular masses extending from a central funnel. Stocks are smaller plutons. Less massive than plutons are sills, which lie parallel to the layers of bedded country rock, and dikes, which cut across those layers.

How do tectonic processes affect magma production? Magmas are produced at mid-ocean ridges, where material rises from the upper mantle and undergoes decompression melting to form basaltic magma, and at subduction zones, where subducting oceanic lithosphere undergoes fluid-enhanced melting to generate magmas of varying composition. Mantle plumes penetrate the interiors of lithospheric plates as "hot spots," resulting in great outpourings of basaltic lavas.

What are the three major types of lavas? Lavas are classified as basaltic (mafic), andesitic (intermediate), and rhyolitic (felsic) on the basis of their silica content. Basaltic lavas are relatively fluid and flow freely; andesitic and rhyolitic lavas are more viscous. Lavas differ from pyroclasts, which are formed by explosive eruptions and vary in size from fine ash particles to house-sized bombs.

How are volcanoes shaped? The chemical composition and gas content of magma are important factors in a volcano's eruptive style and in the shape of the landforms it creates. A shield volcano grows from repeated eruptions of basaltic lava from a central vent. Andesitic and rhyolitic lavas are more viscous and tend to erupt explosively. The erupted pyroclasts may pile up into a cinder cone or cover an extensive area with ash-flow deposits. A stratovolcano is built of alternating layers of lava flows and pyroclastic deposits. The rapid ejection of magma from a large magma chamber, followed by collapse of the chamber's roof, results in a large depression, or caldera. Basaltic lavas can erupt from fissures along mid-ocean ridges as well as on continents, where they flow over the landscape in thin sheets to form flood basalts.

What hazards does volcanism present for human society? Volcanic hazards that can kill people and damage property include pyroclastic flows, tsunamis, lahars, flank collapses, caldera collapses, eruption clouds, and ash falls. Volcanic eruptions have killed about 250,000 people in the past 500 years.

KEY TERMS AND CONCEPTS

andesite (p. 114)
basalt (p. 114)
batholith (p. 120)
bomb (p. 132)
breccia (p. 132)
caldera (p. 133)
country rock (p. 111)
crater (p. 133)
dacite (p. 114)
decompression melting (p. 116)
dike (p. 121)
diorite (p. 113)
felsic rock (p. 113)
fluid-induced melting (p. 124)
fractional crystallization (p. 117)
gabbro (p. 114)
granite (p. 113)
granodiorite (p. 113)
hot spot (p. 126)
intermediate igneous rock (p. 113)
lahar (p. 137)
lava (p. 109)
mafic rock (p. 114)
magma chamber (p. 116)
magmatic differentiation (p. 117)
mantle plume (p. 125)
obsidian (p. 131)
partial melting (p. 115)
pegmatite (p. 122)
peridotite (p. 114)
pluton (p. 120)
porphyry (p. 111)
pumice (p. 131)
pyroclast (p. 111)
pyroclastic flow (p. 132)
rhyolite (p. 113)
shield volcano (p. 133)
sill (p. 120)
stock (p. 120)
stratovolcano (p. 133)
tuff (p. 132)
ultramafic rock (p. 114)
vein (p. 122)
viscosity (p. 114)
volcanic ash (p. 132)
volcano (p. 126)

EXERCISES

1. Why are intrusive igneous rocks coarse-grained and extrusive rocks fine-grained?

2. What kinds of minerals would you find in a mafic igneous rock?

3. What kinds of igneous rocks contain quartz?

4. Name two intrusive igneous rocks with a higher silica content than that of gabbro.

5. How does fractional crystallization lead to magmatic differentiation?

6. Where in the crust, mantle, or core would you find a partial melt of basaltic composition?

7. Why do melts migrate upward?

8. Where on the ocean floor would you find basaltic magmas being extruded?

9. Much of Earth's crustal area, and nearly all of its mantle, is composed of basaltic or ultramafic rocks. Why are granitic and dioritic rocks as plentiful as they are on Earth? Where do the materials that constitute these rocks come from?

10. What is the difference between magma and lava? Describe a geologic situation in which a magma does not form a lava.

11. What are the three major types of volcanic rocks and their intrusive counterparts?

12. Most volcanism occurs near plate boundaries. What type of plate boundary can produce large amounts of rhyolitic lavas?

13. How do scientists predict volcanic eruptions?

14. How would you classify a coarse-grained igneous rock that contains about 50 percent pyroxene and 50 percent olivine?

15. What kind of rock would contain some plagioclase feldspar crystals about 5 mm long "floating" in a dark gray matrix of crystals of less than 1 mm?

16. What differences in crystal size might you expect to find between two sills, one intruded at a depth of about 12 km, where the country rock was very hot, and the other at a depth of 0.5 km, where the country rock was moderately warm?

17. If you were to drill a hole through the crust of a mid-ocean ridge, what intrusive or extrusive igneous rocks might you expect to encounter at or near the surface? What intrusive or extrusive igneous rocks might you expect at the base of the crust?

18. What observations might you make to show that a pluton solidified during fractional crystallization?

19. What processes create the unequal sizes of crystals in porphyries?

20. Water is abundant in the sedimentary rocks and oceanic crust of subduction zones. How would that water affect melting in these zones?

21. Why are eruptions of stratovolcanoes generally more explosive than those of shield volcanoes?

22. Give a few examples of what geologists have learned about Earth's interior by studying volcanoes and volcanic rocks.

23. While on a field trip, you come across a volcanic formation that resembles a field of sandbags. The individual ellipsoidal forms have a smooth, glassy surface texture. What type of lava is this, and what information does this give you about its history?

24. What might be the effects on civilization of a Yellowstone-type caldera eruption, such as the one described in this chapter?

SEDIMENTATION: Rocks Formed by Surface Processes

CHAPTER 6

Black Gold

In 1858, Colonel Edwin Drake, a retired railroad man, was hired by the Seneca Oil Company at an annual salary of $1000. His mission was to look for oil in the Allegheny Mountains of Pennsylvania. Drilling began in August 1859. There were many setbacks—so many that the colonel earned the title "Crazy Drake." At only 16 feet down, groundwater was causing the drill hole to collapse. Drake solved this problem by using a drive pipe, a cast iron pipe into which the drill was lowered.

The drilling went slowly, only 3 feet per day, and at 32 feet down they passed from the soil zone into bedrock. On August 27, on what seemed like any other day on the job, the men drilled to almost 70 feet, and then went home for the day.

The next morning, they found a dark fluid floating on the water in the pipe, and jeers turned to cheers—"We've struck oil!"

We've struck oil!

As crude oil dripped through Drake's fingers, he realized that the trick to finding more of it was to understand its geologic habitat—what kinds of rock it forms in, and where it pools in accumulations large enough to make a good profit. In later years, Drake's successors in Texas realized that oil accumulates at the crests of domelike structures in porous sedimentary rocks, created when the rocks were warped or folded by tectonic processes. They were able to convince their investors that oil, being less dense than water, would "float" to the top of this porous sedimentary rock and accumulate as a pool trapped beneath an overlying layer of impermeable rock. Their strategy worked, and these exploration pioneers quickly learned to drill only the tops of the folds, thus saving capital for further drilling and more discoveries.

At the time of Drake's discovery, the demand for oil was small because its only significant use was in kerosene lamps. It was coal that fueled the industrial revolution in the late 1800s and powered most ships during World War I. It wasn't until the invention of the internal combustion engine that demand for oil increased. Oil replaced coal as the dominant fuel in the late 1950s as the cost of an automobile dropped to within reach of middle-class America. ■

Sedimentary rocks are the primary reservoirs for accumulations of oil and natural gas. Drill rigs can penetrate up to depths in excess of 10 km, though most wells are no deeper than a few kilometers. In the search for new reserves, exploration has moved out onto the submerged edges of continents, where rigs float in water depths of several kilometers. [Dewitt Jones/CORBIS.]

CHAPTER OUTLINE

Weathering and Surface Processes of the Rock Cycle 145

Sedimentary Basins: The Sinks for Sediments 151

Sedimentary Environments 152

Sedimentary Structures 156

Burial and Diagenesis: From Sediment to Rock 159

Classification of Siliciclastic Sediments and Sedimentary Rocks 160

Classification of Chemical and Biological Sediments and Sedimentary Rocks 163

The principles of exploration for oil and gas are simple enough, and the basic strategy has not changed since Drake first figured it out. Four basic factors are important: First, we need sedimentary rocks that lie close to the organic matter from which oil is created—so-called "source rocks." Second, the source rocks must lie close to "reservoir rocks," porous sedimentary rocks into which oil can migrate. Third, we look for "seal rocks," sedimentary rocks that are impermeable to oil (or natural gas) and will not let it escape. Finally, "the trap" forces the oil to accumulate as a pool; very often this involves simple folding of the reservoir and seal rocks. In this manner, the largest oil fields on Earth, such as those in Saudi Arabia, are created. All known oil originates in sedimentary rocks.

As supplies of oil and natural gas begin to diminish, the challenges for geologists increase. These challenges include finding new oil fields as well as squeezing out what is left behind in existing fields. Ultimately, it is the availability of source rocks that limits how much oil can be found. Source rocks were more abundant in some periods of Earth history, and they were formed more easily in certain parts of the world. So there are geologic constraints that we must learn to accept. But we can learn to be smarter about how we explore for what little oil is left, and the need for well-trained geologists has never been greater

Much of Earth's surface, including its seafloor, is covered with sediments. These layers of loose particles have diverse origins. Most sediments are created by weathering of the continental crust. Some are the remains of mineral shells secreted by organisms. Yet others consist of inorganic crystals that precipitated when dissolved chemicals in oceans and lakes combined to form new minerals.

Sedimentary rocks were once sediments, and so they are records of the conditions at Earth's surface when and where the sediments were deposited. Geologists can work backward to infer the sources of the sediments from which the rocks were formed and the kinds of places in which the sediments were originally deposited. The top of Mount Everest, for example, is composed of fossiliferous (fossil-containing) limestones. Because such limestones are formed from carbonate minerals in seawater, Mount Everest must once have been part of an ocean floor!

The analysis used to determine the history of rock formations at the top of Mount Everest applies just as well to ancient shorelines, mountains, plains, deserts, and swamps. In one area, for example, sandstone may record an earlier time when beach sands accumulated along a shoreline that no longer exists. In a bordering area, carbonate reefs may have been laid down along the perimeter of a tropical island. Beyond, there may have been a nearshore area in which the sediments were shallow marine carbonate muds that later became thin-bedded limestone. By reconstructing such environments, we can map the continents and oceans of long ago.

Sedimentary rocks also reveal former plate tectonic events and processes by their presence within or adjacent to volcanic arcs, continental rift valleys, or collisional mountain belts (**Figure 6.1**). In cases in which sediments and sedimentary rocks are derived from the weathering of preexisting rocks, we can form hypotheses about ancient climates and environments. We can also use sedimentary rocks formed by precipitation from seawater to read the history of changes in Earth's climate and seawater chemistry.

The study of sediments and sedimentary rocks has great practical value as well. Oil and gas, our most valuable sources of energy, are found in these rocks. These precious resources are becoming increasingly difficult to find, so it is more important than ever to understand how sedimentary rocks form. As oil and gas decline in abundance,

FIGURE 6.1 Sedimentary rocks can reveal former plate tectonic events and processes. These sedimentary rock layers, exposed in a road cut in Guatemala, originally were deposited as horizontal layers of sand and gravel by streams and landslides. After their deposition, the sediments were broken by a fault that has dropped the layers to the left of the fault down about 3 meters relative to the layers to the right. [Fletcher & Baylis, Photo Researchers.]

FIGURE 6.2 The sedimentary stages of the rock cycle comprise several overlapping processes.

coal—which is a distinct type of sedimentary rock—will be used increasingly to generate energy. Another important energy source that may accumulate in sedimentary rocks is uranium, which is used for nuclear power. Phosphate rock used for fertilizer is also sedimentary, as is much of the world's iron ore. Knowing how these kinds of sediments form helps us to find and use these limited resources.

Finally, because virtually all sedimentary processes take place at or near Earth's surface where we humans live, we can study sedimentary rocks to improve our understanding of our environment.

In this chapter, we will see how geologic surface processes produce sediments and sedimentary rocks. We will describe the compositions, textures, and structures of sediments and sedimentary rocks and examine how they correlate with the kinds of environments in which those sediments and rocks were laid down.

WEATHERING AND SURFACE PROCESSES OF THE ROCK CYCLE

Sediments, and the sedimentary rocks formed from them, are produced by weathering and other surface processes of the rock cycle (discussed in Chapter 4). They form after rocks have been moved from Earth's interior to its surface and uplifted by mountain building and before they are returned to Earth's interior by subduction.

These processes move materials from a *source area,* where sediment particles are created, to a *sink area,* where they are deposited in layers. The path that sediment particles follow from source to sink may be a very long journey—one that involves several important processes resulting from interactions between the plate tectonic and climate geosystems.

Let's look at the role of the Mississippi River in a typical sedimentary process. Tectonic plate movement lifts up rocks in the Rocky Mountains. Rainfall in the mountains—a source area—causes weathering of the rocks there. If precipitation increases in the mountains, weathering also increases. Faster weathering produces more sediment to be released into the river and transported downhill and downstream. At the same time, if the flow in the river also increases because of the higher rainfall, transportation of the sediment down the length of the river will increase. This will increase the volume of sediment delivered to sink areas—sites of deposition, also known as sedimentary basins—in the Mississippi delta and the Gulf of Mexico. And in these sedimentary basins, the sediments will pile up on top of one another—layer after layer—and be buried deep in Earth's crust.

The rock cycle processes that are important in the formation of sedimentary rocks are reviewed in **Figure 6.2** and summarized here.

> Sediments and sedimentary rocks are produced by the surface processes of the rock cycle.

- **Weathering** is the general process by which rocks are broken down at Earth's surface to produce sediment particles. There are two types of weathering. **Physical weathering** takes place when solid rock is fragmented by mechanical processes, such as freezing and thawing or wedging by tree roots (**Figure 6.3**), that do not change its chemical composition. The rubble of broken stone often seen at the tops of mountains and hills is primarily the result of physical weathering. **Chemical weathering** refers to processes by which the minerals in a rock are chemically altered or dissolved. The blurring or disappearance of lettering on old gravestones and monuments is caused mainly by chemical weathering (**Figure 6.4**).

> Physical weathering refers to processes that fragment rock without changing its chemical composition. Chemical weathering refers to processes that change the chemical composition of rock and dissolve it.

- *Erosion* refers to processes that dislodge particles of rock produced by weathering and move them away from the source area. Erosion occurs most commonly when rainwater runs downhill.

- *Transportation* refers to processes by which sedimentary particles are moved to sink areas. It occurs when water currents, wind, or the moving ice of glaciers transport particles to new locations downhill or downstream.

- *Deposition* (also called *sedimentation*) refers to processes by which sedimentary particles settle out as water currents slow, winds die down, or glacier edges melt to form layers of sediment in sedimentary basins. In aquatic environments, particles settle out, chemical precipitates form and are deposited, and the shells of dead organisms are broken up and deposited.

- *Burial* occurs as layers of sediment accumulate in sedimentary basins on top of older, previously deposited sediments, which are compacted and progressively buried deep within the basin. These sediments will remain at depth, as part of Earth's crust, until they are either uplifted again or subducted by tectonic processes.

- *Diagenesis* refers to processes by which pressure, heat, and chemical reactions convert sediments buried within sedimentary basins into sedimentary rocks.

FIGURE 6.4 These early-nineteenth-century gravestones at Wellfleet, Massachusetts, show the results of chemical weathering. The stone on the right is limestone and has become so weathered that it is unreadable. The stone on the left is slate and has remained legible under the same conditions. [Courtesy of Raymond Siever.]

FIGURE 6.3 Plant roots contribute to physical weathering by penetrating fractures and wedging rocks apart. [David R. Frazier/Photo Researchers.]

The Sources of Sediment: Weathering and Erosion

Chemical and physical weathering reinforce each other. Chemical weathering weakens rocks and makes them more susceptible to breakage and the formation of fragments. The smaller the fragments produced by physical weathering, the greater the surface area available for chemical weathering. Together, chemical weathering and physical weathering of rock at the land surface create both solid particles and dissolved products, and erosion carries away these materials. The end products can be classified either as siliciclastic sediments or as chemical and biological sediments.

Siliciclastic Sediments The physical and chemical weathering of preexisting rocks forms *clastic particles* that are transported and deposited as **clastic sediments.** Clastic particles range in size from boulders and pebbles to particles of sand, silt, and clay. They also vary widely in shape. Natural breakage along bedding planes and fractures in the parent rock determine the shapes of boulders, cobbles, and pebbles. Sand grains tend to inherit their shapes from the individual crystals formerly interlocked in the parent rock.

Most clastic sediments are produced by the weathering of rocks composed largely of silicate minerals and so are called **siliciclastic sediments.** The mixture of minerals in siliciclastic sediments varies. Minerals such as quartz resist weathering and thus are found chemically unaltered in siliciclastic sediments. There may also be partly altered fragments of minerals, such as feldspar, that are less resistant to weathering and so less stable. Still other minerals in siliciclastic sediments, such as clay minerals, are newly formed by chemical weathering. Varying intensities of weathering can produce different sets of minerals in sediments derived from the same parent rock. Where weathering is intense, the sediment will contain only clastic particles made of chemically stable minerals, mixed with clay minerals. Where weathering is slight, many minerals that are unstable under land surface conditions will survive as clastic particles in the sediment. Table 6.1 shows three possible sets of minerals in sediments derived from a typical granite outcrop.

> *Siliciclastic sediments are formed by the weathering of silicate rocks.*

FIGURE 6.5 Salts precipitate when water containing dissolved minerals evaporates, as has occurred here in Death Valley, California. [John G. Wilbanks/Agefoto.]

Chemical and Biological Sediments Chemical weathering produces dissolved ions and molecules that accumulate in the waters of soils, rivers, lakes, and oceans. Chemical and biological reactions then precipitate these substances to form chemical and biological sediments. **Chemical sediments** form at or near their place of deposition. The evaporation of seawater, for example, often leads to the precipitation of gypsum or halite (**Figure 6.5**).

Biological sediments also form near their place of deposition, but they are the result of mineral precipitation within organisms as they grow. The abundance of biological sediments depends strongly on climate. Most are restricted to the subtropics and tropics, where carbonate-secreting organisms grow well. After the organisms die, their mineral remains, such as shells or skeletons, accumulate as sediments (**Figure 6.6**).

In shallow marine environments, biological sediments consist of layers of particles, such as whole or fragmented shells of marine organisms. Many different types of organisms, ranging from corals to clams to algae, may contribute their shells. Sometimes the shells are transported, further broken up, and deposited as **bioclastic sediments.** These shallow-water sediments consist predominantly of two calcium carbonate

> *Chemical sediments are formed by the precipitation of products of chemical weathering, usually at or near their place of deposition.*

> *Biological sediments are formed from the remains of organisms that precipitate minerals to form shells or skeletons.*

TABLE 6.1	Minerals Present in Sediments Derived from a Granite Outcrop Under Varying Intensities of Weathering	
Low	Medium	High
Quartz	Quartz	Quartz
Feldspar	Feldspar	Clay minerals
Mica	Mica	
Pyroxene	Clay minerals	
Amphibole		

FIGURE 6.6 One kind of sedimentary rock of biological origin is formed entirely of shell fragments. [John Grotzinger.]

minerals, calcite and aragonite, in variable proportions. Other minerals, such as phosphates and sulfates, are only locally abundant.

In the deep sea, biological sediments are made of the shells of only a few kinds of organisms. These shells are composed predominantly of the calcium carbonate mineral calcite, but silica is precipitated broadly over some parts of the deep sea. Because these biological particles accumulate in very deep water, where agitation by sediment-transporting currents is uncommon, they rarely form bioclastic sediments.

We distinguish between chemical and biological sediments for convenience only; in practice, many chemical and biological sediments overlap. In most of the world, much more rock is fragmented by physical weathering than is dissolved by chemical weathering. Thus, clastic sediments are about 10 times more abundant in Earth's crust than chemical and biological sediments.

Transportation and Deposition: The Downhill Journey to Sedimentation Sites

After clastic particles and dissolved ions have been formed by weathering and erosion, they start their journey to a sedimentary basin. This journey may be very long; for example, it might span thousands of kilometers from the tributaries of the Mississippi River in the highlands of the Rocky Mountains to the swamps of Louisiana.

Most transport agents carry material on a one-way trip downhill. Rocks falling from a cliff, sand carried by a river flowing to the sea, and glacial ice slowly creeping downhill are all responses to gravity. Although winds may blow material from a low elevation to a higher one, in the long run the effects of gravity prevail. When a windblown particle drops into the ocean and settles through the water, it is trapped. It can be picked up again only by an ocean current, which can transport it to and deposit it in another site on the seafloor. Ocean currents transport sediments over a shorter distance than do big rivers on land, and the short transportation distance for chemical and biological sediments contrasts with the much greater distances over which siliciclastic sediments are transported. But eventually, all sediment transportation paths, as simple or complicated as they may be, lead downhill into a sedimentary basin.

Currents as Transport Agents for Clastic Particles

Most sediments are transported by air and water currents. The enormous quantities of all kinds of sediments found in the oceans result primarily from the transporting capabilities of rivers, which annually carry a solid and dissolved sediment load of about 25 billion tons (25×10^{15} g) (Figure 6.7).

Air currents—winds—move materials, too, but in far smaller quantities than rivers or ocean currents. As particles are lifted into the fluid air or water, the current carries them downwind or downstream. The stronger the current—that is, the faster it flows—the larger the particles it can transport.

Most sediments are transported by air and water currents.

Current Strength, Particle Size, and Sorting
Sedimentation starts where transportation stops. For clastic

FIGURE 6.7 Sediments are easily transported by flowing water. In this photo, small ripples of sand in the channel are evidence for sediment transport. [John Grotzinger.]

particles, gravity is the driving force of sedimentation. Particles tend to settle under the pull of gravity. This tendency works against a current's ability to carry a particle. A particle's settling velocity is proportional to its density and to its size. Because all clastic particles have roughly the same density, we use particle size as the best indicator of how quickly a particle will settle. (We will take a more specific look at categories of particle size later in this chapter.) In water, large particles settle faster than small ones. This is also true in air, but the difference is much smaller.

> *The stronger a current is, the larger the particles it can transport.*

As a wind or water current begins to slow, it can no longer keep the largest particles suspended, and those particles settle. As the current slows even more, smaller particles settle. When the current stops completely, even the smallest particles settle. Currents segregate particles in the following ways:

- *Strong currents* (faster than 50 cm/s) carry gravel (which includes boulders, cobbles, and pebbles), along with an abundant supply of smaller particles. Such currents are common in swiftly flowing streams in mountainous terrain, where erosion is rapid. Beach gravels are deposited where ocean waves erode rocky shores.

- *Moderately strong currents* (20–50 cm/s) lay down sand beds. Currents of moderate strength are common in most rivers, which carry and deposit sand in their channels. Rapidly flowing floodwaters may spread sand over the width of a river valley. Waves and currents deposit sand on beaches and in the ocean. Winds also blow and deposit sand, especially in deserts. However, because air is much less dense than water, much higher current velocities are required to move sediments of the same size and density.

- *Weak currents* (slower than 20 cm/s) carry muds composed of the finest clastic particles (silt and clay). Weak currents are found on the floor of a river valley when floodwaters recede slowly or stop flowing entirely. In the ocean, muds are generally deposited some distance from shore, where currents are too slow to keep even fine particles in suspension. Much of the floor of the open ocean is covered with mud particles originally transported by surface waves and currents or by the wind. These particles slowly settle to depths where currents and waves are stilled and, ultimately, all the way to the bottom of the ocean.

As you can see, currents may begin by carrying particles of widely varying sizes, which then become separated as the strength of the current changes. A strong, fast current may lay down a bed of gravel, while keeping sand and mud in suspension. If the current weakens and slows, it will lay down a bed of sand on top of the gravel. If the current then stops altogether, it will deposit a layer of mud on top of the sand bed. This tendency for variations in current velocity to segregate sediments according to size is called **sorting.** A well-sorted sediment consists mostly of particles of a uniform size. A poorly sorted sediment contains particles of many sizes (**Figure 6.8**).

> *Variations in current strength result in the sorting of sediments by particle size.*

As pebbles and large sand grains are being transported by water and air currents, they become abraded. *Abrasion* affects particles in two ways: it reduces their size, and it rounds off their

Well-sorted sand

Poorly sorted sand

FIGURE 6.8 As the strength of currents changes, sediments are sorted according to particle size. The relatively homogeneous group of sand grains on the left is well sorted; the group on the right is poorly sorted. [Bill Lyons.]

FIGURE 6.9 Abrasion during transportation reduces the size and angularity of clastic particles. Particles become more rounded and slightly smaller as they are transported, although the general shape of the particle may not change significantly.

rough edges (Figure 6.9). As these particles are transported, they tumble and strike one another or rub against the underlying rock. These effects apply mostly to the larger particles; smaller sand grains and silt undergo little abrasion.

Particles are generally transported intermittently rather than steadily. A river may transport large quantities of sand and gravel when it floods, but will drop them as the flood recedes, only to pick them up again and carry them even farther in the next flood. Likewise, strong winds may carry large amounts of dust for a few days, then die down and deposit the dust as a layer of sediment. The strong tidal currents along some ocean margins may transport broken shell fragments to places farther offshore and drop them there.

The total time it takes for clastic particles to be transported may be many hundreds or thousands of years, depending on the distance to the final sedimentary basin and the number of stop-offs along the way. Clastic particles eroded by the headwaters of the Missouri River in the mountains of western Montana, for example, take hundreds of years to travel the 3200 km down the Missouri and Mississippi rivers to the Gulf of Mexico.

Oceans as Chemical Mixing Vats

The driving force of chemical and biological sedimentation is precipitation rather than gravity. Chemical substances dissolved in water during weathering are carried along with the water. Materials such as dissolved calcium ions are part of the water solution itself, so gravity cannot cause them to settle out. As dissolved materials flow down rivers, they ultimately enter the ocean.

Oceans may be thought of as huge chemical mixing vats. Rivers, rain, wind, and glaciers constantly bring in dissolved materials. Smaller quantities of dissolved materials enter the oceans through hydrothermal chemical reactions between seawater and hot basalt at mid-ocean ridges. The oceans lose water continuously by evaporation at the surface. The inflow and outflow of water to and from the oceans are so exactly balanced that the amount of water in the oceans remains constant over such geologically short times as years, decades, or even centuries. Over a time scale of thousands to millions of years, however, the balance may shift. During the Pleistocene ice ages, for example, significant quantities of seawater were converted into glacial ice, and sea level was drawn down by more than 100 m.

The entry and exit of dissolved materials, too, are balanced. Each of the many dissolved components of seawater participates in some chemical or biological reaction that eventually precipitates it out of the water and onto the seafloor. As a result, the ocean's **salinity**—the total amount of dissolved substances in a given volume of water—remains constant. Totaled over all the oceans of the world, precipitation balances the total inflow of dissolved material from continental weathering processes and from hydrothermal activity at mid-ocean ridges—yet another way in which the Earth system maintains balance.

We can better understand this chemical balance by considering the element calcium. Calcium is a component of the most abundant biological precipitate formed in the oceans, calcium carbonate ($CaCO_3$). Calcium dissolves when limestone and silicate minerals containing calcium, such as some feldspars and pyroxenes, are weathered on land, and that calcium is brought to the ocean as calcium ions (Ca^{2+}). There, a wide variety of marine organisms take up the calcium ions and combine them with carbonate ions (CO_3^{2-}), also present in seawater, to form their calcium carbonate shells. The calcium that entered the ocean as dissolved ions leaves it as solid sediment particles when the organisms die and their shells settle to the seafloor and accumulate there as calcium carbonate sediments. Ultimately, the calcium carbonate sediments will be buried and transformed into limestone. The chemical balance that keeps the levels of calcium dissolved in the ocean constant is thus controlled in part by the activities of organisms.

Nonbiological mechanisms also maintain chemical balance in the oceans. For example, sodium ions (Na^+) brought into the oceans react chemically with chloride ions (Cl^-) to form the precipitate sodium chloride (NaCl). This happens when evaporation raises the concentrations of sodium and chloride ions past the point of saturation.

> Oceans contain abundant dissolved substances produced by chemical weathering, which can precipitate to form chemical or biological sediments.

As we saw in Chapter 4, solutions crystallize minerals when they become so saturated with dissolved materials that they can hold no more. The intense evaporation required to crystallize salt takes place in warm, shallow arms of the sea.

SEDIMENTARY BASINS: The Sinks for Sediments

As we have seen, the currents that move sediments across Earth's surface generally flow downhill. Therefore, sediments tend to accumulate in depressions in Earth's crust. Such depressions are formed by *subsidence,* in which a broad area of the crust sinks (subsides) relative to the surrounding crust. Subsidence is induced partly by the weight of sediments on the crust but is caused mainly by tectonic forces.

Sedimentary basins are regions of at least 10,000 km² where the combination of sedimentation and subsidence has formed thick accumulations of sediments and sedimentary rock. Sedimentary basins are Earth's primary sources of oil and gas. Commercial exploration for these resources has helped us better understand the deep structure of sedimentary basins and of the continental lithosphere.

Rift Basins and Thermal Subsidence Basins

When plate separation begins within a continent, subsidence results from the stretching, thinning, and heating of the underlying lithosphere by the tectonic forces that are causing the separation (**Figure 6.10**). A long, narrow rift develops. Hot, ductile mantle material rises, melts, and fills

1. A rift develops as hot mantle materials well up and the lithosphere stretches and thins.

2. Seafloor spreading begins.

3. Evaporites, deltaic sediments, and carbonates are deposited.

4. These deposits are then buried by accumulation of further sediments and undergo diagenesis.

FIGURE 6.10 Sedimentary basins are formed where continents separate.

the space created by the thinned lithosphere and crust, initiating the eruption of basaltic lavas in the rift zone. Such **rift basins** are deep, narrow, and long, with thick successions of sedimentary rocks and extrusive and intrusive igneous rocks. The rift valleys of East Africa, the Rio Grande Valley, and the Jordan Valley in the Middle East are examples of rift basins.

At later stages of plate separation, when rifting has led to seafloor spreading and the newly separated continents are drifting away from each other, basin subsidence continues through the cooling of the lithosphere that was thinned and heated during the earlier rifting stage (see Figure 6.10). Cooling leads to an increase in the density of the lithosphere, which in turn leads to its subsidence below sea level. Because cooling of the lithosphere is the main process creating basins at this stage, they are called **thermal subsidence basins.** Sediments from erosion of the adjacent land fill the basin nearly to sea level along the edge of the continent, creating a **continental shelf.**

The continental shelves off the Atlantic coasts of North and South America, Europe, and Africa are good examples of thermal subsidence basins. These basins began to form when the supercontinent Pangaea split apart about 200 million years ago and the North American and South American plates separated from the Eurasian and African plates. The continental shelf continues to receive sediments for a long time because the trailing edge of the drifting continent subsides slowly and because the continents provide a tremendous area from which sediments can be derived. The load of the growing mass of sediment further depresses the crust, so the basins can receive still more material from the land. As a result of this continuous subsidence and sediment supply, continental shelf deposits can accumulate in an orderly fashion to thicknesses of 10 km or more.

Flexural Basins

A third type of basin develops at convergent plate boundaries where one plate pushes up over the other. The weight of the overriding plate causes the underlying plate to bend or flex down, producing a **flexural basin.** The Mesopotamian Basin in Iraq is a flexural basin formed when the Arabian Plate collided with and was subducted beneath the Eurasian Plate. The enormous oil reserves in Iraq (second only to Saudi Arabia's) owe their size to having the right ingredients in this important flexural basin. In effect, oil that had formed in the rocks now beneath the Zagros Mountains in Iran was squeezed out, forming several great pools of oil with volumes larger than 10 billion barrels.

Plate separation produces rift basins. As separation continues, plate cooling forms thermal subsidence basins. Flexural basins are generated where plates converge and one plate bends down under the weight of the other.

SEDIMENTARY ENVIRONMENTS

Between the source area where sediments are formed and the sedimentary basin where they are buried and converted to sedimentary rocks, sediments travel a path through many sedimentary environments. A **sedimentary environment** is a geographic location characterized by a particular combination of climate conditions and physical, chemical, and biological processes (Figure 6.11). Important characteristics of sedimentary environments include the following:

- The type and amount of water (ocean, lake, river, arid land)
- The type and strength of transport agents (water, wind, ice)
- The topography (lowland, mountain, coastal plain, shallow sea, deep sea)
- Biological activity (precipitation of shells, growth of coral reefs, churning of sediments by worms and other burrowing organisms)
- The tectonic settings of sediment source areas (volcanic arc, continental collision zone) and sedimentary basins (rift, thermal subsidence, flexural)
- The climate (cold climates may form glaciers; arid climates may form deserts where minerals precipitate)

Consider the beaches of Hawaii, famous for their unusual green sands, which are a result of their distinct sedimentary environment. Hawaii is a volcanic island made of olivine-bearing basalt, which is released during weathering. Rivers transport the olivine to the beach, where waves approaching and breaking on the shore, and the resulting currents that develop, concentrate the olivine and remove fragments of basalt to form olivine-rich sand deposits.

Sedimentary environments are often grouped by location: on continents, near shorelines, or in the ocean. This very general subdivision highlights the processes that give sedimentary environments their distinct identities.

A sedimentary environment is a geographic location characterized by a particular combination of climate conditions and physical, chemical, and biological processes.

Continental Environments

Sedimentary environments on continents are diverse due to the wide variation in temperature and rainfall on the surface of the land. These environments are built around lakes, rivers, deserts, and glaciers (see Figure 6.11).

- *Lake environments* include inland bodies of fresh or saline water in which the transport agents are relatively small waves and moderate currents.

FIGURE 6.11 Multiple factors interact to create sedimentary environments.

Chemical sedimentation of organic matter and carbonates may occur in freshwater lakes. Saline lakes such as those found in deserts evaporate and precipitate a variety of *evaporite* minerals, such as halite. The Great Salt Lake in Utah is an example.

- *Alluvial environments* include the channel of a river, its borders, and the flat valley floor on either side of the channel that is covered by water when the river floods (the floodplain). Rivers are present on all the continents except Antarctica, so alluvial deposits are widespread. Organisms are abundant in the muddy flood deposits and produce organic sediments that accumulate in swamplands adjacent to river channels. Climates vary from arid to humid. An example is the Mississippi River and its floodplains.

- *Desert environments* are arid. Wind and the rivers that flow intermittently through deserts transport sand and dust. The dry climate inhibits abundant organic growth, so organisms have little effect on the sediments. Desert sand dunes are an example of such an environment.

- A *glacial environment* is dominated by the dynamics of moving masses of ice and is characterized by a cold climate. Vegetation is present but has little effect on sediments. At the melting border of a glacier, meltwater streams form a transitional alluvial environment.

> Continents offer diverse sedimentary environments because factors such as temperature and rainfall vary widely.

Shoreline Environments

The dynamics of waves, tides, and river currents on sandy shores dominate shoreline environments (see Figure 6.11). Shoreline environments include the following:

- *Deltas,* where rivers enter lakes or the sea
- *Tidal flats,* where extensive areas exposed at low tide are dominated by tidal currents
- *Beaches,* where the strong waves approaching and breaking on the shore distribute sediments on the beach, depositing strips of sand or gravel

In most cases, the sediments that accumulate are siliciclastic. Organisms affect these sediments mostly by burrowing into them. However, in some tropical and subtropical settings, sediment particles, particularly carbonate sediments, may be of biological origin. These biological carbonate sediments are also subject to transport by waves and tidal currents.

> Shorelines offer sedimentary environments that are dominated by the movement of water.

Marine Environments

Marine environments are usually classified by water depth, which determines the kinds of currents that are present (see Figure 6.11). Alternatively, they can be classified on the basis of distance from land.

- *Continental shelf environments* are located in the shallow waters off continental shores, where sedimentation is controlled by relatively gentle currents. Sediments may be composed of either siliciclastic particles or biological carbonate particles, depending on how much siliciclastic sediment is supplied by rivers and on the abundance of carbonate-producing organisms. Sedimentation may also be chemical if the climate is arid and an arm of the sea becomes isolated.

- *Organic reefs* are composed of carbonate structures, formed by carbonate-secreting organisms, built up on continental shelves or on oceanic volcanic islands.

- *Continental margin and slope environments* are found in the deeper waters at and off the edges of the continents, where sediments are deposited by turbidity currents. A *turbidity current* is a turbulent submarine avalanche of sediment and water that moves downslope. Sediments deposited by turbidity currents are almost always siliciclastic, except at sites where organisms produce a lot of carbonate sediments. In this case, continental margin and slope sediments may be rich in carbonates.

- *Deep-sea environments* are found far from continents, where the waters are much deeper than the reach of waves and tidal currents. These environments include the lower portion of the continental slope, which is built up by turbidity currents traveling far from continental margins; the abyssal plains of the deep sea, which accumulate carbonate sediments provided mostly by the skeletons of plankton; and the mid-ocean ridges.

> Marine sedimentary environments vary with water depth and distance from land.

We have seen that sedimentary environments can be defined by location. They can also be categorized according to the kinds of sediments found in them or according to the dominant sediment formation process. Grouping in this manner produces two broad classes: siliciclastic sedimentary environments and chemical and biological sedimentary environments.

Siliciclastic versus Chemical and Biological Sedimentary Environments

Siliciclastic sedimentary environments are those dominated by siliciclastic sediments. They include all of the

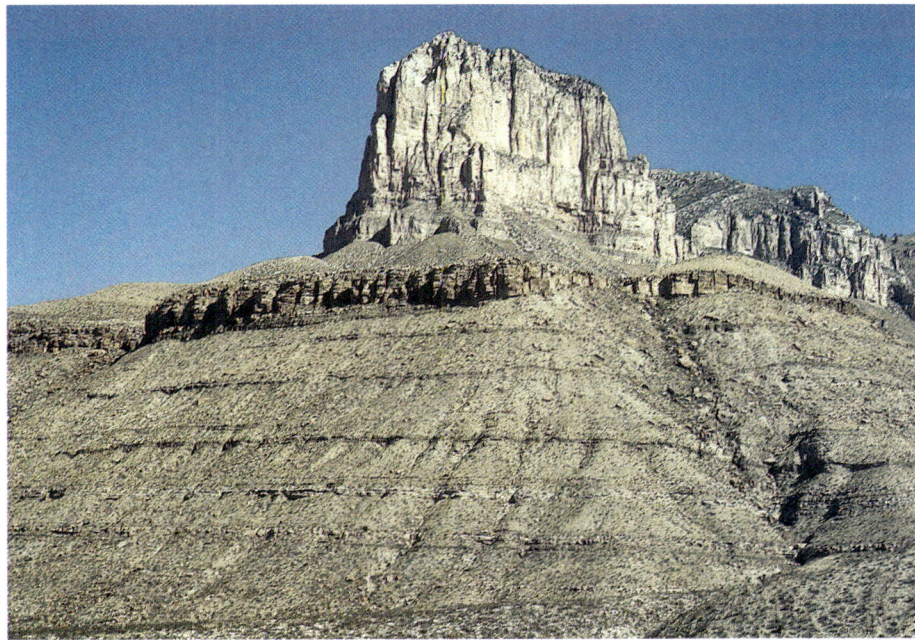

FIGURE 6.12 Sedimentary rocks exposed at El Capitan, in the Guadalupe Mountains of New Mexico, were formed in an ancient ocean about 260 million years ago. The lower slopes of the mountains contain siliciclastic sedimentary rocks formed in deep-sea environments. The overlying cliffs of El Capitan are limestone and dolostone, which formed in a shallow sea when carbonate-secreting organisms died, leaving their shells to form a reef. [John Grotzinger.]

continental sedimentary environments as well as those shoreline environments that serve as transitional zones between continental and marine environments. They also include those marine environments of the continental shelf, continental margin, and deep seafloor where siliciclastic sands and muds are deposited (Figure 6.12).

Chemical and biological sedimentary environments are characterized principally by chemical and biological precipitation (Table 6.2). By far the most abundant are *carbonate environments,* marine settings where calcium carbonate, mostly secreted by organisms, is the main sediment. Hundreds of species of mollusks and other invertebrate organisms, as well as calcareous (calcium-containing) algae, secrete carbonate shell materials. Various populations of these organisms live at different depths, both in quiet areas and in places where waves and currents are strong. As they die, their shells accumulate to form sediments.

Except for those of the deep sea, carbonate environments are found mostly in the warmer tropical and subtropical regions of the oceans, where carbonate-secreting organisms flourish. These environments include organic

TABLE 6.2	Major Chemical and Biological Sedimentary Environments	
Environment	Agent of Precipitation	Sediments
SHORELINE AND MARINE		
Carbonate (reefs, platforms, deep sea, etc.)	Shelled organisms, some algae; inorganic precipitation from seawater	Carbonate sands and muds, reefs
Evaporite	Evaporation of seawater	Gypsum, halite, other salts
Siliceous (deep sea)	Shelled organisms	Silica
CONTINENTAL		
Evaporite	Evaporation of lake water	Halite, borates, nitrates, carbonates, other salts
Swamp	Vegetation	Peat

reefs, carbonate sand beaches, tidal flats, and shallow carbonate platforms. In a few places, carbonate sediments form in cooler waters that are supersaturated with carbonate—waters that are generally below 20°C, such as in some regions of the Antarctic Ocean south of Australia. Carbonate sediments in cool waters are formed by a very limited group of organisms that mainly secrete calcite shell materials.

Siliceous environments are unique deep-sea sedimentary environments named for the remains of silica shells deposited in them. The planktonic organisms that secrete silica grow in surface waters where nutrients are abundant. When they die, their shells settle to the deep seafloor and accumulate as layers of siliceous sediments.

An *evaporite environment* is created when the warm seawater of an arid inlet or arm of the sea evaporates more rapidly than it can mix with seawater from the open ocean. The degree of evaporation and the length of time it has proceeded control the salinity of the evaporating seawater and thus the kinds of chemical sediments formed. Evaporite environments also form in lakes lacking river outlets. Such lakes may produce sediments of halite, borate, nitrates, and other salts.

SEDIMENTARY STRUCTURES

Sedimentary structures include all kinds of features formed at the time of deposition. Sediments and sedimentary rocks are characterized by **bedding,** or *stratification,* which occurs when layers of sediment, or *beds,* with different particle sizes or compositions are deposited on top of one another. These beds range from only millimeters or centimeters thick to meters or even many meters thick. Most bedding is horizontal, or nearly so, at the time of deposition. Some types of bedding, however, form at a high angle relative to the horizontal.

> Bedding is produced when layers of sediment with different particle sizes or compositions are deposited on top of one another.

Cross-Bedding

Cross-bedding consists of beds deposited by wind or water and inclined at angles as much as 35° from the horizontal (**Figure 6.13**). Cross-beds form when sediment particles are deposited on the steeper, downcurrent (lee) slopes of sand dunes on land or of sandbars in rivers and under the sea (**Figure 6.14**). Cross-bedding in wind-deposited sand dunes may be complex as a result of rapidly changing wind directions. Cross-bedding is common in sandstones and is also found in gravels and some carbonate sediments. Cross-bedding is easier to see in sandstones than in sands, which must be excavated to see a cross section.

> Cross-bedding consists of layers of sediments deposited at an angle to the horizontal.

FIGURE 6.13 Cross-bedding in a desert environment: Navajo Sandstone, Zion National Park, southwestern Utah. The varying angles of cross-bedding in this sandstone are due to changes in wind direction at the time the sand dunes were deposited. [Peter Kresan.]

Graded Bedding

Graded bedding is most abundant in continental slope and deep-sea sediments deposited by dense, muddy turbidity currents, which hug the bottom topography of the ocean as they move downhill. Each bed progresses from large particles at the bottom to small particles at the top.

FIGURE 6.14 Cross-beds form when sediment particles are deposited on the steeper, downcurrent (lee) slope of a sand dune or sandbar.

FIGURE 6.15 Ripples. (*left*) Ripples in modern sand on a beach. [John Grotzinger.] (*right*) Ancient ripple-marked sandstone. [John Grotzinger/Ramón Rivera-Moret/MIT.]

As the current progressively slows, it drops progressively smaller particles. The grading indicates a weakening of the current that deposited the particles. A graded bed comprises one set of sediment particles, normally ranging from a few centimeters to several meters thick, that formed a horizontal or nearly horizontal layer at the time of deposition. Accumulations of many individual graded beds can reach a total thickness of hundreds of meters. A graded bed formed as a result of deposition by a turbidity current is called a *turbidite*.

> *Graded bedding is formed when particle sizes within a bed progress from large at the bottom of the bed to small at the top.*

Ripples

Ripples are very small ridges of sand or silt whose long dimension is at right angles to the current. They form low, narrow ridges, usually only a centimeter or two high, separated by wider troughs. These sedimentary structures are common in both modern sands and ancient sandstones (**Figure 6.15**). Ripples can be seen on the surfaces of windswept dunes, on underwater sandbars in shallow streams, and under the waves at beaches. Geologists can distinguish the symmetrical ripples made by waves moving back and forth on a beach from the asymmetrical ripples formed by currents moving in a single direction over river sandbars or windswept dunes (**Figure 6.16**).

> *Ripples are very small ridges of sand or silt formed by currents of water or wind.*

Bioturbation Structures

In many sedimentary rocks, the bedding is broken or disrupted by roughly cylindrical tubes a few centimeters in diameter that extend vertically through several beds. These sedimentary structures are remnants of burrows and tunnels excavated by clams, worms, and other marine organisms that live on the bottom of the sea. These organisms churn

FIGURE 6.16 The shapes of ripples on beach sand, produced by the back-and-forth movements of waves, are symmetrical. Ripples on dunes and river sandbars, produced by the movement of a current in one direction, are asymmetrical.

FIGURE 6.17 Bioturbation structures. This rock is crisscrossed with fossilized tunnels originally made by organisms burrowing through the mud. [John Grotzinger/Ramón Rivera-Moret/Harvard Mineralogical Museum.]

Bedding Sequences

Bedding sequences are built of interbedded and vertically stacked layers of different sedimentary rock types. A bedding sequence might consist of cross-bedded sandstone, overlain by bioturbated siltstone, overlain in turn by rippled sandstone—in any combination of thicknesses for each rock type in the sequence.

Bedding sequences help geologists reconstruct how sediments were deposited and so provide insight into the history of geologic processes and events that occurred at Earth's surface long ago. Figure 6.18 shows a bedding sequence typically formed by rivers. A river lays down sediments as its channel meanders back and forth across the valley floor. Thus, the lower part of each sequence contains the beds deposited in the deepest part of the channel, where the current was strongest. The middle part contains the beds deposited in the shallower parts of the channel, where the current was weaker, and the upper part contains the beds deposited on the floodplain. Typically, a bedding sequence formed in this manner will consist of sediment particles that grade upward from large to small.

Most bedding sequences consist of a number of small-scale subdivisions. In the example shown in Figure 6.18, the basal layers of the bedding contain cross-bedding. These layers are overlain by more cross-bedded layers, but the cross-beds are smaller in scale. Horizontal bedding occurs at the top of the

and burrow through muds and sands—a process called **bioturbation**. They ingest the sediment, digest the bits of organic matter it contains, and leave behind the reworked sediment, which fills the burrow (Figure 6.17). From bioturbation structures, geologists can determine the behavior of organisms that burrowed in the sediment. Since the behavior of burrowing organisms is controlled partly by environmental factors, such as the strength of currents or the availability of nutrients, bioturbation structures help us reconstruct past sedimentary environments.

> Bioturbation is the disruption of sediments by burrowing organisms.

> Bedding sequences consist of vertically stacked beds of different sedimentary rock types. They provide a history of geologic processes and events that occurred when the beds were formed.

FIGURE 6.18 A typical bedding sequence formed by a meandering river. [USDA-NRCS photo by Jim R. Fortner.]

bedding sequence. Today, computer models are used to analyze how bedding sequences of sands were deposited in alluvial environments. Other types of bedding sequences—which consist of different arrangements of sedimentary structures—tell us about different sedimentary environments.

BURIAL AND DIAGENESIS: From Sediment to Rock

Most of the siliciclastic particles produced by weathering and erosion on land end up deposited in various sedimentary basins in the world's oceans. A smaller amount of siliciclastic sediment is deposited in sedimentary basins on land. Most chemical and biological sediments are also deposited in ocean basins, although some are deposited in continental basins containing lakes and swamps.

Burial

Once sediments reach the ocean floor, they are trapped there. The deep seafloor is the ultimate sedimentary basin and, for most sediments, their final resting place. As the sediments are buried under new layers of sediments, they are subjected to increasingly high temperatures and pressures as well as chemical changes.

Diagenesis

After sediments are deposited and buried, they are subject to **diagenesis**—the many physical and chemical changes that result from the increasing temperatures and pressures as they are buried ever deeper in Earth's crust (**Figure 6.19**). These changes continue until the sediment or sedimentary rock is either exposed to weathering or metamorphosed by more extreme heat and pressure.

Diagenesis includes the processes—physical and chemical—that change sediments to sedimentary rocks.

Compaction by burial squeezes out water.

Compaction

50–60% water

↓

10–20% water

Lithification

Precipitation or addition of new minerals cements sediment particles.

Cementation

Lithification

Different sediments result in different sedimentary rocks.

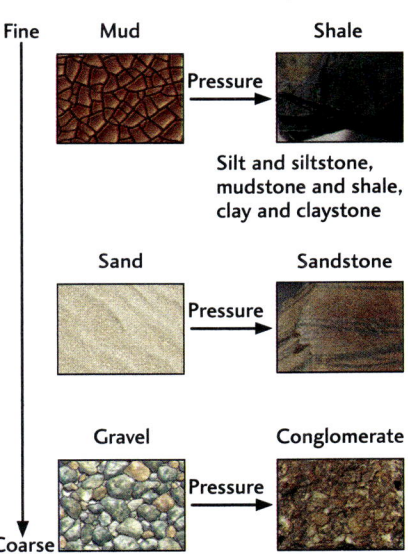

Fine → Mud → Pressure → Shale

Silt and siltstone, mudstone and shale, clay and claystone

Sand → Pressure → Sandstone

Gravel → Pressure → Conglomerate

Coarse

Organic matter

↓ Pressure

Heat to 90°–120°C → Oil and gas

Heat to 90°–120°C → Coal

Sediments are buried, compacted, and lithified at shallow depths in Earth's crust.

FIGURE 6.19 Diagenesis is the process of physical and chemical change that converts sediments into sedimentary rocks. [*shale:* John Grotzinger/Ramón Rivera-Moret/Harvard Mineralogical Museum; *sandstone, conglomerate, coal:* John Grotzinger/Ramón Rivera-Moret/MIT; *coal and gas:* John Woolsey.]

> *Diagenesis is the set of physical and chemical processes that change sediments after they are deposited and buried.*

Temperature increases with depth in Earth's crust at an average rate of 30°C for each kilometer of depth. At a depth of 4 km, buried sediments may reach 120°C or more, the temperature at which certain types of organic matter may be converted to oil and gas. Pressure also increases with depth—on average, about 1 atmosphere for each 4.4 meters of depth. This increased pressure is responsible for the compaction of buried sediments.

Buried sediments are also continuously bathed in groundwater full of dissolved minerals, which can precipitate in the pores between the sediment particles and bind them together—a chemical change called **cementation.** Cementation decreases **porosity,** the percentage of a rock's volume consisting of open pores between particles. In some sands, for example, calcium carbonate is precipitated as calcite, which acts as a cement that binds the grains and hardens the resulting mass into sandstone (**Figure 6.20**). Other minerals, such as quartz, may cement sands, muds, and gravels into sandstone, mudstone, and conglomerate.

The major physical diagenetic change is **compaction,** a decrease in the volume and porosity of a sediment. Compaction occurs as sediment particles are squeezed closer together by the weight of overlying sediment.

> *Cementation, the precipitation of minerals that fill in the pores between sediment particles and bind them together, and compaction, the reduction of pore space by squeezing particles together, result in lithification, the transformation of sediment into sedimentary rock.*

Sands are fairly well packed during deposition, so they do not compact much. However, newly deposited muds, including carbonate muds, are highly porous. Often, more than 60 percent of these sediments consists of water in pore spaces. As a result, muds compact greatly after burial, losing more than half their water.

Both cementation and compaction result in **lithification,** the hardening of soft sediment into rock.

CLASSIFICATION OF SILICICLASTIC SEDIMENTS AND SEDIMENTARY ROCKS

We can now use our knowledge of sedimentation to classify sediments and their lithified counterparts, sedimentary rocks. As we have seen, the major divisions are the siliciclastic sediments and sedimentary rocks and the chemical and biological sediments and sedimentary rocks. Siliciclastic sediments and sedimentary rocks constitute more than three-fourths of the total mass of all types of sediments and sedimentary rocks in Earth's crust (**Figure 6.21**). We therefore begin with them.

Classification by Particle Size

Siliciclastic sediments and rocks are categorized primarily by particle size (Table 6.3):

- *Coarse-grained:* gravel and conglomerate
- *Medium-grained:* sand and sandstone
- *Fine-grained:* silt and siltstone; mud, mudstone, and shale; clay and claystone

We classify siliciclastic sediments and rocks on the basis of their particle size because it distinguishes them by one of the most important conditions of sedimenta-

FIGURE 6.20 This photomicrograph of sandstone shows quartz grains (white and gray) cemented by calcite (brightly colored and variegated) introduced after deposition. [Peter Kresan.]

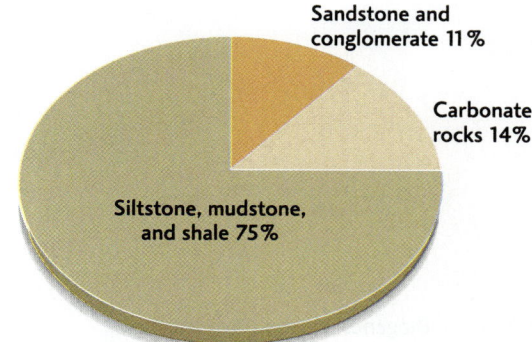

FIGURE 6.21 The relative abundances of the major sedimentary rock types. In comparison with these three types, all other sedimentary rock types—including evaporites, cherts, and other chemical sedimentary rocks—exist in only minor amounts.

TABLE 6.3 Major Classes of Clastic Sediments and Sedimentary Rocks

Particle Size	Sediment	Rock
COARSE	**GRAVEL**	
Larger than 256 mm	Boulder	
256–64 mm	Cobble	Conglomerate
64–2 mm	Pebble	
MEDIUM		
2–0.062 mm	**SAND**	Sandstone
FINE	**MUD**	
0.062–0.0039 mm	Silt	Siltstone
Finer than 0.0039 mm	Clay	Mudstone (blocky fracture)
		Shale (breaks along bedding)
		Claystone

tion: current strength. As we have seen, the larger the particle, the stronger the current needed to move and deposit it. This relationship between current strength and particle size is the reason like-sized particles tend to accumulate in sorted beds. In other words, most sand beds do not contain pebbles or mud, and most muds consist only of particles finer than sand.

Of the various types of siliciclastic sediments and sedimentary rocks, the fine-grained clastics are by far the most abundant—about three times more common than the coarser-grained clastics. The abundance of the fine-grained siliciclastics, which contain large amounts of clay minerals, is due to the chemical weathering of the large quantities of feldspar and other silicate minerals in Earth's crust into clay minerals. We turn now to a consideration of each of the three types of siliciclastic sediments and sedimentary rocks in more detail (**Figure 6.22**).

Coarse-Grained Siliciclastics: Gravel and Conglomerate

Gravel is the coarsest siliciclastic sediment, consisting of particles larger than 2 mm in diameter and including pebbles, cobbles, and boulders (see Table 6.3). **Conglomerates** are the lithified equivalents of gravel (Figure 6.22a). Pebbles, cobbles, and boulders are easy to study and identify because of their large size. Their size can tell us the strength of the currents that transported them. In addition, their composition can tell us about the nature of the distant terrain where they were produced.

(a) Conglomerate

(b) Sandstone

(c) Shale

FIGURE 6.22 Clastic sedimentary rocks. [*conglomerate and sandstone:* John Grotzinger/Ramón Rivera-Moret/MIT; *shale:* John Grotzinger/Ramón Rivera-Moret/Harvard Mineralogical Museum.]

There are relatively few environments—mountain streams, rocky beaches with high waves, and glacier meltwaters—in which currents are strong enough to transport pebbles. Strong currents also carry sand, and we almost always find sand between pebbles. Some of it is deposited with the gravel, and some infiltrates the spaces between fragments after the gravel is deposited. Pebbles and cobbles become rounded very quickly by abrasion in the course of transport on land or in water.

Conglomerates are sedimentary rocks formed from lithified coarse-grained sediments, equivalent in particle size to gravel.

Medium-Grained Siliciclastics: Sand and Sandstone

Sand consists of medium-sized particles, ranging from 0.062 to 2 mm in diameter (see Table 6.3). These sediments can be moved by moderate currents, such as those of rivers, waves at shorelines, and the winds that blow sand into dunes. Sand particles are large enough to be seen with the naked eye, and many of their features are easily discerned with a low-power magnifying glass. The lithified equivalent of sand is **sandstone** (Figure 6.22b).

Sandstones are sedimentary rocks formed from lithified sand.

Sizes and Shapes of Sand Grains Medium-sized siliciclastic particles—sand grains—are subdivided into fine, medium, and coarse grains. The average size of the grains in any one sandstone can be an important clue to both the strength of the current that carried them and the sizes of the crystals eroded from the parent rock. The range and relative abundances of the various sizes are also significant. If all the grains are close to the average size, the sand is well sorted. If many grains are much larger or smaller than the average, the sand is poorly sorted. The degree of sorting can help us distinguish, for example, between sands of beaches (which tend to be well sorted) and muddy sands deposited by glaciers (which tend to be poorly sorted). The shapes of sand grains can also be important clues to their origin. Sand grains, like pebbles, are abraded and rounded during transport. Angular grains imply short transport distances; rounded ones indicate long journeys down a large river system.

Mineralogy of Sands and Sandstones Siliciclastics can be further subdivided by mineralogy, which can help identify the parent rocks. Thus, there are quartz-rich and feldspar-rich sandstones. Some sands are bioclastics; they are formed from material such as carbonate that was originally precipitated as shells, but then was broken and transported by currents. Thus, the mineral composition of sands and sandstones indicates the source areas that were eroded to produce the sand grains. Sodium- and potassium-rich feldspars with abundant quartz, for example, might indicate that the sediments were eroded from a granitic terrain. Other minerals, as we will see in Chapter 7, might indicate metamorphic parent rocks.

The mineral content of sands and sandstones also indicates the plate tectonic setting of the parent rock. Sandstones containing abundant fragments of mafic volcanic rocks, for example, might indicate that the sand grains were derived from a volcanic arc at a subduction zone.

In addition to particle size, siliciclastic sedimentary rocks can be classified by the sorting, shape, and mineralogy of the component particles.

Fine-Grained Siliciclastics

The finest-grained siliciclastic sediments and sedimentary rocks are the silts and siltstones; the muds, mudstones, and shales; and the clays and claystones. All of them consist of particles that are less than 0.062 mm in diameter, but the sediments vary widely in their range of grain sizes and mineral compositions. Fine-grained sediments are deposited by the gentlest currents, which allow the finest particles to settle slowly to the bottom in quiet waves.

Silt and Siltstone **Siltstone** is the lithified equivalent of **silt**, a siliciclastic sediment in which most of the grains are between 0.0039 and 0.062 mm in diameter. Siltstone looks similar to mudstone or very fine grained sandstone.

Mud, Mudstone, and Shale **Mud** is a siliciclastic sediment, mixed with water, in which most of the particles are less than 0.062 mm in diameter (see Table 6.3). Thus, mud can be made of silt- or clay-sized sediments or varying quantities of both. This general term is very useful in fieldwork because it is often difficult to distinguish between silt- and clay-sized sediments without a microscope. Muds are deposited by rivers and tides. As a river recedes after flooding, the current slows, and mud settles, some of it containing abundant organic matter. This mud contributes to the fertility of river floodplains. Muds are also left behind by ebbing tides along many tidal flats where wave action is mild. Much of the deep seafloor, where currents are weak or absent, is blanketed by muds.

The fine-grained rock equivalents of muds are mudstones and shales. **Mudstones** are blocky and show poor or no bedding. Bedding may have been well marked when the sediments were first deposited but then lost through bioturbation. **Shales** (Figure 6.22c) are composed of silt plus a significant component of clay, which causes them to break readily along bedding planes. Many muds are more than 10 percent carbonate, forming calcareous mudstones and shales. Black, or organic, shales contain abundant organic matter. Some, called oil shales, contain large quantities of oily organic material, which makes them a potentially important source of oil. (We consider the oil shales in more detail in Chapter 14.)

Clay and Claystone Clay is the most abundant component of fine-grained sediments and sedimentary rocks and consists largely of clay minerals. Clay-sized particles are less than 0.0039 mm in diameter (see Table 6.3). Rocks made up exclusively of clay-sized particles are called **claystones.**

> *Fine-grained siliciclastic sedimentary rocks include siltstones, mudstones, shales, and claystones.*

CLASSIFICATION OF CHEMICAL AND BIOLOGICAL SEDIMENTS AND SEDIMENTARY ROCKS

Chemical and biological sediments and sedimentary rocks can be classified by their chemical composition (Table 6.4). Geologists distinguish between chemical and biological sediments not only for convenience, but also to emphasize the importance of organisms as the chief mediators of biological sedimentation. Both kinds of sediments can tell us about chemical conditions in the ocean, the predominant environment of sedimentation. Chemical sedimentation occurs in some lakes as well as in oceans, but lake sediments account for only a very small fraction of chemical sediments. Chemical sediments are less abundant than biological sediments.

Biological Sediments: Carbonate Sediments and Rocks

Carbonate sediments and **carbonate rocks** form mostly from the accumulation and lithification of carbonate minerals that are secreted by organisms. In addition to calcite, most carbonate sediments contain aragonite, a less stable form of calcium carbonate. Some organisms precipitate calcite, some precipitate aragonite, and some precipitate both.

Most carbonate sediments of shallow marine environments are bioclastic. They were originally secreted as shells by organisms living near the surface or on the bottom. After they die, the organisms break apart, producing shells or fragments of shells that constitute individual particles of carbonate sediment. These sediments are found from the coral reefs of the Pacific and Caribbean to the shallow carbonate platforms of the Bahamas. Carbonate sediments are most accessible for study in these spectacular vacation spots, but the deep sea is where most carbonate sediments are deposited today.

Most of the carbonate sediments deposited on the abyssal plains of the deep sea are derived from the calcite shells of *foraminifers,* tiny single-celled organisms that live

> *Carbonate sediments and sedimentary rocks form from the accumulation and lithification of carbonate minerals that are secreted by organisms.*

TABLE 6.4 Classification of Biological and Chemical Sediments and Sedimentary Rocks

Sediment	Rock	Chemical Composition	Minerals
BIOLOGICAL			
Sand and mud (primarily bioclastic)	Limestone	Calcium carbonate ($CaCO_3$)	Calcite, aragonite
Siliceous sediment	Chert	Silica (SiO_2)	Opal, chalcedony, quartz
Peat, organic matter	Organics	Carbon compounds; carbon compounded with oxygen and hydrogen	(Coal, oil, gas)
No primary sediment (formed by diagenesis)	Phosphorite	Calcium phosphate ($Ca_3[PO_4]_2$)	Apatite
CHEMICAL			
No primary sediment (formed by diagenesis)	Dolostone	Calcium-magnesium carbonate ($CaMg[CO_3]_2$)	Dolomite
Iron oxide sediment	Iron formation	Iron silicate; oxide (Fe_2O_3); limonite, carbonate	Hematite, siderite
Evaporite sediment	Evaporite	Calcium sulfate ($CaSO_4$); sodium chloride (NaCl)	Gypsum, anhydrite, halite, other salts

in surface waters, and other organisms that secrete calcium carbonate. When the organisms die, their shells and skeletons settle to the seafloor and accumulate there as sediments.

Reefs are moundlike or ridgelike organic structures composed of the carbonate skeletons and shells of millions of organisms. In the warm seas of the present, reefs are built mainly by corals, but hundreds of other organisms, such as algae, clams, and snails, also contribute. In contrast to the soft, loose sediments produced in other carbonate environments, the reef forms a rigid, wave-resistant structure of solid calcite and aragonite that is built up to and slightly above sea level (Figure 6.23). The solid calcite and aragonite of the reef are produced directly by the carbonate-cementing action of the organisms; there is no soft sediment stage.

Carbonate rocks are abundant because of the large amounts of calcium and carbonate present in seawater, which organisms can convert directly into shells. Calcium is supplied by weathering of feldspars and other minerals in igneous and metamorphic rocks. Carbonate is derived from carbon dioxide in the atmosphere. Calcium and carbonate also come from the easily weathered limestone on the continents. During burial and diagenesis,

1 The Bahamas are a carbonate platform system in the Atlantic Ocean east of Florida.

2 Reefs are built in warm, shallow seas by organisms such as corals that precipitate calcium carbonate.

3 Within the coral reef lagoon, growth of carbonate-secreting organisms, including algae and mollusks as well as corals, is rapid, and carbonate sediment forms quickly,...

4 ...whereas in the open ocean outside the reef, sedimentation is much slower.

5 If the sea level rises, the reef continues to grow toward the light at sea level ... and lagoon sedimentation outpaces sedimentation in the open ocean.

6 Eventually, a carbonate platform grows, with steep sides falling away to the open ocean.

FIGURE 6.23 Marine organisms create carbonate platform systems. [*left:* NASA; *middle:* Joseph R. Melanson/Aerials Only Photographs; *right:* Stephen Frink/Index Stock Imagery.]

(a) Limestone

(b) Gypsum

(c) Halite

(d) Chert

FIGURE 6.24 Chemical and biological sedimentary rocks: (a) limestone, lithified from carbonate sediments; (b) gypsum and (c) halite, marine evaporites that crystallize out of shallow seawater basins; and (d) chert, made up of siliceous sediments. [John Grotzinger/Ramón Rivera-Moret/Harvard Mineralogical Museum.]

carbonate sediments react with water to form a new suite of carbonate minerals. In all these processes, the minerals precipitated are either calcium carbonates (calcite or aragonite) or calcium-magnesium carbonate (dolomite).

The dominant biological sedimentary rock lithified from carbonate sediments is **limestone,** which is composed mainly of calcium carbonate ($CaCO_3$) in the form of the mineral calcite (Figure 6.24a; see Table 6.4). Limestone is formed from carbonate sands and muds and, in some cases, ancient reefs (see Figure 6.12). Although reefs are constructed mainly by corals today, they were constructed by other carbonate-secreting organisms—such as a now-extinct variety of mollusk—at earlier times in Earth's history (Figure 6.25).

> *Limestones are sedimentary rocks lithified from carbonate sediments. They are composed mainly of calcite.*

Another abundant carbonate rock is **dolostone,** made up of the mineral dolomite, which is composed of calcium-magnesium carbonate (see Table 6.4). Dolostones are diagenetically altered carbonate sediments and limestones. The mineral dolomite does not form as a primary precipitate from ordinary

FIGURE 6.25 Limestone made from a reef constructed by now-extinct clams (rudists) in the Cretaceous Shuiba formation, Sultanate of Oman. [John Grotzinger.]

> **Dolostone is a sedimentary rock made of the mineral dolomite, which forms by diagenetic alteration of carbonate sediments or limestones.**

seawater, and no organisms secrete shells of dolomite. Instead, the original calcite or aragonite of a carbonate sediment is converted into dolomite after deposition. Some calcium ions in the calcite or aragonite are exchanged for magnesium ions from seawater (or magnesium-rich groundwater) slowly passing through the pores of the sediment. This exchange converts the calcium carbonate mineral, $CaCO_3$, into calcium-magnesium carbonate, $CaMg(CO_3)_2$.

Chemical Sediments: Products of Evaporation

Evaporite sediments and **evaporite rocks** are chemically precipitated from evaporating seawater and, in some cases, lake water.

Marine Evaporites Marine evaporites are chemical sediments and sedimentary rocks formed by the evaporation of seawater. These sediments and rocks contain minerals formed by the crystallization of sodium chloride (halite), calcium sulfate (gypsum and anhydrite), and other combinations of ions commonly found in seawater. As evaporation proceeds and the seawater becomes more concentrated, the minerals crystallize in a set sequence. As dissolved ions precipitate to form each mineral, the evaporating seawater changes in composition.

> **Marine evaporites contain minerals that are precipitated from evaporating seawater.**

Seawater has the same composition in all the oceans, which explains why marine evaporites are so similar the world over. No matter where seawater evaporates, the same sequence of minerals always forms. The study of evaporite sediments also shows us that the composition of the world's oceans has stayed more or less constant over the past 1.8 billion years. Before that time, however, the precipitation sequence may have been different, indicating that the composition of seawater may have been different.

The great volume of many marine evaporites, some hundreds of meters thick, shows that they could not have formed from the small amount of water that could be held in a small, shallow bay or pond. A huge amount of seawater must have evaporated. The way in which such large quantities of seawater evaporate is very clear in bays or arms of the sea that meet the following conditions (**Figure 6.26**):

- The freshwater supply from rivers is small.
- Connections to the open sea are constricted.
- The climate is arid.

In such locations, water evaporates steadily, but the connections allow seawater to flow in to replenish the evaporating waters of the bay. As a result, those waters stay at a constant volume but become more saline than the open ocean. The evaporating bay waters remain more or less constantly supersaturated and steadily deposit evaporite minerals on the floor of the evaporite basin.

FIGURE 6.26 A marine evaporite environment of the past. The drier climate of the Miocene epoch made the Mediterranean Sea shallower than it is today, and its restricted connection to the open ocean created conditions suitable for evaporite formation. As seawater evaporated, gypsum precipitated to form evaporite sediments. A further increase in salinity led to the crystallization of halite. (The basin depth is greatly exaggerated in this drawing.)

1 Salt water entered the Mediterranean through a narrow channel.

2 Evaporation removed more water than was replaced by freshwater inflow.

3 As the basin became more saline, gypsum and halite precipitated, forming evaporite sediments.

As seawater evaporates, the first precipitates to form are the carbonates. Continued evaporation leads to the precipitation of gypsum, or calcium sulfate ($CaSO_4 \cdot 2H_2O$) (Figure 6.24b). By the time gypsum precipitates, almost no carbonate ions are left in the water. Gypsum is the principal component of plaster of Paris and is used in the manufacture of wallboard, which lines the walls of most new houses.

After still further evaporation, the mineral halite, or sodium chloride (NaCl)—one of the most common chemical sediments precipitated from evaporating seawater—starts to form (Figure 6.24c). Halite, as you may remember from Chapter 4, is table salt. Deep under the city of Detroit, Michigan, beds of salt laid down by an evaporating arm of an ancient ocean are commercially mined.

> Gypsum and halite are common marine evaporites.

In the final stages of evaporation, after the sodium chloride is gone, magnesium and potassium chlorides and sulfates precipitate. The salt mines near Carlsbad, New Mexico, contain commercial quantities of potassium chloride. Potassium chloride is often used as a substitute for table salt by people with certain dietary restrictions.

This sequence of mineral precipitation from seawater has been studied in the laboratory and is matched by the bedding sequences found in certain natural salt formations. Most of the world's evaporites consist of thick sequences of dolomite, gypsum, and halite and do not contain the final-stage precipitates. Many do not go even as far as halite. The absence of the final stages indicates that the water did not evaporate completely, but was replenished by normal seawater as evaporation continued.

Nonmarine Evaporites Evaporite sediments also form in arid-region lakes that typically have few or no river outlets. In such lakes, evaporation controls the lake level, and incoming minerals derived from chemical weathering accumulate as sediments. The Great Salt Lake is one of the best known of these lakes. In the dry climate of Utah, evaporation has more than balanced the inflow of fresh water from rivers and rain. As a result, the concentrations of dissolved ions in the lake make it one of the saltiest bodies of water in the world—eight times more saline than seawater. Sediments form when these ions precipitate.

In arid regions, small lakes may precipitate unusual salts, such as borates (compounds of the element boron), and some become alkaline. The water in this kind of lake is poisonous. Economically valuable deposits of borates and nitrates (minerals containing the element nitrogen) are found in the sediments beneath some of these lakes.

> Minerals precipitate and accumulate as sediments in nonmarine environments such as lakes as well as in oceans.

Other Biological and Chemical Sediments

Carbonate minerals secreted by organisms are the principal source of biological sediments, and minerals precipitated from evaporating seawater are the principal source of chemical sediments. However, there are several less common biological and chemical sediments that are locally abundant. They include chert, coal, phosphate rock, iron ore, and the organic carbon-rich sediments that produce oil and gas. The role of biological versus chemical processes in forming these sediments is variable.

Siliceous Sediments: Sources of Chert One of the first sedimentary rocks to be used for practical purposes by our prehistoric ancestors was **chert,** which is made up of silica (SiO_2) (Figure 6.24d). Early hunters used it for arrowheads and other tools because it could be chipped and shaped to form hard, sharp implements. A common name for chert is *flint;* the two terms are virtually interchangeable. The silica in most cherts is in the form of extremely fine crystalline quartz. Some geologically young cherts consist of opal, a less well crystallized form of silica.

> Chert is a sedimentary rock formed from siliceous sediments, which are generally precipitated biologically.

Like calcium carbonate sediments, many siliceous sediments are precipitated biologically, as mineral shells secreted by planktonic organisms that settle to the deep seafloor after they die and accumulate as layers of sediment. After these sediments are buried by later sediments, they are cemented into chert. Chert may also form as nodules and irregular masses replacing carbonate in limestones and dolostones.

Organic Sediments: Sources of Coal, Oil, and Natural Gas Coal is a biological sedimentary rock composed almost entirely of organic carbon formed by the diagenesis of swamp vegetation. Vegetation may be preserved from decay and accumulate as a rich organic material called *peat,* which contains more than 50 percent carbon. If peat is ultimately buried, it may be transformed into coal. Coal is classified as an *organic sedimentary rock,* a class that consists entirely or partly of organic carbon–rich deposits formed by the diagenesis of once-living material that has been buried.

> Coal is a sedimentary rock formed from organic sediments. Oil and natural gas are fluids that are also created by diagenesis of organic matter.

In both lake and ocean waters, the remains of algae, bacteria, and other microscopic organisms may accumulate in sediments as organic matter that can be transformed into oil and natural gas. However, *oil* and *natural gas* are fluids that are not normally classed with sedimentary rocks. They can be considered organic sediments, however, because they form by the diagenesis of organic material in the pores of sedimentary rocks. Deep burial changes the organic matter originally deposited along with inorganic sediments into a fluid that then escapes to porous rock formations and becomes trapped there. As noted earlier in this chapter, oil and gas are found mainly in sandstones and limestones.

SUMMARY >>

What are the major processes that form sedimentary rock? Weathering produces the particles that compose siliciclastic sediments and the dissolved ions that precipitate to form biological and chemical sediments. Currents of water and air and the movement of glaciers transport the sediments to their ultimate resting place in a sedimentary basin. Sedimentation (also called deposition) is the settling out of particles or precipitation of minerals to form layers of sediments. Lithification and diagenesis harden the sediments into sedimentary rock.

What are the two major types of sediments? Sediments and the sedimentary rocks that form from them can be classified as siliciclastic or chemical and biological. Siliciclastic sediments form from fragments of parent rock produced by physical and chemical weathering and are transported to sedimentary basins by water, wind, or ice. Chemical and biological sediments originate from ions dissolved in and transported by water. Through chemical and biological reactions, these ions are precipitated from solution to form sediments.

How are the major kinds of siliciclastic sediments and chemical and biological sediments classified? Siliciclastic sediments and sedimentary rocks are classified by the size of their particles. The major classes, in descending particle size order, are gravels and conglomerates; sands and sandstones; silts and siltstones; muds, mudstones, and shales; and clays and claystones. This classification method emphasizes the importance of the strength of the current that transported the sediments. Chemical and biological sediments and sedimentary rocks are classified on the basis of their chemical composition. The most abundant of these rocks are the carbonate rocks: limestone and dolostone. Limestone is made up largely of biologically precipitated calcite. Dolostone is formed by the diagenetic alteration of limestone. Other chemical and biological sediments include evaporites, siliceous sediments such as chert, and peat and other organic matter that is transformed into coal, oil, and gas.

KEY TERMS AND CONCEPTS >>

bedding (p. 156)
bedding sequence (p. 158)
bioclastic sediment (p. 147)
biological sediment (p. 147)
bioturbation (p. 158)
carbonate rock (p. 163)
carbonate sediment (p. 163)
cementation (p. 160)
chemical and biological sedimentary environment (p. 155)
chemical sediment (p. 147)
chemical weathering (p. 146)
chert (p. 167)
clastic sediment (p. 147)
clay (p. 163)
claystone (p. 163)
coal (p. 167)
compaction (p. 160)
conglomerate (p. 161)
continental shelf (p. 152)
cross-bedding (p. 156)
diagenesis (p. 159)
dolostone (p. 165)
evaporate rock (p. 166)
evaporite sediment (p. 166)
flexural basin (p. 152)
graded bedding (p. 156)
gravel (p. 161)
limestone (p. 165)
lithification (p. 160)
mud (p. 162)
mudstone (p. 162)
physical weathering (p. 146)
porosity (p. 160)
reef (p. 164)
rift basin (p. 152)
ripple (p. 157)
salinity (p. 150)
sand (p. 162)
sandstone (p. 162)
sedimentary basin (p. 151)
sedimentary environment (p. 152)
sedimentary structure (p. 156)
shale (p. 162)
siliciclastic sediment (p. 147)
siliciclastic sedimentary environment (p. 154)
silt (p. 162)
siltstone (p. 162)
sorting (p. 149)
thermal subsidence basin (p. 152)
weathering (p. 146)

EXERCISES >>

1. What processes change sediment into sedimentary rock?

2. How do siliciclastic sedimentary rocks differ from chemical and biological sedimentary rocks?

3. How and on what basis are the siliciclastic sedimentary rocks classified?

4. What kinds of sedimentary rocks are formed by the evaporation of seawater?

5. Define a sedimentary environment, and name three siliciclastic environments.

6. Explain how plate tectonic processes control the development of sedimentary basins.

7. Name two kinds of carbonate rocks and explain how they differ.

8. How do organisms produce or modify sediments?

9. In what two kinds of sedimentary rocks are most oil and gas found?

10. If you drilled an oil well into the bottom of a sedimentary basin that is 1 km deep and another that is 5 km deep, which would have the higher pressures and temperatures? Oil turns into natural gas at high basin temperatures. In which well would you expect to find more gas?

11. You are looking at a cross section of a rippled sandstone. What sedimentary structure would tell you the direction of the current that deposited the sand?

12. You discover a bedding sequence that has a conglomerate at the base; grades upward to a sandstone and then to a shale; and finally, at the top, grades to a limestone of cemented carbonate sand. What changes in the sediment's source area or in the sedimentary environment could have been responsible for this sequence?

13. How can you use the size and sorting of sediment particles to distinguish between sediments deposited in a glacial environment and those deposited in a desert?

14. Where are reefs likely to be found?

DEFORMATION AND METAMORPHISM

CHAPTER 7

The Inspiration of Marble

The use of marble in architecture dates back thousands of years to ancient Egyptian and Mesopotamian cultures. The Parthenon, built in the sixth century B.C., was one of the first buildings constructed entirely of marble. From miles away, incoming ships sailing through the Mediterranean could see the white marble columns of the most famous temple in Athens shining like a beacon of light. The Parthenon still stands today, a testament to the resilience of one of the world's most luxurious building materials. Capable of bearing immense weight, marble was ideally suited for monolithic columns and supporting structures in public, private, and religious buildings.

The word *marble* originates from the Greek *marmaros*, meaning "a snow white and spotless stone" (although not all marble is white). And because no other stone offers the opportunity for refinement quite like marble, it has been the medium of choice for some of the world's greatest sculptors. The most famous, Michelangelo, carved *David* from a single block of marble. Michelangelo preferred to use Carrara marble, which is still the stone preferred by the best sculptors today. Stone from the Carrara region, in the Apennines of Italy, was employed in Rome for architectural purposes in the time of Augustus, but the finer varieties, best suited to the needs of sculptors, were not discovered until later. The purest varieties of Carrara marble have a snow-white color.

> *I found a city of brick and left it a city of marble.*
> Caesar Augustus (64 B.C.–A.D. 14), Roman emperor

In the United States, marble production has been a valued enterprise since George Yule, a mining engineer, discovered marble deposits and recognized their value. The town of Marble, Colorado, was incorporated in 1899 and began its history by producing the finest pure white marble in the world. The quarry, operated until 1941, is most famous for producing the blocks of marble used for the Lincoln Memorial and the Tomb of the Unknowns. Marble was a typical "boom-and-bust" mining town, and the quarry was closed in 1941 at the advent of World War II. At the time it closed, it had the largest marble finishing mill of its kind in the world, employing over 1000 people. With the quarry gone, Marble's population dropped to just 50 residents.

Michelangelo's *David*. [Alinari Archives/CORBIS.]

CHAPTER 7 Deformation and Metamorphism

CHAPTER OUTLINE

Mapping Geologic Structure 173

How Rocks Deform 175

Basic Deformation Structures 176

Styles of Continental Deformation 181

Causes of Metamorphism 185

Types of Metamorphism 187

Metamorphic Textures 188

Regional Metamorphism and Metamorphic Grade 192

Plate Tectonics and Metamorphism 195

Today, marble is increasingly a part of our daily lives. Modern technologies have enabled home designers to take advantage of its beauty and elegance at a fraction of what was paid even 20 years ago. Hewn from Earth's crust, cut with diamond saws, and polished like glass, thin slabs of marble are routinely installed as kitchen and bathroom countertops.

If you call on your local neighborhood marble dealer, you will discover that not all the "marble" being sold is really marble. Most of what may be on display is actually various types of granite, an igneous intrusive rock, or gneiss, a metamorphic rock formed from granite or sedimentary rocks. Granite and gneiss are extremely popular because, unlike true marble, they are formed of silicate minerals, which are much harder and better resist scratching and staining. Other types of "marble" include onyx, a sedimentary rock formed in ancient caves, and travertine, a sedimentary rock formed in thermal springs. These types of "marble" are softer than true marble, and require special care, but will add a touch of elegance in the right place.

What is it that gives marble its strength and beauty? To find out, we must look at the transformations deep in Earth's crust that produce marble and other metamorphic rocks.

During the rock cycle, rocks may be subjected to tectonic forces great enough to tilt, bend, or fracture them. They may also be subjected to temperatures and pressures great enough to cause changes in their mineralogy, texture, or chemical composition. **This chapter examines deformation and metamorphism, the processes that cause these kinds of changes in continental crust.**

Deformation and metamorphism are both intimately associated with plate tectonic processes. The steady relative motion between two rigid lithospheric plates causes deformation along the plate boundary. Within continents, however, that deformation can be "smeared out" across a plate boundary in zones hundreds or even thousands of kilometers wide. Within these broad zones, rocks are deformed by **folding** and **faulting** (Figure 7.1).

Folds in rocks are like folds in clothing. Just as cloth pushed together from opposite sides bunches up in folds, layers of rock slowly compressed by tectonic forces in the crust can be pushed into folds (Figure 7.1a). Tectonic forces can also cause a rock formation to break and slip on both sides of the fracture, forming a fault (Figure 7.1b).

Patterns of deformation also fit into the larger framework of the plate tectonic movements involved in **orogeny,** or "mountain making." Tectonic forces build mountains through the folding and faulting of rock layers, often with accompanying volcanism, especially at ocean-continent convergences. These same enormous forces transform the texture and composition of the rocks themselves. Great belts of metamorphic rock, stretching

(a)

(b)

FIGURE 7.1 (a) An outcrop of originally horizontal rock layers bent into folds by compressive tectonic forces. [Phil Dombrowski.]

(b) An outcrop of once-continuous rock layers displaced along small faults by tensional tectonic forces. [Tom Bean.]

across a region, are often found where continents collided and built mountains. In the cores of the major mountain chains of the world, from the Appalachians to the Alps, we find long belts of regionally metamorphosed and deformed rocks that parallel the lines of folds and faults in the mountains.

MAPPING GEOLOGIC STRUCTURE

Faults and folds are examples of the basic features geologists observe and map to understand crustal deformation. To better understand this process, we need information about the geometry of these rock formations. The best place to find this information is an *outcrop,* where the solid rock that underlies the ground surface—the *bedrock*—is exposed (not obscured by soil or loose boulders). Figure 7.1a is a picture of an outcrop showing sedimentary beds bent into a fold. Often, however, folded rocks are only partly exposed in an outcrop and can be seen only as an inclined layer (Figure 7.2). The orientation of that layer is an important clue we can use to piece together a picture of the overall geologic structure. Two measurements describe the orientation of a rock layer exposed at an outcrop: the strike and the dip of the layer's surface.

Measuring Strike and Dip

The **strike** is the compass direction of a rock layer as it intersects with a horizontal surface. The **dip,** which is measured at right angles to the strike, is simply the amount of tilting—

FIGURE 7.3 The strike and dip of a rock layer define its orientation at a particular place. Strike is the compass direction of a rock layer as it intersects with a horizontal surface; dip is the angle of steepest descent of the rock layer from the horizontal, measured at right angles to the strike. Here the strike is east-west and the dip is 45° to the south. [After A. Maltman, *Geological Maps: An Introduction.* New York: Van Nostrand Reinhold, 1990, p. 37.]

the angle at which the rock layer inclines from the horizontal. Figure 7.3 shows how the strike and the dip are measured in the field. A geologist might describe the outcrop in this figure as "a bed of coarse-grained sandstone striking east-west and dipping 45° south." Strike and dip can also be used to map

Measurements of strike and dip provide information on the orientation of rock layers, fault planes, and other geologic surfaces.

FIGURE 7.2 Dipping limestone and shale beds on the coast of Somerset, England. Children are walking along the strike of beds that dip to the left at an angle of about 15°. [Chris Pellant.]

Geologic Maps

Geologic maps represent the rock formations exposed at Earth's surface. Each rock formation is assigned a particular color on the map, usually keyed to the rock's type and age (**Figure 7.4**). Many different rock formations may be exposed in highly deformed regions, so geologic maps can be very colorful! Softer rocks, such as mudstones, are more easily eroded than harder rocks, such as limestones. Consequently, rock types can exert a strong influence on the topography of the land surface and the exposure of rock formations. The important relationships between geology and topography can be made clear by plotting the geologic structure of a region on a base map that also shows topographic contours (see Figure 7.4).

Geologic Cross Sections

Once a region is mapped, the two-dimensional map must be interpreted in terms of the three-dimensional geologic structure. How can the deformed shapes of the rock layers be reconstructed, even when erosion has removed parts of a formation? The process is like putting together a three-dimensional jigsaw puzzle with some of the pieces missing. Common sense and intuition play important roles, as do basic geologic principles.

To piece together the puzzle, geologists construct **geologic cross sections**—diagrams showing the features that would be visible if vertical slices were made through part of the crust. Small cross sections can sometimes be directly observed in the vertical faces of cliffs, quarries, and road cuts. Cross sections spanning much larger areas can be constructed from the information on a geologic map, including the strikes and dips observed at outcrops.

Figure 7.4 shows a geologic map and geologic cross sections of an area where sedimentary rocks, originally horizontal, were bent into a series of folds and eroded into a set of linear ridges and valleys. We will explore some of the geologic relationships seen in this map later in this chapter. But first we will investigate the basic processes by which rocks deform.

> *Geologic maps and geologic cross sections can be constructed to illustrate the three-dimensional geologic structure of a region.*

FIGURE 7.4 A geologic map and cross section make up a two-dimensional representation of a three-dimensional geologic structure.

HOW ROCKS DEFORM

Rocks deform in response to the tectonic forces acting on them. Whether they will respond to those forces by folding, faulting, or some combination of the two depends on the orientation of the forces, the rock type, and the physical conditions (such as temperature and pressure) during deformation.

Brittle and Ductile Behavior of Rocks in the Laboratory

Early geologists were baffled by the question of how rocks, which seem so strong and rigid, could be distorted into folds or broken along faults by tectonic forces. The question was answered in the mid-1900s, when geologists used big hydraulic rams to bend and break small samples of rock. Engineers had invented such machines to measure the strength of concrete and other building materials, but geologists modified them to discover how rocks deform at pressures and temperatures high enough to simulate physical conditions deep in Earth's crust.

In one such experiment, a compressive (squeezing) force was applied by pushing down with a hydraulic ram on one end of a small cylinder of marble, while at the same time maintaining the force of confining pressure on the cylinder. (Confining pressure is the squeezing your body feels as you dive deeper under water; it increases steadily with depth in proportion to the weight of the overlying mass.) Under low confining pressures, equivalent to those found at shallow depths in Earth's crust, the marble sample deformed only a small amount until the compressive force on its end was increased to the point that the entire sample suddenly broke (**Figure 7.5**). This experiment showed that marble behaves as a **brittle** material at low confining pressures. Repeating the experiment under high confining pressures, equivalent to those that often accompany metamorphism, produced a different result: the marble sample slowly and steadily deformed into a shortened, bulging shape without fracturing. Marble thus behaves as a pliable or **ductile** material at the high confining pressures found deep in the crust.

Rocks behaving as brittle materials respond to deformation by fracturing, whereas rocks behaving as ductile materials respond to deformation by deforming without fracturing.

Other experiments showed that when marble was heated to temperatures as high as those that accompany metamorphism, it acted as a ductile material at a lower confining pressure—just as heating wax changes it from a hard material that can break into a soft one that flows. The

FIGURE 7.5 Results of laboratory experiments conducted to discover how a rock—in this case, marble—is deformed by compressive forces. [M. S. Patterson, Australian National University.]

researchers concluded that the particular marble they were working with would deform by faulting at depths shallower than a few kilometers, but by folding at the greater crustal depths where metamorphism generally occurs.

Brittle and Ductile Behavior in Earth's Crust

Natural conditions in Earth's crust cannot be reproduced exactly in the laboratory. Tectonic forces are applied over millions of years, whereas laboratory experiments are rarely conducted for more than a few hours, or at most a few weeks. Nevertheless, laboratory experiments can help us interpret what we see in the field. Geologists keep the following points in mind as they map crustal folds and faults:

- The same rock can be brittle at shallow depths (where temperatures and pressures are relatively low) and ductile deep in the crust (where temperatures and pressures are higher). Metamorphism is usually accompanied by ductile deformation.

- Rock type affects deformation. In particular, the old, hard igneous and metamorphic rocks that form the bulk of continents act as brittle materials, fracturing along fault planes in earthquakes, while the younger, softer sedimentary rocks that overlie them act as ductile materials, folding gradually.

Brittle behavior is more likely with lower temperatures and pressures, harder rock types, more rapid deformation, and tensional rather than compressive forces.

- A rock formation that would behave as a ductile material if deformed slowly may behave as a brittle material if deformed more rapidly. (Think of Silly Putty, which flows as a ductile clay when you squeeze it slowly but breaks into pieces when you pull it apart very quickly.)

- Rocks break more easily when subjected to tensional (pulling and stretching) forces than when subjected to compressive forces. Sedimentary rock formations that will deform by folding during compression will often break along faults when subjected to tension.

BASIC DEFORMATION STRUCTURES

The simple geometric concepts and measurements described earlier in this chapter (and a rich vocabulary) are used to classify features such as faults and folds into different types of deformation structures.

Tectonic Forces Determine Types of Deformation Structures

The types of faults and folds we see depend on the main tectonic forces in the upper crust, which are produced mostly by horizontal movements of the plates. The tectonic forces that act at plate boundaries are thus predominantly *horizontally directed*. Horizontal plate movements exert three types of tectonic forces:

Horizontal plate movements exert three types of tectonic forces: tensional forces (which stretch and pull rocks apart), compressive forces (which squeeze and shorten rocks), and shearing forces (which push rocks in opposite directions).

- **Tensional forces,** which stretch and pull rock formations apart, dominate at divergent boundaries, where plates move away from each other.

- **Compressive forces,** which squeeze and shorten rock formations, dominate at convergent boundaries, where plates move toward each other.

- **Shearing forces,** which push two sides of a rock formation in opposite directions, dominate at transform-fault boundaries, where plates slide past each other.

Faults

A fault, as we saw in Chapter 1, is a fracture surface across which rock formations have been displaced. We can measure the orientation of a fault plane by its strike and dip, just as we do for other geologic surfaces (see Figure 7.3). The movement of the block of rock on one side of the fault with respect to that on the other side can be described by a *slip direction* and by the total displacement, or *offset*. For a small fault, such as the one pictured in Figure 7.1b, the offset might be only a couple of meters, whereas the offset along a major transform fault, such as the San Andreas fault, can amount to hundreds of kilometers (**Figure 7.6**).

Faults are classified by their slip direction (**Figure 7.7**). A **dip-slip fault** is one on which there has been relative movement of the blocks up or down the dip of the fault plane. A **strike-slip fault** is one on which the movement has been horizontal, parallel to the strike of the fault plane. When movement is along the strike and simultaneously up or down, the result is an *oblique-slip fault*. Dip-slip faults are caused by compressive or tensional forces, whereas strike-slip faults are the work of shearing forces.

FIGURE 7.6 View of the San Andreas fault, showing the northwestward movement of the Pacific Plate with respect to the North American Plate. The map shows a formation of volcanic rocks 23 million years old that has been displaced by 315 km. The fault runs from top to bottom (dashed line) near the middle of the photograph. Note the offset of the stream (Wallace Creek) by 130 m as it crosses the fault. [John S. Shelton.]

Basic Deformation Structures 177

Dip-slip faulting is caused by tensional or compressive forces that move blocks up or down the dip of the fault plane.

Normal faulting is caused by tensional forces that stretch a rock and tend to pull it apart

Reverse faulting is caused by compressive forces that squeeze and shorten a rock.

A thrust fault is a reverse fault with a shallow-dipping fault plane.

Strike-slip faulting is caused by horizontal shearing forces that move blocks left-laterally or right-laterally.

Left-lateral strike-slip fault

Right-lateral strike-slip fault

Oblique-slip faulting is caused by a combination of forces, in this case left-lateral shearing with tension.

FIGURE 7.7 The orientation of tectonic forces determines the style of faulting. Dip-slip faulting (parts a–c) is caused by tensional and compressive forces. Strike-slip faulting (parts d–f) is caused by shearing forces.

An oblique-slip fault results from shearing in combination with either compression or tension.

These fault types require further classification because the movement can be up or down, or right or left. A dip-slip fault is called a **normal fault** if the rocks above the fault plane move downward relative to the rocks below the fault plane, extending the structure horizontally (Figure 7.7a). A dip-slip fault is called a *reverse fault* if the rocks above the fault plane move upward relative to the rocks below, causing a shortening of the structure (Figure 7.7b) —the reverse of what geologists have (somewhat arbitrarily) chosen as "normal." A **thrust fault** is a low-angled reverse fault; that is, one with a dip of less than 45°, so that the overlying block moves mainly horizontally (Figure 7.7c). When subjected to horizontal compression, brittle rocks of the continental crust usually break along thrust

FIGURE 7.8 The Red Canyon fault scarp cuts across a road in Hebgen Lake, Montana. [I. J. Witkind/USGS.]

> *A fault is a surface across which rock formations have been displaced. Dip-slip faults show movement of blocks up (reverse fault) and down (normal fault) the dip of the fault plane. Strike-slip faults show horizontal movement of blocks parallel to the strike of the fault plane.*

faults with dips of about 30° or less, rather than along more steeply dipping reverse faults.

A strike-slip fault is a *left-lateral fault* if an observer on one side of the fault sees that the block on the opposite side has moved to the left (Figure 7.7d). It is a *right-lateral fault* if the block on the opposite side appears to have moved to the right (Figure 7.7e). As you can tell from the stream offset in Figure 7.6, the San Andreas fault is a right-lateral transform fault.

We can recognize faults in the field in several ways. A fault may form a *scarp* (small cliff) that marks the trace of the fault across the ground surface (**Figure 7.8**). If the offset has been large, as it is for transform faults such as the San Andreas, the rock formations currently facing each other across the fault usually differ in type and age. When movements are smaller, offset features can be observed and measured. (As an exercise, try to match up the beds offset by the small-scale fault in Figure 7.1b.) In establishing the time of faulting, we apply a simple rule: a fault must be younger than the youngest rocks that it cuts (the rocks had to be there before they could break!) and older than the oldest undisrupted formation that covers it.

Folds

Folding is a common form of deformation observed in layered rocks. Folds occur when an originally planar structure, such as a sedimentary bed, has been warped into a curved structure. The bending can be produced by either horizontally or vertically directed forces in the crust, just as either pushing together the opposite edges of a piece of paper or pushing up or down on one side or the other can fold it.

Like faults, folds come in all sizes. In many mountain systems, majestic, sweeping folds can be traced over many kilometers. On a much smaller scale, very thin sedimentary beds can be crumpled into folds a few centimeters long (**Figure 7.9**). The bending can be gentle or

FIGURE 7.9 Small-scale folds in sedimentary beds of anhydrite (light) and shale (dark), West Texas. [John Grotzinger/Ramón Rivera-Moret/Harvard Mineralogical Museum.]

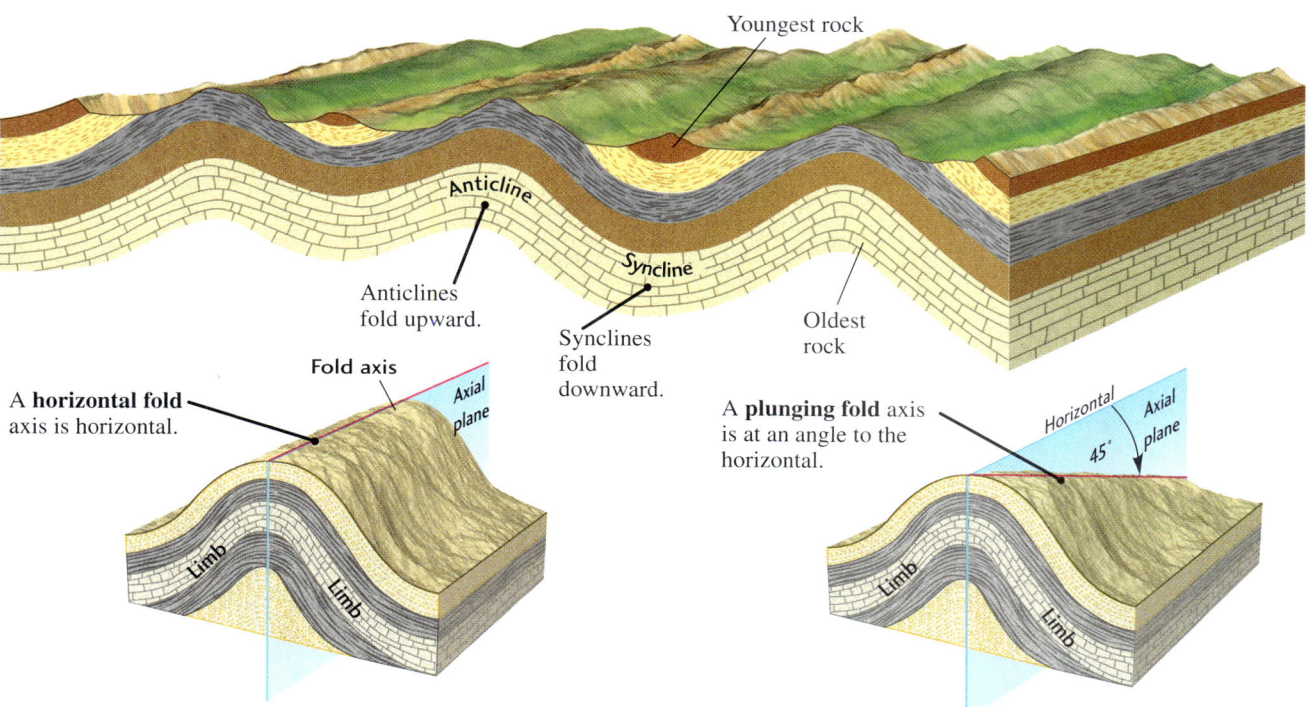

FIGURE 7.10 Rock folding is described by the direction of folding and by the orientation of the fold axis and the axial plane.

severe, depending on the magnitude of the applied forces, the length of time that they were applied, and the ability of the rocks to resist deformation.

Layered rocks that fold upward into arches are called **anticlines;** those that fold downward into troughs are called **synclines** (Figure 7.10). The two sides of a fold are its *limbs*. The *axial plane* is an imaginary surface that divides a fold as symmetrically as possible, with one limb on either side of the plane. The line made by the lengthwise intersection of the axial plane with the rock layers is the *fold axis*. A symmetrical horizontal fold has a horizontal fold axis and a vertical axial plane with limbs dipping symmetrically away from the axis.

Deformed layers of rock may be folded upward into anticlines or downward into synclines.

Folds rarely stay horizontal, however. Follow the axis of any fold in the field, and sooner or later the fold dies out or appears to plunge into the ground. If the axis of a fold is not horizontal, it is called a *plunging fold.* Figure 7.11 diagrams the geometry of plunging anticlines and plunging synclines. In eroded mountain belts, a zigzag pattern of outcrops may appear in the field after erosion has removed much of the surface rock from the folds. The geologic map of Figure 7.4 shows this characteristic pattern.

FIGURE 7.11 The geometry of plunging folds. Note the converging pattern of the layers where they intersect the land surface, which you can also see in Figure 7.4.

Symmetrical folds have limbs dipping symmetrically away from the axial plane.

Asymmetrical folds have beds in one limb that dip more steeply than those in the other.

Overturned folds have limbs that dip in the same direction, but one limb has been tilted beyond the vertical.

FIGURE 7.12 As deformation increases, folds are pushed into asymmetrical shapes. [(*left*) courtesy of Cleet Carlton/Golden Gate Photo; (*center*) courtesy of Mark McNaught; (*right*) John Grotzinger.]

Folds can sometimes be pushed into asymmetrical shapes, with one limb dipping more steeply than the other (**Figure 7.12**). This occurs when the direction of the deformation force is oblique to the layering of the beds. Such *asymmetrical folds* are common. When the deformation is intense and one limb has been tilted beyond the vertical, the fold is called an *overturned fold*. Both limbs of an overturned fold dip in the same direction, but the order of the layers in the bottom limb is precisely the reverse of their original sequence—that is, older rocks are on top of younger rocks.

Observations in the field seldom provide geologists with complete information. Bedrock may be obscured by overlying soils, or erosion may have removed much of the evidence of former structures. So geologists search for clues they can use to work out the relationship of one bed to another. For example, in the field or on a geologic map, an eroded anticline would be recognized by a strip of older rocks forming a core bordered on both sides by younger rocks dipping away from the core. These relationships are illustrated in Figures 7.4 and 7.11. An eroded syncline would show as a core of younger rocks bordered on both sides by older rocks dipping toward the core.

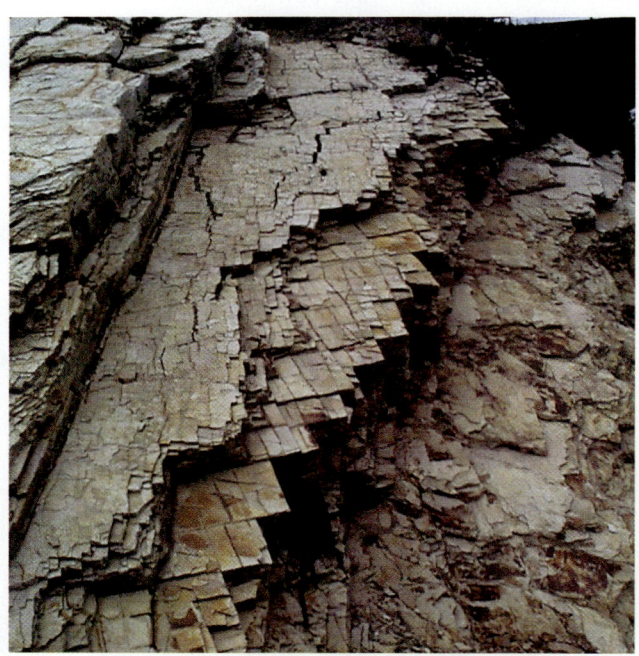

FIGURE 7.13 Intersecting sets of joints have made this rock vulnerable to weathering. (The backpack is shown for scale.) [Courtesy of Peter Kaufman.]

Joints

We have seen how rock deformation depends on tectonic forces and the conditions under which those forces are applied. Some layers crumple into folds, and some fracture. A fracture that has displaced the rock formations on either side is called a fault. A second type of fracture is a **joint**—a crack along which there has been no appreciable movement.

Joints are found in almost every outcrop, and they create surfaces that are vulnerable to weathering. If two or more sets of joints intersect, weathering may cause a rock formation to break into large columns or blocks (**Figure 7.13**). Some joints are caused by tectonic forces. Like any other easily broken material, brittle rocks fracture more easily at flaws or weak spots when they are subjected to pressure. These flaws can be tiny cracks, fragments of other materials, or even fossils. Regional forces—compressive, tensional, or shearing—may leave a set of joints as their imprint long after they have vanished.

> *Joints are fracture surfaces along which there has been no significant movement.*

STYLES OF CONTINENTAL DEFORMATION

If we look closely enough, we can find all the basic deformation structures in any zone of continental deformation. But when we view continental deformation at the scale of the plate boundary, we see distinctive patterns of faulting and folding that relate directly to the tectonic forces causing the deformation. **Figure 7.14** depicts the deformation styles typical of the three main types of tectonic forces.

(a) **Tensional tectonics**: Extension of continental crust produces normal faults with high dip angles in the upper crust that flatten with depth, forming curved fault surfaces.

Normal faulting

Ductile lower crust

(b) **Compressive tectonics**: Compression of continental crust occurs on thrust faults with low dip angles.

Fold and thrust belt

Thrust faulting

(c) **Shearing tectonics**: Shearing of continental crust occurs on a nearly vertical strike-slip fault. The case shown here is for a right-lateral fault.

Strike-slip faulting

A left bend in the fault results in local compression.

A right bend in the fault results in local extension.

FIGURE 7.14 The orientation of tectonic forces determines the style of continental deformation. [After John Suppe, *Principles of Structural Geology.* Upper Saddle River, N.J.: Prentice Hall, 1985.]

Tensional Tectonics

In brittle crust, the tensional forces that produce normal faulting may split a plate apart, resulting in a *rift valley*—a long, narrow trough formed by a block of rock that has dropped downward relative to its two flanking blocks along nearly parallel, steeply dipping normal faults (**Figure 7.15**). The rift valleys of East Africa, the rifts of mid-ocean ridges, the Rhine Valley, and the Red Sea are well-known examples. As we saw in Chapter 6, these structures form basins that fill with sediments eroded from the mountains of the rift walls as well as with volcanic rocks extruded from tensional cracks in the crust.

> Tensional deformation of the lithosphere produces rift valleys bounded by normal faults.

The Basin and Range province, which is centered on the Great Basin of Nevada and Utah, is a good example of a region defined by many adjacent rift valleys. This region, which is now more than 800 km wide, has been stretched and extended in a northwest-southeast direction by a factor of two during the last 15 million years. Here normal faulting has created an immense landscape of eroded, rugged mountains and smooth, sediment-filled valleys, some covered with recent volcanic rocks. This tensional deformation, which appears to be caused by upwelling convection currents beneath the Basin and Range province, continues today.

Extension of the shallow continental crust usually produces normal faults with high dip angles, typically 60° or more. However, below a depth of about 20 km, crustal rocks are hot enough to behave as ductile materials, and the deformation occurs by stretching rather than by fracturing. This change in rock behavior causes the dip of the faults to flatten with increasing depth, which results in normal faults with curved surfaces, as shown in Figure 7.14a. The crustal blocks moving along these curved faults are tilted back as the stretching continues.

Compressive Tectonics

In subduction zones (see Figure 3.8c, d), oceanic lithosphere slips beneath an overriding plate along a huge thrust fault, or *megathrust*. The world's largest earthquakes, such as the great Sumatra earthquake of December 26, 2004, which generated a disastrous tsunami that killed more than 300,000 people, are caused by sudden slips on megathrusts. Thrusting is also the most common type of faulting in continental regions undergoing tectonic compression. During mountain building, sheets of crust may glide over one another for tens of kilometers along nearly horizontal thrust faults, forming *overthrust* structures (**Figure 7.16**).

When two continents collide, the crust can be compressed across a wide zone, resulting in spectacular episodes of orogeny. During such collisions, the brittle basement rocks ride over one another by thrust faulting while the more ductile overlying sedimentary rocks compress into a series of great folds, forming a *fold and thrust belt* (Figure 7.14b).

> Compressive deformation of continental crust produces thrust faults and results in orogeny.

The ongoing collisions of Africa, Arabia, and India with the southern margin of the Eurasian continent have created fold and thrust belts from the Alps to the Himalaya, many of which are still active. The great oil reservoirs of the Middle East are located in anticlines and other structural traps formed by this deformation. Compression across western North America, caused by its westward movement during the opening of the Atlantic Ocean, created the fold and

FIGURE 7.15 A rift valley results from tensional forces that produce normal faulting. The African Plate, on which Egypt rides, and the Arabian Plate, bearing Saudi Arabia, are drifting apart. The tensional forces have created a rift valley, filled by the Red Sea. The diagram shows parallel normal faults bounding the rift valley in the crust beneath the sea. [NASA/TSADO/Tom Stack.]

FIGURE 7.16 The Keystone thrust fault of southern Nevada is a large-scale overthrust sheet of a kind found in California and southern Nevada. Compressive forces have detached a sheet of rock layers (D, C, B) and thrust it a great distance horizontally over a section of the same rock layers (D, C, B, A). [Photo by John S. Shelton.]

thrust belt of the Canadian Rockies. The Valley and Ridge province of the Appalachian Mountains is an ancient fold and thrust belt that dates back to the collisions that created the supercontinent Pangaea.

Shearing Tectonics

A transform fault is a strike-slip fault that forms a plate boundary. Transform faults such as the San Andreas can offset geologic formations by long distances (see Figure 7.6), but as long as they stay aligned with the direction of relative plate motion, the blocks on either side can slide past each other without much internal deformation. Long transform faults are rarely straight, however, so deformation patterns along these faults can be much more complicated. The faults may have bends and jogs that change the tectonic forces acting across portions of the plate boundary from shearing to compression or tension. Those forces, in turn, cause secondary faulting and folding (see Figure 7.14c).

A good example of how this works can be found in southern California, where the right-lateral San Andreas fault bends first to the left and then to the right as one moves along it (in either direction; see Figure 7.6). The segments of the fault on both sides of this "Big Bend" are aligned with the direction of relative plate movement, so the blocks slip past each other in simple strike-slip faulting. Within the Big Bend, however, the change in the fault geometry causes the blocks to push against each other, which causes thrust faulting to the south of the fault. This thrusting has raised mountains to elevations exceeding 3000 m and, during the last half-century, has produced a series of destructive ruptures, including the 1994 Northridge earthquake, which caused more than $40 billion in damage to Los Angeles (see Chapter 13).

In the Salton Trough at the southern end of the San Andreas, the convergent boundary between the Pacific and North American plates jogs to the right in a series of steps. Within these jogs, the plate boundary is subjected to

> *Shearing deformation produces strike-slip faults, but bends in these faults produce more complicated deformation patterns.*

tensional forces, and normal faulting has formed rift valleys that are volcanically active, rapidly subsiding, and filling with sediments. The extension of the Salton Trough occurs within 200 km of the Big Bend compression, demonstrating how variable the tectonics along continental transform faults can be!

We have now described the major types of structural changes that occur when a rock is deformed. When temperatures get high enough, however, rocks undergo changes in their mineral composition, chemical composition, and texture as well as in their structure. These are the changes that are characteristic of metamorphic rocks. Although metamorphism is usually accompanied by ductile deformation, it can also occur without deformation, such as when hot magmas are injected into surrounding rocks. Let's turn next to the process of metamorphism and the types of changes that occur when rocks are subjected to high temperatures and pressures.

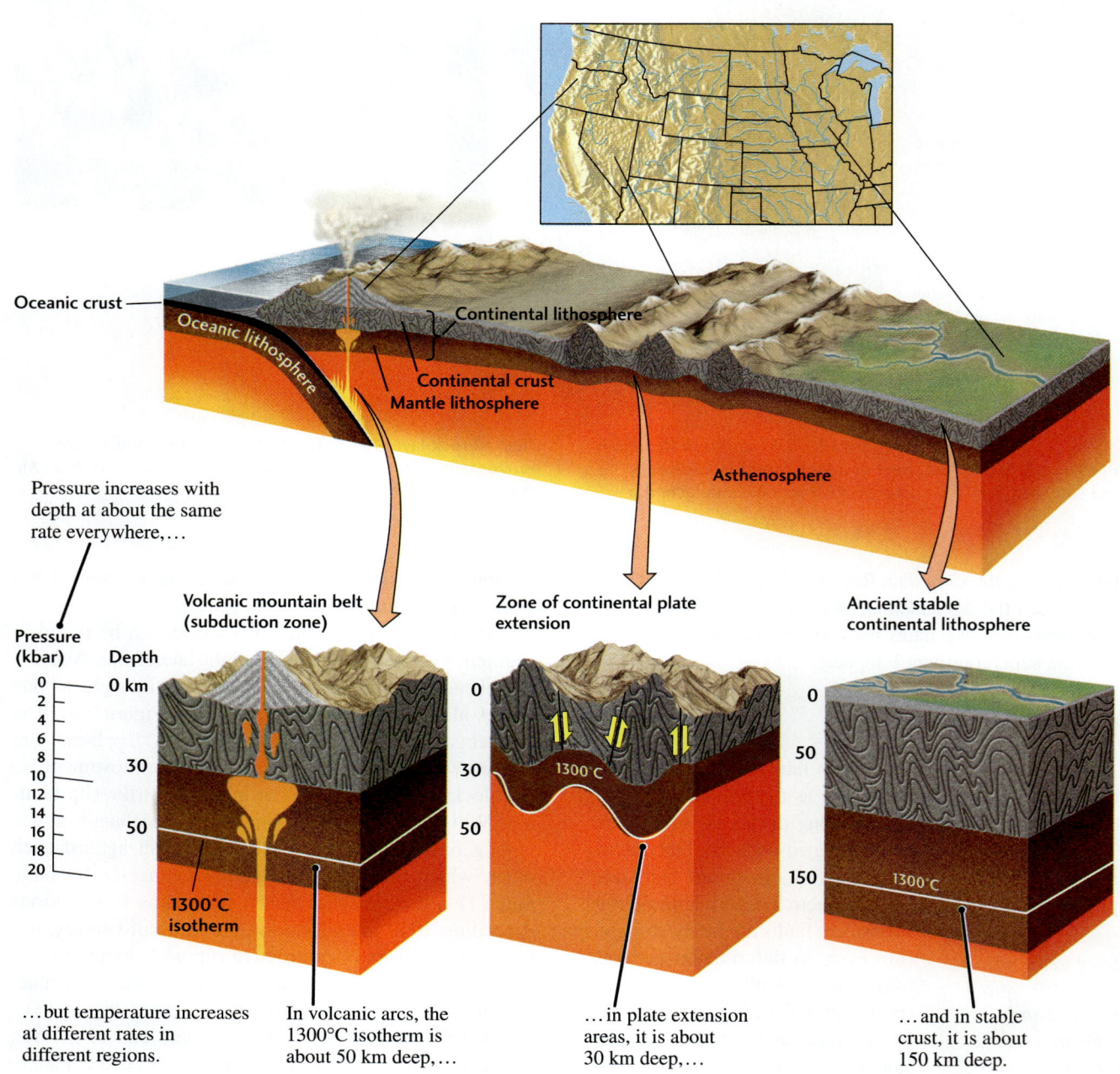

FIGURE 7.17 Pressure and temperature increase with depth in all regions, as shown in these cross sections of a volcanic arc, a continental region, and a region of ancient stable continental lithosphere.

CAUSES OF METAMORPHISM

We are all familiar with some of the ways in which heat and pressure can transform materials. Cooking batter in a waffle iron heats up the batter and also puts pressure on it, transforming it into a rigid solid. In similar ways, rocks change as they encounter high temperatures and pressures. Deep in Earth's crust, tens of kilometers below the surface, temperatures and pressures are high enough to transform rock without being high enough to melt it. Increases in temperature and pressure and changes in the chemical environment can alter the mineral composition and crystalline textures of rock, *even though it remains solid all the while*. The result is **metamorphic rock,** which has undergone changes in mineralogy, texture, chemical composition, or all three.

When a rock is subjected to significant changes in temperature or pressure, it will, given enough time—short by geologic standards but usually a million years or more—undergo changes in its mineralogy and texture. A limestone filled with fossils, for example, may be transformed into a white marble in which no trace of fossils remains. The mineral and chemical composition of the rock may be unaltered, but its texture may have changed drastically, from small calcite crystals to large, interlocked calcite crystals that erase such former features as fossils. Shale, a well-bedded sedimentary rock so fine-grained that no individual crystal can be seen with the naked eye, may become a schist, in which the original bedding becomes hidden by large crystals of mica. In this case, both mineralogy and texture have changed, but the overall chemical composition of the rock has remained the same.

> *Metamorphic rocks are formed when rocks are subjected to significant changes in temperature or pressure.*

Sediments and sedimentary rocks are products of Earth's surface environments, whereas igneous rocks are products of the magmas that originate in the lower crust and mantle. Metamorphic rocks, even those exposed at Earth's surface, are mainly the products of processes acting on rocks at depths ranging from the upper to the lower crust. Most of these rocks are formed at depths of 10 to 30 km, in the middle to lower half of the crust. But metamorphism can also occur at Earth's surface. We can see metamorphic changes, for example, in the baked surfaces of soils and sediments just beneath volcanic lava flows.

The heat and pressure in Earth's interior and its fluid composition are the three principal factors that drive metamorphism. In much of Earth's crust, temperature increases at a rate of 30°C per kilometer of depth, although that rate varies considerably among different regions (Figure 7.17). Thus, at a depth of 15 km, the temperature will be about 450°C—much higher than the

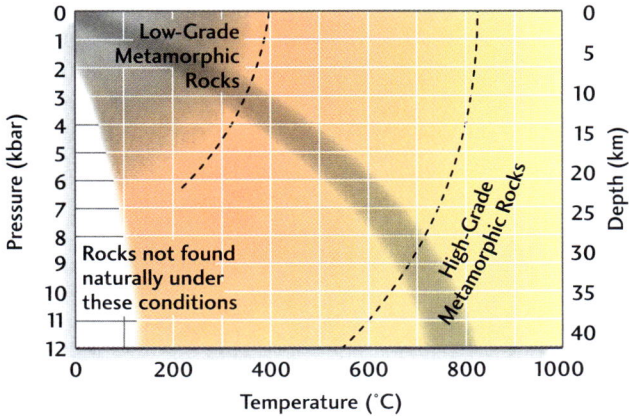

FIGURE 7.18 Temperatures, pressures, and depths at which low- and high-grade metamorphic rocks form. The dark band shows the rates at which temperature and pressure increase with depth over much of the continents.

average temperature at Earth's surface, which ranges from 10° to 20°C in most regions. The contribution of pressure is the result of vertically oriented forces exerted by the weight of overlying rocks as well as horizontally oriented forces developed as the rocks are deformed. The average pressure at a depth of 15 km amounts to about 4000 times the pressure at the surface.

High as these temperatures and pressures may seem, they are only in the middle range of conditions for metamorphism, as Figure 7.18 shows. A rock's *metamorphic grade* reflects the intensity of the temperatures and pressures it was subjected to during metamorphism. We refer to metamorphic rocks formed under the lower temperatures and pressures of shallower crustal regions as *low-grade* rocks and those formed at the higher temperatures and pressures of deeper zones as *high-grade* rocks. As the grade of metamorphism changes, the assemblages of minerals within metamorphic rocks also change.

The Role of Temperature

Heat greatly affects a rock's chemical composition, mineralogy, and texture. For example, plate tectonic processes may move sediment and rock from Earth's surface to its interior, where temperatures are higher. As the rock adjusts to the new temperature, its atoms and ions recrystallize, linking up in new arrangements and creating new mineral assemblages. Many new crystals will grow larger than the crystals in the original rock.

The increase in temperature with increasing depth is called the *geothermal gradient*. The geothermal gradient varies depending on plate tectonic setting, but on average it is about 30°C per kilometer of depth. In areas where plate extension has thinned the continental lithosphere, such as in Nevada's Great Basin, the geothermal gradient is *steep* (for example, 50°C per kilometer of depth). In areas where the continental lithosphere is old and thick, such as beneath central North America, the geothermal gradient is *shallow* (for example, 20°C per kilometer of depth) (see Figure 7.17).

Because different minerals crystallize and remain stable at different temperatures, we can use a rock's mineral composition as a kind of *geothermometer* to gauge the temperature at which it formed. For example, as sedimentary rocks containing clay minerals are buried deeper and deeper, the clay minerals begin to recrystallize and form new minerals, such as micas. With additional burial at greater depths—and temperatures—the micas become unstable and begin to recrystallize into new minerals, such as garnet.

Plate tectonic processes such as subduction and continental collision, which transport rocks and sediments into the hot depths of the crust, are the mechanisms that form most metamorphic rocks. In addition, limited metamorphism may occur where rocks are subjected to elevated temperatures near igneous intrusions. The heat is locally intense but does not penetrate deeply; thus, the intrusions can metamorphose the surrounding country rock, but the effect is localized.

> *The geothermal gradient is the rate at which temperature increases with depth in Earth's crust; it varies depending on plate tectonic setting.*

The Role of Pressure

Pressure, like temperature, changes a rock's chemical composition, mineralogy, and texture. Solid rock is subjected to two basic kinds of pressure, also called *stress:*

1. *Confining pressure* is a general force applied equally in all directions, like the pressure a swimmer feels under water. Just as a swimmer feels greater confining pressure when diving to greater depths, a rock descending to greater depths in Earth's interior is subjected to progressively increasing confining pressure.

2. *Directed pressure* is force exerted in a particular direction, as when you squeeze a ball of clay between your thumb and forefinger. Directed pressure, or *differential stress,* is usually concentrated within particular zones or along discrete planes.

The compressive force exerted where plates converge is a form of directed pressure, and it results in deformation of the rocks near the plate boundary. Heat reduces the strength of a rock, so directed pressure is likely to cause severe folding and other forms of ductile deformation, as well as metamorphism, in mountain belts where temperatures are high. Rocks subjected to differential stress may be severely distorted, becoming flattened in the direction the force is applied and elongated in the direction perpendicular to the force.

The minerals in a rock under pressure may be compressed, elongated, or rotated to line up in a particular direction, depending on the kind of stress applied to the rock. Thus, directed pressure guides the shape and orientation of the new crystals formed as minerals recrystallize under the influence of both heat and pressure. During the recrystallization of micas, for example, the crystals grow with the planes of their sheet silicate structures aligned perpendicular to the directed stress. The rock may develop a banded pattern as minerals of different compositions are segregated into separate planes.

> *Confining pressure increases with depth in Earth's interior. Confining pressure is constant in all directions, whereas directed pressure is force exerted in a particular direction.*

Marble, which we discussed at the opening of this chapter, owes its remarkable strength to this recrystallization process. Marble starts out as the sedimentary rock limestone, which is then heated to very high temperatures that cause it to recrystallize. The original minerals and crystals become reoriented and tightly interlocked during recrystallization to form a very strong structure with no planes of weakness, such as fractures.

> *Metamorphic recrystallization may align minerals in a particular direction depending on the kind of stress applied to the rock.*

The pressure to which rock is subjected deep in Earth's interior is related to both the thickness and the density of the overlying rocks. Pressure is usually recorded in *kilobars* (1000 bars, abbreviated kbar) and increases at a rate of 0.3 to 0.4 kbar per kilometer of depth (see Figure 7.17). One bar is approximately equivalent to the pressure of air at Earth's surface.

The Role of Fluids

Metamorphism can significantly alter a rock's mineralogy by introducing or removing chemical components that are soluble in water. Hydrothermal fluids carry dissolved carbon dioxide as well as chemical substances—such as sodium, potassium, silica, copper, and zinc—that are soluble in hot water under pressure. As hydrothermal solutions percolate up to the shallower parts of the crust, they react with the rocks they penetrate, changing their chemical and mineral compositions and sometimes completely replacing one mineral with another without changing the rock's texture. This kind of change in a rock's composition by fluid transport of chemical substances into or out of the rock is called **metasomatism.** Many valuable deposits of copper, zinc, lead, and other metallic ores are formed by this kind of chemical substitution, as we saw in Chapter 4.

Where do these chemically reactive fluids originate? Although most rocks appear to be completely dry and to

Metasomatism is a change in a metamorphic rock's composition due to interactions with fluids.

have extremely low porosity, they characteristically contain water in minute pores (the spaces between grains). This water comes not from the pores of sedimentary rocks—from which most of the water is expelled during diagenesis—but rather from chemically bound water in clays. Water forms part of the crystalline structures of metamorphic minerals such as micas and amphiboles. The carbon dioxide dissolved in hydrothermal fluids is derived largely from sedimentary carbonates—limestones and dolostones.

TYPES OF METAMORPHISM

Geologists can duplicate metamorphic conditions in the laboratory and determine the precise combinations of pressure, temperature, and chemical composition under which transformations might take place. But to understand how any particular combination of conditions relates to the geology of metamorphism—when, where, and how these conditions came about in Earth's interior—we categorize metamorphic rocks on the basis of their geologic origins (**Figure 7.19**).

Regional Metamorphism

Regional metamorphism, the most widespread type, takes place where both high temperatures and high pressures are imposed over large parts of the crust. We use this term to distinguish this type of metamorphism from more localized transformations near igneous intrusions or faults. Regional metamorphism is a characteristic feature of convergent plate boundaries. It occurs in volcanic arcs, such as the Andes of South America, and in the cores of mountain chains produced by continent-continent collisions, such as the Himalayas of central Asia. These mountain belts are often linear features, so zones of regional metamorphism are often linear in their distribution. In fact, geologists usually interpret regionally extensive belts of metamorphic rocks as representing sites of former mountain chains that were eroded over millions of years, exposing the rocks at Earth's surface.

Regional metamorphism takes place where both high temperatures and high pressures are imposed over large parts of the crust. It is characteristic of convergent plate boundaries.

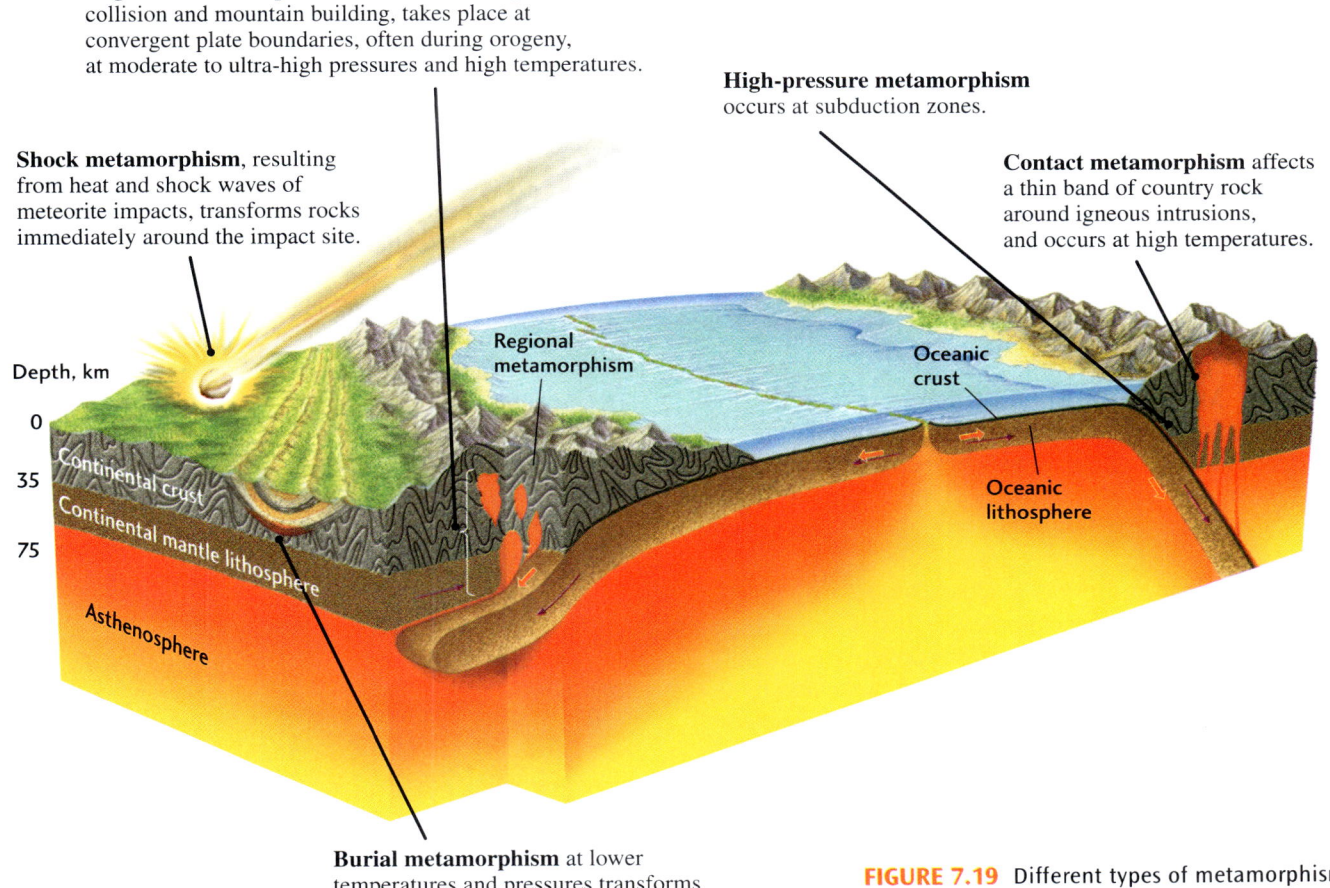

Regional metamorphism, from continental plate collision and mountain building, takes place at convergent plate boundaries, often during orogeny, at moderate to ultra-high pressures and high temperatures.

Shock metamorphism, resulting from heat and shock waves of meteorite impacts, transforms rocks immediately around the impact site.

High-pressure metamorphism occurs at subduction zones.

Contact metamorphism affects a thin band of country rock around igneous intrusions, and occurs at high temperatures.

Burial metamorphism at lower temperatures and pressures transforms sedimentary rocks.

FIGURE 7.19 Different types of metamorphism occur in different plate tectonic settings.

> Contact metamorphism is the localized transformation of rock adjacent to igneous intrusions, mainly as a result of high temperatures.

Some regional metamorphic belts are created by high temperatures and moderate to high pressures near the volcanic arcs formed where subducted plates sink deep into the mantle. Regional metamorphism under very high pressures and temperatures takes place deeper in the crust along boundaries where colliding continents deform rock and raise high mountain belts. In both cases, rocks are typically transported to great depths in Earth's crust, then uplifted and eroded at Earth's surface.

Contact Metamorphism

In **contact metamorphism,** the heat from an igneous intrusion metamorphoses the immediately surrounding rock. This type of localized transformation normally affects only a thin region of country rock along the zone of contact. In many contact metamorphic rocks, especially at the margins of shallow intrusions, the mineral and chemical transformations are largely related to the high temperature of the magma. Pressure effects are important only where the magma is intruded at great depths. Here, the pressure results not from the intrusion forcing its way into the country rock, but from the presence of regional confining pressure. Contact metamorphism by volcanic deposits is limited to very thin zones because lavas cool quickly at the surface and their heat has little time to penetrate deep into the surrounding rocks and cause metamorphic changes.

Other Types of Metamorphism

There are several other types of metamorphism that produce smaller amounts of metamorphic rock. Some of these types are extremely important in helping geologists understand conditions deep within Earth.

Burial Metamorphism Recall from Chapter 6 that when sedimentary rocks are gradually buried during subsidence of the crust, they slowly heat up as they come into equilibrium with the temperature of the crust around them. In the process, diagenesis alters their mineralogy and texture. Diagenesis grades into *burial metamorphism,* or low-grade metamorphism that is caused by the progressive increase in pressure exerted by the growing layers of overlying sediments and sedimentary rocks and by the increase in heat associated with increased depth of burial.

> Burial metamorphism is caused by the increase in temperatures and pressures as sediments are gradually buried deeper in the crust.

Depending on the local geothermal gradient, burial metamorphism typically begins at depths of 6 to 10 km, where temperatures range between 100° and 200°C and pressures are less than 3 kbar. This fact is of great importance to the oil and gas industry, which defines its "economic basement" as the depth where low-grade metamorphism begins. Oil and gas wells are rarely drilled below this depth because temperatures above 130°C convert organic matter trapped in sedimentary rocks into carbon dioxide rather than crude oil and natural gas.

High-Pressure and Ultra-High-Pressure Metamorphism Metamorphic rocks formed by *high-pressure metamorphism* (at 8 to 12 kbar) and *ultra-high-pressure metamorphism* (at pressures greater than 28 kbar) are rarely exposed at Earth's surface for us to study. These rocks are rare because they form at such great depths that it takes a very long time for them to be recycled back to the surface. Most high-pressure metamorphic rocks form in subduction zones as sediments scraped from subducting oceanic plates are plunged to depths of over 30 km, where they experience pressures of up to 12 kbar.

> Rocks produced by high-pressure and ultra-high-pressure metamorphism are rare at Earth's surface because they form at great depths.

Unusual metamorphic rocks once located at the base of Earth's crust can sometimes be found at Earth's surface. These rocks—called *eclogites*—indicate pressures of greater than 28 kbar, suggesting depths of over 80 km. Such rocks form at moderate to high temperatures, ranging from 800° to 1000°C. In a few cases, these rocks contain *microscopic diamonds,* indicative of pressures greater than 40 kbar and depths greater than 120 km!

Shock Metamorphism *Shock metamorphism* occurs when a meteorite collides with Earth. Upon impact, the energy represented by the meteorite's mass and velocity is transformed into heat and shock waves that pass through the impacted country rock. The country rock can be shattered and even partially melted to produce *tektites.* The smallest tektites look like droplets of glass. In some cases, quartz is transformed into *coesite* and *stishovite,* two of its high-pressure forms.

METAMORPHIC TEXTURES

Metamorphism imprints new textures on the rocks that it alters (**Figure 7.20**). The texture of a metamorphic rock is determined by the sizes, shapes, and arrangement of its constituent crystals. Some metamorphic rock textures depend on the particular kinds of minerals formed under metamorphic conditions. Variation in grain size is also important. In general, the size of crystals or grains increases as metamorphic grade increases. Each textural variety of metamorphic rock tells us something about the

FIGURE 7.20 Directed pressure on rocks containing platy minerals causes foliation.
[(a) Marli Miller; (b) S. Dobos.]

metamorphic process that created it. In this section, we examine two major textural classes of metamorphic rocks.

Foliated Rocks

The most prominent textural feature of regionally metamorphosed rocks is **foliation,** a set of flat or wavy parallel cleavage planes produced by deformation under directed pressure. These foliation planes may cut through the bedding of the original sedimentary rock at any angle or be parallel to the bedding (see Figure 7.20a). In general, as the grade of regional metamorphism increases, foliation becomes more pronounced.

> Foliation is created by the alignment of platy minerals under directed pressure.

A major cause of foliation is the presence of minerals with a platy crystal habit, chiefly the micas and chlorite. The planes of all the platy crystals are aligned parallel to the foliation, an alignment called the *preferred orientation* of the minerals (see Figure 7.20b). As platy minerals crystallize, their preferred orientation is usually perpendicular to the main direction of the forces squeezing the rock during metamorphism. Crystals of preexisting minerals may contribute to the foliation by rotating until they also lie parallel to the developing foliation plane.

Minerals with an elongate, needlelike crystal habit also tend to assume a preferred orientation during metamorphism: these crystals, too, normally line up parallel to the foliation plane. Rocks that contain abundant amphiboles, typically metamorphosed mafic volcanics, have this kind of texture.

The **foliated rocks** are classified according to four main criteria:

1. Metamorphic grade
2. Crystal size
3. Type of foliation
4. Banding

FIGURE 7.21 Foliated rocks are classified based on metamorphic grade, crystal size, type of foliation, and banding.

[*slate, phyllite, schist, gneiss:* John Grotzinger/Ramón Rivera-Moret/Harvard Mineralogical Museum; *migmatite:* Kip Hodges.]

Figure 7.21 shows examples of the major types of foliated rocks. In general, foliation progresses from one texture to another with increasing metamorphic grade. In this progression, as temperature and pressure increase, a shale may metamorphose first to a slate, then to a phyllite, then to a schist, then to a gneiss, and finally to a migmatite.

Slate Slates are the lowest grade of foliated rocks. These rocks are so fine-grained that their individual minerals cannot be seen easily without a microscope. They are commonly produced by the metamorphism of shales or, less frequently, of volcanic ash deposits. Slates usually range from dark gray to black, colored by small amounts of organic material originally present in the parent shale. Slate splitters learned long ago to recognize foliation planes and use them to make thick or thin slates for roofing tiles and blackboards. We still use flat slabs of slate for flagstone walks in parts of the country where slate is abundant.

Phyllite The **phyllites** are of a slightly higher grade than the slates but are similar in character and origin. They tend to have a more or less glossy sheen resulting from crystals of mica and chlorite that have grown a little larger than those of slates. Phyllites, like slates, tend to split into thin sheets, but less perfectly than slates.

Schist At low grades of metamorphism, the crystals of platy minerals are generally too small to be seen, foliation planes are closely spaced, and layers are very thin. As rocks are subjected to higher temperatures and pressures, the platy crystals grow large enough to be visible to the naked eye, and the minerals may tend to segregate into lighter and darker bands. This parallel arrangement of platy minerals produces the coarse, wavy foliation known as *schistosity*, which characterizes **schists.** Schists, which are intermediate-grade rocks, are among the most abundant metamorphic rock types. They contain more than 50 percent platy minerals, mainly the micas muscovite and biotite. Schists may contain thin layers of quartz, feldspar, or both, depending on the quartz content of the parent shale.

Gneiss Even coarser foliation is shown by high-grade **gneisses,** light-colored rocks with coarse bands of segregated light and dark minerals throughout the rock. The banding of gneisses into light and dark layers results from the segregation of lighter-colored quartz and feldspar from darker-colored amphiboles and other mafic minerals. Gneisses are coarse-grained, and the ratio of granular to platy minerals is higher than it is in slate or schist. The result is poor foliation and thus little tendency to split. Under conditions of high pressure and temperature, the mineral assemblages of the lower-grade rocks containing micas and chlorite change into new assem-

> *Slates, phyllites, schists, and gneisses are the principal types of foliated metamorphic rocks formed at progressively higher temperatures and pressures.*

blages dominated by quartz and feldspars, with lesser amounts of micas and amphiboles.

Temperatures higher than those necessary to produce gneiss may begin to melt the country rock. The result is a mixture of igneous and metamorphic rock called *migmatite*.

Granoblastic Rocks

Granoblastic rocks are nonfoliated metamorphic rocks composed mainly of crystals that grow in equant (equidimensional) shapes, such as cubes and spheres, rather than in platy or elongate shapes. These rocks result from metamorphic processes, such as contact metamorphism, in which deformation is absent, so foliation does not occur. Granoblastic rocks include hornfels, quartzite, marble, greenstone, amphibolite, and granulite (**Figure 7.22**). All granoblastic rocks, except hornfels, are defined by their mineralogy rather than their texture because all of them have a homogeneous granular texture.

> *Granoblastic rocks are nonfoliated metamorphic rocks composed mainly of crystals that grow in equant shapes.*

Hornfels is a high-temperature contact metamorphic rock of uniform grain size that has undergone little or no deformation. Hornfels has a granular texture overall, even though it commonly contains pyroxene, which makes elongate crystals, and some micas. It is not foliated, and its platy or elongate crystals are oriented randomly.

Quartzites are very hard, white rocks derived from quartz-rich sandstones. Some quartzites are homogeneous, unbroken by preserved bedding or foliation (Figure 7.22a). Others contain thin bands of slate or schist, relics of former interbedded layers of clay or shale.

Marbles are the metamorphic products of heat and pressure acting on limestones and dolomites. Some white, pure marbles, such as the famous Italian Carrara marbles prized by sculptors, show a smooth, even texture of interlocked calcite crystals of uniform size. Other marbles show irregular banding or mottling from silicate and other mineral impurities in the original limestone (Figure 7.22b). The Lincoln Memorial in Washington, D.C., samples the variety of marbles that exist in nature: white Colorado marble is used for the exterior, pink Tennessee marble for the floor, and Alabama marble for the ceiling.

Greenstones are metamorphosed mafic volcanic rocks. Many of these low-grade metamorphic rocks form when mafic lavas and ash deposits react with percolating seawater or other solutions. Large areas of the seafloor are covered with basalts that have been slightly or extensively altered in this way at mid-ocean ridges. An abundance of chlorite gives these rocks their greenish cast.

Amphibolite is a rock made up of amphibole and plagioclase feldspar. It is typically the product of medium- to high-grade metamorphism of mafic volcanic rocks.

(a) Quartzite

(b) Marble

FIGURE 7.22 Granoblastic (nonfoliated) metamorphic rocks: (a) quartzite [Breck P. Kent]; (b) marble [Diego Lezama Orezzoli/CORBIS].

Foliated amphibolites can be produced when deformation occurs.

The high-grade metamorphic rock *granulite*, which is also referred to as *granofels*, has a homogeneous granular texture. It is a medium- to coarse-grained rock in which the crystals are equant and show only faint

> *Granoblastic rocks include hornfels, quartzite, marble, greenstone, amphibolite, and granulite.*

FIGURE 7.23 Garnet porphyroblasts in a schist matrix. The minerals in the matrix are continuously recrystallized as pressure and temperature change and therefore grow to only a small size. In contrast, porphyroblasts grow to large size because they are stable over a broad range of pressures and temperatures. [Chip Clark.]

foliation at most. It is formed by the metamorphism of shale, impure sandstone, and many kinds of igneous rock.

Porphyroblasts are new metamorphic minerals that may grow into large crystals surrounded by a much finer grained matrix of other minerals (**Figure 7.23**). These large crystals are found in rocks formed both by contact and by regional metamorphism. They grow as the chemical components of the matrix are reorganized and thus replace parts of the matrix. Porphyroblasts vary in size, ranging from a few millimeters to several centimeters in diameter. Garnet and staurolite are two common minerals that form porphyroblasts, but many others are also found.

Table 7.1 summarizes the textural classes of metamorphic rocks and their main characteristics.

REGIONAL METAMORPHISM AND METAMORPHIC GRADE

Metamorphic rocks form under a wide range of conditions, and their mineralogies and textures are clues to the pressures and temperatures in the crust where and when they formed. Geologists who study the formation of metamorphic rocks constantly seek to determine the intensity and character of metamorphism more precisely than is indicated by a designation of "low grade" or "high grade." To make these finer distinctions, geologists "read" minerals as though they were pressure gauges and thermometers. These techniques are best illustrated by their application to regional metamorphism.

Mineral Isograds: Mapping Zones of Change

When we study a broad belt of regionally metamorphosed rocks, we can see many outcrops, some showing one set of

TABLE 7.1 Classification of Metamorphic Rocks by Texture

Classification	Characteristics	Rock Name	Typical Parent Rock
Foliated	Distinguished by slaty cleavage, schistosity, or gneissic foliation; mineral grains show preferred orientation	Slate Phyllite Schist Gneiss	Shale, sandstone
Granoblastic (nonfoliated)	Granular, characterized by coarse or fine interlocking grains; little or no preferred orientation	Hornfels Quartzite Marble Argillite Greenstone Amphibolite[a] Granulite[b]	Shale, volcanics Quartz-rich sandstone Limestone, dolomite Shale Basalt Shale, basalt Shale, basalt
Porphyroblastic	Large crystals set in fine matrix	Slate to gneiss	Shale

[a]Typically contains much amphibole, which may show alignment of long, needle-like crystals.
[b]High-temperature, high-pressure rock.

minerals, some showing others. Different zones within the belt may be distinguished by *index minerals,* abundant minerals that all formed under a limited range of temperatures and pressures (Figure 7.24). For example, a zone of unmetamorphosed shales may lie next to a zone of weakly metamorphosed slates (Figure 7.24a). As we move from the shale zone into the slate zone, a new index mineral—chlorite—appears, marking the point at which we move into a new zone with a higher metamorphic grade. If laboratory studies have determined the temperature and pressure at which the index mineral forms, we can draw conclusions about the conditions that existed when the rocks in the zone were formed.

We can make a map of the boundaries between zones where one metamorphic grade changes to another. Geologists define these boundaries by drawing lines called *isograds* that plot the transition from one zone to the next. Isograds are used in Figure 7.24a to show a series of mineral assemblages produced by the regional metamorphism of shale. A pattern of isograds tends to follow the deformation features of a region, as outlined by folds and faults. An isograd based on a single index mineral, such as the chlorite isograd shown here, is a good approximate measure of metamorphic pressure and temperature.

Because isograds reveal the pressures and temperatures at which the minerals in a regional metamorphic belt were formed, the isograd sequence in one belt may differ from that in another. The reason for this difference is that pressure and temperature do not increase at the same rate in all plate tectonic settings. As we discussed earlier in this chapter, pressure increases more rapidly

> Isograds show where one mineral zone, containing rocks of one metamorphic grade, changes to another.

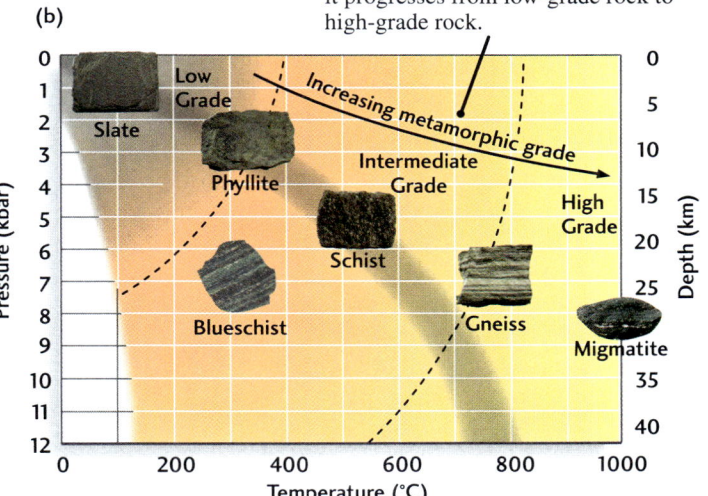

FIGURE 7.24 Index minerals define the different metamorphic zones within a belt of regional metamorphism. (a) Map of New England, showing mineral zones based on index minerals found in rocks metamorphosed from slate. (b) Rocks produced by the metamorphism of slate at different temperatures and pressures. [slate, phyllite, schist, gneiss: John Grotzinger/Ramón Rivera-Moret/Harvard Mineralogical Museum; blueschist: courtesy of Mark Cloos; migmatite: Kip Hodges.] (c) Changes in the mineral composition of metamorphic rock derived from slate define its metamorphic facies and are used to indicate its metamorphic grade.

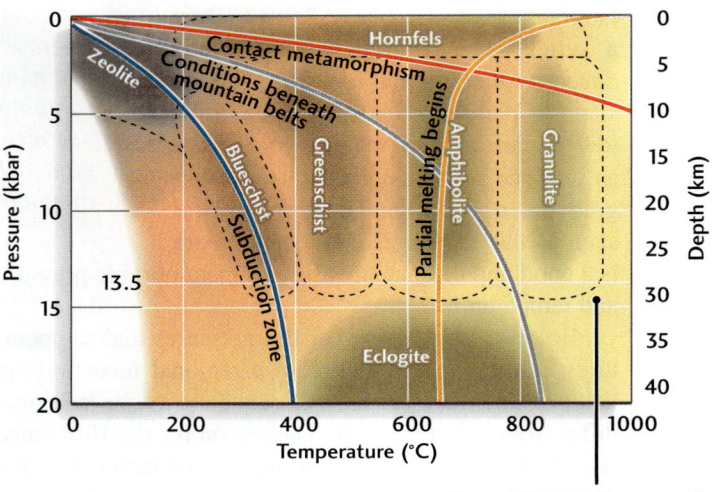

FIGURE 7.25 Metamorphic facies correspond to particular combinations of temperature and pressure, which correspond to particular plate tectonic settings.

Dotted lines indicate overlapping boundaries of metamorphic facies.

than temperature in some settings and more slowly in others (see Figures 7.17 and 7.18).

Metamorphic Facies

We can plot this information about the grades of metamorphic rock in a regional metamorphic belt—derived from parent rocks of many different chemical compositions—on a graph of temperature and pressure (**Figure 7.25**). **Metamorphic facies** are groupings of rocks of various mineral compositions formed under different grades of metamorphism from different parent rocks. By delineating metamorphic facies, we can be more specific about the grades of metamorphism observed in rocks. Two essential points characterize the concept of metamorphic facies:

1. Different kinds of metamorphic rocks form from parent rocks of different composition at the same grade of metamorphism.

2. Different kinds of metamorphic rocks form at different grades of metamorphism from parent rocks of the same composition.

Table 7.2 lists the major minerals of the metamorphic facies produced from shale and basalt. Because parent rocks vary so greatly in composition, there are no sharp boundaries between metamorphic facies (see Figure 7.25). Perhaps the most important reason for analyzing

Metamorphic facies are groupings of rocks of various mineral compositions formed under different grades of metamorphism from different parent rocks.

TABLE 7.2	Major Minerals of Metamorphic Facies Produced from Parent Rocks of Different Composition	
Facies	Minerals Produced from Shale Parent	Minerals Produced from Basalt Parent
Greenschist	Muscovite, chlorite, quartz, albite	Albite, epidote, chlorite
Amphibolite	Muscovite, biotite, garnet, quartz, albite, staurolite, kyanite, sillimanite	Amphibole, plagioclase feldspar
Granulite	Garnet, sillimanite, albite, orthoclase, quartz, biotite	Calcium-rich pyroxene, calcium-rich plagioclase feldspar
Eclogite	Garnet, sodium-rich pyroxene, quartz/coesite, kyanite	Sodium-rich pyroxene, garnet

metamorphic facies is that they give us clues to the tectonic processes responsible for metamorphism.

Metamorphic Grade and Parent Rock Composition

The kind of metamorphic rock that results from a given grade of metamorphism depends partly on the mineral composition of the parent rock. The metamorphism of slate, as shown in Figure 7.24c, reveals the effects of pressure and temperature on parent rock that is rich in clay minerals, quartz, and perhaps some carbonate minerals. The metamorphism of mafic volcanic rock, composed predominantly of feldspars and pyroxene, follows a different course (**Figure 7.26**).

In the regional metamorphism of a basalt, for example, the lowest-grade rocks characteristically contain various *zeolite* minerals. The silicate minerals in the zeolite class contain water in cavities within their crystal structure. Zeolite minerals form at very low temperatures and pressures. Rocks that include this group of minerals are thus placed in the zeolite grade.

Overlapping with the zeolite grade is a higher grade of metamorphosed mafic volcanic rocks, the **greenschists**, whose abundant minerals include chlorite. Next are the amphibolites, which contain large amounts of amphiboles. The granulites, coarse-grained rocks containing pyroxene and calcium plagioclase, constitute the highest grade of metamorphosed mafic volcanics. Rocks of the greenschist, amphibolite, and granulite grades are also formed during metamorphism of sedimentary rocks, as shown in Figure 7.24c.

> *The kind of metamorphic rock that results from a given grade of metamorphism depends partly on the mineral composition of the parent rock.*

The pyroxene-bearing granulites are the products of high-grade metamorphism in which the temperature is high and the pressure is moderate. The opposite conditions, in which the pressure is high and the temperature moderate, produce rocks of the **blueschist** grade from parent rock of various starting compositions, from mafic volcanic rocks to shaley sedimentary rocks. The name comes from the abundance of glaucophane, a blue amphibole, in these rocks. Still another metamorphic rock, formed at extremely high pressures and moderate to high temperatures, is eclogite, which is rich in garnet and pyroxene.

PLATE TECTONICS AND METAMORPHISM

The concept of metamorphic grade can inform us of the maximum pressure or temperature to which a metamorphic rock has been subjected, but it says nothing about where the rock encountered those conditions. Nor does it tell us anything about how the rock was **exhumed,** or transported back to Earth's surface. It is important to understand that most metamorphism is a dynamic process, not a static event. Plate tectonics provides the framework within which we can understand the dynamic processes of metamorphism.

> *P-T paths are distinctive histories of temperatures and pressures that are characteristic of particular plate tectonic settings.*

Each metamorphic rock has a distinctive history of changing conditions of temperature and pressure that is reflected in its texture and mineral composition. This history is called a metamorphic *P-T path*. The P-T path can be a sensitive recorder of many important factors that influence metamorphism—such as sources of heat, which change temperatures, and rates of tectonic transport (burial and exhumation), which change pressures. Thus, P-T paths are characteristic of particular plate tectonic settings.

Ocean-Continent Convergence

The P-T path of the rock assemblages that form when oceanic lithosphere is subducted beneath a plate carrying a continent on its leading edge is shown in **Figure 7.27**a. Thick sediments eroded from the continent rapidly fill the deep-sea trench that forms at the subduction zone. As it descends, the cold slab of oceanic lithosphere stuffs the region below the inner wall of the trench (the wall closer to the continent) with these sediments, as well as with deep-sea sediments and shreds of ophiolite suites (see Figure 5.12) scraped off the descending plate. Regions of this sort, located between the volcanic arc on the continent and the

FIGURE 7.26 Changes in the mineral composition of mafic rock during metamorphism indicate its metamorphic grade.

FIGURE 7.27 P-T paths indicate the trajectory of rocks during metamorphism at (a) an ocean-continent convergence and (b) a continent-continent convergence. The different P-T paths of rocks formed in these different plate tectonic settings indicate differences in geothermal gradients. Rocks transported to similar depths—and pressures—beneath mountain belts become much hotter than rocks at an equivalent depth in subduction zones.

trench offshore, are enormously complex and variable. The rocks formed there are all highly folded, intricately faulted, and metamorphosed. Blueschist-grade metamorphism, which is characteristic of high pressures and low temperatures, is found in these regions because the material may be carried relatively rapidly to depths as great as 30 km, where recrystallization occurs in the environment of the still-cool subducting slab. Typical P-T paths for blueschist-grade metamorphic rocks reveal this initial downward transport in the subduction zone. During exhumation, the trajectory reverses itself.

Continent-Continent Convergence

When a continent collides with another continent, a wide zone of intense deformation develops where the continents grind together, as shown in Figure 7.27b. The intense deformation that occurs during orogeny results in a much-thickened continental crust in the collision zone, often producing high mountains. As the lithosphere thickens, the deep parts of the continental crust heat up and undergo varying grades of metamorphism. Millions of years afterward, when erosion has stripped off the surface layers of the mountains, their cores are exposed at the surface, providing a rock record of the metamorphic processes that formed the schists, gneisses, and other metamorphic rocks.

P-T paths for metamorphic rocks produced by continent-continent collision indicate initial transport deep within the collision zone. Geologists generally interpret this initial trajectory as indicating the burial of rocks beneath high mountains during orogeny. P-T paths then indicate transport of metamorphic rocks back to Earth's surface as the buried rocks are uplifted and exhumed.

Exhumation: The Link Between Metamorphism and Climate

Forty years ago, plate tectonic theory provided a ready explanation for how metamorphic rocks could be produced by seafloor spreading, subduction, and continent-continent collision. By the mid-1980s, the study of P-T paths provided a more highly resolved picture of the specific tectonic mechanisms involved in the deep burial of rocks. At the same time, however, it surprised geologists by providing an equally well resolved image of the subsequent, and often very rapid, uplift and exhumation of these deeply buried rocks. Since the time of this discovery, geologists have been searching for exclusively tectonic mechanisms that could bring these rocks back to Earth's surface so quickly.

An important new idea is that climate, and not tectonic processes alone, drives the flow of rocks from the deep crust to the shallow crust through the process of rapid erosion. Thus, plate tectonics—which acts through orogeny—and climate—which acts through weathering and erosion—*interact* to control the flow of metamorphic rocks to Earth's surface. After decades of emphasis on solely tectonic explanations for regional and global geologic processes, it now seems that two apparently unrelated processes—metamorphism and erosion—may be linked in an elegant way. As one geologist exclaimed: "Savory the irony should the metamorphic muscles that push mountains to the sky be driven by the pitter patter of tiny raindrops."

By increasing erosion rates, climate may help drive the flow of metamorphic rocks from deep in Earth's crust to Earth's surface.

SUMMARY

What are geologic maps and cross sections? A geologic map is a model of the rock formations exposed at Earth's surface. A geologic cross section is a diagram representing the geologic features that would be visible if a vertical slice were made through part of the crust.

What do laboratory experiments tell us about the way rocks deform? Laboratory studies show that rocks subjected to tectonic forces may behave as brittle materials or as ductile materials. These behaviors depend on temperature and pressure, the type of rock, the speed of deformation, and the orientation of tectonic forces.

What are the basic deformation structures that geologists observe in the field? The most important geologic structures in rock formations that result from deformation are folds and faults. Fractures are known as faults if rocks are displaced across the fracture surface and as joints if no displacement is observed.

What kinds of forces produce these deformation structures? Faults and folds are produced primarily by horizontally directed forces at plate boundaries. Horizontal tensional forces at divergent boundaries produce normal faults, horizontal compressive forces at convergent boundaries produce thrust faults, and horizontal shearing forces at transform-fault boundaries produce strike-slip faults. Folds are usually formed in layered rock by compressive forces, especially in regions where continents collide.

What are the causes of metamorphism? Metamorphism—alteration in the chemical composition, mineralogy, or texture of solid rock—is caused by increases in pressure and temperature and by reactions with chemical components introduced by migrating fluids. As pressures and temperatures deep within the crust increase as a result of plate tectonic processes, the chemical components of the parent rock rearrange themselves into a new set of minerals that are stable under the new conditions. Rocks metamorphosed at relatively low temperatures and pressures are referred to as low-grade rocks; those metamorphosed at high temperatures and pressures are high-grade rocks. Chemical components may be added to or removed from a rock during metamorphism, usually by migrating fluids.

What are the various types of metamorphism? The major types of metamorphism are regional metamorphism, during which large areas are metamorphosed by high pressures and temperatures generated during mountain building, and contact metamorphism, during which country rock close to an igneous intrusion is transformed primarily by the heat of the intruding magma. Less common types are burial metamorphism, during which sedimentary rocks are altered by increases in pressure and temperature as they are buried deeper in the crust; high-pressure and ultra-high-pressure metamorphism, which occur at great depths, as when sediments are subducted; and shock metamorphism, which results from meteorite impacts.

What are the chief kinds of metamorphic rocks? Metamorphic rocks fall into two major textural classes: foliated rocks (displaying foliation, a pattern of parallel cleavage planes resulting from a preferred orientation of crystals) and granoblastic, or nonfoliated, rocks. The kinds of rocks produced by metamorphism depend on the composition of the parent rock and the grade of metamorphism. Regional metamorphism of a shale leads to zones of foliated rock of progressively higher grade, from slate to phyllite, schist, gneiss, and finally migmatite. Among granoblastic rocks, marble is derived from the metamorphism of limestone, quartzite from quartz-rich sandstone, and greenstone from basalt. Hornfels is the product of contact metamorphism of fine-grained sedimentary rocks and other types of rock containing an abundance of silicate minerals.

What do metamorphic rocks reveal about the conditions under which they were formed? Mineral zones can be mapped with isograds defined by the first appearance of an index mineral. The presence of an index mineral can indicate the temperature and pressure under which the rocks in the zone were formed. According to the concept of metamorphic facies, rocks of the same metamorphic grade may differ because of variations in the chemical composition of the parent rock, whereas rocks metamorphosed from the same parent rock may vary because they were subjected to different grades of metamorphism.

How do metamorphic rocks relate to plate tectonic processes? During both subduction and continent-continent collision at convergent plate boundaries, rocks and sediments are pushed to great depths in Earth's crust, where they are subjected to increasing pressures and temperatures that result in metamorphism. Metamorphic P-T paths provide insight into the plate tectonic settings where these rocks are metamorphosed. In the case of ocean-continent convergences, P-T paths indicate subduction of rocks and sediments to sites with high pressures and relatively low temperatures. In continent-continent collision zones, rocks are pushed down to depths where pressure and temperature are both high. In both settings, the P-T paths show that after the rocks experienced the maximum pressures and temperatures, they were pushed back up to shallow depths. This process of exhumation may be driven by weathering and erosion at Earth's surface as well as by plate tectonic processes.

KEY TERMS AND CONCEPTS >>

anticline (p. 179)
blueschist (p. 195)
brittle (p. 175)
compressive force (p. 176)
contact metamorphism (p. 188)
dip (p. 173)
dip-slip fault (p. 176)
ductile (p. 175)
exhumation (p. 195)
faulting (p. 172)
folding (p. 172)
foliated rocks (p. 189)
foliation (p. 189)
geologic cross section (p. 174)
geologic map (p. 174)
gneiss (p. 190)
granoblastic rocks (p. 191)
greenschist (p. 195)
joint (p. 181)
metamorphic facies (p. 194)
metamorphic rock (p. 185)
metasomatism (p. 186)
normal fault (p. 177)
orogeny (p. 172)
phyllite (p. 190)
regional metamorphism (p. 187)
schist (p. 190)
shearing force (p. 176)
slate (p. 190)
strike (p. 173)
strike-slip fault (p. 176)
syncline (p. 179)
tensional force (p. 176)
thrust fault (p. 177)

EXERCISES >>

1. Is the small fault on the left side of Figure 7.1b a normal fault or a thrust fault? Estimate the fault's offset, expressing your answer in meters.

2. The motion of the San Andreas fault has offset the stream channel in Figure 7.6 by 130 m. Geologists have determined that this channel is 3800 years old. What is the slip rate along the San Andreas fault at this site, expressed in millimeters per year?

3. Show that a left jog in a right-lateral strike-slip fault will produce compression, whereas a right jog in a right-lateral strike-slip fault will produce extension. Write a similar rule for left-lateral strike-slip faults.

4. At what depths in Earth do metamorphic rocks form? What happens if temperatures get too high?

5. What does preferred orientation refer to in a metamorphic rock? Think about how the alignment of minerals relates to deformation.

6. You have mapped an area of metamorphic rocks, such as the region in Figure 7.24, and have observed mineral zones marked by north-south isograd lines running from sillimanite in the east to chlorite in the west. Were metamorphic temperatures higher in the east or in the west?

7. Subduction zones are generally characterized by high pressure–low temperature metamorphism. In contrast, continent-continent collision zones are marked by moderate pressure–high temperature metamorphism. Which type of plate boundary has a higher geothermal gradient? Explain.

TIMING THE GEOLOGIC RECORD

CHAPTER 8

The Abyss of Time

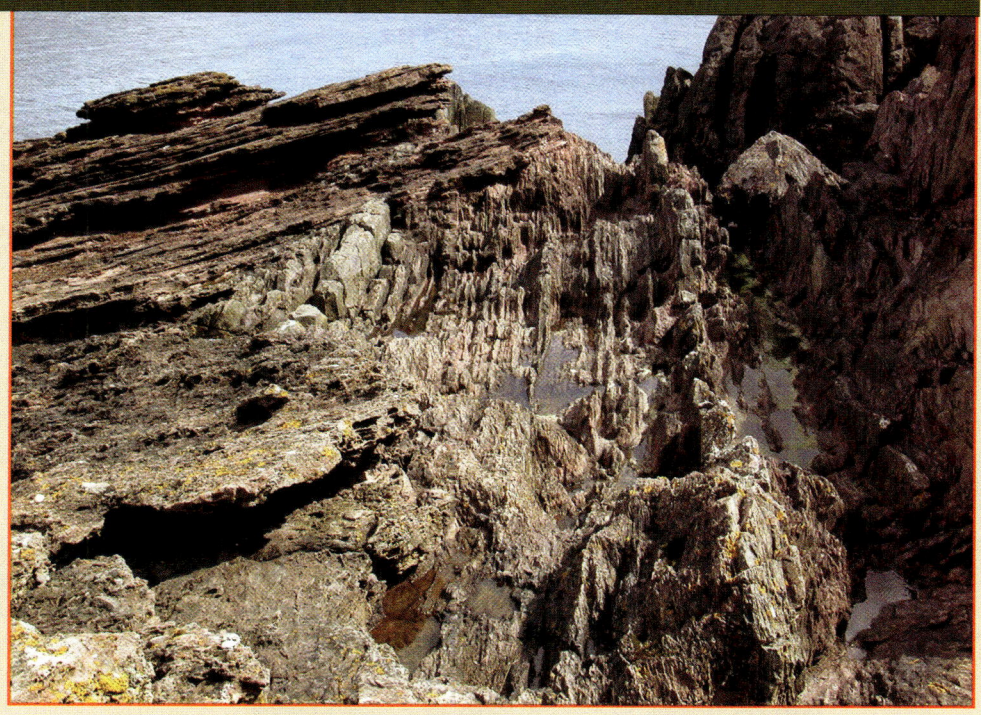

On a clear June day in 1788, the geologist James Hutton sailed along the Berwickshire coast of southeastern Scotland with a mathematician friend, John Playfair, and Sir James Hall, looking for evidence of Earth's ancient past. They rounded Siccar Point and landed at the base of an impressive rock formation, soon to become one of the most famous sites in the study of geology. Nearly vertical layers of gray slate (schistus) rose up in a sheer cliff face, only to terminate abruptly against nearly horizontal beds of red sandstone. In the words of Playfair,

We felt ourselves necessarily carried back to the time when the schistus on which we stood was yet at the bottom of the sea, and when the sandstone before us was only beginning to be deposited, in the shape of sand or mud, from the waters of a superincumbent ocean. An epoch still more remote presented itself, when even the most ancient of these rocks, instead of standing upright in vertical beds, lay in horizontal planes at the bottom of the sea, and was not yet disturbed by that immeasurable force which has burst asunder the solid pavement of the globe.... The mind seemed to grow giddy by looking so far into the abyss of time.

> *We find no vestige of a beginning, no prospect of an end.*
> — James Hutton

How deep is that abyss? Hutton was a gentleman farmer and sometime philosopher who had carefully observed how water and ice, year by year, wore down the highlands a few rocks at a time and rivers washed away, grain by grain, the soils of his farmlands. He had realized that these erosional processes, which laid sediments down in the sea, must be balanced by the deformation and uplift of sedimentary layers to rejuvenate the landscape in a continuing rock cycle.

It was obvious to Hutton—though not to many others of his day—that the contorted geology of the British Isles could not have been created by such slow processes in the mere thousands of years of recorded human history. After all, Hadrian's Wall, built to protect the Roman emperor's English settlements from rowdy Scottish tribes, hardly looked the worse for wear despite 16 centuries of exposure to the elements. What much greater expanse of time must nature have taken to create the skewed, cross-cutting formations at Siccar Point?

Siccar Point, southeastern Scotland, where gently dipping beds of red sandstone rest on near-vertical beds of gray slate. [Clifford E. Ford, University of Edinburgh.]

CHAPTER OUTLINE

Reconstructing Geologic History from the Stratigraphic Record 201

The Geologic Time Scale: Relative Ages 206

Measuring Absolute Time with Radioactive Clocks 209

The Geologic Time Scale: Absolute Ages 211

In his great treatise, *Theory of the Earth*, published in the same year as his expedition with Playfair, Hutton presented the evidence for his principle of uniformitarianism (introduced in Chapter 2). He described the indefinite span of geologic time in a memorable line: "we find no vestige of a beginning, no prospect of an end."

Spurred by Hutton's vision, geologists set out to explore Earth's history through ever more careful and precise readings of the rock record. They eventually discovered that Earth did indeed have a beginning, and a violent one at that. The methods they invented to measure geologic time are described in this chapter; what they found is the subject of the next.

In this chapter, we will learn how geologists have plumbed the abyss of time by finding order in the geologic record and using "clocks in rocks" to date the events that have occurred throughout Earth's 4.56-billion-year history.

Geologic processes occur on time scales that range from seconds (meteorite impacts, volcanic explosions, earthquakes) to tens of millions of years (the recycling of oceanic lithosphere) and even billions of years (the tectonic evolution of continents). If we are careful enough, we can measure the rates of short-term processes, such as beach erosion or the seasonal variations in the transport of sediments by rivers, in a few years. Precise surveying can monitor the slow movements of glaciers (meters per year), and with the new GPS instruments depicted in Figure 3.14, we can track the even slower movements of tectonic plates (centimeters per year). Historical documents can provide certain types of geologic data, such as the dates of major earthquakes or volcanic eruptions, from hundreds or, in some cases, thousands of years ago.

However, the record of human observation is far too short for the study of many slow geologic processes. In fact, it's not even long enough to capture some types of rapid but infrequent events; for example, we have never witnessed a meteorite impact as big as the one that left the crater shown in Figure 2.7. We must rely instead on the geologic record—the information preserved in rocks that have survived erosion and subduction. Almost all oceanic crust older than 200 million years has been subducted back into the mantle, so most of Earth's history is documented only in the older rocks of the continents. Geologists can

(a)

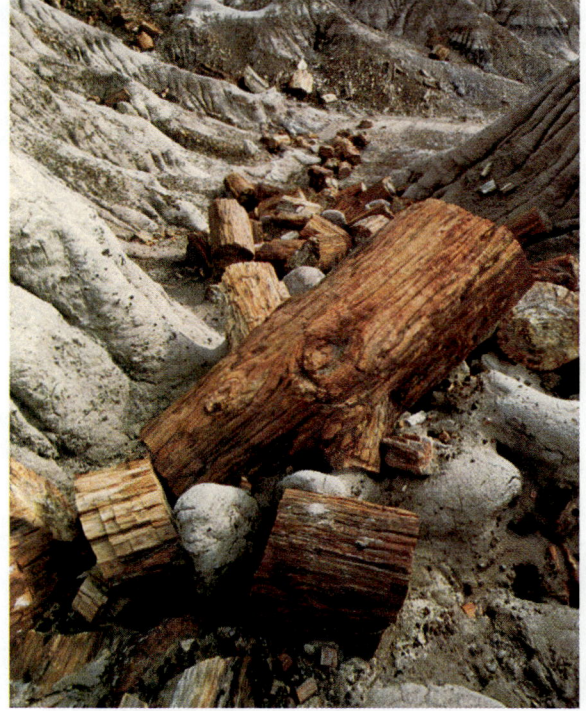

(b)

FIGURE 8.1 Fossils are traces of living organisms preserved in the geologic record. (a) Ammonite fossils, ancient examples of a large group of invertebrate organisms that are now largely extinct. Their sole representative in the modern world is the chambered nautilus. [Chip Clark.] (b) Petrified Forest, Arizona. These ancient logs are millions of years old. Their substance was completely replaced by silica, which preserved all the original details of form. [Tom Bean.]

reconstruct subsidence from the record of sedimentation; uplift from the erosion of rock layers; and deformation from faults, folds, and metamorphic rocks. But to measure the pace of these processes and understand their common causes, we must be able to date the geologic record.

Geologists speak carefully about time. Dating means measuring the **absolute age** of an event in the geologic record—the number of years elapsed from the event until now. Before the twentieth century, no one knew much about absolute ages; geologists could determine only whether one event was older or younger than another —their **relative ages.** They could say, for instance, that fish bones were first deposited in marine sediments before mammal bones appeared in land sediments, but they couldn't tell how many millions of years ago the first fish or mammals appeared. As we will see in this chapter, powerful methods have been invented for estimating absolute ages and developing a precise and detailed geologic time scale.

RECONSTRUCTING GEOLOGIC HISTORY FROM THE STRATIGRAPHIC RECORD

Philosophers have struggled with the notion of time throughout human history, but until fairly recently, they had very little data. The first geological observations pertaining to the question of deep time came in the mid-seventeenth century from the study of fossils. A fossil is an artifact of life preserved in the geologic record (**Figure 8.1**). However, few people living in the seventeenth century would have understood this definition. Most thought that the seashells and other lifelike forms found in rocks dated from Earth's beginnings or grew there spontaneously.

In 1667, the Danish scientist Nicolaus Steno, who was working for the royal court in Florence, Italy, demonstrated that the peculiar "tongue stones" found in certain Mediterranean sediments were essentially identical to the teeth of modern sharks. He concluded that tongue stones *really were* ancient shark teeth preserved in the rocks, and more generally, that fossils were the remains of ancient life deposited with the sediments. To convince people of his ideas, Steno wrote a short but brilliant book about the geology of Tuscany in which he laid the foundation for the modern science of **stratigraphy**— the study of *strata* (layers) in rocks.

> Stratigraphy is the study of layers, or strata, of sedimentary rock.

Principles of Stratigraphy

Geologists still use Steno's principles to interpret sedimentary strata. Two of his basic rules are so simple they seem obvious to us today (**Figure 8.2**):

1. The **principle of original horizontality** states that sediments are deposited under the influence of gravity as nearly horizontal beds. Observations in a wide variety of sedimentary environments support this principle. If we find folded or faulted strata, we know that the layers were deformed by tectonic forces after the sediments were deposited.

2. The **principle of superposition** states that each layer of an undisturbed sedimentary sequence is younger than the one beneath it and older than the one above it. A new layer cannot be deposited beneath an existing layer. Thus, strata can be vertically ordered in time from the

1 Sediments are deposited in horizontal layers and slowly change into sedimentary rock.

2 If left undisturbed by tectonic processes, the youngest layers remain on the top and the oldest on the bottom.

FIGURE 8.2 Layers of sedimentary rock in Marble Canyon, part of the Grand Canyon, illustrate Steno's principles. The Grand Canyon was cut by the Colorado River through what is now northern Arizona, revealing layers that record millions of years of geologic history. Stratigraphy is the study of sedimentary sequences such as this one. [Fletcher and Baylis/Photo Researchers.]

lowest (oldest) bed to the uppermost (youngest) bed. A chronologically ordered set of strata is known as a **stratigraphic succession.**

> *Sediments are deposited sequentially in horizontal layers with the oldest layers at the bottom and the youngest layers at the top.*

We can apply Steno's principles in the field to determine whether one sedimentary layer is older than another. At each outcrop within a region, there are distinct **formations,** groups of rock layers that can be identified throughout the region by their physical properties. Some formations consist of a single rock type, such as limestone. Others are made up of thin, interlayered beds of different kinds of sediments, such as sandstone and shale. However they vary, each formation comprises a distinctive set of rock layers that can be recognized and mapped as a unit. By piecing together the relative ages of formations exposed in different outcrops, we can sort them into chronological order and thus construct the stratigraphic succession of a region—at least in principle.

In practice, there were two problems with this strategy. First, geologists almost always found gaps in a region's stratigraphic succession, indicating time intervals that had gone entirely unrecorded. Some were short, such as periods of drought between floods; others lasted for millions of years—for example, periods of regional tectonic uplift when thick sequences of sedimentary rocks were removed by erosion. Second, it was difficult to determine the relative ages of two formations that were widely separated in space; stratigraphy alone couldn't determine whether a sequence of mudstones in, say, Tuscany was older, younger, or the same age as a similar sequence in England. It was necessary to expand Steno's ideas about the biological origin of fossils to solve these problems.

Fossils as Recorders of Geologic Time

In 1793, only 5 years after Hutton visited Siccar Point, William Smith, a surveyor working on the construction of canals in southern England, recognized that fossils could help geologists order the relative ages of sedimentary rocks. Smith was fascinated by the variety of fossils, and he collected them from the rock strata exposed along canal cuts. He observed that different layers contained different sets of fossils, and he was able to tell one layer from another by the characteristic fossils in each. He established a general order for the sequence of fossil assemblages and strata, from lowest (oldest) to uppermost (youngest) rock layers. Regardless of the location, Smith could predict the stratigraphic position of any particular layer or formation in any outcrop in southern England, basing his prediction on its fossil assemblages. This stratigraphic ordering of the fossils of animal species (fauna) produces a sequence known as a *faunal succession*.

Smith's **principle of faunal succession** states that the sedimentary rock strata in an outcrop contain fossils in a definite sequence. The same sequence can be found in outcrops at other locations, and so strata from one location can

FIGURE 8.3 Fossils can be used to correlate rock layers in different outcrops.

Characteristic assemblages of fossils can be used to identify strata of the same age in different locations.

be matched to strata in another location. Figure 8.3 shows such a composite stratigraphic succession for two outcrops.

Using faunal successions, Smith was able to identify formations of the same age in different outcrops. By noting the vertical order in which the formations were found in each place, he compiled a composite stratigraphic succession for the entire region. His compilation showed how the complete succession would have looked if the formations at different levels in all the various outcrops could have been brought together at a single spot.

The geologists who followed in Steno's and Smith's footsteps described and catalogued hundreds of fossils and their relationships to modern organisms, establishing the new science of paleontology—the historical study of ancient life-forms. The most common fossils they found were the shells of invertebrate animals. Some were similar to clams, oysters, and other living shellfish; others represented strange species with no living examples, such as the trilobites (Figure 8.4). Less common were the bones of vertebrates, such as mammals, birds, and the huge extinct reptiles they called dinosaurs. Plant fossils were found to be abundant in some rocks, particularly in coal beds, where leaves, twigs, branches, and even whole tree trunks could be recognized. Fossils were not found in intrusive igneous rocks—no surprise, since any biological material would have been destroyed when the rocks melted—nor were they found in highly metamorphosed rocks, where any remains of organisms would have been distorted beyond recognition.

By the beginning of the nineteenth century, paleontology had become the single most important source of information about geologic history. Charles Darwin studied paleontology as a young scientist, and he collected many unusual fossils on his famous voyage aboard the *Beagle* (1831–1836). During this world-circling tour, he also studied many unfamiliar animal and plant species in their native habitats. Darwin pondered what he had seen for a couple of decades and then, in 1859, proposed his theory of evolution, which revolutionized the science of biology and provided a sound theoretical framework for paleontology. If organisms have evolved progressively over time, then the fossils in each layer must represent the organisms living when that layer was deposited.

Unconformities: Gaps in the Geologic Record

In putting together the stratigraphic succession of a region, we often find places in the geologic record where a formation is missing. Either no rock was ever deposited, or it was eroded away before the next strata were laid down. The surface between two layers that were laid down with a time gap between them—the boundary along which the two existing formations meet—is called an **unconformity** (Figure 8.5). An unconformity, like a sedimentary sequence, represents the passage of time.

An unconformity may imply that tectonic forces raised the land above sea level, where erosion removed some rock layers. Alternatively, the unconformity could have been produced by the erosion of newly exposed land as the sea level fell. As we will see in Chapter 10, global sea level can be lowered by hundreds of meters during ice ages, when water from the oceans is withdrawn to form continental ice sheets near the poles.

Unconformities are classified according to the relationships between the layers above and below them. An unconformity in which the upper set of layers overlies an erosional surface developed on an undeformed, still-horizontal lower set of layers is a *disconformity* (see Figure 8.5). Sea level drops caused by ice ages and broad tectonic uplifts often create disconformities. In a *nonconformity*, the upper layers overlie metamorphic or igneous rocks (see Figure 8.10 for an example). An *angular unconformity* is one in which the upper layers overlie lower layers that have been folded by tectonic processes and then eroded to a more or less even plane. In an angular unconformity, the two sets of layers have bedding planes that are not parallel. The photograph at the opening

An unconformity indicates a gap in time between layers in a stratigraphic succession.

FIGURE 8.4 Trilobites, an extinct life-form, preserved as fossils in rocks about 365 million years old found in Ontario, Canada. [William E. Ferguson.]

TIME 1
Beneath the sea, sedementary beds accumulate in layers A–D.

TIME 2
Later, tectonic forces cause uplift of the beds above sea level, exposing them to erosion.

TIME 3
Erosion strips away layer D and part of C, leaving an irregular surface of hills and valleys.

TIME 4
Subsidence below sea level allows a new layer, E, to be deposited over C. The irregular surface of C is preserved as an unconformity.

FIGURE 8.5 An unconformity is a surface between two rock layers representing a layer that never formed or was eroded away. The type of unconformity created through uplift and erosion, followed by subsidence and another round of sedimentation, is called a disconformity.

FIGURE 8.6 The Great Unconformity in the Grand Canyon, Arizona, is an angular unconformity between the horizontal Tapeats Sandstone above and the steeply dipping Precambrian Wapatai Shale below. The Wapatai Shale is part of the Precambrian Grand Canyon beds; the Tapeats Sandstone is Cambrian in age. [GeoScience Features Picture Library.]

of this chapter shows the famous angular unconformity that Hutton discovered at Siccar Point. **Figure 8.6** depicts another dramatic angular unconformity found near the bottom of the Grand Canyon. The formation of an angular unconformity by tectonic processes is illustrated in **Figure 8.7**.

Cross-Cutting Relationships

Other disturbances of the layering of sedimentary strata also provide clues for determining the relative ages of rocks. Recall that dikes can cut through sedimentary layers; sills can be intruded parallel to bedding planes (see Chapter 6); and faults can displace bedding planes, dikes, and sills as they shift blocks of rock (see Chapter 7). These *cross-cutting relationships* can be used to establish the relative ages of igneous intrusions or faults within the stratigraphic succession. Because the deformation or intrusion events must have taken place after the affected sedimentary layers were deposited, those structures must be younger than the rocks they cut (**Figure 8.8**). If the intrusions or fault

FIGURE 8.7 An angular unconformity is a surface that separates two sets of layers whose bedding planes are not parallel. This sequence of drawings shows how such a surface can form.

FIGURE 8.8 Cross-cutting relationships allow us to establish the relative ages of igneous intrusions or faults within a stratigraphic succession.

Cross-cutting relationships can be used to determine the relative ages of igneous intrusions, faults, and sedimentary strata.

displacements are eroded and planed off at an unconformity and then overlaid by younger sedimentary beds, we know that those structures are older than the younger strata.

THE GEOLOGIC TIME SCALE: Relative Ages

Early in the nineteenth century, geologists began to apply Steno's and Smith's stratigraphic principles to outcrops all over the world. The same distinctive fossils were discovered in similar formations on many continents. Moreover, faunal successions from different continents often displayed the same changes in fossil assemblages. By matching up faunal successions and using cross-cutting relationships, geologists could determine the relative ages of rock formations on a global basis. By the end of the century, they had pieced together a worldwide history of geologic events—a **geologic time scale.**

Divisions of Geologic Time

The geologic time scale divides Earth's history into intervals marked by distinctive sets of fossils, and it places the boundaries of those intervals at times when those sets of fossils changed abruptly (**Figure 8.9**). The basic intervals of this time scale are the **eras:** the Paleozoic (from the Greek *paleo,* meaning "old," and *zoi,* meaning "life"), the Mesozoic ("middle life"), and the Cenozoic ("new life").

The eras are subdivided into **periods,** usually named for the geographic locality in which the formations representing them were first or best described, or for some distinguishing characteristic of the formations. The Jurassic period, for example, is named for the Jura mountain range of France and Switzerland, and the Carboniferous period is named for the coal-bearing sedimentary rocks of Europe and North America. The Paleogene and Neogene periods of the Cenozoic are two exceptions; these Greek names mean "old origin" and "new origin," respectively.

Periods are further subdivided into **epochs,** such as the Miocene and Pliocene epochs of the Neogene period (see Figure 8.9). Today we are living in the Holocene ("completely new") epoch of the Neogene period in the Cenozoic era.

Distinctive fossil assemblages mark the intervals of the geologic time scale.

Interval Boundaries Mark Mass Extinctions

Many of the major boundaries in the geologic time scale represent **mass extinctions**—short intervals during which a large proportion of the species living at the time simply disappeared from the fossil record, followed by the blossoming of many new species. These abrupt changes in the

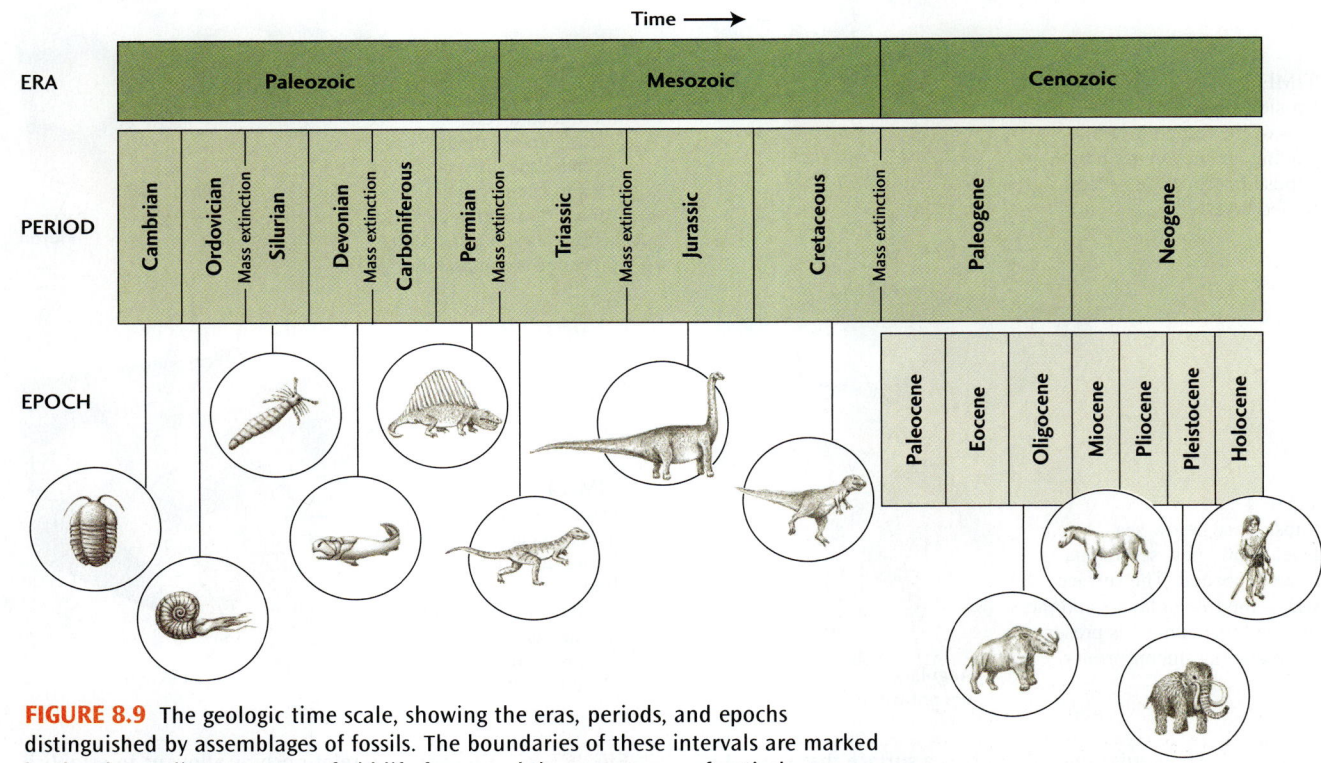

FIGURE 8.9 The geologic time scale, showing the eras, periods, and epochs distinguished by assemblages of fossils. The boundaries of these intervals are marked by the abrupt disappearance of old life-forms and the appearance of entirely new ones. The five most dramatic mass extinctions are indicated. Note that this diagram shows only the relative ages of the intervals. Time progresses from left to right.

faunal succession were a great mystery to the geologists who discovered them. Darwin's theory of evolution explained how new species could evolve, but what had caused the mass extinctions?

In some cases, we think we know. The mass extinction at the end of the Cretaceous, which killed off 75 percent of the living species, including all the dinosaurs, was almost certainly the result of a large meteorite impact that darkened the atmosphere and plunged Earth's climate into many years of bitter cold. This disaster marks the end of the Mesozoic era and the beginning of the Cenozoic.

In other cases, we are still not sure. The largest mass extinction, at the end of the Permian period, which defines the Paleozoic-Mesozoic boundary, eliminated nearly 95 percent of all living species, but the cause of this event is the subject of debate, as we will see in Chapter 9. The timing of this mass extinction precisely corresponds to a huge eruption of flood basalts in Siberia 251 million years ago, but why this particular eruption had such a devastating effect on the biosphere (while others almost as large did not) remains a mystery. Perhaps the eruption injected enough material into the atmosphere to trigger catastrophic climate change. The extreme events that cause mass extinctions are the subject of very active geologic research.

Mass extinctions, in which a large proportion of living species disappeared from the fossil record, mark many of the major boundaries in the geologic time scale.

The Grand Canyon Sequence: An Exercise in Relative Dating

The sequence of rocks exposed in the Grand Canyon and other parts of the Colorado Plateau illustrates how relative dating works. These beds record a long history of sedimentation in a variety of environments, sometimes on land and sometimes under the sea. By matching the rock sequences exposed at different localities, geologists have constructed a stratigraphic succession that spans both the Paleozoic and Mesozoic eras (**Figure 8.10**).

The lowest—and therefore oldest—rocks exposed at the Grand Canyon are dark igneous and metamorphic rocks that make up the Vishnu group of formations. Above the Vishnu are the younger Grand Canyon beds. Though these sedimentary rocks contain evidence of early life, they do not contain the shelly fossils distinctive of the Cambrian and later periods, so they are called *Precambrian* rocks.

A nonconformity separates the Vishnu and Grand Canyon beds. This structure indicates a period of deformation accompanying metamorphism of the Vishnu, and then a period of erosion, before the deposition of the Grand Canyon beds. The tilting of the Grand Canyon beds from their originally horizontal position shows that they, too, were folded after deposition and burial.

An angular unconformity divides the Grand Canyon beds from the overlying horizontal Tapeats Sandstone (see Figure 8.6). This unconformity indicates a long period of erosion after the lower rocks had been tilted. The Tapeats Sandstone and Bright Angel Shale can be dated as Cambrian by their fossils, many of which are trilobites.

Above the Bright Angel Shale is a group of horizontal limestone and shale formations (Muav Limestone, Temple Butte Limestone, Redwall Limestone) that represent a long history of marine sedimentation from the late Cambrian period to the end of the Mississippian period. There are so many time gaps represented by disconformities in these rocks that less than 40 percent of the Paleozoic is actually represented by rock strata.

The next set of strata, high up on the canyon wall, is the Supai group of formations (Pennsylvanian and Permian), which contain fossils of land plants like those found in the coal beds of North America and other continents. Overlying the Supai is the Hermit, a sandy red shale.

Continuing up the canyon wall, we find another continental deposit, the Coconino Sandstone, which contains vertebrate animal tracks. At the top of the cliffs at the canyon rim are two more formations of Permian age: the Toroweap, made mostly of limestone, overlain by the Kaibab, a massive layer of sandy and cherty limestone. These two formations record subsidence below sea level and the deposition of marine sediments. Above the Kaibab Limestone and the canyon rim itself, but exposed within Grand Canyon National Park, is the Moenkopi formation, a red sandstone of Triassic age—the first appearance of rocks from the Mesozoic era in this stratigraphic succession.

The stratigraphic succession at the Grand Canyon, though picturesque and informative, represents an incomplete picture of Earth's history. Younger periods of geologic time are not preserved, and we must travel to nearby locations in Utah, such as Zion Canyon and Bryce Canyon, to fill in this more recent history. At Zion Canyon, we find equivalents of the Kaibab and Moenkopi, which allow us to correlate this stratigraphic succession with the one at the Grand Canyon and establish a link. In contrast to the Grand Canyon, however, the Zion strata extend upward to Jurassic time, including ancient sand dunes represented by the Navajo Sandstone (see Figure 6.13). If we travel a bit farther, we find that the Navajo Sandstone also occurs in Bryce Canyon, but that at this location the strata extend still farther upward, to the Wasatch formation of the Paleogene period.

The correlation of strata among these three areas of the Colorado Plateau shows how widely separated localities—each with an incomplete record of geologic time—can be pieced together to build a composite record of Earth's history.

The geologic history of the Colorado Plateau can be reconstructed from overlapping stratigraphic sequences exposed in the Grand Canyon, Zion, and Bryce Canyon national parks.

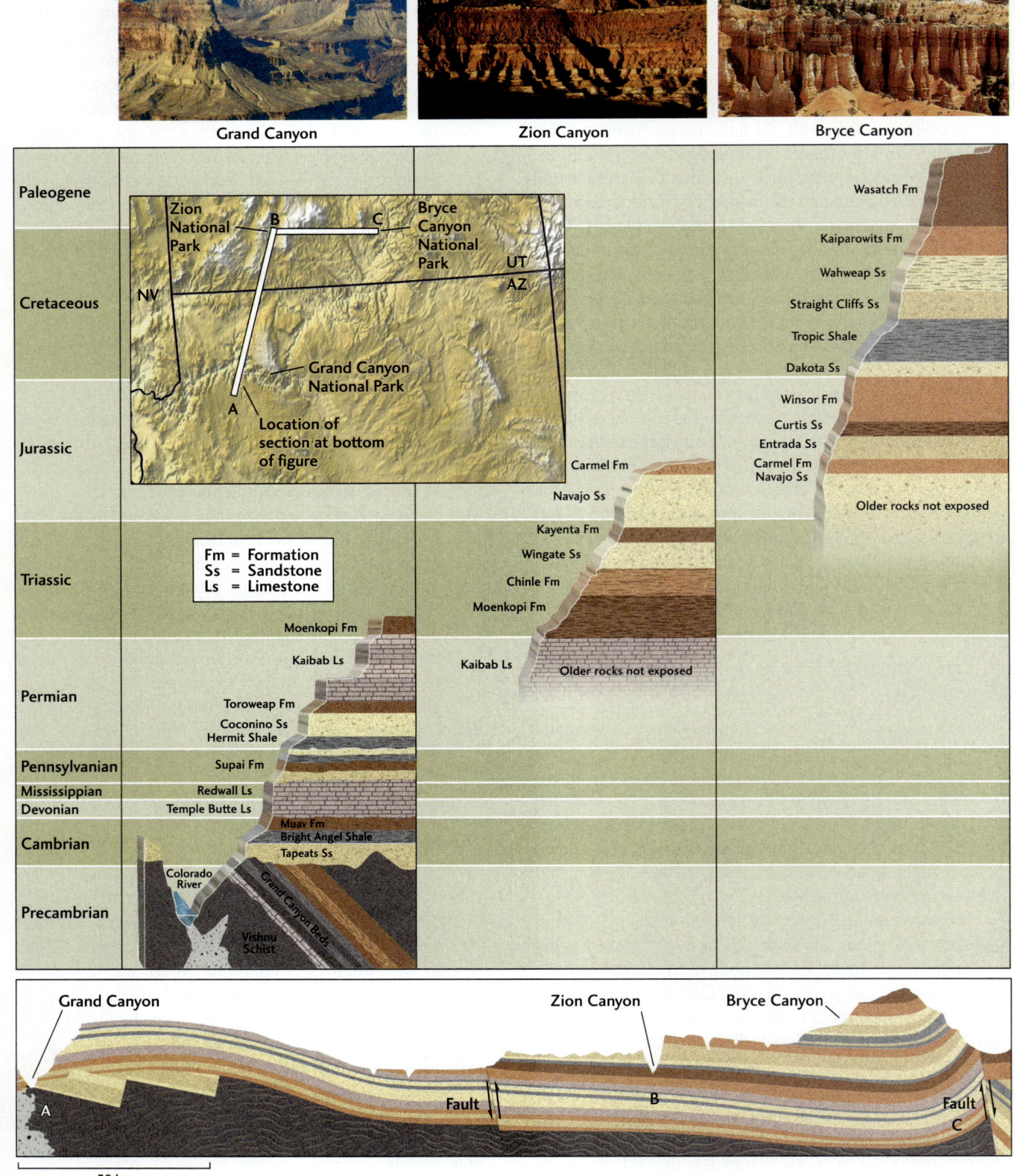

FIGURE 8.10 Stratigraphic sequence of the Colorado Plateau, reconstructed from strata exposed in Grand Canyon, Zion Canyon, and Bryce Canyon National Parks. [*Grand Canyon:* John Wang/Photo Disc/Getty Images; *Zion Canyon:* David Muench/CORBIS; *Bryce Canyon:* Tim Davis/Photo Researchers.]

MEASURING ABSOLUTE TIME WITH RADIOACTIVE CLOCKS

The geologic time scale based on studies of stratigraphy and faunal successions is a relative time scale. It tells us whether one formation or fossil assemblage is older than another, but not how long the eras, periods, and epochs were in actual years. Estimates of how long it takes mountains to erode and sediments to accumulate suggested that most geologic periods had lasted for millions of years, but geologists of the nineteenth century did not know whether the duration of a specific period was 10 million years, 100 million years, or even longer.

They did know that the geologic time scale was incomplete. The earliest period of geologic history recorded by faunal successions was the Cambrian, when animal life, in the form of shelly fossils, suddenly appeared in the geologic record. Many rock formations were clearly older than the Cambrian period because they occurred below Cambrian rocks in the stratigraphic succession. But these formations contained no recognizable fossils, so there was no way to determine their relative ages, and all such rocks were lumped into the general category "Precambrian." What fraction of Earth's history was locked up in these cryptic rocks? How old was the oldest Precambrian rock?

These questions sparked a huge debate in the latter half of the nineteenth century. Physicists and astronomers argued for a maximum age of less than 100 million years, but most geologists regarded this age as much too young, even though there were no precise data to back them up.

The Discovery of Radioactivity

In 1896, a major advance in physics paved the way for reliable and accurate measurements of absolute ages. Henri Becquerel, a French physicist, discovered radioactivity in uranium. Within a year, the French chemist Marie Sklodowska-Curie discovered and isolated a new and highly radioactive element, radium.

In 1905, the physicist Ernest Rutherford suggested that the absolute age of a rock could be determined by measuring the decay of radioactive elements found in rocks. He calculated the age of one rock by measuring its uranium content. This was the start of **isotopic dating,** the use of naturally occurring radioactive elements to determine the ages of rocks. Within a decade of Rutherford's first attempt, geologists were able to show that some Precambrian rocks were billions of years old.

In 1956, the geologist Clare Paterson measured the decay of uranium in meteorites to determine that the solar system—and, by implication, Earth—was formed 4.56 billion years ago. That age has been modified by less than 10 million years since Paterson's original measurement, so we might say that he completed the discovery of geologic time.

Radioactive Isotopes: The Clocks in Rocks

How do we use radioactivity to determine the age of a rock? Recall that the nucleus of an atom consists of protons and neutrons (see Chapter 4). For a given element, the number of protons is constant, but the number of neutrons can vary among different *isotopes* of the same element. Most isotopes are stable, but the nucleus of a *radioactive* isotope can spontaneously disintegrate, emitting particles and transforming the atom into an atom of a different element. We call the original atom the *parent* and the product of decay its *daughter.*

A useful element for isotopic dating is rubidium, which has 37 protons and two naturally occurring isotopes: rubidium-85, which has 48 neutrons and is stable, and rubidium-87, which has 50 neutrons and is radioactive. A neutron in the nucleus of a rubidium-87 atom can spontaneously eject an electron, changing to a proton that stays in the nucleus. The parent rubidium atom thus forms a daughter strontium-87 atom, with 38 protons and 49 neutrons (Figure 8.11).

A parent isotope decays into a daughter isotope at a constant rate. The rate of radioactive decay is measured by the isotope's **half-life**—the time required for one-half of the original number of parent atoms to be transformed into daughter atoms. For instance, the half-life of rubidium-87 is 47 billion years, meaning that this is the time it takes for half of the rubidium in a rock sample to decay to strontium-87. Of course, we must know the initial amount of the daughter that was present in a rock to calculate the isotopic age. In the case of rubidium-strontium dating, the initial amount of strontium-87 can be estimated from the amount of strontium-86, a stable isotope that is not the product of radioactive decay and therefore does not change as the mineral ages.

Radioactive isotopes make good clocks because half-life does not vary with the

> *Radioactive isotopes decay at a constant rate to form daughter isotopes. The age of a rock sample can be determined by measuring how much decay has occurred.*

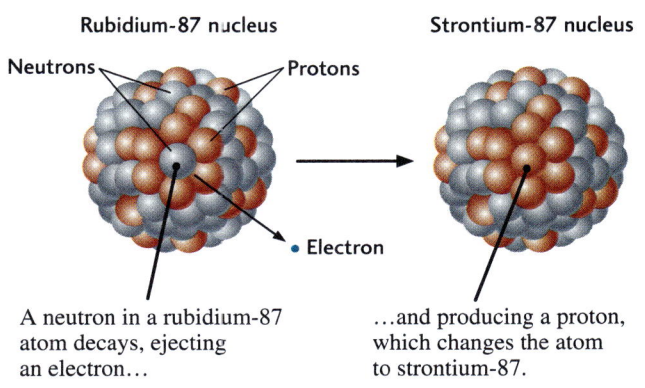

FIGURE 8.11 The radioactive decay of rubidium to strontium.

FIGURE 8.12 The number of radioactive atoms of any element in any mineral declines at a fixed rate over time. This rate of decay is given by the half-life of the isotope.

changes in temperature, pressure, chemical environment, or other factors that can accompany geologic processes on Earth or other planets. So when atoms of a radioactive isotope are created anywhere in the universe, they start to act like a ticking clock, steadily transforming from one type of atom to another at a fixed rate. At the end of the first half-life, the number of parent atoms has decreased by a factor of two; at the end of the second half-life, by a factor of four; at the end of the third half-life, by a factor of eight, and so forth (**Figure 8.12**).

We can measure the ratio of parent and daughter atoms with a mass spectrometer, which can detect even minute quantities of isotopes, and determine how much of the daughter has been produced from the parent. Knowing the half-life, we can then calculate the time elapsed since the radioactive clock began to tick.

The isotopic age of a rock corresponds to the time since the isotopic clock was "reset" when the isotopes were locked into the minerals of the rock. This locking usually occurs when a mineral crystallizes from a magma or recrystallizes during metamorphism. During crystallization, the number of daughter atoms in a mineral is not necessarily reset to zero, so the initial amount must be taken into account when calculating isotopic age.

Many other complications make isotopic dating a tricky business. A mineral can lose daughter isotopes by weathering or can be contaminated by fluids circulating in the rock. Metamorphism of igneous rocks can reset the isotopic composition of minerals to a date much later than the crystallization age.

Isotopic Dating Methods

To determine the ages of rocks, geologists use a number of naturally occurring radioactive isotopes, each of which has its own half-life (Table 8.1). Isotopic dating is possible only if a measurable number of parent and daughter atoms remain in the rock. For example, if a rock is very old and the decay rate is fast, almost all the parent atoms will already have been transformed. In that case, we could determine that the isotopic clock had run down, but we would have no way of knowing how long ago it stopped. Thus, isotopes that decay slowly over billions of years, such as rubidium-87, are most useful in measuring the ages of older rocks, whereas those that decay rapidly, such as carbon-14, can only be used to date younger rocks.

One of the most precise dating methods for old rocks is based on the decay of two related isotopes: the decay of uranium-238 to lead-206 and the decay of uranium-235 to lead-207. Isotopes of the same element behave similarly in the chemical reactions that alter rocks because the chemistry of an element depends mainly on its atomic number, not its atomic mass. The two uranium isotopes have different half-lives, however (see Table 8.1), so together they provide a consistency check that helps geologists account for the problems of weathering, contamination, and metamorphism discussed above.

The lead isotopes from single crystals of zircon—a crustal mineral with a relatively high concentration of

TABLE 8.1 Major Radioactive Elements Used in Isotopic Dating

Isotopes		Half-Life of Parent (years)	Effective Dating Range (years)	Examples of Minerals and Materials That Can Be Dated
Parent	Daughter			
Uranium-238	Lead-206	4.4 billion	10 million–4.6 billion	Zircon, apatite
Uranium-235	Lead-207	0.7 billion	10 million–4.6 billion	Zircon, apatite
Potassium-40	Argon-40	1.3 billion	50,000–4.6 billion	Muscovite, biotite, hornblende
Rubidium-87	Strontium-87	47 billion	10 million–4.6 billion	Muscovite, biotite, potassium feldspar
Carbon-14	Nitrogen-14	5730	100–70,000	Wood, charcoal, peat, bone and tissue, shells and other calcium carbonates

uranium—can be used to date the oldest rocks on Earth with an uncertainty of less than 1 percent. These formations turn out to be about 4 billion years old. Individual zircon grains with ages as great as 4.4 billion years have been found in sediments, indicating that Earth had a continental crust within 150 million years of its formation!

Crustal minerals more than 4 billion years old have been dated using uranium isotopes.

Carbon-14, which has a half-life of 5700 years, is especially useful for dating fossil bone, shell, wood, and other organic materials in sediments less than a few tens of thousands of years old. Carbon is an essential element in the living cells of all organisms. As green plants grow, they continuously incorporate carbon into their tissues from carbon dioxide in the atmosphere. When a plant dies, it stops absorbing carbon dioxide. At that moment, the ratio of carbon-14 to the stable carbon isotopes in the plant is identical to that in the atmosphere. Thereafter, the ratio of carbon-14 in the dead tissue decreases. Nitrogen-14, the daughter isotope of carbon-14, is a gas and thus leaks from the fossil material, so it cannot be measured to determine the time that has elapsed since the plant died. We can, however, estimate the absolute age by comparing the ratio of carbon-14 left in the plant material with the ratio in the atmosphere at the time the plant died.

THE GEOLOGIC TIME SCALE: Absolute Ages

Armed with isotopic dating techniques, geologists of the twentieth century were able to nail down the absolute ages of the key events on which their predecessors had based the geologic time scale. More important, they were able explore the early history of the planet recorded in Precambrian rocks. **Figure 8.13** presents the results of this century-long effort as a ribbon of geologic time.

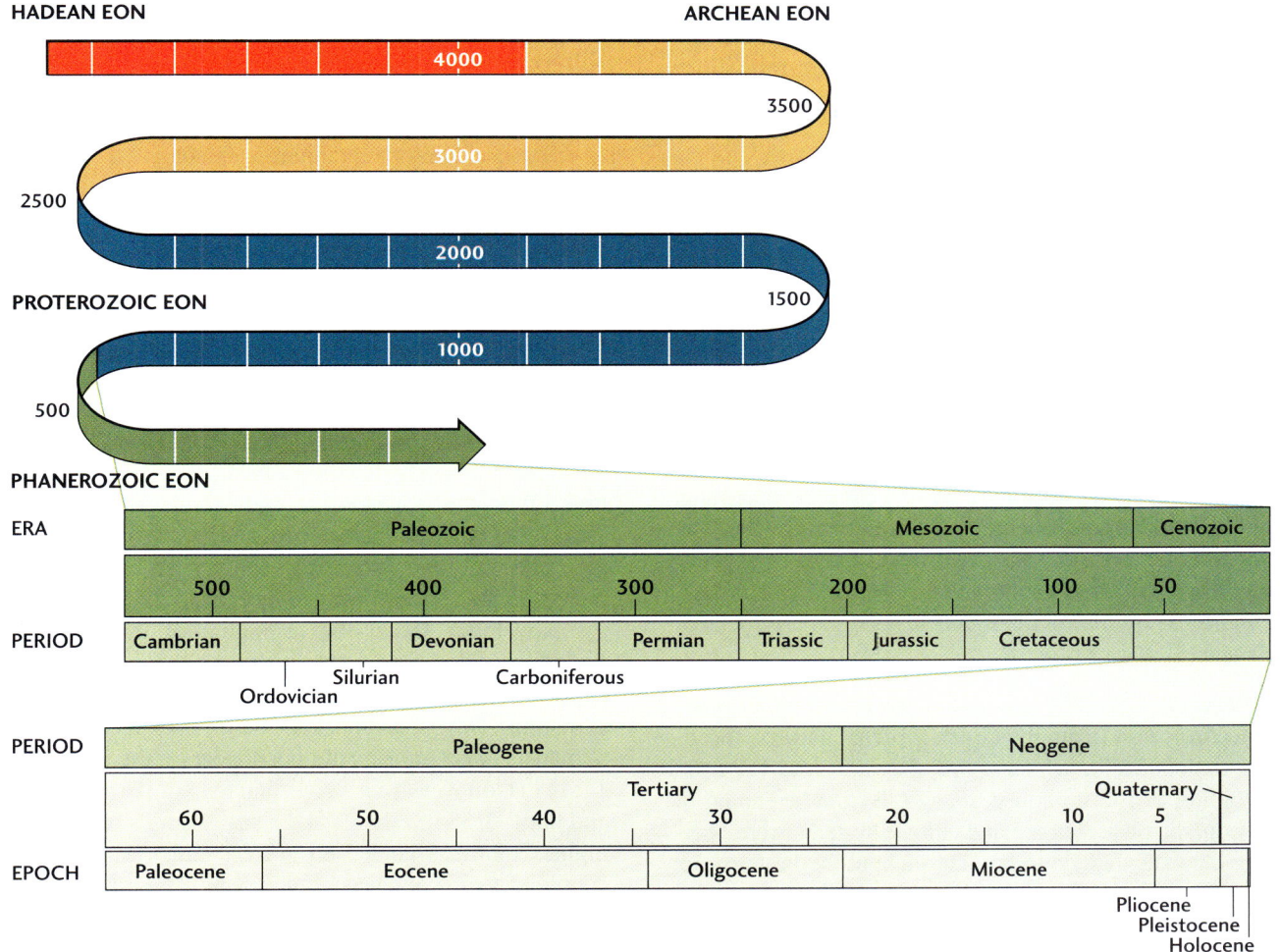

FIGURE 8.13 The ribbon of geologic time shows the complete geologic time scale. All numbers are ages in millions of years ago. (The Tertiary and Quaternary periods are older divisions that have been largely replaced by the Paleogene and Neogene periods, but are still sometimes used by geologists.)

The assignment of absolute ages to the geologic time scale reveals great differences in the time spans of the geologic periods. The Cretaceous period (80 million years) turns out to be more than three time longer than the Neogene (only 23 million years), and the Paleozoic era (291 million years) lasted longer than the Mesozoic and Cenozoic eras combined. The biggest surprise is the Precambrian, which had a duration of over 4000 million years—almost nine-tenths of Earth's history!

Eons: The Longest Intervals of Geologic Time

To represent the rich history of the Precambrian, a division of the geologic time scale longer than the era, called the **eon**, was introduced. Four eons, based on the isotopic ages of terrestrial rocks and meteorites, are now recognized.

Hadean Eon The earliest eon, whose name comes from *Hades* (the Greek word for "hell"), began with the formation of Earth 4.56 billion years ago and ended about 3.9 billion years ago. During its first 660 million years, Earth was bombarded by chunks of material from the early solar system. Although very few rock formations survived this violent period, individual mineral grains 4.4 billion years old have been discovered. There is also evidence that some liquid water existed on Earth's surface at about this time, suggesting that the planet cooled rapidly. In Chapter 9, we will explore this early phase of Earth's history in more detail.

Archean Eon The name of the next eon comes from *archaios* (the Greek word for "ancient"). Rocks of Archean age range from 3.9 billion to 2.5 billion years old. The geodynamo and the climate system were established during the Archean eon, and felsic crust accumulated to form the first stable continental masses. The processes of plate tectonics were probably operating, although perhaps substantially differently from the way they did later in Earth's history. Life, in the form of primitive single-celled microorganisms, became established, as indicated by the fossils found in sedimentary rocks of this age.

Proterozoic Eon The last part of the Precambrian is the Proterozoic eon (from the Greek *proteros* and *zoi*, meaning "earlier life"), which spans the time interval from 2.5 billion to 542 million years ago. By the beginning of this eon, the plate tectonic and climate systems were operating much as they do today. Throughout the Proterozoic, organisms that produced oxygen as a waste product (as plants do today) increased the amount of oxygen in Earth's atmosphere. Although life remained soft-bodied, some organisms evolved into sophisticated creatures with cells containing nuclei. The increase in atmospheric oxygen to nearly present-day concentrations toward the end of the Proterozoic may have encouraged single-celled organisms to evolve into multicellular algae and animals, which are preserved in the late Proterozoic fossil record.

Phanerozoic Eon The start of the Phanerozoic eon is marked by the first appearance of shelly fossils at the beginning of the Cambrian period, now dated at 542 million years ago. The name of this eon—from the Greek *phaneros* and *zoi* ("visible life")—certainly fits, because it comprises all three eras recognized in the fossil record: the Paleozoic (542 million to 251 million years ago), the Mesozoic (251 million to 65 million years ago), and the Cenozoic (65 million years ago to the present).

Precambrian time is divided into three eons: the Hadean, the Archean, and the Proterozoic. The Phanerozoic, a fourth eon, comprises time since the beginning of the Cambrian period.

Perspectives on Geologic Time

In the dusty sheep country of far western Australia stands a small promontory of ancient red rocks called the Jack Hills (**Figure 8.14**). Geologists have pulverized truckloads of these rocks to isolate a few sand-sized crystals of zirconium silicate, a very hard mineral nearly as indestructible as diamond. Using the uranium-lead isotopic dating technique, they have found one small crystal fragment with an age of 4.4 billion years—the oldest piece of our planet ever discovered. How can we relate to such a mind-boggling span of time?

Imagine compressing the 4.56 billion years of Earth history into a single year, starting with the formation of Earth on January 1 and ending right now at midnight on December 31. Within the first week, Earth was organized into core, mantle, and crust. The oldest zircon grain from the Jack Hills crystallized on January 13. The first primitive organisms appeared in mid-March. By mid-June, stable continents had developed, and throughout the summer and fall, the biological activity of evolving life increased the concentration of oxygen in the atmosphere. On November 18, at the beginning of the Cambrian period, complex organisms, including those with shells, appeared. On December 11, reptiles evolved, and late on Christmas Day, the dinosaurs became extinct. Modern humans, *Homo sapiens*, did not appear on the scene until 11:42 P.M. on New Year's Eve, and the last ice age did not end until 11:58 P.M. Three and a half seconds before midnight, Columbus landed on a West Indies island, and a couple of tenths of a second ago, you were born.

SUMMARY >>

How do we know whether one rock is older than another? We can determine the relative ages of rocks by studying the stratigraphy, fossils, and cross-cutting rela-

FIGURE 8.14 The outcrop in the Jack Hills of western Australia from which geologists have dated zircon grains as old as 4.4 billion years. [Bruce Watson, Rensselaer Polytechnic Institute.] *Inset:* Zircon crystals ($ZrSiO_4$) from the Hadean eon extracted from the Jack Hills. [Photo Chd GFDL.]

tionships of rock formations observed at outcrops. An undeformed sequence of sedimentary rock layers will be horizontal, with each layer younger than the layers beneath it and older than the ones above it. In addition, the fossils found in each layer reflect the organisms that were present when that layer was deposited. Knowing the faunal succession makes it easier to spot unconformities, which indicate intervals of time missing in the stratigraphic record.

How was a global geologic time scale created? By using fossils to match up rocks in outcrops around the world, geologists compiled a composite stratigraphic sequence from which they developed a geologic time scale. The use of isotopic dating allowed them to assign absolute dates to the eons, eras, periods, and epochs that constitute the time scale. Isotopic dating is based on the decay of radioactive isotopes, in which unstable parent atoms are transformed into stable daughter atoms at a constant rate. By measuring the amounts of parent and daughter atoms, geologists can calculate the absolute ages of rocks. The isotopic clock starts ticking when radioactive isotopes are locked into minerals as igneous rocks crystallize or metamorphic rocks recrystallize.

Why is the geologic time scale important to geology? The geologic time scale enables us to reconstruct the chronology of events that have shaped the planet. The time scale has been instrumental in studying plate tectonics and in estimating the rates of geologic processes too slow to be monitored directly. The creation of the relative time scale paralleled the development of paleontology and the theory of evolution. The assignment of absolute ages to the time scale revealed that Earth is 4.56 billion years old.

KEY TERMS AND CONCEPTS >>

absolute age (p. 201)
eon (p. 212)
epoch (p. 206)
era (p. 206)
formation (p. 202)
geologic time scale (p. 206)
half-life (p. 209)
isotopic dating (p. 209)
mass extinction (p. 206)
period (p. 206)
principle of faunal succession (p. 202)
principle of original horizontality (p. 201)
principle of superposition (p. 201)
relative age (p. 201)
stratigraphic succession (p. 202)
stratigraphy (p. 201)
unconformity (p. 203)

EXERCISES

1. Many fine-grained muds are deposited at a rate of about 1 centimeter per 1000 years. At this rate, how long would it take to accumulate a stratigraphic sequence half a kilometer thick?

2. How many formations can you count in the geologic cross section of the Grand Canyon in Figure 8.10 ? How many are the same formations observed in Zion Canyon? Are any of the formations observed in both the Grand Canyon and Bryce Canyon cross sections?

3. What type of unconformity would a broad tectonic uplift followed by subsidence be likely to produce? What type of unconformity might separate young flat-lying sediments from older metamorphosed sediments?

4. In studying an area of tectonic compression, a geologist discovers a sequence of older, more deformed sedimentary rocks on top of a younger, less deformed sequence, separated by an angular unconformity. What tectonic process might have created the angular unconformity?

5. Mass extinctions have been dated at 444 million years, 416 million years, and 359 million years. How are these events expressed in the geologic time scale of Figure 8.13?

6. A geologist discovers a distinctive set of fish fossils that dates from the Devonian period within a low-grade metamorphic rock. The rubidium-strontium isotopic age of the rock is determined to be only 70 million years. Give a possible explanation for the discrepancy.

7. Explain why the last eon of geologic history is named the Phanerozoic.

8. At the present rate of seafloor spreading, the entire seafloor is recycled every 200 million years. Assuming that this rate has been constant, calculate how many times the seafloor has been recycled since the end of the Archean eon.

9. Is carbon-14 a suitable isotope for dating geologic events in the Pliocene epoch?

HISTORY OF EARTH

CHAPTER 9

Geologists on the Moon

Harrison "Jack" Schmitt is the only geologist to have done fieldwork on another celestial body. He was the lunar module pilot aboard the *Apollo 17* spacecraft that landed in the narrow Taurus-Littrow Valley, situated in the ancient highlands on the southeastern rim of Mare Serenitatis—the Moon's lava-filled Sea of Serenity. The mission was near-perfect from its spectacular midnight launch by a 393-foot-tall Saturn V rocket on December 7, 1972, until its splashdown in the South Pacific on December 19.

During their 75 hours in Taurus-Littrow, Schmitt and the mission commander, Gene Cernan, traversed 30 km of the valley floor in a "Moon buggy," investigating the volcanic rocks that partially fill the valley, the boulders that had rolled into the valley from the surrounding mountains, and the impact-generated soils that cover the valley floor. They set up an intricate series of geophysical experiments that measured the properties of the soils, the crust, and the deep lunar interior, and they collected 110 kg of lunar rocks, more than any other Apollo mission. Isotopic dating of the rock samples they and other astronauts returned to Earth demonstrated that the lunar highlands were formed very early in the history of the solar system—before 4.4 billion years ago—and have a common ancestry with the silicate materials that constitute our own planet. As we will see in this chapter, the Apollo samples provided the key to unlock the early geologic history not only of the Earth-Moon system, but also of Mercury, Venus, and Mars.

Schmitt was trained at Caltech and at Harvard, where he received his Ph.D. in geology in 1964. He was recruited into NASA's first group of scientist-astronauts in 1965 while working at the U.S. Geological Survey's Astrogeology Center at Flagstaff, Arizona. He and astrogeologist Eugene Shoemaker, the first director of the Flagstaff center, played key roles in training Apollo crews to be competent geological observers when

they were in lunar orbit and on the lunar surface. Throughout the Apollo program, Schmitt participated in the examination and evaluation of the returned lunar samples and helped the crews with their scientific reports.

Jack Schmitt was the last person to leave footprints in the lunar soil. Gene Shoemaker died in a car accident in July 1997 while prospecting for impact craters in central Australia. Shoemaker's ashes were carried aboard the *Lunar Prospector* space probe, an orbiter that mapped the Moon's surface in great detail for over a year. On July 31, 1999, the spacecraft was deliberately crashed into Shoemaker crater near the lunar south pole in a search for ancient ice deposits. None were detected. To date, Shoemaker, the founder of astrogeology, is the only person to have been buried on another celestial body. ■

Geologist-astronaut Harrison "Jack" Schmitt, *Apollo 17* lunar module pilot, uses an adjustable scoop to retrieve samples of lunar rock on the rim of Camelot crater in the Taurus-Littrow Valley. [NASA.]

CHAPTER OUTLINE

Origin of the Solar System 216

Early Earth: Formation of a Layered Planet 218

Diversity of the Planets 221

Evolution of the Continents 228

Biological Events in Earth's History 239

Today, the solar system is a well-ordered place, with planets moving in stately orbits around the Sun. But the crater-marked surface of the Moon and the occasional meteorite that crashes through Earth's atmosphere remind us of a more disorganized, chaotic time when the solar system was young and Earth's environment was much less hospitable.

Some of the most fascinating scientific questions concern the birth of our planet and its early evolution. How did its rocky mass come together and differentiate into a core, mantle, and crust? Why does Earth's surface, with its blue oceans and wandering continents, look so different from those of its planetary neighbors? And how did that surface evolve and become populated with the myriad species that now constitute our biosphere?

Geologists can draw from many lines of evidence to answer these questions. The rocks of continents preserve a record of geologic processes more than 4 billion years old, and even more ancient materials have been collected from meteorites and by space exploration. Astronauts trained in geology have explored the lunar surface, mapping outcrops and collecting dust and rock samples for analysis back on Earth. A new generation of observing platforms such as the Hubble Space Telescope have allowed us to peer into the deeper reaches of the universe.

In this chapter, we will explore the 4.5-billion-year history of Earth from the beginning of the solar system to the present day. We will see how Earth and the other planets formed around the Sun and how they differentiated into layered bodies. Then we'll see how tectonic processes began, and how they led to the formation of continents. Finally, we'll chart the evolution of life by examining the geologic record preserved in the ancient rocks of those continents.

ORIGIN OF THE SOLAR SYSTEM

The search for the origins of the universe—and of our own small part of it—goes back to the earliest recorded mythologies. Today, the generally accepted scientific explanation is the Big Bang theory, which holds that our universe began about 13.7 billion years ago with a cosmic "explosion." Before that moment, all matter and energy were compacted into a single, inconceivably dense point. Although we know little of what happened in the first fraction of a second after time began, astronomers have a general understanding of the billions of years that followed. In a process that still continues, the universe has expanded and thinned out to form the galaxies and stars. Geology explores the latter third of that time: the past 4.5 billion years, during which our *solar system*—the star that we call the Sun and the planets that orbit it—formed and evolved. In particular, geologists look to the formation of the solar system to understand the formation of Earth and the Earthlike planets.

Evidence from astronomy tells us that the universe began with the Big Bang: a cosmic explosion of matter and energy that expanded from a single densely packed point.

The Nebular Hypothesis

In 1755, the German philosopher Immanuel Kant suggested that the origin of the solar system could be traced to a rotating cloud of gas and fine dust, an idea called the **nebular hypothesis.** We now know that outer space beyond our solar system is not as empty as we once thought. Astronomers have recorded many clouds of the type that Kant surmised, and they have named them *nebulae* (plural of the Latin word for "fog" or "cloud"). They have also identified the materials that form these clouds. The gases are mostly hydrogen and helium, the two elements that make up all but a small fraction of our Sun. The dust-sized particles are chemically similar to materials found on Earth.

How could our solar system take shape from such a cloud? This diffuse, slowly rotating mass contracted under the force of gravity (**Figure 9.1**). Contraction, in turn, accelerated the rotation of the particles (just as ice skaters spin more rapidly when they pull in their arms), and the faster rotation flattened the cloud into a disk.

The Sun Forms

Under the pull of gravity, matter began to drift toward the center of the cloud, accumulating into a protostar, the precursor of our present Sun. Compressed under its own weight, the material in the proto-Sun became dense and hot. The internal temperature of the proto-Sun rose to millions of degrees, at which point nuclear fusion began. The Sun's nuclear fusion, which continues today, is the same nuclear reaction that occurs in a hydrogen bomb. In both cases, hydrogen atoms, under intense pressure and at high temperature, combine (fuse) to form helium. Some mass is converted into energy in the process. The Sun releases some of that energy as sunshine; an H-bomb releases it as an explosion.

The Planets Form

Although most of the matter in the original nebula was concentrated in the proto-Sun, a disk of gas and dust, called the

FIGURE 9.1 The nebular hypothesis explains the formation of the solar system.

solar nebula, remained to envelop it. The temperature of the solar nebula rose as it flattened into a disk. It became hotter in the inner region, where more of the matter accumulated, than in the less dense outer regions. Once formed, the disk began to cool, and many of the gases condensed; that is, they were transformed to their liquid or solid state, just as water vapor condenses into droplets on the outside of a cold glass and water solidifies into ice when it cools below the freezing point. Gravitational attraction caused the dust and condensing material to clump together (accrete) into small, kilometer-sized chunks, or *planetesimals.* In turn, these planetesimals collided and stuck together, forming larger, Moon-sized bodies (see Figure 9.1). In a final stage of cataclysmic impacts, a few of these larger bodies—with their larger gravitational attraction—swept up the others to form the planets in their present orbits. Planetary formation happened rapidly, probably within 10 million years after the condensation of the nebula.

As the planets formed, those in orbits close to the Sun and those in orbits farther from the Sun developed in markedly different ways. Thus, the composition of the inner planets is quite different from that of the outer planets.

> The planets, including Earth, formed from the gas and dust of a nebula that enveloped the proto-Sun.

Inner Planets The four inner planets, in order of closeness to the Sun, are Mercury, Venus, Earth, and Mars (**Figure 9.2**). They are also known as the Earthlike, or *terrestrial,* planets. In contrast to the outer planets, the four inner planets are small and are made up of rocks and metals. They formed close to the Sun, where conditions were so hot that most of the *volatile* materials (those that most easily become gases) boiled away. Radiation and matter streaming from the Sun—the solar wind—blew away most of the hydrogen, helium, water, and other light gases and liquids on these planets. The inner planets formed from the dense matter that was left behind, which included the rock-forming silicates as well as metals such as iron and nickel. From isotopic dating of the meteorites that occasionally strike Earth and are believed to be remnants of this preplanetary process, we know that the inner planets began to accrete about 4.56 billion years ago (see Chapter 8). Computer simulations indicate that they would have grown to planetary size in a remarkably short time—perhaps as quickly as 10 million years or less.

> The four inner planets formed primarily from dense materials such as silicate minerals, iron, and nickel.

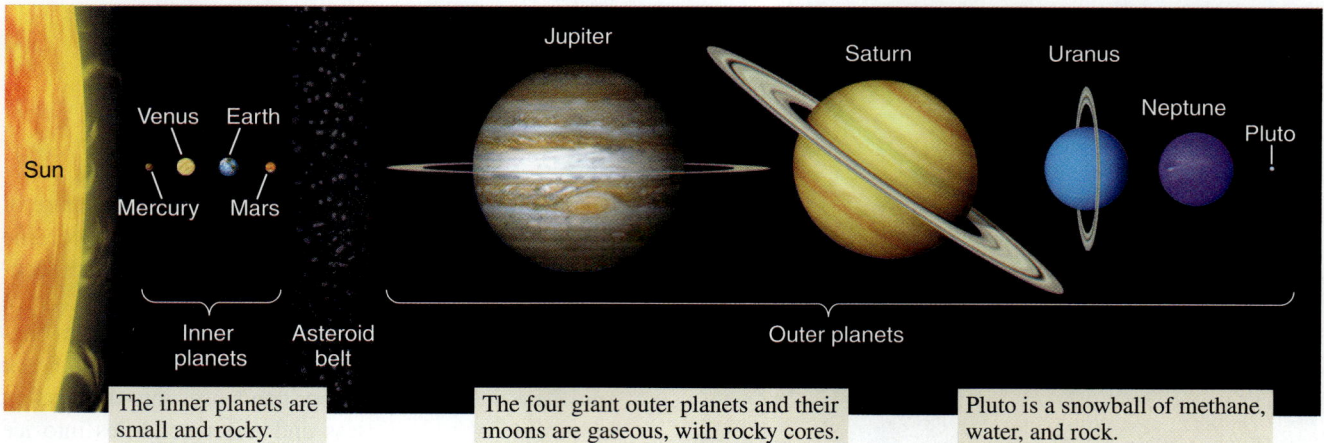

The inner planets are small and rocky.

The four giant outer planets and their moons are gaseous, with rocky cores.

Pluto is a snowball of methane, water, and rock.

FIGURE 9.2 The solar system. This diagram shows the relative sizes of the planets as well as the asteroid belt separating the inner and outer planets. Though considered one of the nine planets since its discovery in 1930, Pluto was demoted from that status by the International Astronomical Union in 2006. With this revision, there are only eight true planets, not nine.

Giant Outer Planets Most of the volatile materials swept from the region of the terrestrial planets were carried to the cold outer reaches of the solar system to form the giant outer planets—Jupiter, Saturn, Uranus, and Neptune—and their satellites. The giant planets were big enough, and their gravitational attraction strong enough, to enable them to hold onto the lighter nebular material. Thus, although they have rocky and metal-rich cores, they, like the Sun, are composed mostly of hydrogen and helium and the other light materials of the original nebula.

Small Bodies of the Solar System

Not all of the material from the solar nebula ended up in planets. Some planetesimals collected between the orbits of Mars and Jupiter to form the *asteroid belt* (see Figure 9.2). This region now contains more than 10,000 **asteroids** with diameters larger than 10 km and about 300 larger than 100 km. The biggest is Ceres, which has a diameter of 930 km. Most **meteorites**—bolides from outer space that strike Earth—are tiny pieces of asteroids ejected from the asteroid belt during collisions with one another. Astronomers originally thought the asteroids were the remains of a large planet that had broken apart early in the history of the solar system, but it now appears they are pieces that never coalesced into a planet, probably due to the gravitational influence of Jupiter.

Another important group of small, solid bodies is the *comets,* aggregations of dust and ice that condensed in the cooler outer reaches of the solar nebula. Most of the comets—there are probably many millions of them with diameters larger than 10 km—orbit the Sun far beyond the outer planets, forming concentric "halos" around the solar system. Occasionally, collisions or near misses throw one of them into an orbit that penetrates the inner solar system. We can then observe it as a bright object with a tail of gases blown away from the Sun by the solar wind. Comets are intriguing to geologists because they provide clues about the more volatile components of the solar nebula, including water and carbon-rich compounds, which they contain in abundance.

EARLY EARTH: Formation of a Layered Planet

We know that Earth is a layered planet with a core, mantle, and crust surrounded by a fluid ocean and a gaseous atmosphere (see Chapter 2). How did Earth evolve from a hot, rocky mass into a living planet with continents, oceans, and a pleasant climate? The answer lies in **differentiation:** the transformation of random chunks of primordial matter into a body whose interior is divided into concentric layers that differ from one another both physically and chemically. Differentiation occurred early in Earth's history, when the planet got hot enough to melt.

Earth Heats Up and Melts

Although Earth probably started out as an accretion of planetesimals and other remnants of the solar nebula, it did not retain this form for long. To understand Earth's present layered structure, we must return to the time when Earth was still subject to violent impacts by planetesimals and

Early Earth: Formation of a Layered Planet

larger bodies. As these objects crashed into the primitive planet, most of their energy of motion was converted into heat—another form of energy—and that heat caused melting. A planetesimal colliding with Earth at a typical velocity of 15 to 20 km/s would deliver as much energy as 100 times its weight in TNT. The impact energy of a body the size of Mars colliding with Earth would be equivalent to exploding several trillion 1-megaton nuclear bombs (a single one of which would destroy a large city), enough to eject a vast amount of debris into space and to generate enough heat to melt most of what remained of Earth.

Many scientists now think that such a cataclysm did occur during the middle to late stages of Earth's accretion. A giant impact by a Mars-sized body created a shower of debris from both Earth and the impacting body and propelled it into space. The Moon aggregated from that debris (**Figure 9.3**). According to this theory, Earth re-formed as a body with an outer molten layer hundreds of kilometers thick—a *magma ocean*. The huge impact sped up Earth's rotation and changed its axis, knocking it from vertical with respect to Earth's orbital plane to its present 23° inclination. All this occurred about 4.5 billion years ago, between the beginning of Earth's accretion (4.56 billion years ago) and the formation of the oldest Moon rocks brought back by the Apollo astronauts (4.47 billion years old).

Another source of heat that contributed to melting early in Earth's history was radioactivity. When radioactive elements such as uranium disintegrate spontaneously, they emit heat. Though present in only small amounts, radioactive elements have continued to keep Earth's interior hot.

> **Collision with a Mars-sized body ejected debris that formed the Moon and melted the outer part of Earth.**

Differentiation of Earth's Core, Mantle, and Crust

As a result of the tremendous impact energy absorbed during Earth's formation, its entire interior was heated to a "soft" state in which its components could move around. Heavy material sank to become the core, releasing gravitational energy and causing more melting, and lighter material floated to the surface and formed the crust. The rising lighter matter brought interior heat to the surface, where it

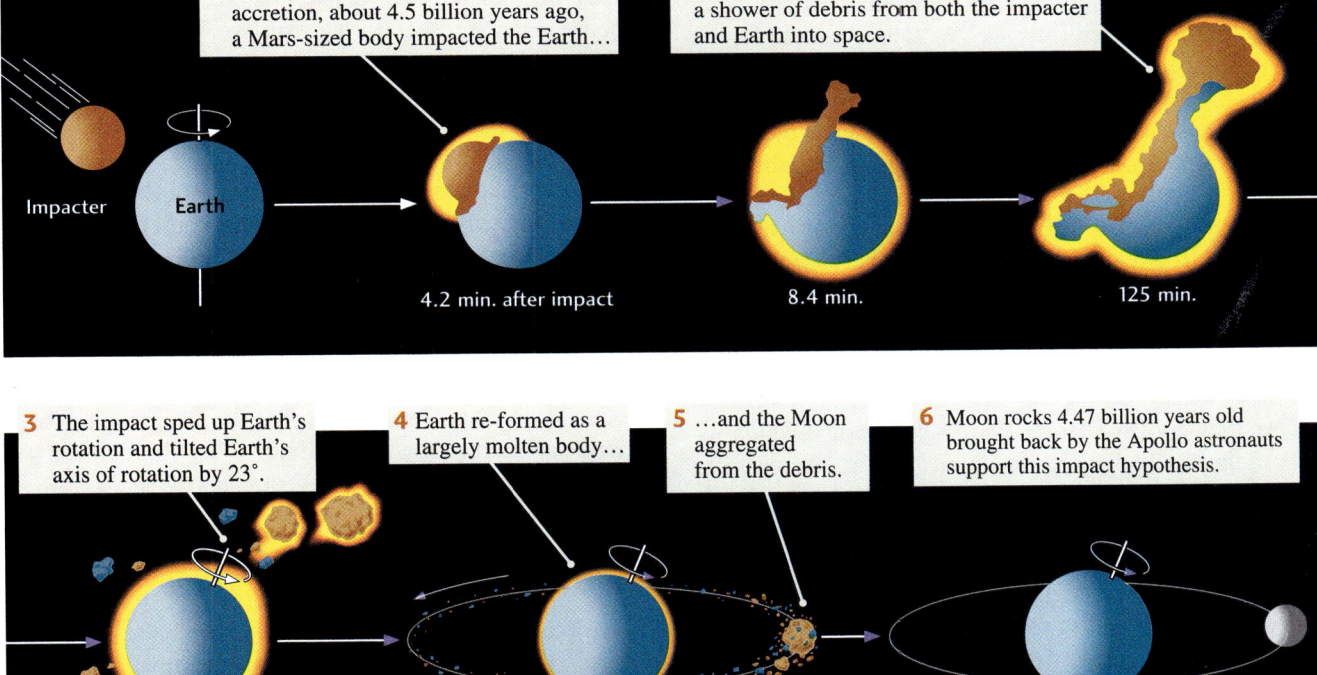

FIGURE 9.3 Computer simulation of the impact of a Mars-sized body on Earth.
[*Solid-Earth Sciences and Society*. Washington, D.C.: National Research Council, 1993.]

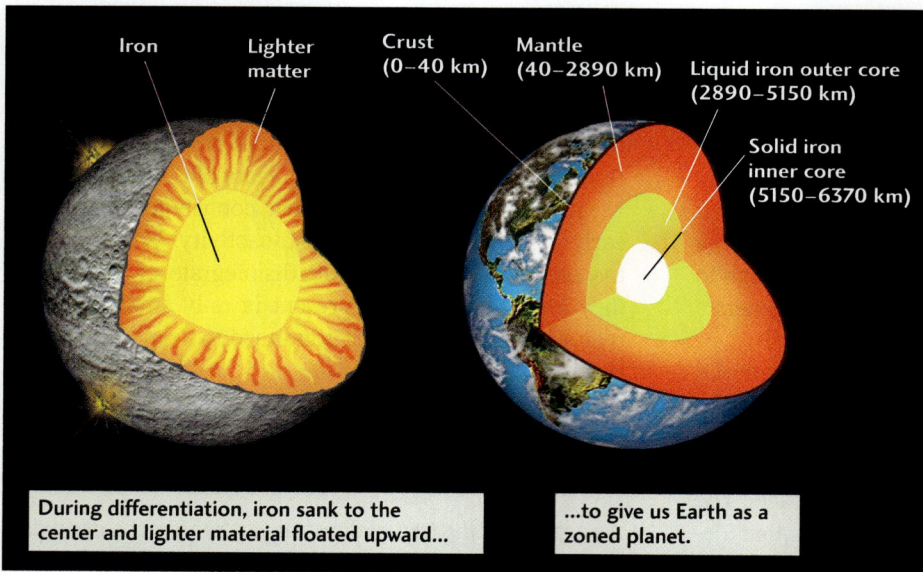

FIGURE 9.4 The differentiation of early Earth resulted in a zoned planet with three main layers.

could radiate into space. In this way, Earth differentiated into a zoned planet with three main layers: a central core, a mantle, and an outer crust (Figure 9.4).

Earth's Core Iron, which is denser than most of the other elements, accounted for about a third of the primitive planet's material. Most of this iron and other heavy elements, such as nickel, sank to form a central core, which begins at a depth of about 2900 km. By probing the core with seismic waves, scientists have found that it is molten on the outside but solid in a region called the inner core, which extends from a depth of about 5200 km to Earth's center at about 6400 km. Today the inner core is solid because the pressures at the center are too high for iron to melt.

In the young, hot Earth, iron and nickel sank to form the core, while the less dense materials floated to the surface, forming a primitive crust.

Earth's Crust Other molten materials were less dense, so they floated toward the surface of the magma ocean. There they cooled to form Earth's solid crust, which today ranges in thickness from about 7 km in the oceans to about 40 km on the continents. We know that oceanic crust is constantly generated by seafloor spreading and recycled back into the mantle by subduction. In contrast, the continental crust began to accumulate early in Earth's history from silicates of relatively low density with a felsic composition and low melting temperatures.

The zircon grains recently found in western Australia, dated to about 4.4 billion years ago (see Chapter 8), are the oldest terrestrial material yet discovered. Chemical analysis indicates that the samples formed near the surface under relatively cool conditions and in the presence of water. This finding suggests that Earth had cooled enough for a crust to exist only 100 million years after the planet re-formed following the giant impact that formed the Moon.

Earth's Mantle Between the core and the crust lies the mantle, the layer that forms the bulk of the solid Earth. The mantle is made up of the material left in the middle zone after most of the denser material sank and the less dense material rose toward the surface. It is about 2900 km thick and consists of ultramafic silicate rocks containing more magnesium and iron than crustal silicates do. Convection in the mantle removes heat from Earth's interior (see Chapter 2).

Because the mantle was hotter early in Earth's history, it was probably convecting more vigorously than it does today. Some form of plate tectonics may have been operating even then, although the "plates" were probably much smaller and thinner, and the tectonic features were probably very different from the linear mountain belts and long mid-ocean ridges we now see on Earth's surface. Some scientists think that Venus today provides an analog for these long-vanished processes on Earth. We will compare tectonic processes on Earth and Venus shortly.

Earth's Oceans and Atmosphere Form

The oceans and atmosphere can be traced back to the "wet birth" of Earth itself. The planetesimals that aggregated into our planet contained ice, water, and other volatiles, such as nitrogen and carbon. Originally, these volatiles were locked up in minerals carried by the aggregating planetesimals. As Earth differentiated, water vapor and other gases were freed, carried to the surface by magmas, and released through volcanic activity.

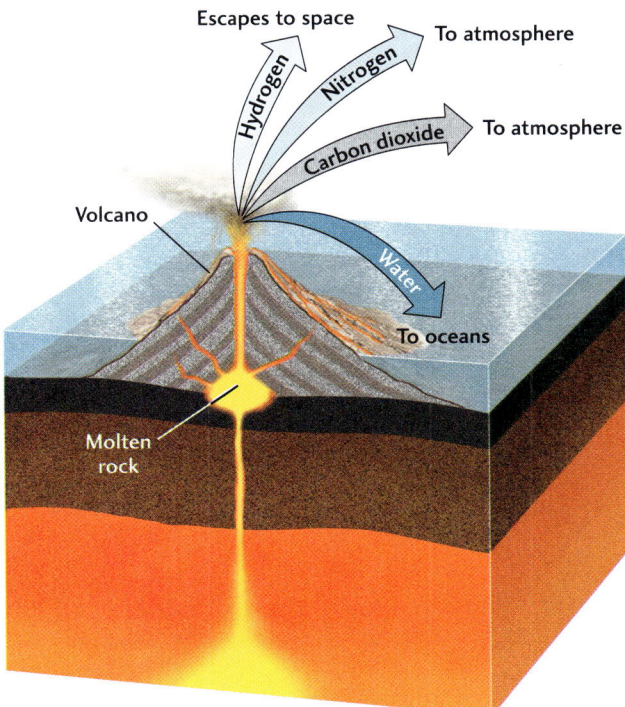

FIGURE 9.5 Early volcanic activity contributed enormous amounts of water vapor, carbon dioxide, and other gases to the atmosphere and oceans. Later, photosynthetic microbes removed carbon dioxide and added oxygen to the primitive atmosphere. Hydrogen, because it is lighter, escaped into space.

The gases spewed from volcanoes around 4 billion years ago probably consisted of the same substances that are expelled from present-day volcanoes (though not necessarily in the same relative abundances): primarily hydrogen, carbon dioxide, nitrogen, water vapor, and a few others (Figure 9.5). Almost all of the hydrogen escaped to outer space, while the heavier gases enveloped the planet. Some of the air and water may also have come from volatile-rich matter from the outer solar system that struck the planet after it formed. The comets, for example, are composed largely of frozen water (ice) plus frozen carbon dioxide and other gases. Countless comets may have bombarded Earth early in its history, contributing water and gases to the early oceans and atmosphere. The early atmosphere lacked the oxygen that makes up 21 percent of the atmosphere today. Oxygen did not enter the atmosphere until oxygen-producing organisms evolved, as described later in this chapter.

DIVERSITY OF THE PLANETS

By about 4.4 billion years ago, in less than 200 million years since its origin, Earth had become a fully differentiated planet. The core was still hot and mostly molten, but the mantle was fairly well solidified, and a primitive crust and continents had begun to develop. Oceans and atmosphere had formed, and the geologic processes that we observe today had been set in motion. But what about the early history of the other terrestrial planets?

Planetary Differentiation and Layering

Information transmitted from space probes indicates that the four terrestrial planets have all undergone differentiation into layered structures with iron-nickel cores, silicate mantles, and an outer crust. Venus is close to Earth in mass and size (Table 9.1), and its core seems to be about the same size as Earth's. Though Mars is considerably smaller than Earth—only about one-tenth of Earth's mass—the relative size of its core appears to be similar, about half its surface radius.

> Earth's oceans and atmosphere developed when water and gases were released from its interior during differentiation. Water and gases may also have arrived with comets that bombarded Earth after it had already formed.

> The four terrestrial planets have all undergone differentiation into layered structures with iron-nickel cores, silicate mantles, and an outer crust.

TABLE 9.1	Characteristics of the Terrestrial Planets and Earth's Moon				
	Mercury	Venus	Earth	Mars	Earth's Moon
Radius (km)	2440	6052	6371	3390	1737
Mass (Earth = 1)	0.06 (3.3×10^{23} kg)	0.81 (4.9×10^{24} kg)	1.00 (6.0×10^{24} kg)	0.11 (6.4×10^{23} kg)	0.01 (7.3×10^{22} kg)
Mean density (g/cm³)	5.43	5.24	5.52	3.93	3.34
Orbit period (Earth days)	88	225	365	687	27
Distance from Sun (millions of km)	58	108	150	228	
Moons	0	0	1	2	0

Mercury's average density is nearly as great as Earth's, even though it is a much smaller planet. Accounting for differences in interior pressure (remember, higher pressures increase density), scientists have surmised that Mercury's iron-nickel core must make up about 70 percent of its mass, a record for solar system planets (Earth's core is only one-third of its mass). Perhaps Mercury lost part of its silicate mantle in a giant impact. Alternatively, the Sun could have vaporized part of its mantle during an early phase of intense radiation. Scientists are still debating these hypotheses.

As stated earlier, the Moon appears to have formed from remnants of Earth's collision with a Mars-sized body (see Figure 9.3). In bulk, the Moon's materials are lighter than Earth's, probably because the heavier matter of the giant impacting body remained embedded in Earth. The lunar core is therefore small, comprising only about 20 percent of the lunar mass.

The Man in the Moon: A Planetary Time Scale

If you look at the face of the Moon through binoculars on a clear night, you will see two distinct types of terrain: rough areas that appear light-colored with lots of big craters, and smooth, dark areas, usually circular in shape, where craters are small or nearly absent (**Figure 9.6**). The light-colored regions are the mountainous *lunar highlands*,

FIGURE 9.6 The Moon has two types of terrain: the lunar highlands, with many craters, and the lunar lowlands, or maria, with few craters. The maria appear darker due to the presence of widespread basalts that flowed across their surface over 3 billion years ago. The highlands are lighter because the abundant craters reflect sunlight better. [NASA.]

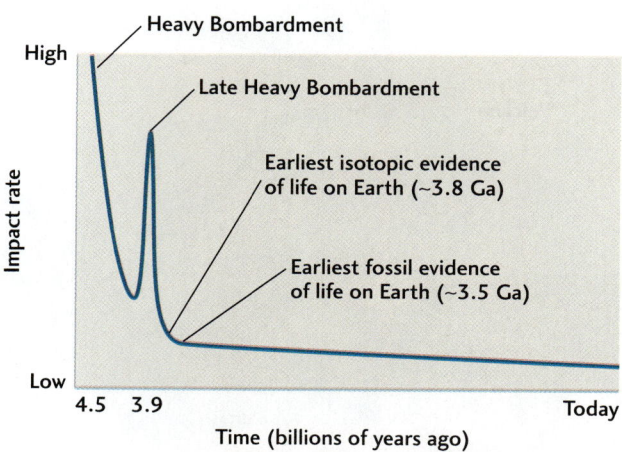

FIGURE 9.7 The number of impacts varies over time. The planets were formed by a chain of collisions that tapered off over the first 500 million years of their evolution. However, there was a later phase, known as the Late Heavy Bombardment, that peaked around 3.9 billion years ago. (Ga, billion years ago.)

which cover about 80 percent of the surface. The dark regions are low-lying plains called *lunar maria*, which is Latin for "seas," because they looked like seas to early Earth-bound observers.

The heavily cratered lunar surface we see today is that of a very old, geologically dead body, dating back to a period early in the history of the solar system known as the **Heavy Bombardment**, when crater-forming impacts were very frequent (**Figure 9.7**). In this way, the Moon contrasts with Earth, where most surfaces older than about 500 million years have been obliterated through the combined activities of the plate tectonic and climate systems.

> The Moon's heavily cratered surface indicates that it is geologically dead; no active geologic processes are changing its surface features.

In preparation for the Apollo missions to the Moon, geologists such as Gene Shoemaker (**Figure 9.8**) developed a relative time scale for the formation of lunar surfaces based on the following simple principles:

- Craters are absent on a new geologic surface; older surfaces have more craters than younger surfaces.

- Impacts by small bodies are more frequent than impacts by large bodies; thus, older surfaces have larger craters.

- More recent impact craters cross-cut or cover older craters.

By applying these principles and by mapping the numbers and sizes of craters, geologists showed that the lunar highlands are older than the maria. They interpreted the maria to be basins formed by the impacts of asteroids or comets that were subsequently flooded with basalts,

FIGURE 9.8 Astrogeologist Eugene Shoemaker (at left, pointing with hammer) leads an astronaut training trip on the rim of Meteor Crater, Arizona, in May 1967. (An aerial view of Meteor Crater is shown in Figure 2.7.) [U.S. Geological Survey.]

which "repaved" the basins. They were able to assign different parts of the Moon to geologic intervals analogous to those in the relative time scale worked out by nineteenth-century geologists for Earth.

By applying the isotopic dating methods described in Chapter 8 to rock samples brought back by the Apollo astronauts, geologists were able to calibrate an absolute time scale for the Moon's geologic intervals. Sure enough, the highlands turned out to be very ancient (4.4 billion to 4.0 billion years old) and the maria younger (4.0 billion to 3.2 billion years old). **Figure 9.9** plots these ages on the ribbon of geologic time.

FIGURE 9.9 By calibrating the relative time intervals from crater counting with the absolute ages of lunar rocks, geologists have constructed a geologic time scale for the terrestrial planets. (Ma, million years ago.)

The youth of the maria turned out to be a puzzle, however. The best computer simulations of the Heavy Bombardment indicate that it should have been over rather quickly, perhaps in a few hundred million years or even less. Why, then, did some of the biggest impacts observed on the Moon—those that formed the maria—occur so late in lunar history?

The simulations missed an important event. The rate at which large objects struck the Moon did decrease quickly, as the simulations predicted, but then spiked up again in a period known as the *Late* Heavy Bombardment, which occurred between about 4.0 billion and 3.8 billion years ago (see Figure 9.7). The explanation of this event is still controversial, but it looks as though small changes in the orbits of Jupiter and Saturn about 4 billion years ago (caused by their mutual gravitational interactions as they settled into their present orbits) perturbed the orbits of the asteroids. Some of the asteroids were thrown into the inner solar system, where they collided with the Moon and the terrestrial planets, including Earth. The Late Heavy Bombardment explains why so few rocks on Earth have ages greater than 3.9 billion years. It is this event that marks the end of the Hadean eon and the beginning of the Archean eon (see Figure 9.9).

> *Impacts during the Late Heavy Bombardment created the lunar maria.*

Mercury: The Ancient Planet

The topography of Mercury is poorly understood. *Mariner 10* was the first and only spacecraft to visit Mercury when it flew by the planet in March 1974. It mapped less than half the planet, and we have little idea of what is on the other side.

The *Mariner 10* mission confirmed that Mercury has a geologically dormant, heavily cratered surface, the oldest of all the terrestrial planets (**Figure 9.10**). This lifeless planet has no atmosphere. Between the large old craters lie younger plains, which are probably volcanic, like the lunar maria. Unlike Earth and Venus, Mercury shows very few features that are clearly due to tectonic forces having reshaped its surface.

In many respects, the face of Mercury seems very similar to that of Earth's Moon. The two bodies are similar in size and mass, and most of their geologic activity took place within the first billion years of their histories. There is one interesting difference, however. Mercury's face has several scars marked by scarps nearly 2 km high and up to 500 km long (**Figure 9.11**). Such features are common on Mercury, but rare on Mars and absent on the Moon. These cliffs appear to have resulted from horizontal compression of Mercury's brittle crust, which formed enormous thrust faults (see Chapter 7). Some scientists think they formed during the cooling of the planet's crust immediately after its formation.

> *The heavy cratering of Mercury's surface indicates that it is ancient and has not been reshaped by tectonic activity.*

FIGURE 9.10 Mercury has an ancient, geologically dormant surface similar to that of Earth's Moon. [NASA/JPL/Northwestern University.]

On August 3, 2004, the first new mission to Mercury in 30 years was launched successfully. If all goes well, *Messenger* will arrive and enter an orbit around Mercury in September 2009. *Messenger* will provide information about Mercury's surface composition, its geologic history, and its core and mantle, and it will search for evidence of ice and other frozen gases, such as carbon dioxide, at the poles.

FIGURE 9.11 The prominent scarp that snakes across this image is thought to have formed as Mercury was compressed, possibly as the planet cooled following its formation. Note that the scarp must be younger than the craters it offsets. [NASA/JPL/Northwestern University.]

Venus: The Volcanic Planet

Venus is our closest planetary neighbor, often brightly visible just before sunset. The entire planet is shrouded in a dense fog of carbon dioxide, water vapor, and sulfuric acid, which prevents scientists from studying its surface with ordinary telescopes and spacecraft cameras. Not much was known about Venus until the early 1990s, when the *Magellan* spacecraft mapped its surface, using radio waves (radar) to penetrate the fog (**Figure 9.12**).

FIGURE 9.12 A comparison of the solid surfaces of Venus, Earth, and Mars, all at the same scale. The topography of Mars, which shows the greatest range, was measured in 1998 and 1999 by a laser altimeter aboard the orbiting *Mars Global Surveyor* spacecraft. That of Venus, which shows the smallest range, was measured from 1990 to 1993 by a radar altimeter aboard the orbiting *Magellan* spacecraft. Earth's topography, which is intermediate in range and dominated by continents and oceans, has been synthesized from altimeter measurements of the land surface, ship-based measurements of ocean depth, and gravity field measurements of the seafloor surface from Earth-orbiting spacecraft. [Greg Neumann/MIT/GSFC/NASA.]

> *Venus is a geologically active planet with a convecting mantle like Earth's. Its surface shows evidence of tectonic activity, especially volcanism.*

The images that *Magellan* returned to Earth showed that Venus is a geologically active planet with mountains, plains, volcanoes, and rift valleys. The lowland plains of Venus—the green regions in Figure 9.12—have far fewer craters than the Moon's youngest maria, indicating that they must be younger still. Estimates of their age range between 1600 million and 300 million years.

These young plains are dotted with hundreds of thousands of small volcanic domes 2 to 3 km across and perhaps 100 m or so high, which formed over places where Venus's crust got very hot. There are larger, isolated volcanoes as well, up to 3 km high and 500 km across, similar to the shield volcanoes of the Hawaiian Islands (**Figure 9.13**a). *Magellan* also observed unusual circular features called *coronae* that appear to result from blobs of

(a)

(b)

FIGURE 9.13 Many surface features on Venus were formed by ongoing geologic activity. (a) Maat Mons, a volcanic mountain that may be up to 3 km high and 500 km across. (b) Volcanic features called coronae are not observed on any other planet except Venus. The visible lines that define the coronae are fractures, faults, and folds produced when a large blob of hot lava collapsed like a fallen soufflé. Each corona is a few hundred kilometers across. [Images from NASA/USGS.]

hot lava that rose, created a large bulge or dome in the surface, and then sank, collapsing the dome and leaving a wide ring that looks like a fallen soufflé (Figure 9.13b).

Because Venus has so much evidence of widespread volcanism, it has been called the Volcanic Planet. Venus has a convecting mantle like Earth's, in which hot material rises and cold material sinks (see Figure 2.16), but unlike Earth, it does not appear to have thick plates of rigid lithosphere. Instead, only a thin crust of frozen lava overlies the convecting mantle. As the convection currents push and stretch the surface, the crust breaks up into flakes or crumples like a rug, and blobs of hot lava bubble up to form large landmasses and volcanic deposits. Scientists have called this type of geology *flake tectonics*. When Earth was younger and hotter, it is possible that flakes, rather than plates, were the main expression of its tectonic activity.

Mars: The Red Planet

Of all the planets, Mars has a surface most similar to Earth's. Although Venus is closer to Earth in size and mass (see Table 9.1), Mars has features suggesting that liquid water once flowed across its surface, and liquid water may still exist in its deep subsurface. And where there is water, there may be living organisms. No other planet in our solar system has as much chance of harboring extraterrestrial life as Mars.

The abundance of iron oxide minerals on the surface of Mars makes it the Red Planet. Iron oxide minerals are common on Earth and tend to form where weathering of iron-bearing silicates occurs. We now know that many other minerals common on Earth, such as olivine and pyroxene, which form in basalt, are also present on Mars. But there are other relatively unusual minerals, such as sulfates, that record an earlier, wetter phase of the Martian climate, when liquid water may have been stable.

The topography of Mars shows a greater range of highs and lows than that of Earth or Venus (see Figure 9.12). Olympus Mons, at 25 km high, is a giant, recently active volcano—the tallest mountain in the solar system (**Figure 9.14**a). The Vallis Marineris canyon, 4000 km long and averaging 8 km deep, stretches the distance from New York to Los Angeles and is five times deeper than the Grand Canyon (Figure 9.14b).

Like the Moon and Mercury, Mars has ancient cratered terrain that preserves a record of the Late Heavy Bombardment and is therefore likely to be older than 3.9 billion years (see Figure 9.9). But younger surfaces are also widespread on Mars, occupying much of its northern hemisphere. Some are paved with volcanic lavas, as on Venus, whereas others are covered with sediments and landslide deposits. The

> Parts of the Martian surface consist of ancient cratered terrain, but other parts are covered with lavas and sediments.

(a)

(b)

FIGURE 9.14 (a) Olympus Mons is the tallest volcano in the solar system, with a summit almost 25 km above the surrounding plains. Encircling the volcano is an outward-facing scarp 550 km in diameter and several kilometers high. Beyond the scarp is a moat filled with lava, most likely derived from Olympus Mons. [NASA/USGS.] (b) Vallis Marineris is the longest (4000 km) and deepest (up to 10 km) canyon in the solar system. It is five times deeper than the Grand Canyon. In this image, the canyon is exposed as series of fault basins whose sides have partially collapsed (such as at upper left), leaving piles of rock debris. The walls of the canyon are 6 km high here! The layering of the canyon walls suggests deposition of sedimentary or volcanic rocks prior to faulting. [ESA/DLR/FU Berlin.]

FIGURE 9.15 *Spirit* (*right*), one of the Mars Exploration Rovers, is about the size of a golf cart. *Spirit* is standing next to a twin of *Sojourner*, a rover that was sent to Mars in 1997. [NASA/JPL.]

surface, taking pictures, performing experiments, and analyzing the chemical composition of the rocks they have come across. (One of the authors of this textbook, John Grotzinger, is a member of the Mars Exploration Rovers science team.) Combined with detailed mapping of the planet by the *Mars Express* orbiter, the data from these "robotic geologists" have shown us that at least some—and perhaps many—of the younger surfaces are underlain by sedimentary rocks.

One of the most exciting findings has been the discovery of a vertical succession of sedimentary rocks by the rover *Opportunity* at a site called Endurance crater (**Figure 9.16**), the first of its kind on another planet. These sediments accumulated in a shallow lake or sea, and the sulfate minerals they contain precipitated as water evaporated. This site has provided information similar to stratigraphic successions observed on Earth (compare, for example, Figure 6.18).

> *Evidence that liquid water once flowed on the Martian surface includes sulfide mineral deposits, deep canyons, and sedimentary sequences that accumulated in a shallow lake or sea.*

The data collected by the Mars Exploration Rovers and orbiting space probes provide compelling evidence for liquid water on the Martian surface at some point in the planet's history. But many mysteries remain. How much water was there? How long did it last? Did it ever rain, or was all the evidence produced by groundwater leaking to the surface? Did the water last long enough, and have the right composition, to allow life to get started? Only one thing is certain at this point: more missions will be required to answer these questions.

largest volcanoes, including Olympus Mons, appear to have erupted during the last 100 million years and may still be active.

Our understanding of surface processes on Mars dramatically improved when two golf-cart-sized Mars Exploration Rovers, named *Spirit* and *Opportunity*, landed on Mars in January 2004 (**Figure 9.15**). They have since traveled more than 18 kilometers in total across the Martian

FIGURE 9.16 A sedimentary sequence exposed along the flank of Endurance crater, photographed by the rover *Opportunity*. (a) An interpretive drawing showing each stage in the history of the outcrop. (b) The vertical succession of layers in the outcrop preserves an excellent record of early Martian environments. [NASA/JPL/Cornell.]

EVOLUTION OF THE CONTINENTS

In contrast to the cratered face of the Moon, the exterior of our home planet preserves no vestiges of its fiery beginning. Evidence of the Heavy Bombardment has been obliterated by the resurfacing processes of the plate tectonic and climate systems. As we saw in Chapter 3, nearly two-thirds of Earth's surface—all of its oceanic crust—was created by seafloor spreading over the past 200 million years, an interval spanning a mere 4 percent of Earth's history. The story of all earlier events is entirely contained in the continental crust.

Although the oldest crustal minerals are older than the Late Heavy Bombardment (see Chapter 8), indicating that some type of continental crust existed in the Hadean eon, the main episodes of continent building occurred after this violent period. In this section, we describe the structure of the continents and the processes that formed them—and are still forming them today.

The North American Continent

Continents, like people, show a great variety of features that reflect their parentage and experience over time. Yet, also like people, continents share many similarities in their basic structure and growth patterns. Let's begin by outlining the major features of one particular continent: North America. The long-term tectonic evolution of the continent is reflected in its *tectonic provinces*—large-scale regions formed by a particular set of tectonic processes (**Figure 9.17**).

Canadian Shield The oldest parts of North America's crust, built during the most ancient episodes of deformation, tend to be found in the northern interior of the continent, in central and eastern Canada. This huge region was named the *Canadian Shield* by nineteenth-century geologists because the crystalline rocks that form the upper crust ("basement" rocks) emerge from the surrounding sediments like a shield partially buried in the dirt of a battle-

FIGURE 9.17 The major tectonic provinces of North America reflect the processes that formed the continent.

field. The Canadian Shield has remained largely undisturbed by rifting, continental drift, or continent-continent collision throughout the Phanerozoic, but the entire region has been eroded nearly flat (**Figure 9.18**). It consists primarily of Precambrian granitic and metamorphic rocks, such as gneisses, together with highly deformed and metamorphosed sedimentary and volcanic rocks, and it contains major deposits of iron, gold, copper, diamond, and nickel. Large portions of the shield were formed during the Archean eon, representing one of the oldest records of Earth's history.

> *The Canadian Shield is a vast tectonic province of very old Precambrian granitic and metamorphic rocks that have been eroded nearly flat.*

Interior Platform Extensive flat-lying (*platform*) sediments have been deposited on stable continental crust around the periphery of the Canadian Shield and also near its center, beneath Hudson Bay (see Figure 9.17). The vast low-lying, sediment-covered region south and west of the Canadian Shield, which includes the Great Plains of Canada and the United States, is called the *interior platform*. The Precambrian basement rocks of the interior platform are a continuation of the Canadian Shield, although here they lie under nearly flat layers of Paleozoic sedimentary rocks, typically less than 2 km thick.

Within the interior platform are broad sedimentary basins, roughly circular or oval depressions where the sediments are thicker than in the surrounding areas (**Figure 9.19**). Most of these features are thermal subsidence

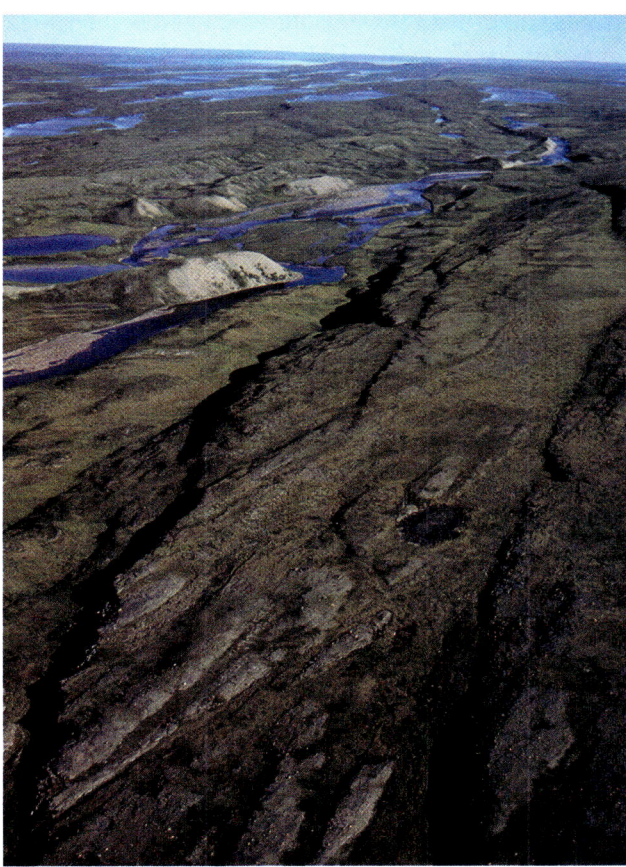

FIGURE 9.18 An aerial view of ancient, eroded metamorphic rocks in Nunavut, Canada, exposed on the surface of the Canadian Shield. [Roy Tanami/Ursus Photography.]

FIGURE 9.19 A map of the interior platform of North America, showing its basin and dome structure. The basins are nearly circular regions of thick sediments, whereas the domes are regions where the sediments are anomalously thin. Basement rocks are exposed on the tops of some domes, such as the Black Hills and the Ozark Dome.

basins; that is, regions that subsided when heated portions of the lithosphere cooled and contracted (see Chapter 6). An example is the Michigan Basin, a circular area of about 200,000 km² that covers most of the Lower Peninsula of Michigan. This basin subsided throughout much of the Paleozoic era and received sediments more than 5 km thick in its central, deepest part. The sandstones and other sedimentary rocks of these basins, laid down under tectonically quiet conditions, have remained unmetamorphosed and only slightly deformed to this day. The interior platform basins contain important deposits of uranium, coal, oil, and natural gas.

In the interior platform, the rocks of the Canadian Shield are overlain by Paleozoic sediments.

Appalachian Fold Belt On the edges of these older tectonic provinces are younger metamorphic belts where most of the present-day mountain chains are found. These mountain chains form elongated topographic features near the margins of the continent. The two main examples are the *Cordillera,* which runs down the western edge of North America and includes the Rocky Mountains, and the *Appalachian fold belt,* which runs southwest to northeast on the continent's eastern margin from Alabama to Newfoundland.

The rock assemblages and structures of the old, eroded Appalachian Mountains resulted from the continent-continent collisions that formed the supercontinent Pangaea 470 million to 270 million years ago. The western side

FIGURE 9.20 The Appalachian Mountain region of the south and central eastern United States, with an aerial view to the northeast and an idealized cross section. The major physiologic subregions extend from west to east in order of increasing intensity of deformation. [After S. M. Stanley, *Earth System History.* New York: W. H. Freeman, 2005. Aerial view from NASA.]

of the Appalachians is bounded by the *Allegheny Plateau,* a region of slightly uplifted, mildly deformed sediments that is rich in coal and oil. Moving eastward, we encounter regions of increasing deformation (**Figure 9.20**):

- The folded and faulted Paleozoic sedimentary rocks of the *Valley and Ridge province*

- The eroded mountains of the *Blue Ridge province,* composed of highly metamorphosed Precambrian and Cambrian rocks, which were thrust as sheets over the sedimentary rocks of the Valley and Ridge province

- The hilly *Piedmont,* containing Precambrian and Paleozoic metamorphosed sedimentary and volcanic rocks, intruded by granite and thrust over Blue Ridge rocks

- The modern *coastal plain,* made up of relatively undisturbed sediments of Jurassic age and younger, underlain by rocks similar to those of the Piedmont

> The Appalachian fold belt was formed by continent-continent collisions during the assembly of Pangaea.

North American Cordillera Some of the highest peaks on the continent are found in the mountainous Cordillera, a relatively young region that extends the length of North America from Alaska to Guatemala. Across its middle section, between San Francisco and Denver, the Cordilleran system is about 1600 km wide and includes several different tectonic provinces: the Coast Ranges along the Pacific Ocean; the lofty Sierra Nevada; the Basin and Range province; the high tableland of the Colorado Plateau; and the rugged Rocky Mountains, which end abruptly at the edge of the Great Plains on the stable interior platform (**Figure 9.21**).

The history of the Cordillera involves interactions among the Pacific, Farallon, and North American plates over the past 200 million years. Before the breakup of Pangaea, the Farallon Plate occupied most of the eastern Pacific Ocean. As North America moved westward, most of this plate's oceanic lithosphere was subducted eastward under the continent. The westward margin of the continent swept up island arcs and continental fragments, and the subduction zone eventually swallowed portions of the Pacific-Farallon spreading center, which converted

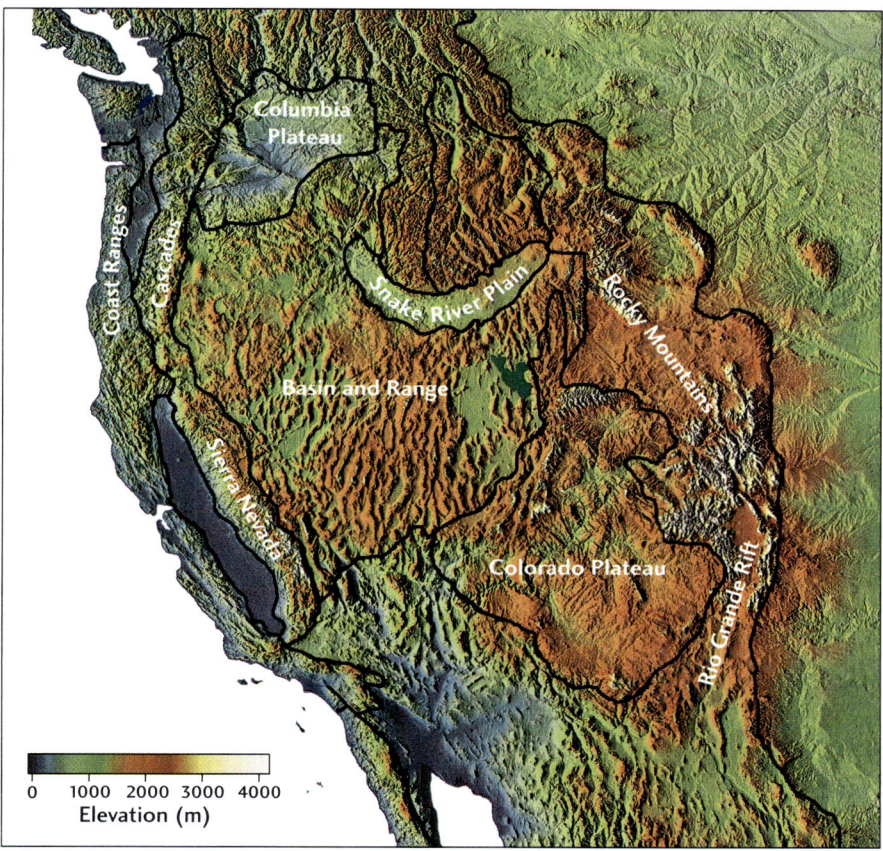

FIGURE 9.21 Topography of the North American Cordillera in the western United States. Computer manipulation of digitized elevation data produced this color shaded-relief map in which the major geologic provinces of the area are clearly visible, as if illuminated by a light source low in the west.

FIGURE 9.22 The interaction of the western coast of North America with the shrinking Farallon Plate as it was progressively subducted beneath the North American Plate, leaving the present-day Juan de Fuca and Cocos plates as small remnants. Large solid arrows show the present-day direction of relative movement between the Pacific and North American plates. [After W. J. Kious and R. I. Trilling, *This Dynamic Earth: The Story of Plate Tectonics*. Washington, D.C.: U.S. Geological Survey, 1996.]

the convergent boundary into the modern San Andreas transform-fault system (Figure 9.22). Today, all that is left of the Farallon Plate are small remnants, including the Juan de Fuca and Cocos plates, which are still subducting beneath North America.

The *Basin and Range* province developed through the uplift and stretching of the crust in a northwest-southeast direction (see Figure 9.21). This extension began with the heating of the lithosphere by upwelling convection currents about 15 million years ago and continues to the present day (see Chapter 7). It has resulted in a wide zone of normal faulting extending from southern Oregon to Mexico and from eastern California to western Texas. The Basin and Range province is volcanically active and contains extensive hydrothermal deposits of gold, silver, copper, and other valuable metals.

The *Colorado Plateau* seems to be an island of stability that has experienced no major extension or compression since Precambrian time (see Figure 9.21). The broad uplift of the plateau has allowed the Colorado River to cut through flat-lying rock formations, creating the Grand Canyon. Geologists believe that this uplift was caused by the same type of lithospheric heating that led to stretching of the crust in the Basin and Range province.

> The North American Cordillera was formed during the subduction of the Farallon Plate beneath the North American continent.

Coastal Plain The Atlantic coastal plain and the offshore continental shelf (see Figure 9.17) are sediment-covered regions underlain by crust that was extended by normal faulting during the Triassic-Jurassic splitting of North America from Africa and the subsequent opening of the modern North Atlantic Ocean about 180 million years ago. The opening rift valleys formed sedimentary basins that trapped a thick series of nonmarine sediments. In the early Cretaceous period, the deeply eroded, sloping surface of the Atlantic coastal plain and continental shelf began to subside, accumulating sediments as much as 5 km thick.

The coastal plain and continental shelf of the Gulf of Mexico are similar to their Atlantic equivalents, separated from them by the Florida Peninsula, a large carbonate platform. The Mississippi, Rio Grande, and other rivers that drain the interior of the North American continent have delivered enough sediments to fill a trough some 10 to 15 km deep running parallel to the Gulf Coast. This region contains exceptionally rich reservoirs of oil and natural gas.

> The Atlantic coastal plain and continental shelf were areas of rifting during the breakup of Pangaea and have since subsided.

Continents Around the World

Each continent has its own distinctive features, but a general pattern becomes evident when continental geology is viewed on a global scale (Figure 9.23). Continental shields and platforms make up the most stable parts of the continental lithosphere, called **cratons,** and contain the eroded remnants of ancient deformed rocks. The North American

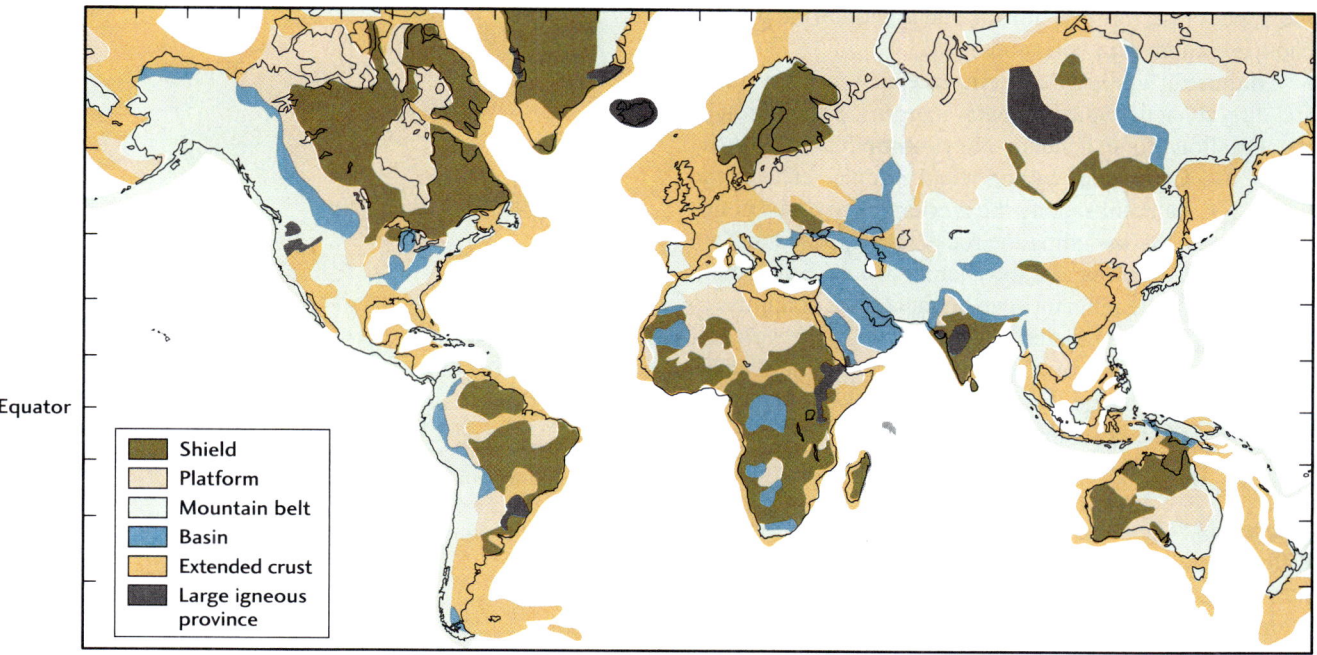

FIGURE 9.23 A global map of the continents, showing the major tectonic provinces. [W. Mooney/USGS.]

craton, for instance, comprises the Canadian Shield and the interior platform (see Figure 9.17).

Around these cratons are elongated mountain belts that were formed by later episodes of compressive deformation. The youngest orogenic (mountain-building) systems, such as the North American Cordillera, are found along the *active margins* of continents, where tectonic activity caused by relative plate movement continues to deform the continental crust.

The *passive margins* of continents—those that are attached to oceanic crust as part of the same plate and thus are not near plate boundaries—are zones of extended crust, stretched during the rifting that broke older continents apart and initiated seafloor spreading. This rifting often occurred along older mountain belts, such as the Appalachian fold belt.

> Continents are built by orogenic activity around stable parts of the continental lithosphere called cratons.

The current distribution of continental tectonic provinces and their ages is like a giant puzzle in which the original pieces have been rearranged and reshaped by continental rifting, continental drift, and continent-continent collisions over billions of years. Only the past 200 million years of plate movements can be reliably determined from existing oceanic crust (see Chapter 3). Earlier plate movements must be inferred from the indirect evidence found in continental rocks. From this evidence, we can gain insights into several key questions about continental evolution: What geologic processes built the continents we see today? How do these processes fit into the theory of plate tectonics? Can plate tectonics explain the formation of the cratons?

We will illustrate the answers to these questions using the evolution of North America as our prime example, starting with its youngest provinces along its west coast and working backward in time to the Canadian Shield.

How Continents Grow

Over the 4-billion-year history of the continents, new crust has been added to them at an average rate of about 2 km^3 per year. In the modern plate tectonic system, two basic processes work together to form new continental crust: magmatic addition and accretion.

Magmatic Addition The process of magmatic differentiation of low-density, silica-rich rock in Earth's mantle and *vertical transport* of this buoyant, felsic material from the mantle to the crust is called **magmatic addition.**

Most new continental crust is born in subduction zones from magmas formed by fluid-induced melting of the subducting lithospheric slab and the mantle wedge above the slab (see Chapter 5). These magmas, which are of basaltic to andesitic composition, migrate toward the surface, pooling in magma chambers near the base of the crust. Here they incorporate crustal materials and differentiate further to form the felsic magmas that migrate into the upper crust, forming dioritic and granodioritic plutons capped by andesitic volcanoes.

Magmatic addition can emplace new crustal material directly at active continental margins. Subduction of the Farallon Plate beneath North America during the Cretaceous period, for example, created the batholiths along the western edge of the continent, including the rocks now exposed in Baja California and the Sierra Nevada. Subduction of the remnant Juan de Fuca Plate continues to add new material to the crust in the volcanically active Cascade Range of the Pacific Northwest, just as subduction of the Nazca Plate is building up the crust in the Andes of South America.

> *Magmatic addition occurs primarily in subduction zones as mantle material undergoes fluid-induced melting and silica-rich magmas rise into the crust.*

Accretion The integration of crustal material previously differentiated from mantle material into existing continental masses by *horizontal transport* during plate movements is called **accretion**.

Much of this new crust is produced far away from continents, in volcanic island arcs where one oceanic plate subducts beneath another. Over time, these island arcs can merge into thick sections of silica-rich crust, such as those found today in the Philippines and other island groups of the southwestern Pacific. Plate movements transport these buoyant fragments of crust horizontally across the globe and eventually attach them to active continental margins.

Geologic evidence for accretion can be found on the active margins of North America. In the Pacific Northwest and Alaska, the crust consists of a mix of odd pieces—island arcs, seamounts (extinct underwater volcanoes) and remnants of thickened oceanic plateaus, old mountain ranges and other slivers of continental crust—that were plastered onto the leading edge of the continent as it moved across Earth's surface (**Figure 9.24**). These pieces are sometimes referred to as *accreted terrains*.

The geologic arrangement of accreted terrains can be chaotic. Adjacent terrains can contrast sharply in their assemblages of rock types, the nature of their folding and faulting, and their history of magmatism and metamorphism. Geologists often find fossils indicating that these blocks originated in different environments and at different times from those of the surrounding area. For example, a terrain comprising ophiolite suites (pieces of oceanic crust) that contain deep-water fossils might be surrounded by remnants of island arcs and continental fragments containing shallow-water fossils of a completely different age. The boundaries between terrains are almost always major faults that have undergone substantial slipping, although the nature of the faulting is often difficult to discern.

> *Accretion occurs when plate movement attaches a fragment of buoyant crust to the edge of a continent.*

FIGURE 9.24 Much of the Cordillera of western North America has been formed by accretion over the past 200 million years. Wrangellia, for example, is a former island arc that was transported to its present location from 5000 km away in the Southern Hemisphere. [After D. R. Hutchison, "Continental Margins." *Oceanus* 35 (Winter 1992–1993): 34–44; modified from work of D. G. Howell, G. W. Moore, and T. J. Wiley.]

FIGURE 9.25 When plates bearing continents collide, the continental crust can break into multiple thrust fault sheets stacked one above the other.

How Continents Are Modified

The Cordillera of western Canada includes many terrains accreted during the drift of North America since the breakup of Pangaea. The geology of this youthful part of the continent looks nothing like that of the ancient Canadian Shield, which lies directly east of the Cordillera. In particular, the accreted terrains do not show the high degree of melting or the high-grade metamorphism that characterize the Precambrian crust of the shield. Why such a difference?

The answer involves *orogeny:* mountain building by folding, faulting, magmatism, and metamorphism. Orogenic processes have repeatedly modified the edges of cratons throughout their long history. Most periods of mountain building (orogenies) result from plate convergence. Orogenies can result when a continent rides forcefully over subducting oceanic crust, as in the Andean orogeny now under way in South America, but the most intense orogenies are caused by the collision of two or more continents.

As we saw in Chapter 7, when two continents collide, thrust faulting caused by the convergence can stack the upper part of the crust into multiple thrust fault sheets tens of kilometers thick, deforming and metamorphosing the rocks they contain (**Figure 9.25**) and creating fold and thrust belts (see Chapter 7). Horizontal compression throughout the crust can double its thickness, causing the rocks in the lower crust to melt. This melting can generate huge amounts of granitic magma, which rises to form extensive batholiths in the upper crust.

Orogenies modify continents by metamorphosing and thickening continental crust and causing the lower crust to melt.

The Alpine-Himalayan Orogeny Orogeny can be seen in action today along the great chains of high mountains that stretch from Europe through the Middle East and across Asia, known collectively as the *Alpine-Himalayan belt*. The breakup of Pangaea sent the continental crust of Africa, Arabia, and India northward, causing the Tethys Ocean to close as its lithosphere was subducted beneath Eurasia (see Figure 3.16).

The Himalayas, the world's highest mountains, are the most spectacular result of this modern episode of continent-continent collision (**Figure 9.26**). They were

FIGURE 9.26 The collision between India and Eurasia has produced many spectacular tectonic features, including large-scale faulting and uplift. [After P. Molnar and P. Tapponier, "The Collision Between India and Eurasia." *Scientific American* (April 1977): 30.]

formed from overthrust slices of the old northern portion of India stacked one atop the other, as in Figure 9.25. The horizontal compression and formation of fold and thrust belts also thickened the crust north of India, causing the uplift of the huge Tibetan Plateau, which now has a crustal thickness of 60 to 70 km (almost twice the thickness of normal continental crust) and stands nearly 5 km above sea level. Further compression has pushed China and Mongolia eastward, out of India's way, like toothpaste squeezed from a tube. Most of this sideways movement has taken place along the major strike-slip faults shown in Figure 9.26.

Paleozoic Orogenies During the Assembly of Pangaea

If we go farther back in geologic time, we find abundant evidence of older orogenies caused by the earlier episodes of plate convergence that produced the supercontinent Pangaea. For example, at least three distinct orogenies were responsible for the Paleozoic deformation now exposed in the eroded Appalachian fold belt of the eastern United States (**Figure 9.27**).

The supercontinent Rodinia began to break up toward the end of the Proterozoic eon, spawning several paleocontinents (see Figure 3.16). One was the large continent of Gondwana. Two of the others were *Lauren-*

(a) Middle Cambrian (510 Ma)
After the breakup of Rodinia, the continent of Laurentia straddled the equator, and its southern side was a passive continental margin, bounded on the south by the Iapetus Ocean.

(b) Late Ordovician (450 Ma)
The island arc built up by the southward-directed subduction of Iapetus lithosphere collided with Laurentia in the middle to late Ordovician, causing the Taconic orogeny.

(d) Late Mississippian (340 Ma)
The collision of Gondwana with Laurussia began with the Variscan orogeny in what is now central Europe…

(e) Upper Pennsylvanian (300 Ma)
… and continued along the margin of the North American craton with the Appalachian orogeny. During this terminal cataclysm, Siberia converged with Laurussia in the Ural orogeny to form Laurasia, while the Hercynian orogeny created new mountain belts across Europe and northern Africa.

FIGURE 9.27 Paleogeographic reconstructions of the present North Atlantic region, showing the sequence of orogenic events that resulted from the assembly of Pangaea. (Ma, million years ago.) [Ronald C. Blakey, Northern Arizona University, Flagstaff.]

tia, which included the North American craton and Greenland, and *Baltica*, comprising the lands around the Baltic Sea (Scandinavia, Finland, and the European part of Russia). In the Cambrian period, Laurentia was north of the proto-Atlantic or *Iapetus Ocean* (in Greek mythology, Iapetus was the father of Atlantis), Baltica lay off to the southeast, and Gondwana was thousands of kilometers to the south.

An island arc built up by the southward-directed subduction of oceanic lithosphere collided with Laurentia in the middle to late Ordovician (470 million to 440 million years ago), causing the first episode of mountain building: the *Taconic orogeny*. (You can see some of the rocks accreted and deformed during this period if you drive the Taconic State Parkway, which runs east of the Hudson River for about 160 km north of New York City.) The second episode began when Baltica and a connected set of island arcs began to collide in the Devonian, causing the *Caledonian orogeny* in Europe and the *Acadian orogeny* in North America.

The grand finale in the assembly of Pangaea was the collision of the behemoth landmass of Gondwana with Laurasia and Baltica, by then joined into a continent named *Laurussia* (see Figure 9.27). The collision began around 340 million years ago with the *Variscan orogeny* in what is now central Europe and continued along the margin of the North American craton with the *Appalachian orogeny* (320 million to 270 million years ago). This latter phase of assembly pushed Gondwanan crust over Laurentia, lifting the Blue Ridge into a mountain chain that may have been as high as the modern Himalaya and causing much of the deformation now seen in the Appalachian fold belt.

The Wilson Cycle Geologists have documented orogenies even farther back in time; for example, during the assembly of the supercontinent Rodinia in the late Proterozoic eon. Some of the best evidence is found at the eastern and southern margins of the Canadian Shield in a broad belt known as the Grenville province, where new crustal material was added to the continent about 1.1 billion to 1.0 billion years ago. Geologists have discovered orogenic belts of similar age on continents worldwide, providing a general picture of how Rodinia came together.

From our brief look at the history of eastern North America, we can infer that the edges of many cratons have experienced multiple episodes of deformation in a general plate tectonic cycle that comprises four main phases (**Figure 9.28**):

1. Rifting during the breakup of a supercontinent

2. Passive margin cooling and sediment accumulation during seafloor spreading and ocean opening

3. Active magmatic addition and accretion during subduction and ocean closure

4. Orogeny during the continent-continent collision that forms the next supercontinent

This idealized sequence of events was named the **Wilson cycle** after the Canadian pioneer of plate tectonics J. Tuzo Wilson, who first recognized its importance in the evolution of continents.

(c) Early Devonian (400 Ma)
The collision of Laurentia with the continent of Baltica caused the Caledonian orogeny and formed Laurussia. The southward continuation of the convergence caused the Acadian orogeny.

(f) Early Permian (270 Ma)
The end product of these episodes of continental convergence was the supercontinent of Pangaea.

> The Wilson cycle is the sequence of continental tectonic events that occur during the assembly and breakup of supercontinents and the opening and closing of ocean basins.

FIGURE 9.28 The Wilson cycle describes the plate tectonic movements that open and close ocean basins and their effects on the continents.

The geologic data suggest that the Wilson cycle has operated throughout the Proterozoic and Phanerozoic eons (Figure 9.29), resulting in the formation of at least two supercontinents prior to Rodinia. One of these supercontinents (*Columbia*) formed about 1.9 billion to 1.7 billion years ago. An even earlier one, whose assembly marks the transition from the Archean to the Proterozoic eon, formed about 2.7 billion to 2.5 billion years ago. Did the Wilson cycle also operate in the Archean eon? That question is still being debated by geologists.

Formation of Cratons

Every continental craton contains regions of ancient lithosphere that have been stable (i.e., undeformed) since the Archean eon (4.0 billion to 2.5 billion years ago). As we have seen, deformation has occurred at the edges of these stable landmasses and new crust has subsequently accreted around them. But how were these central parts of the cratons created in the first place?

We know that Earth was a hotter planet 4 billion years ago. Therefore, mantle convection during the Archean was more vigorous, although we don't know to what extent the movements of rocks on Earth's surface resembled modern plate tectonic processes. We do know that a silica-rich continental crust existed at this early stage in Earth's history. The Acasta gneiss, in the northwestern part of the Canadian Shield, has been dated to 4.0 billion years ago, and its rocks look very similar to modern gneisses (Figure 9.30). Formations as old as 3.8 billion years have been found on many continents; most are metamorphic rocks evidently derived from even older continental crust.

Cratons, the most stable regions of today's continents, were formed in the Archean by the accretion of smaller rafts of lithosphere.

FIGURE 9.29 The ribbon of geologic time, showing some important events in the history of the continents. All numbers are ages in millions of years ago.

In the early part of the Archean eon, the continental lithosphere was very mobile. It may have been organized in small rafts that were rapidly pushed together and torn apart by intense tectonic activity—a version of the flake tectonics that appears to be happening on Venus today. The first crustal regions with long-term stability began to form around 3.3 billion to 3.0 billion years ago. In North America, the oldest surviving example is the central Slave province in northwestern Canada, where the Acasta gneiss is found, which stabilized around 3 billion years ago. Geologists have been able to show that this stabilization process involved not only the continental crust, but also chemical changes to the mantle portion of the continental lithosphere.

By the end of the Archean, 2.5 billion years ago, enough continental lithosphere had been stabilized in cratons to allow the formation of larger and larger continents by magmatic addition and horizontal accretion. Plate tectonics was probably operating much as it does today. It is at about this time that we see the first evidence of major continent-continent collisions and the assembly of supercontinents. From this point onward in Earth's history, the evolution of the continents was governed by the plate tectonic processes of the Wilson cycle.

BIOLOGICAL EVENTS IN EARTH'S HISTORY

As the continents began to take shape during the Archean eon, so did early forms of life. We don't understand how life got started on Earth, but several lines of evidence—fossils as well as chemical traces of life—indicate that single-celled microorganisms, or **microbes,** were flourishing by 3.5 billion years ago. According to one evolutionary hypothesis, these primitive organisms, and all that came after, descended from a single "universal ancestor."

Not much is known about the earliest forms of life, but they must have been composed of carbon-rich compounds, like all subsequent organic substances, and they must have contained *genes*—molecules with instructions for growth and reproduction. Biologists have learned how to use the genetic information contained in living organisms to understand which forms of life are most closely related to one another. This knowledge has allowed them to organize

FIGURE 9.30 The Acasta gneiss in the Archean Slave craton of northwestern Canada, dated to 4.0 billion years ago, is the oldest rock formation so far discovered. It demonstrates that continental crust existed on Earth's surface at the beginning of the Archean eon. [Courtesy of Sam Bowring, Massachusetts Institute of Technology.]

FIGURE 9.31 The universal tree of life shows how all organisms are related to one another. Organisms are subdivided into three great domains: the Bacteria, Archaea, and Eukaryota. These domains are all descended from a universal common ancestor. All three domains are dominated by microorganisms. Note that animals appear at the tip of the eukaryote branch. (Ma, million years ago.)

the hierarchy of ancestors and descendants into a *universal tree of life,* with the hypothesized universal ancestor at its root. The tree has three main branches, called *domains:* Bacteria, Archaea, and Eukaryota (**Figure 9.31**).

Early Life

Sometime early in Earth history, probably during the Archean eon, the microbes split into these three branches. The Bacteria and Archaea appear to have evolved first; all of their descendants have remained single-celled microorganisms. The Eukaryota, thought to be the youngest branch, have a more complex cellular structure, which includes a cell nucleus that contains the genetic material. Eukaryotes evolved from small, single-celled organisms into larger, multicellular organisms—an essential step in the evolution of animals and plants. We have identified fossils of multicellular organisms only in rocks younger than 1 billion years. It therefore appears that microbes were the only organisms on Earth for at least 2.5 billion years!

Precambrian microbes, like those living today, were tiny, only a few microns in size (1 micron = 10^{-6} m). The traces of individual microbes preserved in rocks are therefore called **microfossils.** Needless to say, such features were much harder to find than the macroscopic fossils of shells, bones, and twigs

> *Microfossils and chemical traces of life indicate that microbes were alive on Earth's surface as early as 3.5 billion years ago.*

(a)

(b)

FIGURE 9.32 (a) Abundant microfossils are well preserved in the 2.1-billion-year-old Gunflint formation of southern Ontario, Canada. (b) Early Archean (3.4 billion years old) stromatolites in the Warrawoona formation, Western Australia. The conical shapes suggest that the microbial mats that formed these rocks may have grown toward the sunlight. [Images courtesy of H. J. Hofmann.]

(a) Modern stromatolites in Shark Bay, Australia, grow in the intertidal zone.

(b) Ancient stromatolites (over 1 billion years old) in cross section form columns, northern Siberia, Russia.

(c) A cross section of this stromatolite reveals layering similar to that seen in ancient stromatolites.

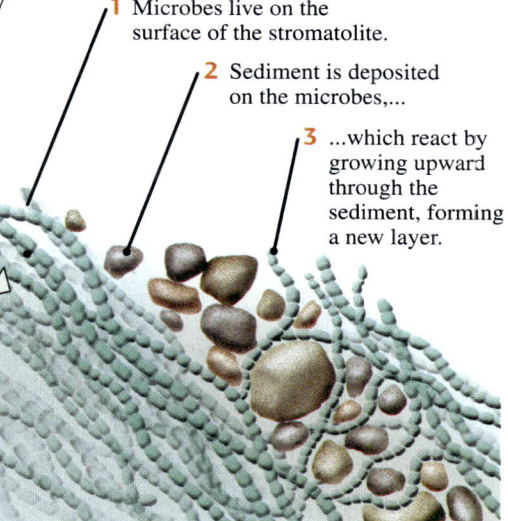

1. Microbes live on the surface of the stromatolite.
2. Sediment is deposited on the microbes,...
3. ...which react by growing upward through the sediment, forming a new layer.

FIGURE 9.33 Stromatolites are sedimentary features that result from the interaction of microbes with their environment. [Images from John Grotzinger.]

used by geologists to study the evolution of animals and plants during the Phanerozoic eon (recall that *Phanerozoic* means "visible life"). In fact, the first microfossils were not discovered until 1954, in the 2.1-billion-year-old Gunflint formation of southern Canada (Figure 9.32a). Well-preserved microfossils have also been found in the 3.2-billion-year-old Fig Tree formation of South Africa. Tiny threads similar in size and appearance to modern microorganisms occur in formations in Western Australia as old as 3.5 billion years, though the interpretation of these older features as microfossils remains controversial.

Some of the best evidence for life in the early Archean comes not from microfossils, but from macroscopic structures called *stromatolites:* sedimentary features with distinctive laminations formed by microbial communities as matted sheets and domes in lakes and tidal flats. Microbial communities can be observed building stromatolites today in intertidal environments such as Shark Bay, Western Australia (Figure 9.33). Stromatolites are common in Precambrian continental cratons and have been identified in sedimentary rocks of early Archean age (Figure 9.32b). Other types of indirect evidence include chemical traces of early life, such as the ratios of carbon isotopes found in some early Archean rocks, which show values that can be produced only by biological processes.

Origin of Earth's Oxygenated Atmosphere

As we learned earlier in this chapter, Earth's early atmosphere was dominated by carbon dioxide released from Earth's interior by volcanism. Our current oxygen-rich atmosphere was produced by early life through the basic metabolic process of **photosynthesis.** Through this process, organisms such as green plants, algae, and certain types of bacteria use energy from sunlight to convert water and carbon dioxide to carbohydrates, such as sugar, and oxygen. The oxygen is released to the atmosphere, and the

FIGURE 9.34 The ribbon of geologic time, showing major events in the history of life. (Ma, million years ago.)

sugar is stored as an energy source for future use by the organism.

The rise of atmospheric oxygen occurred in two main steps, separated by more than a billion years (**Figure 9.34**). The first major increase began with the evolution of the first group of microorganisms capable of photosynthesis, the *cyanobacteria,* perhaps as early as 2.7 billion years ago. The oxygen produced by cyanobacteria reacted with iron dissolved in seawater, causing iron oxide minerals, such as magnetite and hematite, and silica-rich minerals, such as chert and iron silicates, to precipitate and sink to the seafloor. These minerals accumulated in thin, alternating layers of distinctive sediments, called *banded iron formations* (**Figure 9.35**a). This process continued until most of the dissolved iron was used up, allowing oxygen to accumulate in the atmosphere.

Atmospheric oxygen concentrations reached an initial plateau around 2.1 billion to 1.8 billion years ago, when the first eukaryote fossils, of a type of algae, entered the geologic record (Figure 9.35b). The large size of these organisms—at least 10 times larger than anything that came before them—is thought to be a consequence of the oxygen increase. This time also marks the first appearance of *red beds,* unusual river deposits of sandstones and shales bound together by iron oxide cement, which gives them their red color (Figure 9.35c).

After eukaryotic algae came on the scene, not much happened for over a billion years. Then, about 600 million years ago, atmospheric oxygen rose dramatically, almost to its modern level (see Figure 9.34). The reason for this second increase is still not understood, though it may be related to the enrichment of the deep sea with oxygen, perhaps by an increase in the burial of organic carbon by sedimentation. In any case, the consequences were dramatic: the first really large animals suddenly appeared, and all the modern groups of animals evolved shortly thereafter, ushering in the Phanerozoic eon with its wonderfully complex multicellular organisms.

> Earth's atmosphere was oxygenated by early life in two main steps separated by more than a billion years.

Biological Events in Earth's History   243

(a)

(b)

(c)

FIGURE 9.35 Unusual sedimentary rocks and new, larger organisms mark the rise of oxygen in the atmosphere between 2.7 billion and 2.1 billion years ago. (a) Banded iron formation. [Pan Terra.] (b) These fossils of *Grypania*, a type of eukaryotic algae, are visible to the naked eye. [Courtesy of H. J. Hofmann.] (c) Red beds are mostly sandstones and shales partly cemented by iron oxide minerals. [John Grotzinger.]

Radiation of Life: The Cambrian Explosion of Animals

Perhaps the most remarkable geobiological event in Earth's history, aside from the origin of life itself, was the sudden appearance of large animals with skeletons at the end of Precambrian time (**Figure 9.36**). This rapid development of new types of organisms—what biologists call an *evolutionary radiation*—had such an extraordinary effect on the fossil record that its culmination 542 million years ago is used to mark the most profound boundary of

Namacalathus

Hallucigenia

Trilobites

FIGURE 9.36 Fossils that record the Cambrian explosion. Precambrian organisms such as *Namacalathus* (*left*) were the first organisms to use calcite in making shells. These organisms became extinct at the Precambrian-Cambrian boundary, paving the way for a strange new group of organisms, including *Hallucigenia* (*center*) and the more familiar trilobites (*right*). In each example, the fossils are shown on top and the reconstructed organism is shown on the bottom. [*left, top:* John Grotzinger; *left, bottom:* W. A. Watters; *center, top:* National Museum of Natural History/Smithsonian Institution; *center, bottom:* Chase Studio/Photo Researchers; *right, top:* courtesy of Musée cantonal de géologie, Lausanne. Photo by Stéphane Ansermat; *right, bottom:* Chase Studio/Photo Researchers.]

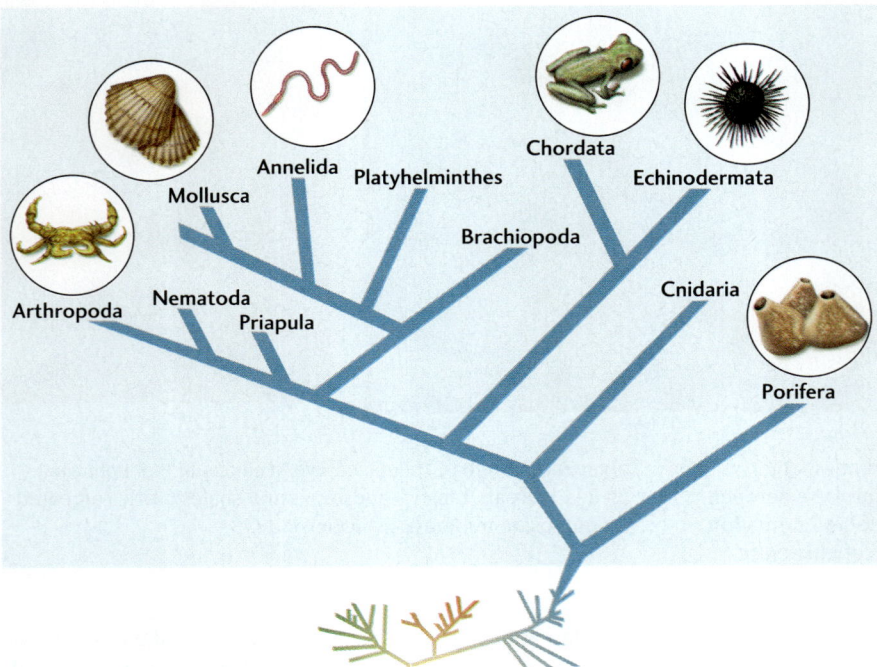

FIGURE 9.37 Every major group of animals alive today originated during the early Cambrian radiation, also known as the Cambrian explosion.

the geologic time scale: the beginning of the Phanerozoic eon, the Paleozoic era, and the Cambrian period (see Chapter 8 and Figure 9.34).

The radiation of animals during the early Cambrian was so fast that it is often called the **Cambrian explosion.** Every major animal group that exists on Earth today, as well as a few more that have since become extinct, originated during the Cambrian explosion. **Figure 9.37** shows the tree of life for the major groups of animals; all of the branches on this tree (*phyla*) originated during the Cambrian explosion.

Fossil as well as living organisms are classified according to shared characteristics, such as shape. Modern animal phyla, for example, are differentiated based on whether their bodies are symmetrical, or are segmented, or have skeletons, to name just a few criteria. What allowed these early animals to develop such complex body shapes so rapidly, and therefore to become so diverse, is still not fully understood. Systematic change in organisms over many generations is referred to as **evolution.** Evolution is driven by **natural selection,** the process by which populations of organisms adapt to new environments. The theory of *evolution by natural selection* states that, over many generations, individuals with the most favorable traits are most likely to survive and reproduce, passing those traits on to their offspring. This process can lead eventually to the emergence of new species.

One hypothesis for the Cambrian explosion is that the genes of these early animals may have changed in a way that made it possible for them to exceed some sort of evo-

lutionary barrier. The stage was set by the development of multicellularity in late Precambrian time (**Figure 9.38**), which opened up new evolutionary possibilities. For example, the development of skeletons may have been an important trigger: once one group of animals had evolved hard parts, the others had to as well, or they would have been eliminated through competition.

The right genes were probably not enough to account for such a major radiation, however. The dramatic envi-

> *All major groups of animals on Earth today originated in the early Cambrian period during a rapid evolutionary radiation called the Cambrian explosion.*

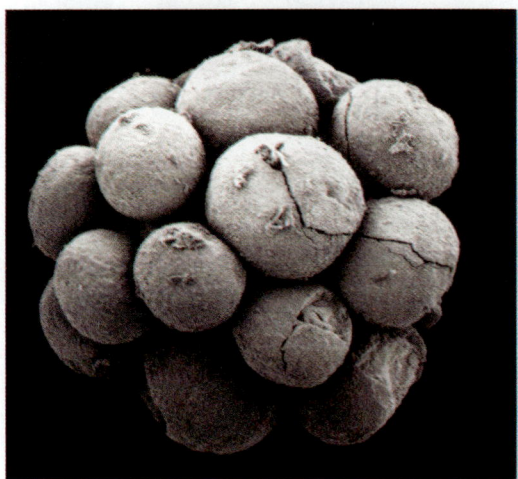

FIGURE 9.38 A fossilized animal embryo from the latest part of Precambrian time. Such fossils show that multicellular animals had evolved before the Precambrian-Cambrian boundary and are the ancestors of the animals that evolved during the Cambrian explosion. This cluster of cells is about 100 microns in diameter. [Courtesy Shuhai Xiao, Virginia Tech.]

1. Animal diversity increased dramatically during the Cambrian radiation.
2. Animal diversity decreased dramatically during the end-Permian extinction.
3. The end-Cretaceous extinction included the demise of dinosaurs.

FIGURE 9.39 The diversity of animal fossils reveals both mass extinctions and radiations. This graph shows the number of "shelly" animal families found in the fossil record during the last 600 million years; each family comprises many species. During a radiation, such as the Cambrian explosion, the number of new animal families increases. During a mass extinction, such as occurred at the end of the Permian period, the number of families decreases. (Ma, million years ago.)

ronmental transformations that occurred near the end of Precambrian time, including major changes in climate and ocean chemistry as well as the increase in atmospheric oxygen, may have set the evolutionary stage for the Cambrian explosion.

Extinction of Life: The Demise of the Dinosaurs

Another dramatic event in Earth's history was the extinction of the dinosaurs at the end of the Cretaceous period (about 65 million years ago). This event defines the *Cretaceous-Tertiary boundary* of the geologic time scale and also marks the end of the Mesozoic era (see Figure 8.13).

When groups of organisms are no longer able to adapt to changing environmental conditions or compete with more successful groups of organisms, they become extinct. An interval when many groups of organisms become extinct at the same time is called a *mass extinction* (see Chapter 8). The Cretaceous-Tertiary boundary represents one of the greatest mass extinctions in Earth's history (Figure 9.39). Entire global ecosystems were obliterated, and about 75 percent of all species on Earth, both on land and in the oceans, were extinguished forever. The dinosaurs are only one of several groups that became extinct at the end of the Cretaceous period, but they are certainly the most prominent. Other groups, such as ammonites, marine reptiles, certain types of clams, and many types of plants and plankton, also perished.

We are now virtually certain that the cause of the Cretaceous-Tertiary mass extinction was a gigantic asteroid impact (see Figure 1.23). In 1980, geologists discovered a very thin layer of dust containing the distinctive element *iridium*, which is typically found in extraterrestrial bolides, in sediments deposited at the end of the Cretaceous in Italy (Figure 9.40). This extraterrestrial dust was

FIGURE 9.40 Knife marks a light-colored layer of clay containing both extraterrestrial materials and materials from local rocks at the Chicxulub impact site that accumulated in the Raton Basin of the southwestern United States following the asteroid impact that killed the dinosaurs. Such deposits have been found worldwide. [From David Kring and Daniel Dura, "The Day the World Burned," *Scientific American* (December 2003): 104. © December 2003 by Scientific American. All rights reserved.]

subsequently found at many other locations around the world, but always exactly at the Cretaceous-Tertiary boundary. The geologists argued that the accumulation of this much iridium-bearing dust in such a thin layer would require an asteroid about 10 km in diameter to hit Earth, explode, and send its cosmic detritus across the globe.

In the early 1990s, geologists found the smoking gun: a huge crater, almost 200 km in diameter, that was buried near a town on Mexico's Yucatán Peninsula, called Chicxulub. The material that had been blown out of the Chicxulub crater, including the fragments of the asteroid itself, formed a blanket that covered all of North America and possibly South America as well. Additional, finer-grained material had been lofted into the atmosphere, enveloping Earth and eventually depositing a thin layer of dust around the world.

The immediate aftermath of this impact is the stuff that disaster movies are made of. The blast itself would have been hundreds of millions of times more devastating than the most powerful nuclear weapon ever detonated and 6 million times more powerful than the 1980 eruption of Mount St. Helens. It would have created a hurricane of unimaginable fury and tsunamis as high as 1 km (100 times higher than the Indian Ocean tsunamis of 2004). The sky would have turned black as massive amounts of dust and vapor were ejected into the upper atmosphere and even into space. A global firestorm may have resulted as the flaming fragments from the blast fell back to Earth (**Figure 9.41**).

The impact of a 10-km asteroid 65 billion years ago killed off 75 percent of all species, including all the dinosaurs. This mass extinction marks the Cretaceous-Tertiary boundary in the geologic time scale.

The hellish effects of the impact itself would have been devastating for many organisms. Most organisms near the impact zone would have been instantly wiped out. But worse yet would have been the aftermath over months and years to come, which scientists think led to the actual mass extinction. The high concentration of debris in the atmosphere would have blocked out the Sun, vastly reducing the light energy available for photosynthesis. In addition to the solid debris, poisonous sulfur- and nitrogen-bearing gases would have been injected into the atmosphere, where they would have reacted with water vapor to form toxic sulfuric and nitric acids that would have rained down on Earth. The combination of these two effects, and many others, would have killed most of the plants and other photosynthetic organisms at the base of the food chain. Once the food sources were gone, the dinosaurs would have died off as well.

Global Warming Disaster: The Paleocene-Eocene Mass Extinction

The mass extinction at the Paleocene-Eocene boundary (55 million years ago; see Figure 8.13) is significant in the evolution of life because it paved the way for the mammals, including the primates, to radiate into an important group. But unlike the mass extinction that wiped out the dinosaurs, it had no extraterrestrial cause. Instead, it was the result of abrupt global warming. Earth scientists are very interested in the details of what happened because global warming, this time produced by human activities, may threaten life in the coming decades (see Chapter 14).

FIGURE 9.41 An artist's rendition of the Cretaceous-Tertiary scene after the asteroid impact. [Alfred Kamajian.]

We now believe that the global warming at the end of the Paleocene epoch occurred when the ocean suddenly belched an enormous amount of methane—a very potent greenhouse gas—into the atmosphere. The resulting global warming was the primary cause of death that led to the mass extinction. But where did all that methane come from?

When marine organisms die, they settle to the bottom of the ocean, where they accumulate as organic debris. Some fraction of this carbon-rich debris is buried in sediments under oxygen-poor conditions, where it is consumed by microbes that produce methane (CH_4) as a waste product. That methane accumulates in the pores of seafloor sediments. If the seafloor is as cold as it is in our present climate (about 3°C), the methane combines with water to form a frozen solid (a methane-water ice), which is stored within the sediments. Geologists searching for oil have found layers with abundant methane-water ices in the upper 1500 m of sediments along many continental margins. If temperatures rise by even a few degrees, however, the methane-water ice melts, and the methane is quickly transformed into a gas.

A massive release of methane from the seafloor around 55 million years ago caused abrupt global warming and led to the mass extinction that marks the Paleocene-Eocene boundary.

At the end of the Paleocene epoch, average temperatures in the deep sea increased enough to start this process. Once some of the methane-water ice thawed, methane entered the atmosphere and reinforced the global warming trend. This, in turn, raised temperatures further in the deep sea, which accelerated the rate of thawing. This cycle resulted in a sudden—and catastrophic—release of methane that caused average global temperatures to rise dramatically. As much as 2 *trillion* tons of carbon may have escaped to the atmosphere in this methane over a period as short as 10,000 years or less!

Because methane reacts with oxygen, the release of methane also caused oxygen levels in the oceans to plummet. Marine organisms were essentially suffocated when oxygen concentrations dropped below a critical level. The oxygen decrease and temperature rise were devastating to organisms in the deep sea, and up to 80 percent of bottom feeders, such as clams, became extinct.

Today, the global inventory of all methane deposits is estimated at 10 to 20 trillion tons of carbon present as methane, far more than what was released to cause the Paleocene-Eocene mass extinction. We are adding carbon dioxide to the atmosphere through the burning of fossil fuels, causing the climate to warm significantly. If this trend continues and the oceans warm up, it is possible that the current methane deposits could thaw. We would be wise to pay attention to the lessons of our geologic history.

The Mother of All Mass Extinctions: Whodunit?

The Cretaceous-Tertiary and Paleocene-Eocene extinctions are clear-cut examples of dramatic changes in Earth's environment that caused the catastrophic collapse of ecosystems and led to mass extinction. They were big, but not the biggest.

The extinction at the end of the Permian period was even worse (see Figure 9.39)—95 percent of all species became extinct! Here, there is no smoking gun, and it seems unlikely that something as simple as an asteroid impact could explain how almost every species on Earth was killed. Not surprisingly, the absence of clear-cut evidence has resulted in a long list of hypotheses. Some scientists point to extraterrestrial events, such as a comet impact or an increase in the solar wind. Others argue for events that come from Earth itself, such as an increase in volcanism, the depletion of oxygen in the oceans, or the sudden release of carbon dioxide from the oceans. As in the Paleocene-Eocene extinction, the sudden release of methane from the oceans has also been proposed.

Recently, it has been shown that the mass extinction at the end of the Permian, which occurred 251 million years ago, precisely matches the age of an enormous deposit of flood basalts in Siberia. Flood basalts, as we learned in Chapter 5, are extrusive igneous rocks formed from huge volumes of lava that pour out across the surface of Earth in a relatively short time. In Siberia, volcanic fissures spewed out some 2 million cubic kilometers of basaltic lava, covering an area of 1.6 million square kilometers, about the size of Europe. Geochronological data show that all of this lava poured out within 1 million years or less. It is hard to escape the conclusion that the Permian mass extinction was somehow related to this catastrophic eruption. These tremendous volcanic emissions would have injected enormous amounts of carbon dioxide and sulfur dioxide gases into the atmosphere. Carbon dioxide contributes to global warming, and sulfur dioxide is the principal source of acid rain (see Chapter 1). Both are harmful to life if atmospheric concentrations get too high.

The cause of the greatest mass extinction of all time, which wiped out 95 percent of all species at the end of the Permian period, is still being debated.

Whatever the cause of the Permian mass extinction, one point is clear: just as in the Cretaceous-Tertiary and Paleocene-Eocene mass extinctions, the ultimate cause was the collapse of ecosystems. We know that this collapse occurred, but we don't know exactly how. The message that we should take away from this history lesson is that the past may repeat itself. As we will see in Chapter 14, the decisions that humans make today will inevitably affect our environment—we just don't know exactly how, at least not yet.

SUMMARY

How did our solar system originate? According to the nebular hypothesis, the Sun and its planets formed when a cloud of gas and dust condensed about 4.5 billion years ago. The inner planets, including Earth, differ from the giant outer planets in their composition.

How did Earth form and evolve over time? Earth probably grew by the accretion of colliding planetesimals. Soon after it formed, it was struck by a large body about the size of Mars. Matter ejected into space from both Earth and the impacting body reassembled to form the Moon. The impact generated enough heat to melt much of Earth's outer layers. Radioactivity also contributed to early heating and melting. Heavy matter, rich in iron, sank toward Earth's center, and lighter matter floated upward to form the outer layers that became the crust and the continents. Still lighter gases formed Earth's atmosphere and oceans. In this way, Earth was transformed into a differentiated planet with distinct layers.

What are some major events in the early history of the solar system? The age of the solar system, as determined from isotopic dating of meteorites, is about 4.56 billion years. Earth and the other terrestrial planets had formed within about 10 million years. The impact that formed the Moon occurred about 4.5 billion years ago. Minerals as old as 4.4 billion years have survived in Earth's crust. The lunar maria were formed by impacts during the Late Heavy Bombardment, which peaked around 3.9 billion years ago and marks the end of the Hadean eon on Earth.

How can planetary surfaces be dated? Rocks returned from the surface of the Moon by the Apollo missions have been dated using isotopic methods. The lunar highlands show ages from 4.47 billion to about 4.0 billion years. The lunar maria show ages from 4.0 billion to 3.2 billion years. These isotopic ages allowed geologists to calibrate the relative time scale they had developed by mapping and counting craters.

Do other planets have plate tectonics? Venus is the only planet, other than Earth, that has features indicating tectonic activity resulting from mantle convection. But Venus does not appear to have thick plates. Instead, it has a thin crust of frozen lava that breaks up into flakes or crumples like a rug. A similar process, which geologists refer to as flake tectonics, may have occurred on Earth when it was younger and hotter.

What are the major tectonic provinces of North America? The continent's most ancient crust is exposed in the Canadian Shield. South of the Canadian Shield is the interior platform, where Precambrian basement rocks are covered by layers of Paleozoic sedimentary rocks. These older regions form the continental craton. Around the edges of the craton are elongated mountain chains. The most important of these orogenic belts are the Cordillera, which runs down the western edge of North America, and the Appalachian fold belt, which trends southwest to northeast on its eastern margin. The coastal plain and continental shelf of the Atlantic Ocean and Gulf of Mexico are parts of a passive continental margin that subsided after rifting during the breakup of Pangaea.

How do continents grow? Two plate tectonic processes, magmatic addition and accretion, add crust to continents. Buoyant silica-rich rocks are produced by magmatic differentiation, primarily in subduction zones, and added to the continental crust by vertical transport. Accretion occurs when preexisting crustal material is attached to existing continental masses by horizontal plate movement.

How do orogenies modify continents? Orogenies caused by plate convergence can deform continental crust. Thrust faulting can stack the upper part of the crust into multiple thrust fault sheets tens of kilometers thick, deforming and metamorphosing the rocks they contain. Compression throughout the crust can double its thickness, causing the rocks in the lower crust to melt and form extensive batholiths.

What is the Wilson cycle? The Wilson cycle is a sequence of tectonic events that occur during the assembly and breakup of supercontinents and the opening and closing of ocean basins. It has four main phases: rifting during the breakup of a supercontinent; passive margin cooling and sediment accumulation during seafloor spreading and ocean opening; active magmatic addition and accretion during subduction and ocean closure; and orogeny during continent-continent collision.

What evidence do we have of life on early Earth? The oldest possible fossils on Earth are 3.5 billion years old and appear to be the remnants of microbes. Photosynthetic bacteria had probably evolved by 2.7 billion years ago. Banded iron formations, red beds, and the appearance of eukaryotic algae testify to an initial rise in atmospheric oxygen concentrations by 2.1 billion years ago. A second, more dramatic rise in oxygen concentrations occurred near the end of Precambrian time and might have triggered the evolution of animals.

What is the difference between radiation and extinction? Evolutionary radiation is the rapid development of new types of organisms. When groups of organisms are no longer able to adapt to changing environmental conditions or to compete with more successful groups of organisms, they become extinct. In a mass extinction, many groups of organisms become extinct at the same time. The greatest radiation of animals in Earth's history occurred during the early Cambrian, when all the animal phyla living today evolved. Several mass extinctions have occurred throughout the Phanerozoic eon. A major mass extinction occurred at the end of the Cretaceous period, when an asteroid hit Earth and 75 percent of all species were wiped out. Global warming by release of methane caused a mass extinction at the Paleocene-Eocene boundary. The cause of the greatest mass extinction of all time, which wiped out 95 percent of all species at the end of the Permian period, is still being debated.

KEY TERMS AND CONCEPTS

accretion (p. 234)
asteroid (p. 218)
Cambrian explosion (p. 244)
craton (p. 232)
differentiation (p. 218)
evolution (p. 244)
Heavy Bombardment (p. 222)
magmatic addition (p. 233)
meteorite (p. 218)
microbe (p. 239)
microfossil (p. 240)
natural selection (p. 244)
nebular hypothesis (p. 216)
photosynthesis (p. 241)
solar nebula (p. 217)
Wilson cycle (p. 237)

EXERCISES

1. How and why do the inner planets differ from the giant outer planets?

2. What caused Earth to differentiate into a layered planet, and what was the result?

3. Mercury's average density is less than Earth's, but the relative size of its core is larger than that of Earth's core. How can you explain this?

4. What aspects of the geology of Earth and the Moon are consistent with high impact rates during the Late Heavy Bombardment?

5. Why is the topography of the North American Cordillera higher than that of the Appalachian Mountains? How long ago were the Appalachians at their highest elevation?

6. Describe the tectonic province in which you live.

7. Are the interiors of continents usually younger or older than their margins? Explain your answer using the concept of the Wilson cycle.

8. Two continents collide, thickening the crust from 35 km to 70 km and forming a high plateau. After hundreds of millions of years, the plateau is eroded down to sea level. (a) What kinds of rocks might be exposed at the surface by this erosion? (b) Estimate the crustal thickness after the erosion has occurred. (c) Where in North America has this sequence of events been recorded in surface geology?

9. How many times have the continents been joined in a supercontinent since the end of the Archean eon? Use this number to estimate the typical duration of a Wilson cycle and how fast plate tectonic processes move continents.

10. During an evolutionary radiation, organisms evolve rapidly. What would the geologic record look like if a slow evolutionary radiation occurred during the time of an unconformity? How would you distinguish between a rapid radiation and the effects of an unconformity?

11. Compare and contrast the events that led to the Permian, Cretaceous-Tertiary, and Paleocene-Eocene mass extinctions. Are there particular processes that contributed to these mass extinctions that we, as members of a global society, should be aware of? Which processes may be induced by human activities?

THE CLIMATE SYSTEM AND GLACIATION

CHAPTER 10

The Ice Hunter

Lonnie Thompson has spent more time above 18,000 feet than any other person on Earth. Trekking to the Himalayas and Andes and beyond, he has risked blood clots and temporary blindness in the name of a single pursuit: preserving tens of thousands of years of weather history coded deep in the planet's fast-vanishing glaciers. "No scientist has taken bigger risks to track ancient weather patterns and help us understand the anomaly of current climate trends," says Al Gore.

Thompson stores his prehistoric glacial samples at Ohio State University in vaults kept at subarctic temperatures and studies the dust particles and air trapped within the ice. From this atmospheric evidence, he has reconstructed a meticulous calendar of temperatures dating back 750,000 years. The upshot: "It proves that the warming trends of today are vastly more dramatic than what we've seen over 5,000 years," says Thompson.

> *You could have knocked me over with a feather the day I discovered, firsthand, that glaciers contain a frozen history of the Earth.*

Growing up on a small farm in West Virginia, Thompson studied geology so he could work in the coal industry. But he got sidetracked in grad school, when he examined the first ice core ever extracted by American scientists. "You could have knocked me over with a feather the day I discovered, firsthand, that glaciers contain a frozen history of the Earth," he recalls. Now, in his work at the Byrd Polar Research Center at Ohio State, Thompson possesses a will to survive on par with Lance Armstrong's, defying frostbite and hurricane-force winds. Photographs he has taken provide disturbing views of the world's melting glaciers—including the ice cap on Mount Kilimanjaro, which is expected to disappear entirely by 2015.

Thompson dismisses skeptics who contend that the current warming trend is due to a natural cycle. "Name one who has ever really studied climate or collected data," he says. "I bet you can't." Glaciers, he adds, "have no political agenda. They don't care if you're a Democrat or a Republican. Science is about what is, not what we believe or hope. And it shows that global warming is wiping out invaluable geological archives right before our eyes."

— From *Rolling Stone,* November 3, 2005

Glaciologist Lonnie Thompson at an altitude of 5300 m (17,390 ft) on Tibet's Dasuopu Glacier. Ice coring on this glacier provides evidence of abnormal global warming during the twentieth century. [Lonnie Thompson/Byrd Polar Research Center/Ohio State University.]

CHAPTER OUTLINE

Components of the
Climate System 252

The Greenhouse Effect 260

Climate Variation 262

Glacial Landscapes 267

Pleistocene Glaciations 272

The Record of Ancient
Glaciations 276

No aspect of Earth science is more important to our continued well-being than the study of climate. Throughout geologic time, evolutionary radiations and extinctions of organisms have been closely connected to changes in climate. Even the short history of our own species is deeply imprinted by climate change: agricultural societies began to flourish only about 11,000 years ago, when the harsh climate of the last ice age rapidly transformed into the mild and steady climate of the Holocene epoch. Now a globalized human society based on a petroleum-fueled economy is injecting greenhouse gases into the atmosphere at a phenomenal rate, with potentially dire consequences—global warming, sea level rise, and unfavorable changes in weather patterns. The climate system is a huge, incredibly complex machine, and, like or not, our hands are on the controls. We're in the driver's seat, with pedal to the metal, so we had better understand how the machinery works!

In this chapter, we will examine the main components of the climate system and the ways in which these components interact to produce the climate we live in today. We will look at the evidence for recent global warming and its relationship to changes in the composition of the atmosphere caused by human activities. We will investigate the geologic record of past climate change, particularly the ice ages that spread vast glaciers across the northern continents, and we will see how those glaciations have modified the landscape.

Knowledge of the climate system will equip us to study a wide range of geologic processes that shape the face of our planet—weathering, erosion, sediment transport by wind and water, the interaction of the plate tectonic and climate systems—which will be the topics of the next two chapters. The material presented here will also prepare us for the final subject of this textbook: a geological perspective on the resource needs and environmental impacts of human society.

COMPONENTS OF THE CLIMATE SYSTEM

At any point on Earth's surface, the amount of energy received from the Sun changes daily, yearly, and on longer-term cycles associated with Earth's movement through the solar system. These cyclical variations in the input of solar energy, known as *solar forcing,* cause changes in the surface environment: temperatures rise during the day and fall at night, and they rise in summer and fall in winter. The term **climate** refers to the average conditions at a point on Earth's surface and their variation during these cycles of solar forcing.

Climate is described by daily and seasonal statistics on atmospheric temperature, humidity, cloud cover, rate of rainfall, wind speed, and other weather conditions. Table 10.1 gives an example of temperature statistics for New York City, which include measures of temperature variability (record highs and lows) as well as the mean values. In addition to these common weather statistics, a full scientific description of climate includes the nonatmospheric components of the surface environment, such as soil moisture and streamflow on land, as well as sea surface temperature and the velocity of currents in the ocean.

> Climate is described by the average weather conditions at a point on Earth's surface during daily, yearly, and longer-term cycles of solar forcing.

The **climate system** includes all of the components of the Earth system, and all of the interactions among those

TABLE 10.1 Daily and Seasonal Temperatures (°F) in Central Park, New York City

Data type[a]	January 1	April 1	July 1	October 1
Record high	62	83	100	88
Mean high	39	56	82	69
Mean low	28	39	67	55
Record low	−4	12	52	36

[a]Mean temperatures are for 1970–2004; record temperatures are for 1869–2004.

FIGURE 10.1 Earth's climate system involves complex interactions among many components.

components, that determine how climate behaves in space and time (**Figure 10.1**). The main components of the climate system are the atmosphere, hydrosphere, cryosphere, lithosphere, and biosphere. Each component plays a different role in the climate system, and that role depends on its ability to store and transport energy.

The Atmosphere

Earth's atmosphere is the most mobile and rapidly changing part of the climate system. Like Earth's interior, the atmosphere is layered (**Figure 10.2**). About three-fourths of its mass is concentrated in the layer closest to Earth's surface, the **troposphere,** which has an average thickness of 11 km. The troposphere convects vigorously due to the heating of Earth's surface by the Sun, causing storms and other short-term disturbances in the weather (*tropos* is the Greek word for "turn" or "mix"). Above the troposphere is the **stratosphere,** a dryer layer that extends to an altitude of about 50 km. Above the stratosphere, the atmosphere has no abrupt cutoff; it slowly becomes thinner and fades away into outer space.

The convection patterns in the troposphere (which we'll examine more closely in Chapter 12), combined with Earth's rotation, set up a series of prevailing wind belts. In temperate regions, the prevailing winds have a generally eastward flow, such that they transport a typical parcel of air eastward around the globe in about a month (which is why it takes a few days for storms to

> *The Sun's heat causes convection in the troposphere, which, combined with Earth's rotation, sets up prevailing wind belts that transport heat from the warmer equatorial regions to the cooler polar regions.*

FIGURE 10.2 Layers of the atmosphere, showing the variation of temperature (indicated by the blue line) and pressure with altitude.

blow across the continental United States). The spiral-like circulation of air in these global wind belts also transports heat energy from the warmer equatorial regions to the cooler polar regions.

The atmosphere is a mixture of gases, mainly nitrogen (78 percent by volume in dry air) and oxygen (21 percent by volume in dry air). The remaining 1 percent consists of argon (0.93 percent), carbon dioxide (0.035 percent), and other minor gases (0.035 percent). Water vapor (H_2O) is concentrated near Earth's surface in highly variable amounts (up to 3 percent, but typically about 1 percent). Water vapor and carbon dioxide (CO_2) are the principal greenhouse gases in the atmosphere. If there were no greenhouse gases, incoming solar radiation would easily pass out again through the atmosphere, and Earth's climate would be much colder. We touched on the greenhouse effect in Chapters 1 and 2, and we will discuss it in more detail shortly.

FIGURE 10.3 Two major circulation systems in the oceans. (a) Currents at the surface of the oceans. [U.S. Naval Oceanographic Office.] (b) A schematic representation of thermohaline circulation, which acts like a conveyor belt to transport heat from warm equatorial regions to cool polar regions.

> *The atmosphere is composed mostly of nitrogen gas (78%) and oxygen gas (21%).*

Another minor constituent of the atmosphere is ozone (O_3^+), a highly reactive greenhouse gas produced primarily by the effect of solar radiation on atmospheric oxygen. Most atmospheric ozone is found in the stratosphere, with a maximum concentration at an altitude of 25 to 30 km (see Figure 10.2). As we saw in Chapter 1, this stratospheric ozone layer filters out incoming ultraviolet radiation, protecting the biosphere at Earth's surface from its potentially damaging effects.

The Hydrosphere

The *hydrosphere* comprises all the liquid water on, over, and under Earth's surface, including oceans, lakes, streams, and groundwater. Almost all of that liquid water is in the global oceans; lakes, streams, and groundwater constitute a mere 1 percent. However small, these continental components of the hydrosphere play a vital role in the climate system. They are reservoirs for moisture on land and provide the transport system for returning precipitation and transporting salt and other minerals to the oceans.

> *The hydrosphere comprises all of the liquid water on Earth, most of which is in the oceans.*

Although water circulates more slowly in the oceans than air does in the atmosphere, water can store a much larger amount of heat than air. For that reason, ocean currents transport heat energy very effectively. Prevailing winds blowing across the ocean generate surface currents and large-scale circulation patterns within the ocean basins (**Figure 10.3**a). As in the atmosphere, the most important currents for regulating climate are those that transport heat from equatorial to polar regions. These currents involve vertical as well as horizontal movement.

An example is the Gulf Stream, which flows along the western Atlantic margin, bringing warm water from the Caribbean Sea that warms the climate of the North Atlantic and Europe. In the North Atlantic, that water cools and becomes more salty (because less fresh water enters the sea from rivers at high latitudes than evaporates from the ocean surface). Cool water is denser than warm water, and salty water is denser than less salty water; therefore, this cooler, saltier water sinks. In this way, a subsurface cold current is created that flows southward as part of a global pattern of **thermohaline circulation**—so called because it is driven by differences in temperature and salinity.

> *Ocean currents, like prevailing wind belts, transport heat energy from warmer regions near the equator to cooler regions near the poles.*

On a global scale, thermohaline circulation acts like an enormous conveyor belt running through the oceans that moves heat from the equatorial regions toward the poles (Figure 10.3b). Changes in this circulation pattern can strongly influence the global climate.

The Cryosphere

The ice component of the climate system is called the *cryosphere* (**Figure 10.4**). Thought of as a mineral, ice has some unusual properties. Its melting temperature is very low (0°C), hundreds of degrees below the melting temperatures of silicate minerals. Most minerals are denser in their solid form than in their liquid form, which is why magma rises buoyantly through the lithosphere. Ice, on the

FIGURE 10.4 Several valley glaciers flow together in the St. Elias Mountains, Kluane National Park, Yukon, Canada. [Stephen J. Krasemann/DRK Photo.]

FIGURE 10.5 Iceberg calving at Wrangell-St. Elias National Park, Alaska. Calving occurs when huge blocks of ice break off at the edge of a glacier that has moved to a shoreline. [Tom Bean.]

other hand, is less dense than its liquid form (water), which is why icebergs float on the ocean. And although ice may seem hard, it is much softer than most minerals.

Because ice is so easily deformed, it flows readily downhill like a viscous fluid. *Glaciers* are large masses of ice on land that move under the force of gravity. **Valley glaciers** are rivers of ice that form in the cold heights of mountain ranges and flow down valleys. A valley glacier usually occupies the complete width of a valley and may bury its floor under hundreds of meters of ice (see Figure 10.4). In warmer, low-latitude climates, valley glaciers are found only at the heads of valleys on the highest mountain peaks. Valley glaciers that flow down coastal mountain ranges at high latitudes may terminate at the ocean's edge, where masses of ice break off and form icebergs—a process called *iceberg calving* (**Figure 10.5**).

A **continental glacier** is a thick, slow-moving sheet of ice (sometimes called an *ice sheet*) that covers a large part of a continent or other large landmass. Today, the world's largest continental glaciers overlie much of Greenland and Antarctica, covering about 10 percent of Earth's land surface and storing about 75 percent of the world's fresh water (**Figure 10.6**). In Greenland (Figure 10.6a), 2.6 million cubic kilometers of ice occupy 80 percent of the island's total area. The upper surface of the ice sheet resembles an extremely wide convex lens. At its highest point, in the middle of the island, the ice is more than 3200 m thick. From this central area, the ice surface slopes to the sea on all sides. At the mountain-rimmed coast, the ice sheet breaks up into narrow tongues resembling valley glaciers that wind through the mountains to reach the sea, where icebergs form by calving.

Though very large, the Greenland Glacier is dwarfed by the Antarctic ice sheet, which covers an area of about 13.6 million square kilometers and reaches thicknesses of 4000 m (Figure 10.6b). The total volume of Antarctic ice, about 30 million cubic kilometers, constitutes over 90 percent of the cryosphere. As in Greenland, the ice forms a dome in the center and slopes down to the margins of the continent.

Parts of Antarctica are rimmed by thinner sheets of ice—*ice shelves*—floating on the ocean and attached to the main glacier on land. The best known of these is the Ross Ice Shelf, a

> The cryosphere comprises all of the ice on Earth, most of which is found in the Antarctic continental glacier.

FIGURE 10.6 Topographic maps and cross sections of the two existing continental glaciers. (a) Greenland. [After R. F. Flint, *Glacial and Quaternary Geology*. New York: Wiley, 1971.] (b) Antarctica. [After U. Radok, "The Antarctic Ice." *Scientific American* (August 1985): 100; based on data from the International Antarctic Glaciological Project.]

thick layer of ice about the size of Texas (see Figure 10.6b). Floating ice also includes *sea ice* in the open ocean. During the winter, sea ice typically covers 14 million to 16 million square kilometers of the Arctic Ocean (Figure 10.7) and 17 million to 20 million square kilometers of the Antarctic Ocean, shrinking to about one-third of that area in the summer.

The seasonal exchange of water between the cryosphere and hydrosphere is an important process of the climate system. About one-third of the land surface is covered by seasonal snows, almost entirely (all but 2 percent) in the Northern Hemisphere. Melting snow is the source of much of the fresh water in the hydrosphere. In the U.S. Sierra Nevada and Rocky Mountains, for example, 60 to 70 percent of annual precipitation is snowfall, which is later released as water during spring snowmelt and river runoff.

About one-third of Earth's land surface is covered by seasonal snow, an important source of fresh water when it melts.

FIGURE 10.7 Sea ice is an important part of the cryosphere. This satellite image shows sea ice flowing through the Bering Strait in May 2002. [NASA MODIS satellite.]

In very cold regions, the ground stays frozen throughout the year. Perennially frozen soil, or *permafrost,* today covers as much as 25 percent of Earth's total land area and constitutes an important part of the cryosphere (**Figure 10.8**). Permafrost is defined solely by temperature, not by soil moisture content, overlying snow cover, or location: any rock or soil remaining at or below 0°C for 2 or more years is permafrost. Permafrost covers about 82 percent of Alaska and 50 percent of Canada, as well as great parts of Siberia. Outside the polar regions, it is present in high, mountainous areas such as the Tibetan Plateau.

The Lithosphere

The part of the lithosphere that is most important to the climate system is the land surface, which makes up about 30 percent of Earth's total surface area (**Figure 10.9**) The

FIGURE 10.8 A map of the Northern Hemisphere with the North Pole at its center, showing the distribution of permafrost on the continents. The large area of high mountain permafrost at the top of the map is on the Tibetan Plateau. [T. L. Pewe, Arizona State University.]

FIGURE 10.9 A snapshot of the climate system taken by sensors on several spacecraft, showing cloud cover (in white), variations in sea surface temperature (from the warmest in red to the coolest in dark blue), and land surface properties, including density of vegetation (from the lowest density in brown to the highest in green). [R. B. Husar/NASA Visible Earth.]

composition of the land surface affects the way it absorbs solar energy or releases it to the atmosphere. As the temperature of the land surface rises, more heat energy is radiated back into the atmosphere, and more water evaporates from the land surface. Because evaporation uses considerable energy, it causes the surface to cool. Consequently, soil moisture and other factors that influence evaporation—such as vegetation and the subsurface flow of water—are very important in controlling atmospheric temperatures.

Topography has a direct effect on climate through its influence on atmospheric circulation. Air masses that flow over mountain ranges dump rain on the windward side, creating a rain shadow on the leeward side of the mountains (see Figure 11.4). In fact, the overall asymmetry of the continents—a direct consequence of plate movements—induces hemispheric asymmetries in the global climate system.

Geologists have documented many long-term changes in the climate system due to the rearrangement of continents by plate movements. Changes in the shape of the seafloor due to seafloor spreading induce changes in sea level, and the drift of continents over the poles leads to the growth of continental glaciers. Plate movements can also block ocean currents or open gateways through which they can flow. The emergence of the Isthmus of Panama, for example, closed a gateway connecting the Atlantic and Pacific oceans about 5 million years ago, which may have initiated the Pleistocene glacial cycles we will describe below. If future geologic activity were to close the narrow channel between the Bahamas and Florida through which the Gulf Stream flows, temperatures in western Europe might drop drastically.

The land surface, the part of the lithosphere that is most important to the climate system, absorbs and releases solar energy.

Plate movements can cause long-term climate changes by rearranging continents and ocean basins.

Volcanism in the lithosphere affects climate by changing the composition and properties of the atmosphere. Volcanic eruptions inject tiny liquid drops and solid particles, called *aerosols,* into the stratosphere, blocking solar radiation and temporarily lowering atmospheric temperatures on a global scale. After the massive April 1815 eruption of Mount Tambora in Indonesia, New England suffered through a "year without a summer" in 1816. According to a diarist in Vermont, "no month passed without a frost, nor one without a snow," and crop failures were common. Careful studies have shown that more recent large volcanic eruptions—including those of Krakatau (1883), El Chichón (1982), and Mount Pinatubo (1991)—each produced an average dip of 0.3°C in global temperatures about 14 months after the eruption (local temperature variations can be much larger, of course). Temperatures returned to normal in about 4 years.

The Biosphere

The term *biosphere* was coined in 1875 by Eduard Suess, the geologist who also named Gondwana and first described the exposed continental cratons as "shields." The biosphere comprises all the organisms living at Earth's surface: plants, animals, and microbes, both marine and terrestrial. Life is ubiquitous at Earth's surface, but the amount of life depends on local climate conditions, as we can see from the satellite images of plant concentrations in Figure 1.35.

The amount of energy contained and transported by living organisms is relatively small—less than 0.1 percent of incoming solar energy is used by plants in photosynthesis and thus enters the biosphere. However, the biosphere is strongly coupled to the climate system through its effects on the atmosphere, hydrosphere, and lithosphere. Terrestrial vegetation can affect both atmospheric temperature, because plants absorb solar radiation and release it as heat, and atmospheric moisture, because they soak up groundwater and release it as water vapor. Life also regulates the

The biosphere, which comprises the living organisms of the Earth system, affects climate by modifying the temperature of the land surface and the composition of the atmosphere and oceans.

composition of the atmosphere by taking up or releasing greenhouse gases such as carbon dioxide and methane. Through photosynthesis, marine and terrestrial plants extract carbon dioxide from the atmosphere. Some of it is precipitated in the oceans as shells made of calcium carbonate or buried as organic matter in sediments. The biosphere thus plays a central role in the carbon cycle (discussed in Chapter 14).

Humans, of course, are part of the biosphere, though hardly an ordinary part. Our influence over the biosphere is growing rapidly, and we have become the most active agents of environmental change. As an organized society, we behave in fundamentally different ways from other species. For example, we can study climate change scientifically and modify our actions according to what we have learned.

THE GREENHOUSE EFFECT

The Sun is a yellow star that puts out about half its radiant energy as visible light. The other half is split between infrared waves of lower energy intensity than visible light (which we perceive as heat) and ultraviolet waves of higher energy intensity. At Earth's position in the solar system, the average amount of solar radiation its surface receives throughout the year is 342 watts per square meter of surface area (W/m^2) (1 watt = 1 joule per second; a joule is a unit of energy or heat). In comparison, the average amount of heat flowing out of Earth's deep interior by mantle convection is minuscule, only 0.06 W/m^2. Essentially all the energy driving the climate system ultimately comes from the Sun (**Figure 10.10**).

If the global temperature, averaged over daily and seasonal cycles, is to remain the same, then Earth's surface must radiate energy back into space at a rate of precisely 342 W/m^2. Any less would cause the surface to heat up; any more would cause it to cool down. How is this *radiation balance*—the equilibrium between incoming and outgoing radiant energy—achieved?

A Planet Without Greenhouse Gases

Suppose Earth had no atmosphere at all, but was a rocky sphere like the Moon. Some of the sunlight falling on the surface would be reflected back into space, and some would be absorbed by the rocks, depending on the color of the surface. A perfectly white planet would reflect all the solar energy, whereas a perfectly black planet would absorb it all. The fraction of the solar energy reflected is called the planet's **albedo** (from the Latin word *albus*, meaning "white"). Although the full Moon looks bright to us, the rocks on its surface are mainly dark basalts, so its albedo is only about 7 percent. In other words, the Moon is very nearly black.

The energy radiated by a black body increases rapidly as its temperature increases. A cold bar of iron is black and gives off little heat. If you heat the bar to 100°C, it gives off warmth in the form of infrared waves (like a steam radiator). If you heat the bar to 1000°C, it becomes bright orange, radiating heat at visible wavelengths (like the burner on an electric stove).

A black body exposed to the Sun heats up until its temperature is just the right value for it to radiate the

A balance between incoming solar energy and outgoing radiation maintains the average global temperature of Earth's surface.

1 Solar energy input to Earth's surface is, on average, 342 W/m^2.

2 Heat flowing out of Earth's deep interior is much smaller — only 0.06 W/m^2.

3 To maintain a constant temperature, heat radiating from Earth must balance solar input.

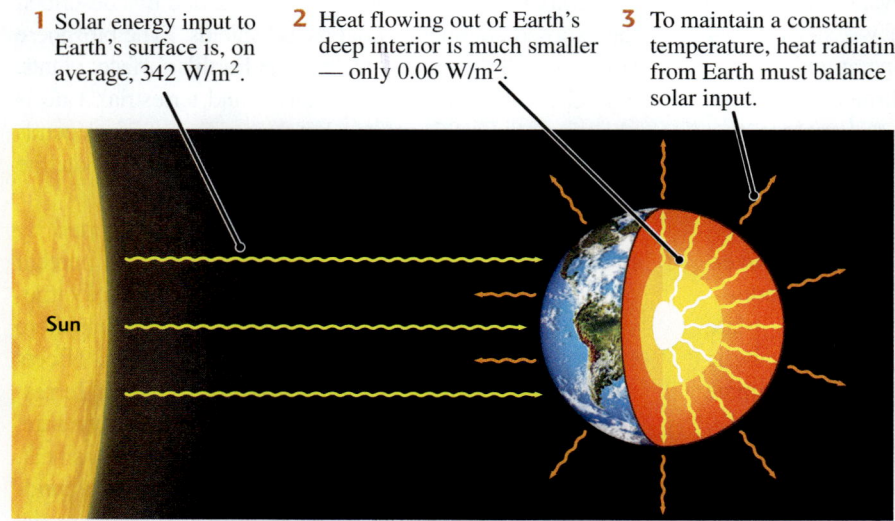

FIGURE 10.10 Earth's energy balance is achieved by the radiation of incoming solar energy back into space. Heat gain from Earth's interior is negligible in comparison to that from solar energy.

incoming solar energy back into space. The same principle applies to a "gray body" like the Moon, except that the reflected energy must be excluded from the radiation balance. And, in the case of rotating bodies like the Moon and Earth, day-and-night cycles must be taken into account. The Moon's daytime temperatures rise to 130°C, and its nighttime temperatures drop to –170°C. Not a pleasant environment!

Earth rotates much faster than the Moon (once per day rather than once per month), which evens out the day-and-night extremes of temperature. Earth's albedo, at about 31 percent, is much higher than the Moon's because Earth's blue oceans, white clouds, and ice caps are more reflective than dark lunar basalts. If our atmosphere did not contain greenhouse gases, the average surface temperature required to balance the absorbed solar radiation would be about –19°C (–2°F), cold enough to freeze all the water on the planet. Instead, Earth's average surface temperature remains a balmy 14°C (57°F). The difference of 33°C is a result of the greenhouse effect.

> *Greenhouse gases in Earth's atmosphere, primarily water vapor, carbon dioxide, methane, and ozone, trap solar energy, thereby helping to maintain Earth's equable temperatures.*

Earth's Greenhouse Atmosphere

A planet with an atmosphere containing greenhouse gases radiates solar energy back into space less efficiently than a planet with no atmosphere. Therefore, its surface temperature must rise to achieve radiation balance. In this way, the atmosphere acts like the glass in a greenhouse, allowing solar energy to pass through, but trapping heat. This trapping of heat is known as the **greenhouse effect.**

Greenhouse gases such as water vapor, carbon dioxide, methane, and ozone absorb energy—both that coming directly from the Sun and that radiated by Earth's surface—and reradiate it as infrared energy in all directions, including downward to the surface. Their net effect is to trap heat within the atmosphere, thereby increasing the temperature at the surface relative to the temperature at higher levels of the atmosphere.

How Earth's atmosphere balances incoming and outgoing radiation is illustrated in Figure 10.11. The incoming solar radiation not directly reflected is absorbed by

FIGURE 10.11 To maintain radiation balance, Earth radiates the sum of the radiation absorbed by its atmosphere and surface (342 W/m^2) back into space. Flows of warm air, moisture, and radiation transport more energy away from Earth's surface (492 W/m^2) than it receives as incoming solar radiation (168 W/m^2); the greenhouse gases reflect most of this energy back to Earth as infrared radiation (324 W/m^2). [IPCC, *Climate Change 2001: The Scientific Basis.*]

Earth's atmosphere and surface. To achieve radiation balance, Earth radiates this same amount of energy back into space as infrared energy. Because of the heat trapped by the greenhouse gases, the amount of energy transported away from Earth's surface, both by radiation and by the flow of warm air and moisture from the surface, is significantly larger than the amount Earth receives as direct solar radiation. The excess is exactly the energy received as Earthward infrared radiation from the greenhouse gases. It is this "back radiation" that causes Earth's surface to be 33°C warmer than it would be if the atmosphere contained no greenhouse gases.

Balancing the System Through Feedbacks

How does the climate system actually achieve the radiation balance illustrated in Figure 10.11? Why does the greenhouse effect yield an overall warming of 33°C and not some larger or smaller amount? The answers to these questions are not simple because they depend on interactions among the many components of the climate system. The most important of those interactions involve feedbacks.

Feedbacks come in two basic types: **positive feedbacks,** in which a change in one component is *enhanced* by the changes it induces in other components, and **negative feedbacks,** in which a change in one component is *reduced* by the changes it induces in other components. Positive feedbacks tend to amplify changes in a system, whereas negative feedbacks tend to stabilize the system against change.

Here are some of the feedbacks within the climate system that can significantly affect the temperatures achieved by radiation balance:

- *Water vapor feedback.* A rise in temperature increases the amount of water vapor that moves from Earth's surface into the atmosphere through evaporation. Water vapor is a greenhouse gas, so this increase enhances the greenhouse effect, and the temperature rises further—a positive feedback.

- *Albedo feedback.* A rise in temperature reduces the accumulation of ice and snow in the cryosphere, which decreases Earth's albedo and increases the energy its surface absorbs. This increased warming of the atmosphere enhances the temperature rise—another example of positive feedback.

- *Radiation feedback.* In "radiative damping," a rise in atmospheric temperature greatly increases the amount of infrared energy radiated back into space, which moderates the temperature rise—a negative feedback. Radiative damping stabilizes Earth's climate against major changes, keeping the oceans from freezing up or boiling off and thus maintaining an equable habitat for water-loving life.

- *Plant growth feedback.* The stimulation of plant growth by increasing atmospheric CO_2 concentrations removes CO_2 from the atmosphere by converting it to carbon-rich organic matter, thus reducing the greenhouse effect—a negative feedback.

> Changes in the climate system can be amplified by positive feedbacks; negative feedbacks can stabilize the climate system against change.

Feedbacks can involve much more complex interactions among components of the climate system. For example, an increase in atmospheric water vapor produces more clouds. Because clouds reflect solar energy, they increase the planetary albedo, which sets up a negative feedback between atmospheric water vapor and temperature. On the other hand, clouds absorb infrared radiation efficiently, so increasing the cloud cover enhances the greenhouse effect, thus providing a positive feedback between atmospheric water vapor and temperature. Does the net effect of clouds produce a positive or negative feedback?

Scientists have found it surprisingly difficult to answer such questions. The components of our climate system are joined through an amazingly complex web of interactions on a scale far beyond experimental control. Consequently, it is often impossible to gather data that isolate one type of feedback from all the others. Scientists must therefore turn to computer models to understand the inner workings of the climate system. The role of computer models in making climate predictions will be discussed in Chapter 14.

CLIMATE VARIATION

Earth's climate varies considerably from place to place—its poles are frigid and arid, its tropics sweltering and humid. Comparable variations in climate can also occur over time. The geologic record shows us that periods of global warmth have alternated with periods of glacial cold many times in the past. This climate variation is erratic; dramatic changes can happen in just a few decades or evolve over time scales of many millions of years.

Some climate variation can be attributed to factors outside the climate system, such as solar forcing and changes in the distribution of land and sea surfaces caused by continental drift. Others result from variations within the climate system itself, such as the growth of continental glaciers that

> Climate variations can be caused by factors outside the climate system, such as solar forcing, or effects within the climate system, such as changes in albedo.

increase Earth's albedo. In this section, we will examine several types of climate variation and discuss their causes, beginning with short-term variations on a regional scale.

Regional Variations in Climate

Local and regional climates show much larger fluctuations in temperature and rainfall than the global averages. Over periods of years to decades, the predominant regional variations result from interactions between atmospheric circulation and the sea and land surfaces. They generally occur in distinct geographic patterns, although their timing and amplitudes can be highly irregular.

One of the best-known examples is an abnormal warming of the eastern Pacific Ocean that occurs every 3 to 7 years and lasts for a year or so. Peruvian fishermen call such an event **El Niño** ("the boy child" in Spanish) because the warming typically reaches the surface waters off the coast of South America around Christmastime. El Niño events can decimate fish populations, which depend on the upwelling of cold water for their nutrient supply, and can thus be disastrous for coastal human populations that depend on fishing for food.

Scientists have shown that El Niño and its flip side, a cooling event called *La Niña* ("the girl child"), are part of a natural variation in the exchange of heat between the atmosphere and the tropical Pacific Ocean (**Figure 10.12**). Normally, atmospheric pressure gradients cause prevailing winds, called trade winds, to blow from east to west along the equator, pushing warm water westward, and this movement of water causes colder water to well up from the ocean depths off Peru. Sporadically, the trade winds weaken or occasionally even reverse direction, cutting off the upwelling and equalizing water temperatures across the tropical Pacific (an El Niño event). At other times, the trade winds strengthen, enhancing the temperature difference between the eastern and western Pacific

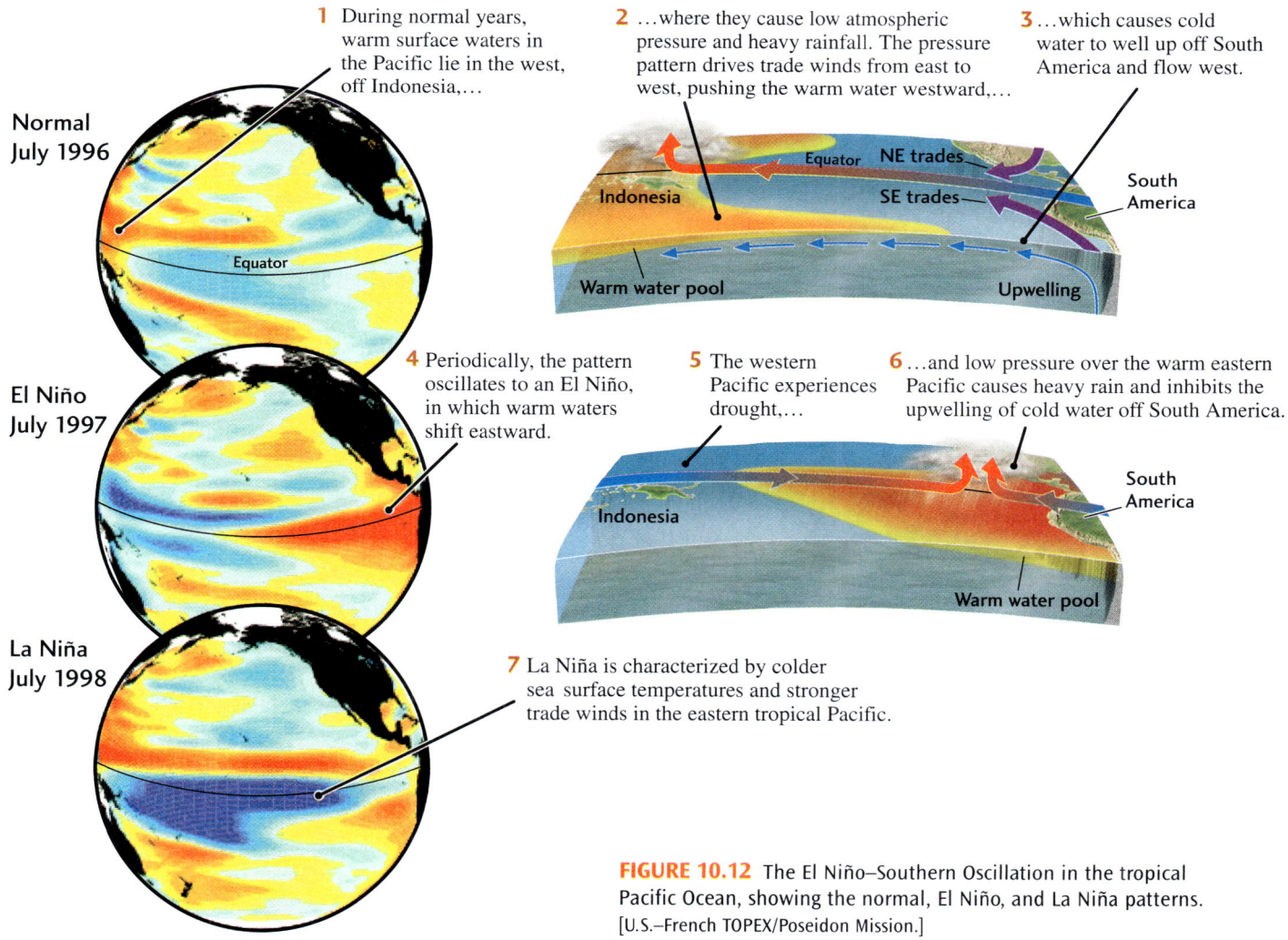

FIGURE 10.12 The El Niño–Southern Oscillation in the tropical Pacific Ocean, showing the normal, El Niño, and La Niña patterns. [U.S.–French TOPEX/Poseidon Mission.]

(a La Niña event). Climate scientists call this regional variation the El Niño–Southern Oscillation.

Besides disrupting fishing in the eastern Pacific, El Niño can trigger changes in wind and precipitation patterns in both the Northern and Southern hemispheres. The strongest El Niño on record, in 1997–1998, generated droughts in Australia and Indonesia, severe typhoons in the Pacific and storms in California, and heavy rains in Peru, Ecuador, and Kenya. According to one estimate, the global disruption in weather patterns and ecosystems may have cost 23,000 lives and caused $33 billion in damage.

Climate scientists have identified similar patterns of weather and climate variation in other regions. One example is the North Atlantic Oscillation, a highly irregular fluctuation in the atmospheric pressure between Iceland and the Azores that has a strong influence on the movement of storms across the North Atlantic and thus affects weather conditions throughout Europe and parts of Asia. A better understanding of these patterns is improving long-range weather forecasting and may provide important information about the regional effects of human-induced climate change.

Twentieth-Century Warming: Human Fingerprints of Global Change?

The most basic device for measuring climate, the thermometer, was invented in the early seventeenth century, and Daniel Fahrenheit set up the first standard temperature scale in 1724. By the mid-nineteenth century, temperatures around the world were being reported by enough meteorological stations on land and on ships at sea to allow accurate estimation of Earth's average annual surface temperature.

The results of tracking global temperatures have become quite interesting. Although there are substantial fluctuations from year to year and decade to decade, the overall trend has been upward (**Figure 10.13**). Between the end of the nineteenth century and the beginning of the twenty-first, the mean surface temperature rose by about 0.6°C, an increase called the **twentieth-century warming** (Figure 10.13a). Fossil-fuel burning and deforestation since the industrial revolution have substantially increased the concentration of CO_2 in the atmosphere during the same interval.

We know that humans are responsible for the CO_2 increase because the carbon isotopes of fossil fuels have a distinctive signature that precisely matches the changing isotopic composition of atmospheric carbon. But how certain can we be that the twentieth-century warming was a direct consequence of the CO_2 increase—that is, a result of an *enhanced greenhouse effect*—and not some other kind of change associated with natural climate variation?

> *Global temperatures fluctuate considerably from year to year, but the increase in temperatures during the twentieth century was abnormal compared with changes in the preceding millennium.*

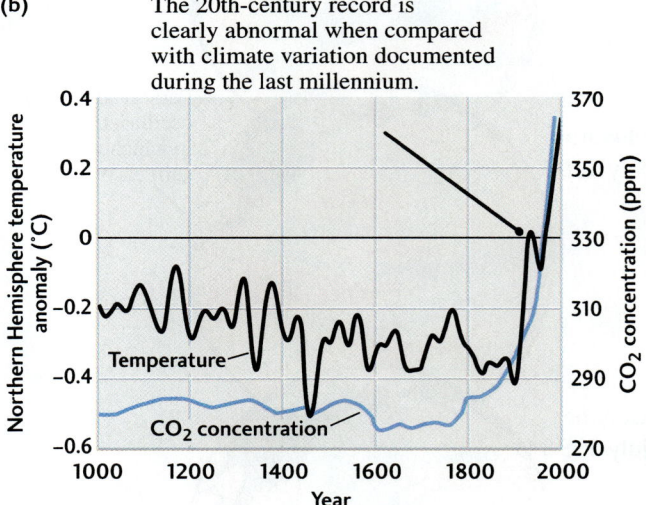

FIGURE 10.13 Earth's climate is getting warmer. A comparison of average annual surface temperature anomaly (black lines) with CO_2 concentrations in the atmosphere (blue lines) shows that the recent warming trend correlates with the increase in CO_2 caused by human activities since the industrial revolution. (a) Earth's average annual surface temperature anomaly, calculated from thermometer measurements, between 1850 and 2000. (b) Average annual surface temperature anomaly inferred for the Northern Hemisphere from tree rings, ice cores, and other climate indicators. In both of these figures, the temperature anomaly is defined as the difference between the observed temperature and the temperature average for the period 1961–1990. [IPCC, *Climate Change 2001: The Scientific Basis.*]

To answer this question, we have to look at a longer history of climate. Although direct temperature measurements are not available before the nineteenth century, climate indicators such as tree rings and ice cores have allowed scientists to reconstruct a temperature record for the Northern Hemisphere over the last millennium (Figure 10.13b). This record shows an irregular but steady global cooling of about 0.2°C in the nine centuries between 1000 and 1900. The fluctuations in mean temperature during any one of these centuries were less than a few tenths of a degree. Against this background, the twentieth-century warming appears to be abnormal.

Based on this observation and a variety of other data, almost all experts on Earth's climate are now convinced that the twentieth-century warming was in large part anthropogenic and that the warming will continue into the twenty-first century as human activities continue to raise the concentrations of greenhouse gases in the atmosphere. We will return to this issue in Chapter 14.

The twentieth-century warming was due primarily to the anthropogenic increase in atmospheric greenhouse gases.

Reconstructing Past Climates

In 1830, the geologist Charles Lyell famously stated that "the present is the key to the past." Modern geologists are fond of turning his wording around to say that the past is the key to the present, and Figure 10.13b illustrates why: knowing how the climate system has changed in the past can help us understand how it might change in the future. For this reason, *paleoclimatology* is one of the hottest fields in geology.

Some of the best data on past climates come from cores of ice recovered by drilling into the thick continental glaciers of Antarctica and Greenland (**Figure 10.14**). The cores show layers produced by annual cycles of ice formation from snow. The age of the ice can be determined by carefully counting the layers, working from the top down, just as the age of wood in a tree can be determined by counting tree rings. The composition of the atmosphere, including the concentrations of greenhouse gases such as carbon dioxide and methane (CH_4), can be measured in tiny bubbles of air trapped at the time the ice formed.

The atmospheric temperature at which each layer in the ice core formed can be estimated by measuring the proportions of two oxygen isotopes found in the ice: oxygen-16, the most common oxygen isotope, and oxygen-18, a heavier and less common form.

Ice cores taken from continental glaciers provide records of atmospheric temperature and composition over the last several hundred thousand years.

FIGURE 10.14 Russian scientists at Vostok Station carefully remove an ice core from a drill. The layers produced by annual cycles of ice formation are visible in the core. Vostok Station, at an altitude of 3500 m near the center of Antarctica (see Figure 10.6b), is an especially grueling place to do research. Its average annual temperature is only −55°C, and the lowest reliably measured temperature on Earth's surface, −89.2°C, was recorded there in 1983. [R. J. Delmas, Laboratoire de glaciologie et géophysique de l'environnement, Centre National de la Recherche Scientifique.]

Water containing the heavier isotope evaporates less readily and condenses more rapidly as the air gets colder. Therefore, the ratio of oxygen-18 to oxygen-16 in the ice is a good indicator of how cold the air was when the snow fell; the lower the proportion, the colder the air.

One of the best ice core records, stretching back more than 400,000 years, has come from drilling of the East Antarctic ice sheet at Russia's Vostok Station (**Figure 10.15**). This record shows a great deal of short-term variation in temperature, but the largest swings correspond to **glacial cycles,** which form a sawtooth pattern of peaks and troughs with an average spacing of about 100,000 years. As you can also see from Figure 10.15, the concentrations of carbon dioxide and methane are higher in ice layers formed at higher temperatures, suggesting that past climate variations were strongly influenced by the greenhouse effect.

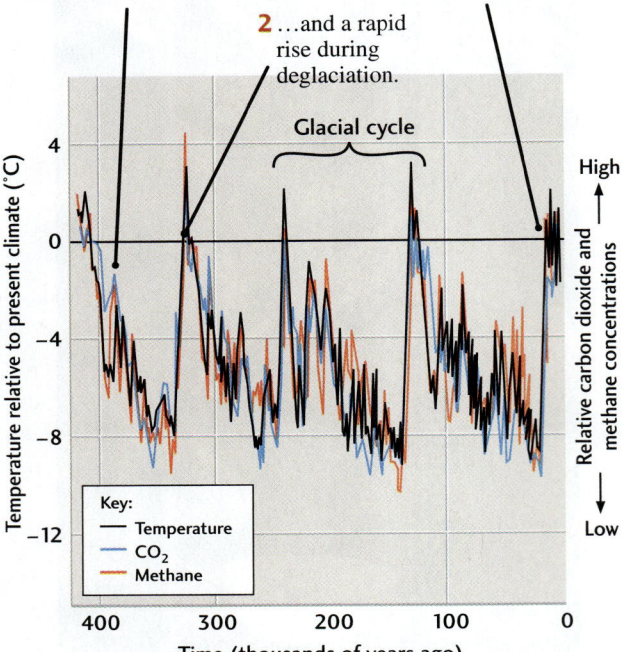

1 There is a decline in both temperature and greenhouse gas concentrations during glacial periods...

2 ...and a rapid rise during deglaciation.

3 Climate has been relatively warm and stable during the last 10,000 years—the Holocene interglacial period.

FIGURE 10.15 Three types of data were recovered from the Vostok Station ice cores, which were drilled to a depth of 3600 m in East Antarctica: temperatures and the concentrations of two important greenhouse gases, carbon dioxide and methane. The temperatures were estimated from oxygen isotopes; the gas concentrations came from measurements of air samples trapped as tiny bubbles within the Antarctic ice. [IPCC, *Climate Change 2001: The Scientific Basis.*]

Glacial Cycles

The most dramatic features observed in the climate record of the last 400,000 years are glacial cycles. In Figure 10.15, we can see that each glacial cycle is marked by a gradual decline of about 6° to 8°C from a warm **interglacial period** to a cold *glacial period,* or **ice age.** In a glacial period, Earth cools, and water is transferred from the hydrosphere to the cryosphere. The amount of sea ice increases, and more snow falls on the continents in winter than melts in summer, increasing the volume and area of the continental glaciers and decreasing the sea level. As the polar ice caps expand into lower latitudes, they reflect more solar energy back into space, and Earth's surface temperatures fall further—an example of albedo feedback. The glacial period ends with a rapid rise in temperature during a short interval of deglaciation. The ice sheets melt, and water is transferred from the cryosphere to the hydrosphere.

During the Pleistocene, Earth's climate has alternated between cold glacial periods and warmer interglacial periods.

During the last ice age, called the *Wisconsin glaciation,* which peaked about 18,000 years ago, the volume of the cryosphere exceeded 100 million cubic kilometers, more than three times the amount of ice on Earth today (**Figure 10.16**). Ice sheets with thicknesses of 2 to 3 km built up over North America, Europe, and Asia. Continental shelves that today are under water were exposed by a drop in sea level, increasing the land area. The Wisconsin glaciation is only the latest in a series of ice ages that we will study later in this chapter.

Temperatures began to drop in the last glacial cycle about 120,000 years ago, reaching their lowest values 21,000 to 18,000 years ago (the Wisconsin *glacial maximum*). Temperatures then rebounded to the warm interglacial climate of the last 11,000 years, corresponding to the Holocene epoch. Three other ice ages, which reached their minimum temperatures about 140,000, 260,000, and 350,000 years ago, are visible in the Vostok record (see Figure 10.15).

The Wisconsin glaciation peaked about 18,000 years ago, when ice sheets up to 3 km thick covered the continents of the Northern Hemisphere.

Short-Term Variations Within Glacial Cycles

The temperatures shown in Figure 10.15 do not vary smoothly over time. Superimposed on the 100,000-year glacial cycles are climate fluctuations of shorter duration, some nearly as large as the changes from glacial to inter-

FIGURE 10.16 The extent of continental glaciers and sea ice in the Northern Hemisphere at the Wisconsin glacial maximum, around 18,000 years ago. The continental shelves were exposed by the lowering of the sea level, exemplified here by the expanded coastline of Florida. [Wm. Robert Johnston.]

glacial periods. Geologists have combined the information from cores in continental and valley glaciers, lake sediments, and deep-sea sediments to reconstruct a decade-by-decade history of short-term climate variations during the last glacial cycle. Here we summarize some of the basic features of this remarkable chronicle.

- During the Wisconsin glaciation, Earth's climate was highly variable, with shorter (1000-year) oscillations between warm and cold occurring within longer (10,000-year) cycles. The most extreme variations appear to have been in the North Atlantic region, where mean local temperatures rose and fell by as much as 15°C. The massive discharges of icebergs and fresh water resulting from the sudden warmings altered thermohaline circulation in the oceans and dumped large amounts of glacial material into deep-sea sediments.

Within glacial cycles, there have been many short-term variations in climate, some marked by abrupt global warming.

- The main phase of warming after the Wisconsin glaciation occurred in the interval between 14,500 and 10,000 years ago. It was not a smooth transition, but involved two main stages, with a pause in deglaciation and a return to cold conditions between 13,000 and 11,500 years ago. The extremely abrupt increases in temperature at 14,500 and 11,500 years ago are perhaps the most astonishing aspect of this jerky transition. Broad regions of Earth experienced almost simultaneous changes from glacial to interglacial temperatures during intervals as short as 30 to 50 years. Evidently, atmospheric circulation can reorganize very rapidly, flipping the entire climate system from one state (glacial cold) to another (interglacial warmth) in less than a human lifetime! This observation raises the possibility that human-induced global climate change could involve abrupt shifts to a new (and unknown) climate state, rather than just a gradual warming.

- The warm interglacial climate of the Holocene, which began about 11,000 years ago, has been much more stable than the previous parts of the glacial cycle; in fact, the Holocene appears to be the longest and most stable warm period over the last 400,000 years. The warmest temperatures occurred at the beginning of this epoch. Geologists have documented regional variations of about 5°C on time scales of 1000 years or so, but the global changes during this period are much smaller, with a total range of only 2°C. Holocene conditions were no doubt favorable for the rapid rise of agriculture and civilization that followed the end of the Wisconsin glaciation. Scientists have speculated that the production of greenhouse gases caused by deforestation, agriculture, and other preindustrial human activities have kept those gases at high enough atmospheric concentrations to prevent the climate from plunging into another ice age.

The Holocene interglacial period has been the longest and most stable warm period during the last 400,000 years.

GLACIAL LANDSCAPES

Glaciation is the most important geologic process of the climate system. Many spectacular features of the landscapes we see today were created by recent episodes of glaciation. In mountainous regions, glaciers have gouged out steep-walled valleys, scraped their surfaces down to bedrock, and plucked huge blocks from their rocky floors. During ice ages, glaciers have pushed across entire continents, carving far more topography than rivers and wind.

The Glacial Budget

For a glacier to form, temperatures must be low enough to keep snow on the ground year-round. These conditions occur at high latitudes, because the Sun's rays strike Earth at low angles there (see Figure 12.4), and at high altitudes, because the atmosphere becomes steadily cooler up to altitudes of about 10 km (see Figure 10.2). Therefore, the height of the *snow line*—the altitude above which snow does not completely melt in summer—generally decreases toward the poles, where snow and ice cover the ground year-round even at sea level. Near the equator, glaciers form only on mountains that are higher than about 5000 m.

The precipitation of snow and the formation of glaciers require moisture as well as cold. Moisture-laden winds tend to drop most of their snow on the windward side of a high mountain range, so the leeward side is likely to be dry and unglaciated. Parts of the high Andes of South America, for instance, lie in a belt of prevailing easterly winds. Glaciers form on the moist eastern slopes, but the dry western side has little snow and ice. In arid climates, glaciers are unlikely to form unless the temperature is so frigid throughout the year that very little snow melts, as in Antarctica.

Glaciers form where temperatures are cold enough that snow covers the ground year-round. Such conditions are found at high latitudes and at high altitudes.

A fresh snowfall is a fluffy mass of loosely packed snowflakes. As these small, delicate ice crystals age on the ground, they shrink and become grains, and the mass of snowflakes compacts to form a dense, granular snow. As new snow falls and buries the older snow, that older, granular snow compacts, eventually producing solid glacial ice as the grains recrystallize and are cemented together. The amount of ice added to the glacier annually is its **accumulation.**

FIGURE 10.17 Accumulation takes place mainly by snowfall over a glacier's colder upper regions. Ablation takes place mainly by melting and iceberg calving in the warmer lower regions. The difference between accumulation and ablation is the glacial budget.

When glaciers accumulate enough ice and become sufficiently massive, they begin to flow downhill, eventually losing ice through ablation.

When ice accumulates to a mass sufficient for gravity to act on it, a glacier is born. Ice, like water, flows downhill under the pull of gravity. The ice moves down a mountain valley or downward from the dome of ice at the center of a continental glacier. In either case, the glacier flows to lower altitudes where temperatures are warmer and it loses ice, a process called **ablation.**

The two mechanisms by which glaciers lose the most ice are melting and iceberg calving (see Figure 10.5), although wind erosion and, in cold climates, transformation of ice directly into water vapor can also contribute. Most of the ablation takes place at the glacier's leading edge. Therefore, even when a glacier is advancing downward or outward from its center, the ice front may be retreating.

The difference between accumulation and ablation, called the *glacial budget,* determines the growth or shrinkage of a glacier (Figure 10.17). When accumulation equals ablation over a long period, the glacier remains a constant size, even as it continues to flow downslope.

Such a glacier accumulates snow and ice in its upper reaches as an equal amount is ablated in its lower reaches. If accumulation exceeds ablation, the glacier grows; if ablation exceeds accumulation, the glacier shrinks.

Glacial budgets vary from year to year. In the past several thousand years, many glaciers have maintained a constant average size, though some show evidence of growth or shrinkage in response to short-term regional climate variations. In the last century, however, mountain glaciers in many low-latitude regions have been shrinking in response to global warming, as described in the opening story of this chapter.

The growth or shrinkage of a glacier is determined by its glacial budget: the difference between accumulation and ablation.

Glacial Erosional Landforms

Ice is a far more efficient agent of erosion than water or wind. A valley glacier only a few hundred meters wide can tear up and crush millions of tons of bedrock in a single year. The ice carries this heavy load of sediment to the ice front, where it is dropped as the ice melts. The total amount of sediment deposited in the world's oceans per year has been several times larger during recent ice ages than during interglacial periods.

As a glacier drags rocks along its base, the rocks scratch or groove the bedrock beneath it. Such abrasions are termed *striations.* The orientation of striations shows

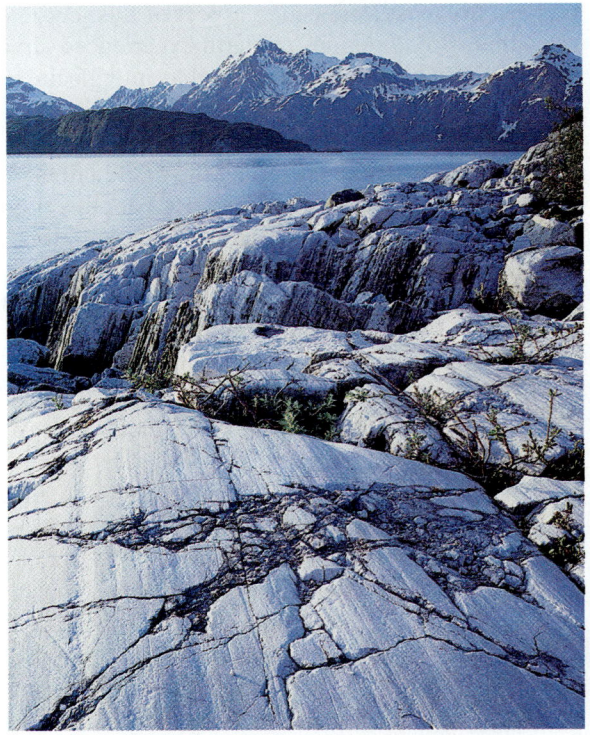

FIGURE 10.18 Glacial striations on bedrock in Glacier Bay National Park, Alaska. Striations are evidence of the direction of ice movement and are especially important clues for reconstructing the movement of continental glaciers. [Carr Clifton.]

us the direction of ice movement—an especially important factor in the study of continental glaciers, which lack obvious valleys. By mapping striations over wide areas formerly covered by a continental glacier, we can reconstruct the glacier's flow patterns (**Figure 10.18**).

A flowing valley glacier carves a series of erosional forms (**Figure 10.19**). At the head of a glacial valley, the plucking and tearing action of the ice tends to carve out an amphitheater-like hollow called a *cirque,* usually shaped like half of an inverted cone. With continued erosion,

1. Before glaciation, a mountain river cuts through a V-shaped valley.

2. During glaciation, cirques and arêtes form. The glacier moving down from its cirque creates a U-shaped valley.

3. When ice melts and retreats, the tributary valley is left as a hanging valley. If the U-shaped valley is at the coastline, it may flood with seawater to form a fjord.

A cirque is the head of a glacial valley, with nearly vertical upper walls and a flat or hollowed-out base.

Arêtes are sharp, jagged crests along an eroded divide.

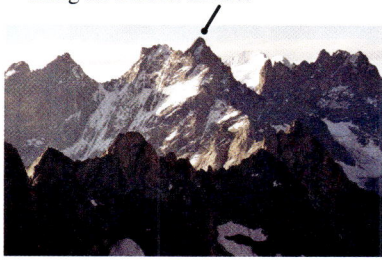

A glacial valley is U-shaped, with steep sides, hanging valleys, and waterfalls.

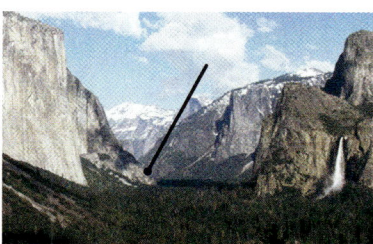

A fjord is a U-shaped valley occupied by an arm of the sea.

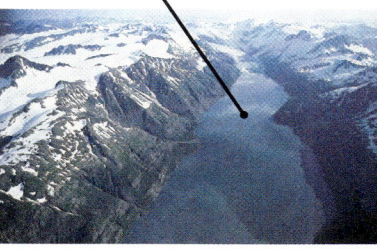

FIGURE 10.19 Erosion by valley glaciers creates distinctive landforms. [*cirque:* Marli Miller; *arêtes:* Julien Beausseron; *glacial valley:* Tom Bean/CORBIS; *fjord:* Steve McCutcheon/Visuals Unlimited.]

cirques at the heads of adjacent valleys gradually meet at the mountaintops, producing sharp, jagged crests called *arêtes* along the divide. As the glacier flows down from its cirque, it excavates a new valley or deepens an existing river valley, creating a characteristic **U-shaped valley** (see Figure 10.19). Glacial valley floors are wide and flat and have steep walls, in contrast to the V-shaped valleys of many mountain rivers (see Chapter 12).

Glaciers and rivers also differ in how their tributaries form junctions. Although the ice surface is level where a tributary glacier joins a main valley glacier, the floor of the tributary valley may be much shallower than that of the deeply carved main valley. When the ice melts, the tributary valley is left as a *hanging valley*—one whose floor lies high above the main valley floor (see Figure 10.19). After the ice is gone and streams occupy the valleys, the junction is marked by a waterfall as the stream in the hanging valley plunges over the steep cliff separating it from the main valley below.

Valley glaciers at coastlines may erode their valley floors far below sea level. When the ice retreats, these steep-walled valleys—which still maintain a U-shaped profile—are flooded with seawater (see Figure 10.19). These arms of the sea carved out by glaciers, called

> As valley glaciers flow, they pick up and transport heavy loads of sediment, carving U-shaped valleys and other characteristic erosional landforms.

FIGURE 10.20 Glacial till is deposited as an end moraine at the ice front, as lateral moraines at the rocky valley walls, and as a ground moraine beneath the ice. Meltwater streams deposit glacial outwash downstream of the ice front. The inset photo shows till deposited during the Pleistocene epoch on the eastern side of the Sierra Nevada in California. Note the wide range of particle sizes and the lack of stratification. The large boulder the person is standing on is an example of a glacial erratic. [Photo by Marli Miller.]

fjords, create the spectacular rugged scenery for which the coasts of Alaska, British Columbia, Norway, and New Zealand are renowned.

Glacial Sedimentary Landforms

Glaciers transport eroded rock materials of all kinds and sizes downstream, eventually depositing them where the ice melts. Ice is a very effective transporter of sediment because the material it picks up does not settle out like the load of sediment carried by a river. It can carry huge blocks many meters in diameter that no other transport agent could budge. Some glacial ice is so full of rock material that it is dark and looks like sediment cemented with ice.

When glacial ice melts, it deposits a poorly sorted, heterogeneous load of boulders, pebbles, sand, and clay. A wide range of particle sizes is the characteristic that differentiates glacial sediment, called *drift,* from the much better sorted material deposited by streams and winds.

Ice-Laid Deposits Some drift is deposited directly by melting ice. This unstratified and poorly sorted sediment is known as **till,** and it may contain all sizes of rock fragments from clay to boulders (**Figure 10.20**). The large boulders often contained in till are called *erratics* because of their seemingly random composition, often very different from that of local rocks.

A **moraine** is an accumulation of rocky, sandy, and clayey material carried by glacial ice or deposited as till. There are many types of moraines, each named for its position with respect to the glacier that formed it (see Figure 10.20). One of the most prominent in size and appearance is an *end moraine,* formed at the ice front. As the ice flows steadily downstream, it brings more and more sediment to its melting edge. The unsorted material accumulates there as a hilly ridge of till.

> When a glacier melts at the ice front, it deposits a poorly sorted sediment called till, forming an end moraine.

Some continental glacial terrains display prominent landforms called *drumlins*—large, streamlined hills of till and bedrock that parallel the direction of ice movement (**Figure 10.21**). Drumlins can form when the sediment-rich layer at the base of a glacier encounters a knob of bedrock or other obstacle and the excess pressure squeezes out water and drops the sediment.

Water-Laid and Windblown Deposits Other deposits of drift are laid down by meltwater from glaciers. Meltwater flowing in tunnels within and beneath the ice and in streams at the ice front may pick up, transport, and deposit some of the drift. Like any other waterborne sediment, this material is stratified and well sorted and may be cross-bedded. Drift that has been picked up and modified, sorted, and distributed by meltwater is called *outwash,* and it often forms broad sedimentary plains downstream of melting glaciers. Strong winds can blow fine-grained material from these *outwash plains* and transport it

FIGURE 10.21 Water-laid glacial deposits. *Drumlins,* from Rochester, New York. [John Shelton.] *Kettle lake,* near headwaters of the Thelon River, Canada. [Galen Rowell/CORBIS.] Pleistocene *varved clay,* from an excavation in Stockholm, Sweden. The light layers are the coarse sediments deposited in a lake during warm seasons. The dark layers are the fine clays deposited when the lake was frozen in winter. [John Shelton.] *Esker,* near Dahlen, North Dakota. [Tom Bean.]

> Outwash is well-sorted sediment deposited by glacial meltwater.

FIGURE 10.22 The Channeled Scablands are unique geological erosion features in eastern Washington State formed by cataclysmic draining of Lake Missoula, a large glacial lake. This aerial photograph shows Dry Falls, a 350-foot-high, 3-mile-wide group of scalloped cliffs created by the flood. [Dave Rahm/Easterbrook Photo and Image Center.]

over long distances; such windblown deposits of glacial sediments are called *loess*.

Deposits of outwash take a variety of forms. *Eskers* are long, narrow, winding ridges of sand and gravel deposited by meltwater flowing in tunnels along the bottom of a melting glacier (see Figure 10.21). *Kames* are small hills of sand and gravel created when drift fills a hole in a glacier and is left behind when the glacier recedes.

Glacial terrains are dotted with **kettles,** hollows or undrained depressions that often have steep sides and may be occupied by ponds or lakes. Kettles are formed when retreating glaciers leave behind isolated blocks of ice in outwash plains. By the time the block has melted completely, the ice front has retreated so far from the region that little outwash reaches the area to fill the hole. The sand and gravel that were deposited as outwash around the block of ice now surround a depression (see Figure 10.21).

The melting of isolated blocks of ice in outwash plains forms kettles.

Glacial meltwater may build up behind deposits of till to form lakes. Some lakes formed by continental glaciers were huge, many thousands of square kilometers in extent. The till dams that created these lakes were sometimes breached and carried away at a later time, causing the lakes to drain rapidly and creating huge floods. In eastern Washington State, an area called the Channeled Scablands (**Figure 10.22**) is covered by broad, dry stream channels, relics of torrential floodwaters draining from Lake Missoula, a large glacial lake, now completely emptied. From the giant ripples, sandbars, and coarse gravels found there, geologists have estimated that this flood discharged 21 million cubic meters of water per second, flowing as fast as 30 m/s. For comparison, the discharge rate of the Mississippi River in full flood is less than 50,000 m³/s.

Valley glaciers may deposit silts and clays on the bottom of a lake formed at the edge of the ice. These deposits are characterized by a series of alternating coarse and fine layers called *varves*, formed by seasonal freezing of the lake surface (see Figure 10.21).

PLEISTOCENE GLACIATIONS

Louis Agassiz, a Swiss geologist, was the first to propose (in 1837) that the glaciers he observed in the Alps must have been much larger and thicker in the geologically recent past. He suggested that during that past ice age, Switzerland was covered by an extensive continental glacier almost as thick as its mountains were tall, similar to the one in Greenland today. Among the evidence he cited was the obvious glacial sculpting of the high Alpine peaks, such as the mighty Matterhorn (**Figure 10.23**). Agassiz's hypothesis was controversial and was not immediately accepted.

Agassiz emigrated to the United States in 1846 and became a professor at Harvard University, where he continued his studies in geology and other sciences. His research took him to many places in the northern parts of Europe and North America, from the mountains of Scandinavia and New England to the rolling hills of the American Midwest. In all these diverse regions, Agassiz saw signs of glacial erosion and sedimentation. In the flat country of the American Great Plains, he observed deposits of glacial drift that reminded him of the end moraines of Swiss valley glaciers (Figure 10.24). The heterogeneous material of the drift, including erratic boulders, convinced him of its glacial origin, and the freshness of the soft sediments indicated that they were deposited in the recent past. The areas covered by this drift were so vast that the ice that deposited them must have been a continental glacier much larger than the ones that now cover Greenland and Antarctica.

Agassiz expanded his ice age hypothesis, proposing that a great continental glaciation had extended the polar ice caps far into regions that now enjoy more temperate climates. For the first time, people began to talk about ice ages.

FIGURE 10.24 Irregular hills alternate with lakes in a terrain of glacial till in Coteau des Prairies, South Dakota. This sculpting provided evidence for the great continental glaciations of the Pleistocene ice ages. [John S. Shelton.]

FIGURE 10.23 The high mountains of the Alps, such as the famous Matterhorn, shown here, were sculpted by a continental glacier nearly as thick as the peaks were tall. These obviously sculpted peaks provided compelling evidence for an ice age in the recent geologic past. [Hubert Stadler/CORBIS.]

The Wisconsin Glaciation

Geologists have determined the ages of the glacial sediments that Agassiz studied by isotopic dating, using carbon-14 in logs buried in the drift. The most recent drift was deposited during the latter part of the Pleistocene epoch (see Chapter 8). Along the east coast of the United States, the southernmost advance of this ice is recorded by the enormous sand and till deposits of the end moraines that form Long Island and Cape Cod. North American geologists named this glacial period after Wisconsin because its effects are particularly well manifested in the glacial terrains of that state. The ice advance during the Wisconsin glaciation reached its maximum 21,000 to 18,000 years ago. Figure 10.16 shows the distribution of ice across North America at the time of the Wisconsin glacial maximum.

The Wisconsin glaciation was a global event (and geologists in various parts of the world have therefore given it their own local names, calling it the Würm glaciation in the Alps, for example). Thick ice sheets moved southward across the northern parts of North America, Europe, and Asia. In the Southern Hemisphere, the Antarctic ice sheet expanded, and the southern tips of South America and Africa were covered with ice.

Glaciation and Sea Level Change

During the Wisconsin glacial maximum, the continents were slightly larger than they are today because the continental shelves surrounding them—some more than 100 km wide—were exposed by a drop in sea level of about 130 m (see Figure 10.16). This drop in sea level was due to the enormous amount of water transferred from the hydrosphere to the cryosphere. The volume of the continental ice sheets grew by about 70 million cubic kilometers, so that they contained three times the amount of ice on Earth today. Rivers extended across the newly emergent continental shelves and began to erode channels in the former seafloor. Early cultures, such as those of prehistoric Egypt, were evolving in the lands beyond the ice sheets, and humans lived in these low coastal plains.

The relationship between sea level change and glaciation illustrates the interactions of the hydrosphere and cryosphere within the climate system. As continental glaciers grow, the volume of the ocean decreases, and sea level falls. When the glaciers melt, sea level rises. If the current global warming melts parts of the remaining continental glaciers in Greenland and Antarctica, the sea level could rise significantly, a problem we will discuss in more detail in Chapter 14.

The Geologic Record of Pleistocene Glaciations

Soon after Agassiz's ice age hypothesis became widely accepted in the mid-nineteenth century, geologists discovered that there had been multiple ice ages during the Pleistocene epoch (1.8 million to 11,000 years ago), with warmer interglacial periods between them. As they mapped glacial deposits in more detail, they became aware of several distinct layers of drift, the lower ones corresponding to earlier ice ages. Between these older layers of glacial material were soils containing fossils of warm-climate plants. These fossils were evidence that the glaciers had retreated as the climate warmed. By the early part of the twentieth century, scientists were convinced that at least four major glaciations had affected North America and Europe during the Pleistocene epoch. In North America, these ice ages, from youngest to oldest, are named after American states where the evidence of glacial advance is best preserved: Wisconsin, Illinois, Kansas, and Nebraska.

There have been multiple glacial cycles during the Pleistocene epoch.

This simple scheme was modified in the late twentieth century, when geologists and oceanographers examining oceanic sediments found fossil evidence of warming and cooling of the oceans. Sediments that had accumulated continuously in undisturbed ocean basins showed a much more complex history of glacial advance and retreat than continental glacial deposits. Pleistocene and earlier ocean sediments contained fossils of foraminifers—small, single-celled marine organisms that secrete shells of calcite ($CaCO_3$). As we learned earlier in this chapter, the ratio of oxygen-18 to oxygen-16 is a sensitive indicator of climate conditions. By analyzing the proportions of these oxygen isotopes in marine sediments around the world, geologists have constructed a record of climate history that spans the entire Pleistocene epoch.

Milankovitch Cycles

The major ups and downs in the marine sediment record (**Figure 10.25**) match the glacial cycles in the ice core record (see Figure 10.15). Why does the climate fluctuate in such a pattern? Changes in the amount of solar energy received by Earth's surface are an obvious possibility. We know that it gets cold in winter because the amount of sunlight that falls at a particular latitude decreases due to the tilt of Earth's axis. Could periods of glacial cold be explained by decreases in the solar energy received by Earth's surface on much longer time scales?

The answer appears to be yes. Small periodic variations in the amount of radiation Earth receives from the

FIGURE 10.25 Changes in global climate over the last 1.8 million years, as inferred from the ratio of oxygen-18 to oxygen-16 in the calcite of foraminifers from deep-sea sediments. The peaks indicate interglacial periods, and the valleys indicate glacial periods. [After L. E. Lisiecki and M. E. Raymo, *Paleoceanography* 20 (2005): 1003.]

(a) Eccentricity (100,000 years)

(c) Precession (23,000 years)

(b) Tilt (41,000 years)

FIGURE 10.26 Three kinds of Milankovitch cycles (here much exaggerated) affect the amount of solar radiation Earth receives. (a) Eccentricity is the degree of ellipticity of Earth's orbit. (b) Tilt is the angle between Earth's axis of rotation and the angle perpendicular to the orbital plane. (c) Precession is the wobble of the axis of rotation. One can imagine this motion by thinking of the wobble of a spinning top.

Sun are caused by **Milankovitch cycles,** periodic variations in Earth's movement named after the Serbian geophysicist who first calculated them in the early twentieth century. Three kinds of Milankovitch cycles can be correlated with climate variation (**Figure 10.26**).

We know that the shape of Earth's orbit around the Sun changes periodically, becoming more circular at some times and more elliptical at others. The degree of ellipticity of Earth's orbit around the Sun is known as *eccentricity*. A nearly circular orbit has low eccentricity, and a more elliptical orbit has high eccentricity (Figure 10.26a). The amount of solar radiation Earth receives varies slightly with eccentricity. The length of one cycle of variation in eccentricity is about 100,000 years.

Milankovitch cycles are small periodic variations in the orientation of Earth's axis and in the shape of its orbit around the Sun.

We also know that the *tilt angle* of Earth's axis changes periodically. Today this angle is 23.5°, but it cycles between 21.5° and 24.5° with a period of about 41,000 years. These variations also slightly change the amount of radiation received from the Sun (Figure 10.26b).

In addition, Earth's axis of rotation wobbles like a top, giving rise to a pattern of variation called *precession* with a time period of about 23,000 years (Figure 10.26c). Precession, too, modifies the amount of radiation Earth receives from the Sun, though by less than variations in eccentricity and tilt.

You can see lots of small ups and downs in the climate record of Figure 10.25, but in the last half-million years, the record reveals a sawtooth pattern of major glacial cycles that looks roughly like this:

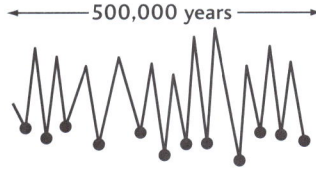

In particular, you can count five glacial maxima (represented by the black dots). This pattern, which is also visible in Figure 10.15, reveals an average time interval between glacial maxima of 100,000 years. The 100,000-year spacing of ice ages closely matches the times of high eccentricity, when Earth receives less radiation, on average, from the Sun—a Milankovitch cycle.

Now let's move backward in time to examine the first half-million years of the record shown in Figure 10.25, from 1.8 to 1.3 million years ago. Again we see many small fluctuations, but major glacial maxima and minima occur more frequently than they do in the later record, as approximated in the following sketch:

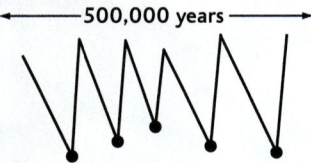

During this period of time, we find 12 glacial maxima, with an average spacing between glacial maxima of about 41,000 years (500,000 years/12 cycles = 41,667 years per cycle). This shorter interval is very close to the 41,000-year pattern of variation in the tilt angle of Earth's axis—another Milankovitch cycle! Like the change in eccentricity, the change in tilt is very small, only about 3° (see Figure 10.26b), but it's evidently enough to trigger ice ages. And, although it's not as obvious, the climate record also shows variations on the 23,000-year time scale of precession.

The periodicity of ice ages can be correlated with that of Milankovitch cycles.

By themselves, the changes in solar radiation caused by Milankovitch cycles are too small to explain the large drops in Earth's average temperature from interglacial to glacial periods (about 10°C), so some type of positive feedback must be operating within the climate system to amplify the solar forcing. The data in Figure 10.15 demonstrate that this feedback involves greenhouse gases. The atmospheric concentrations of these gases precisely track the temperature variations throughout the glacial cycles—warm interglacial periods are marked by high concentrations, cold glacial periods by low concentrations. Although these data clearly indicate the importance of the greenhouse effect in long-term climate variations, exactly how this feedback works has not yet been fully explained.

Other aspects of this story are not yet understood. You will notice in Figure 10.25 that the 41,000-year periodicity dominates the climate record up to about 1 million years ago. Then the highs and lows become more variable, eventually shifting to the 100,000-year periodicity about 700,000 years ago. What caused this transition? Climate scientists are still scratching their heads.

In fact, we don't really know what triggered the Pleistocene ice ages in the first place. The climate record shows that the 41,000-year glacial cycles were not confined to the Pleistocene but extended back at least into the Pliocene epoch (5.3 million to 1.8 million years ago), when Antarctica became covered in ice. The global cooling of Earth's climate that preceded these glaciations began during the Miocene epoch (23 million to 5.3 million years ago). Its cause continues to be debated, although most geologists believe it is related to plate movements and continental drift. According to one hypothesis, the collision of the Indian subcontinent with Eurasia and the resulting orogeny led to an increase in the weathering of silicate rocks, which consumed carbon dioxide from the atmosphere. Other hypotheses are based on changes in oceanic circulation associated with the opening of the Drake Passage between South America and Antarctica (25 million to 20 million years ago) or the closing of the Isthmus of Panama between North and South America (about 5 million years ago). Perhaps the cooling resulted from a combination of these events.

THE RECORD OF ANCIENT GLACIATIONS

The Pleistocene glacial cycles were not unique in Earth's history. Since the early part of the twentieth century, we have known from glacial striations and lithified ancient tills, called **tillites,** that glaciers covered parts of the continents several times in the distant geologic past, long before the Pleistocene. Tillites record major continental glaciations during Permian-Pennsylvanian time, Ordovician time, and at least twice during Precambrian time (**Figure 10.27**).

The Permian-Pennsylvanian glaciation covered much of southern Gondwana, leaving deposits that have been preserved as tillites across much of the Southern Hemisphere. The Ordovician glaciation was more limited in its distribution and is best preserved in northern Africa.

The oldest confirmed glaciation occurred during the Proterozoic eon, about 2.4 billion years ago. Its glacial deposits are preserved in Wyoming, along the Canadian portion of the Great Lakes, in northern Europe, and in South Africa. Some geologists argue for an even older glaciation in the Archean eon about 3 billion years ago, but this interpretation is disputed.

Glaciations in the late Proterozoic eon were more extensive than those of the Pleistocene.

The youngest Proterozoic glaciation, which spanned a period between 750 million and about 600 million years ago, involved several glacial periods separated by warm interglacial periods. Curiously, the reconstruction of paleocontinents indicated that the ice sheets extended much farther south than during the Pleistocene glaciations, perhaps all the way to the equator! This evidence has provoked some geologists to speculate that Earth may have been completely covered by ice, from pole to pole—a bold hypothesis called *Snowball Earth* (see Figure 10.27d). Glacial deposits of this age have been found on every continent (see Figure 10.27c).

According to the Snowball Earth hypothesis, there was ice everywhere—even the oceans were frozen. The

FIGURE 10.27 Ancient glaciations. (a) The first map shows the extent of the Permian-Pennsylvanian glaciation, which occurred more than 350 million years ago. At that time, the continents were assembled as the supercontinent Gondwana, and the ice sheet was situated in the Southern Hemisphere, centered over Antarctica, which is home to a continental glacier today. The second map shows the distribution of Permian-Pennsylvanian glacial deposits today. (b) Permian glacial deposits from South Africa. (c) Late Proterozoic glacial deposits. (d) The development of a late Proterozoic Snowball Earth. Geologists debate the extent to which ice covered the globe, but some think even the oceans became frozen. [Photos by John Grotzinger.]

average global temperatures would have been about –40°C, like those of the Antarctic today. Except for a few warm spots near volcanoes, no life would have survived. How could such an apocalyptic event have occurred? And how could it have ended, returning us to the climate we know today? The answers may lie in the feedbacks that occur within the climate system.

According to one scenario, as Earth initially cooled, ice sheets at the poles spread outward, their white surfaces reflecting more and more sunlight away from Earth.

This increase in Earth's albedo cooled the planet, which further expanded the ice sheets. This self-reinforcing process continued until it reached the tropics, encasing the planet in a layer of ice as much as 1 km thick. This scenario is an example of albedo feedback gone wild.

Earth remained buried in ice for millions of years, but the few volcanoes that poked above the surface slowly pumped carbon dioxide into the atmosphere. When the concentration of carbon dioxide reached a critical level, the temperature rose, the ice melted, and Earth again became a greenhouse. In this scenario, the increase in carbon dioxide represents a negative feedback in the climate system, reversing the cooling process.

The Snowball Earth hypothesis is very controversial, and some geologists disagree with the idea that the oceans may have frozen. Nevertheless, the evidence for glaciation at low latitudes is strong, and the hypothesis serves as an example of the potential of feedbacks in Earth's climate system to produce extreme change. Geologists have their work cut out for them in trying to understand the extremes of Earth's climate system.

SUMMARY >>

What is the climate system? The climate system includes all of the components of the Earth system, and all of the interactions among those components, that determine how climate behaves in space and time. The main components of the climate system are the atmosphere, hydrosphere, cryosphere, lithosphere, and biosphere. Each component plays a role in the climate system that depends on its ability to store and transport energy.

What is the greenhouse effect? For the global temperature, averaged over daily and seasonal cycles, to remain constant, Earth's surface must radiate as much energy back into space at it receives from the Sun. Much of the radiant energy that reaches Earth from the Sun is absorbed by Earth's surface and atmosphere. The warmed land surface radiates heat back into the atmosphere. Carbon dioxide and other greenhouse gases absorb some of this infrared radiation and reradiate it in all directions, including downward to Earth's surface. The net effect is an increase in the temperature at the surface relative to the temperature at higher levels in the atmosphere.

Was the twentieth-century warming caused by human activities? An increase in Earth's average annual surface temperature of about 0.6°C during the twentieth century is correlated with a significant rise in the atmospheric concentrations of CO_2 and other greenhouse gases caused by fossil-fuel burning, deforestation, and other human activities. Most experts on Earth's climate are now convinced that the twentieth-century warming was in part human-induced and that the warming will continue into the twenty-first century as atmospheric concentrations of greenhouse gases continue to rise.

How has Earth's climate changed over time? Natural variations in climate occur on a wide range of scales in both time and space. Some result from factors outside the climate system, such as solar forcing and changes in the distribution of land and sea surfaces caused by continental drift. Others result from variations within the climate system itself. The largest recent climate variations have been the 100,000-year glacial cycles, during which surface temperatures have changed by as much as 6° to 8°C. Within these glacial cycles, there have been climate fluctuations of shorter duration, some as large as the changes from glacial to interglacial periods. The Holocene has been the longest and most stable warm period of the last 400,000 years.

How do glaciers form? Glaciers form where climates are cold enough that snow, instead of melting completely in summer, is transformed into ice. As ice accumulates, the glacier thickens, until it becomes so massive that gravity starts to pull it downhill. The glacial budget is the difference between accumulation and ablation. If accumulation equals ablation, the size of the glacier remains constant. If the amount of ice ablated by melting, iceberg calving, and transformation into water vapor is greater than the amount of ice accumulated in the glacier's upper reaches, the glacier shrinks. Conversely, if accumulation exceeds ablation, the glacier grows.

How do glaciers shape the landscape? Glaciers scrape, pluck, and grind rocks into sizes ranging from boulders to fine dust. Valley glaciers erode cirques and arêtes at their heads, excavate U-shaped and hanging valleys, and create fjords by eroding their valleys below sea level at the coast. Glaciers transport huge quantities of sediment to the ice front, where melting releases them. The sediments may be deposited directly by the melting ice as till or picked up by meltwater and laid down as outwash.

What are ice ages, and what causes them? Widespread glacial drift is evidence that continental glaciers once expanded far beyond the polar regions. Studies of the geologic ages of glacial deposits on land and of sediments on the seafloor show that there have been multiple glacial cycles during the Pleistocene epoch. Each glacial period corresponded to a global lowering of sea level that exposed large areas of continental shelves. During interglacial periods, sea level rose and submerged the shelves. The favored explanation for these glacial cycles is the effect of Milankovitch cycles. These very small periodic variations in the eccentricity of Earth's orbit and the angle of its axis alter the amount of solar radiation received at Earth's surface. These changes have been amplified by positive feedbacks involving concentrations of greenhouse gases. The global cooling that preceded the glacial cycles may have resulted from continental movements that changed oceanic circulation patterns.

KEY TERMS AND CONCEPTS >>

ablation (p. 268)
accumulation (p. 267)
albedo (p. 260)
climate (p. 252)
climate system (p. 252)
continental glacier (p. 256)
El Niño (p. 263)
fjord (p. 271)
glacial cycle (p. 265)
greenhouse effect (p. 261)
ice age (p. 266)
interglacial period (p. 266)
kettle (p. 272)
Milankovitch cycle (p. 275)
moraine (p. 271)
negative feedback (p. 262)
positive feedback (p. 262)
stratosphere (p. 253)
thermohaline circulation (p. 255)
till (p. 271)
tillite (p. 276)
troposphere (p. 253)
twentieth-century warming (p. 264)
U-shaped valley (p. 270)
valley glacier (p. 256)

EXERCISES >>

1. What is a greenhouse gas, and how does it affect Earth's climate?

2. What is the role of continental glaciers in climate variation?

3. Give an example not discussed in this chapter of a positive feedback and a negative feedback in the climate system.

4. Why do scientists think the twentieth-century warming is associated with human activities?

5. What information about glacial cycles has been obtained by studying ice cores?

6. What does glacial shrinkage tell us about the balance between ablation and accumulation?

7. What type of sedimentary deposit marks the farthest advance of a glacier?

8. Why are kettles described as water-laid sedimentary deposits, rather than ice-laid deposits?

9. Why does sea level drop during ice ages?

10. What variation in Earth's orbit correlates most strongly with glacial cycles during the past 500,000 years?

11. Do Milankovitch cycles fully explain the warming and cooling of the global climate during Pleistocene glacial cycles?

THE HYDROLOGIC CYCLE AND GROUNDWATER

CHAPTER 11

Drought in the Southwest

If you live in the state of Arizona, California, Colorado, New Mexico, Nevada, or Utah, chances are you've heard a lot about reducing water use over the past several years. Some cities are imposing a variety of restrictions, such as allowing watering of lawns only on alternate days, banning planting of new grass, shutting off water to decorative fountains, installing low-volume flush toilets and no-water urinals, and increasing fees for excessive water use. In Phoenix, city departments were required to permanently reduce their water use by at least 5 percent. In 2007, Denver increased its water rates by 7 percent, while at the same time offering businesses a rebate on their water bills for improving the efficiency of processes that use water. In Las Vegas, which continues to grow at a rate that is one of the highest in the country, the Southern Nevada Water Authority is predicting that the city will be short 64 million gallons of water a day in 2010.

The Upper Colorado River Basin, which covers a large part of the southwestern United States, has been experiencing a prolonged multiyear drought since October 1999, as we saw in Chapter 1. Lake Powell, one of the West's greatest artificial reservoirs, serves as a very visible gauge for this drought. Lake Powell was full during the summer of 1999, with reservoir storage at 97 percent of its capacity. In 2000, however, when rainfall totals for the year added up to only 30 percent of the average, it became clear that the stage was set for the first year of a possible drought and receding lake levels.

By 2005, Lake Powell had dropped 145 feet from its 1999 high, and water storage was at only 33 percent of reservoir capacity. The last time Lake Powell had been this low was in 1969. Through the spring and early summer of 2005, the water surface elevation increased. By 2007, the lake level had risen by 50 feet, though it is still well below its normal levels. ■

Water levels are low near Lake Powell's Dangling Rope Marina. [David McNew/Getty Images.]

CHAPTER OUTLINE

Flows and Reservoirs   282

Hydrology and Climate   285

Groundwater   290

Human Effects on Groundwater Resources   296

Erosion by Groundwater   297

Water Quality   299

Water Deep in the Crust   301

The level of Lake Powell provides a simple gauge of the drought conditions in the Upper Colorado River Basin. Droughts—temporary decreases in precipitation—are a normal part of natural climate variation in any region, but they are defined by the local climate. In some places, such as rain forests, a drought may mean a month of no rain, whereas in other places, such as the southwestern United States, it may mean annual rainfall of less than 4 inches. A drought is also defined by its effect on the availability of water for intended human uses. In the Southwest, the effects of drought are heightened because of the ever-increasing demands on a limited water supply. The uses of Colorado River water are not limited to the supply of water to major metropolitan areas, but also include irrigation of crops and hydroelectric power generation. As the population of the Southwest increases, so does the demand for water. Eventually, even modest droughts will have major effects on the people of this region.

No one can say with certainty when the current drought in the Southwest will end. One of our greatest concerns as Earth scientists is whether this drought and others have been amplified by the effects of global warming, and whether future droughts will be more frequent and of greater magnitude. A 2007 study from the Lamont-Doherty Earth Observatory indicates that, according to computer models, the current drought in the Southwest is indeed a consequence of global warming. The study suggests that continued drought may force adjustments of the social and economic order from Colorado to California. How water will be allocated is the principal concern: will it go to cities or to agriculture?

Most of us have heard the phrase from Samuel Taylor Coleridge's "Rime of the Ancient Mariner": "Water, water everywhere, Nor any drop to drink." About 70 percent of Earth's surface is covered in water, but only a fraction of that water is available for human consumption. Sometimes it takes a prolonged drought, like the one in the southwestern United States, to remind us how important water is to us.

The United States, one of the heaviest users of water in the world, has been steadily increasing its water consumption for over 100 years. In the 30 years between 1950 and 1980 alone, water use increased over 10 times, from 34 billion gallons a day to about 370 billion gallons a day. Some, but not all, of this increase is a result of pop-

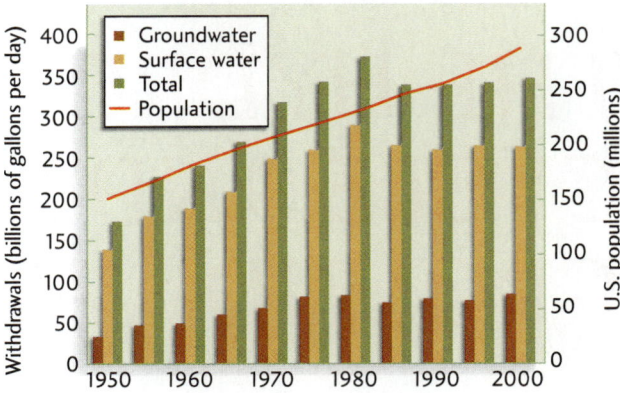

FIGURE 11.1 Estimated use of water, by source, in the United States from 1950 to 2000. [Data from U.S. Geological Survey.]

ulation growth. The good news is that because of increased awareness, water consumption per person in the United States actually fell by about 20 percent between 1980 and 1985 and has remained relatively constant since then despite continued population growth (**Figure 11.1**). Many developed countries have started to emphasize more efficient use of their finite water resources.

Water is also essential to a wide variety of geologic processes. It is important in weathering and erosion, both as a solvent of minerals in rock and soil and as a transport agent that carries away dissolved and weathered materials. Rivers and glacial ice help to shape the landscape of the continents. Water that sinks into Earth's crust forms large reservoirs of groundwater.

We saw in Chapter 10 that water exchanged between the oceans and atmosphere forms a critical link in Earth's climate system. Climate scientists now recognize that understanding the cycling of water on, over, and under Earth's surface is one of the most important steps in climate prediction.

> *Hydrology is the study of the movements and characteristics of water on and under Earth's surface.*

In this chapter, we will study the movements and characteristics of water on and under Earth's surface. This field of study, known as **hydrology,** is becoming more important to all of us as the demand on limited water supplies increases and our need to understand Earth's climate patterns is heightened. To address these concerns, we must understand not only where to find water, but also how water cycles through the Earth system.

FLOWS AND RESERVOIRS

We can see water in Earth's lakes, oceans, and polar ice caps, and we can see water moving over Earth's surface in streams and glaciers. It is harder to see the massive amounts of water stored in the atmosphere and underground, or the flows of water into and out of these storage places. As water evaporates, it moves into the atmosphere as water vapor. As rain sinks into the ground, it becomes **groundwater**—the mass of water that flows beneath Earth's surface.

Each place that stores water is referred to as a **reservoir.** Lake Powell is an example of an artificial reservoir, but Earth also has abundant natural reservoirs. The largest are the oceans, glaciers and polar ice, groundwater, lakes and rivers, the atmosphere, and the biosphere. **Figure 11.2** shows the distribution of water among these reservoirs. Although the total amount of water in rivers and lakes is very small compared with the amount in the oceans, these reservoirs are important to human populations because they contain fresh water. The amount of groundwater is more than 100 times the amount in rivers and lakes, but much of it is unusable because it contains large quantities of dissolved material.

> *Reservoirs are places where water is stored.*

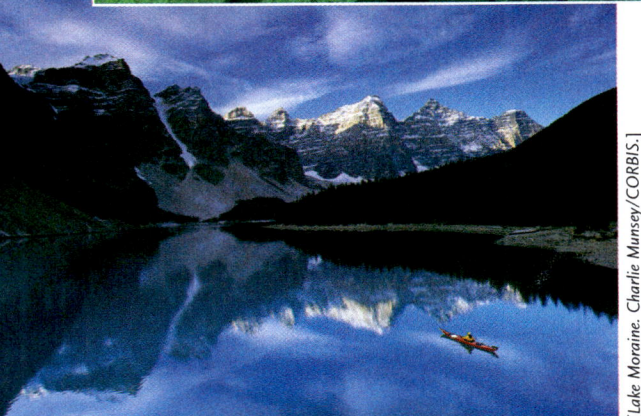

SALT WATER 95.96%

Oceans and seas (1.40×10^9 km^3)

FRESH WATER 4.04%

Glaciers and polar ice 2.97% (4.34×10^7 km^3)

Atmosphere 0.001% (1.5×10^4 km^3)

Lakes and rivers 0.009% (1.27×10^5 km^3)

Underground waters 1.05% (1.54×10^7 km^3)

Biosphere 0.0001% (2×10^3 km^3)

FIGURE 11.2 The distribution of water on Earth. [Data from J. P. Peixoto and M. Ali Kettani, "The Control of the Water Cycle." *Scientific American* (April 1973): 46; E. K. Berner and R. A. Berner, *Global Environment.* Upper Saddle River, N.J.: Prentice Hall, 1996, pp. 2–4.]

Reservoirs gain water from inflows, such as rain and streams running in, and lose water from outflows, such as evaporation and streams running out. If inflow equals outflow, the size of the reservoir stays the same, even though water is constantly entering and leaving it. These flows mean that any given quantity of water spends a certain average time, called the *residence time,* in a reservoir.

How Much Water Is There?

The world's total water supply is enormous—about 1.4 billion cubic kilometers distributed among the various reservoirs. If all of that water covered the land area of the United States, it would submerge the 50 states under a layer of water about 145 km deep. This total is constant, even though the flows from one reservoir to another may vary from day to day, year to year, and century to century. Over these geologically short time intervals, there is neither a net gain or net loss of water to or from Earth's interior nor any significant loss from the atmosphere to outer space.

The Hydrologic Cycle

All water on Earth cycles among the various reservoirs in the oceans, in the atmosphere, and on and under the land surface. The cyclical movement of water—from the ocean to the atmosphere by evaporation, to Earth's surface by precipitation, to streams through runoff and groundwater, and back to the ocean—is the **hydrologic cycle (Figure 11.3)**.

Because organisms use water, small amounts are also stored in the biosphere.

The Sun's heat drives the hydrologic cycle, mainly by evaporating water from the oceans and transporting it as water vapor in the atmosphere. Under the right temperature and humidity conditions, water vapor condenses into the tiny droplets of water that form clouds and eventually fall as rain or snow—together known as **precipitation.** Some of the precipitation that falls on land soaks into the ground by **infiltration,** a process by which water enters rock or soil through cracks or small pores between particles. Part of this groundwater evaporates through the soil surface. Another part moves through the biosphere as it is absorbed by plant roots, carried up to the leaves, and returned to the atmosphere as water vapor. Most of it flows underground, eventually returning to the surface in springs that empty into rivers and lakes.

The precipitation that does not infiltrate the ground runs off the land surface, gradually collecting into streams and rivers as **runoff.** Some runoff may later seep into the ground or evaporate from rivers and lakes, but most of it eventually flows into the oceans.

Snowfall may be converted into ice in glaciers, which return water to the oceans by melting and runoff. A much smaller amount is returned to the atmosphere by transformation from a solid (ice) directly into a gas (water vapor).

Most of the water that evaporates from the oceans returns to them as precipitation. The remainder falls over

FIGURE 11.3 The hydrologic cycle is the movement of water through Earth's crust, atmosphere, oceans, lakes, and streams. The numbers indicate the amount of water (in thousands of cubic kilometers per year) that flows among these reservoirs annually.

land and either evaporates or returns to the ocean as runoff.

Figure 11.3 shows how the total flows among reservoirs balance one another in the hydrologic cycle. The land surface, for example, gains water from precipitation and loses the same amount of water by evaporation and runoff. The ocean gains water from runoff and precipitation and loses the same amount by evaporation. As you can see from Figure 11.3, more water evaporates from the oceans than falls on them as rain. This loss is balanced by the water returned as runoff from the continents. Thus, on a global scale, the size of each reservoir stays constant. Differences in climate, however, produce local variations in the balance among evaporation, precipitation, runoff, and infiltration.

> *The hydrologic cycle, which is driven by the Sun's heat, represents the cyclical flow of water among the oceans, the atmosphere, and the land.*

How Much Water Can We Use?

Almost all the water used by human society is *fresh water*—water that is not salty. Artificial desalination of seawater (the removal of salt) is producing small but steadily growing amounts of fresh water in areas such as the arid Middle East. In the natural world, however, fresh water is supplied only by rain, rivers, lakes, groundwater, and water melted from snow or ice on land. All these waters are ultimately supplied by precipitation. Therefore, the practical limit to the amount of natural fresh water that we can ever envision using is the amount steadily supplied to the continents by precipitation.

> *Our supply of water can be replenished only by precipitation.*

HYDROLOGY AND CLIMATE

For most practical purposes, local hydrology—the amount of water there is in a region and how it flows from one reservoir to another—is more important than global hydrology. The strongest influence on local hydrology is the local climate, especially temperature and precipitation levels. In warm areas where rain falls frequently throughout the year, water supplies—both at the surface and underground—are abundant. In warm arid or semiarid regions, it rarely rains, and water is a precious resource. People who live in icy climates rely on meltwaters from snow and ice. In some parts of the world, seasons of heavy rain, called *monsoons*, alternate with long dry seasons during which water supplies shrink, the ground dries out, and vegetation shrivels.

Wherever we live, climate and the geology of the landscape strongly influence the amounts of water that move from one reservoir to another. Geologists are especially interested in how changes in precipitation and evaporation affect water supplies by altering the amounts of infiltration and runoff, which determine groundwater levels. Over the longer term, climate may affect water supplies in another way: if the sea level rises as a result of global warming, groundwater in low-lying coastal regions may become salty as seawater invades formerly fresh groundwater reservoirs.

Humidity, Rainfall, and Landscapes

Many variations in climate are related to the average temperature of the air and the average amount of water vapor it contains, both of which affect levels of precipitation. The **relative humidity** is the amount of water vapor in the air, expressed as a percentage of the total amount of water the air could hold at the same temperature if it were saturated. When the relative humidity is 50 percent and the temperature is 15°C, for example, the amount of moisture in the air is one-half the maximum amount the air could hold at 15°C.

Warm air can hold much more water vapor than cold air. When unsaturated warm air cools enough, it becomes supersaturated, and some of its water vapor condenses into water droplets. Those condensed water droplets form clouds. We can see clouds because they are made up of visible water droplets rather than invisible water vapor. When enough moisture has condensed into clouds and the droplets have grown too heavy to stay suspended by air currents, they fall as rain.

> *Relative humidity is the amount of water vapor in the air, expressed as a percentage of the amount that air could hold. Warm air can hold more water vapor than cold air.*

Most of the world's rain falls in warm, humid regions near the equator, where both the air and the surface waters of the ocean are warmed by the Sun. Under these conditions, a great deal of ocean water evaporates, resulting in high relative humidity. When water-laden winds from these tropical oceans rise and blow over nearby continents, the air cools and becomes supersaturated. The result is heavy rainfall over the land, even at great distances from the coast.

Unlike tropical climates, polar climates tend to be very dry. The polar oceans and the air above them are cold, so little ocean water evaporates, and the air can hold little moisture. Between the tropical and polar extremes are the temperate climates, where rainfall and temperatures are moderate.

Certain landscape features can alter precipitation patterns. Mountain ranges, for example, form **rain shadows**, areas of low rainfall on their leeward (downwind) slopes. Moisture-filled winds rising over high mountains cool and precipitate on the windward slopes, losing much of

FIGURE 11.4 Rain shadows are areas of low rainfall on the leeward (downwind) slopes of a mountain range.

1. Prevailing winds carry warm air over oceans, where it gathers moisture as water vapor.
2. When moist air encounters mountains, it rises, cools, and condenses, precipitating rain or snow.
3. The result is a rainy windward slope.
4. As the air mass passes over the mountains, the cool air—now depleted of moisture—sinks and warms. Its relative humidity decreases…
5. …and a dry leeward slope, or rain shadow, is formed.

their moisture by the time they reach the leeward slopes (**Figure 11.4**). The air warms again as it drops to lower elevations on the other side of the mountain range. Because the warmer air can hold more moisture, relative humidity declines, decreasing the likelihood of precipitation even more. There is a rain shadow on the eastern side of the Cascade Range of Oregon, where the prevailing winds blow inland from the Pacific Ocean. These moist winds release heavy rainfall on the mountains' western slopes. The eastern slopes, on the other side of the range, are dry and barren.

> Rain shadows are areas of low rainfall that form on the leeward side of a mountain range.

Droughts

Droughts—periods of months or years when precipitation is much lower than normal—can occur in all climates, but arid regions are especially vulnerable to their effects. Lacking replenishment from precipitation, streams may shrink and dry up, artificial reservoirs may evaporate, and the soil may dry and crack while vegetation dies. As human populations grow, demands on reservoirs increase, so a drought can deplete already inadequate water supplies.

The severest drought of the past few decades has affected lands along the southern border of the Sahara Desert (**Figure 11.5**), where tens of thousands of lives have been lost to famine. This long drought has expanded the desert and has effectively destroyed farming and grazing in the area, as we'll see in Chapter 12.

Another prolonged but less severe drought affected most of California from 1987 until February 1993, when torrential rains arrived. During the drought, groundwater and surface reservoirs dropped to their lowest levels in 15 years. Some restrictions on water use were instituted, but a move to reduce the extensive use of water supplies for irrigation encountered strong political resistance from farmers and the agricultural industry.

> Droughts are periods when precipitation is much lower than normal.

FIGURE 11.5 This millet field in Mali, Africa, shows the effects of the 1984–1985 drought. [Frans Lemmens/Alamy.]

Who Should Get Water?

As threats of drought and water shortages loom, the use of water enters the arena of public policy debate. Until recently, most people in the United States have taken their water supply for granted. In the near future, however, many areas of the country will experience water shortages more and more frequently. These shortages will create conflict among several sectors of society—residential, industrial, agricultural, and recreational—over who has the greatest rights to the water supply. Figure 11.6 shows U.S. water use by category in 2000.

In recent years, widely publicized droughts and restrictions on water use in California, Florida, Colorado, and many other places have alerted the public that the nation faces major water shortages. Public concern waxes and wanes, however, as periods of drought and abundant rainfall come and go, and governments are not pursuing long-term solutions with the urgency that they deserve. Here are some facts to ponder:

- A human can survive with about 2 liters of water per day. In the United States, the per capita use for individuals is about 250 liters per day. If industrial, agricultural, and energy production are considered, then the per capita usage rises to about 6000 liters per day.

- Industry uses about 38 percent and agriculture about 43 percent of the water withdrawn from U.S. reservoirs.

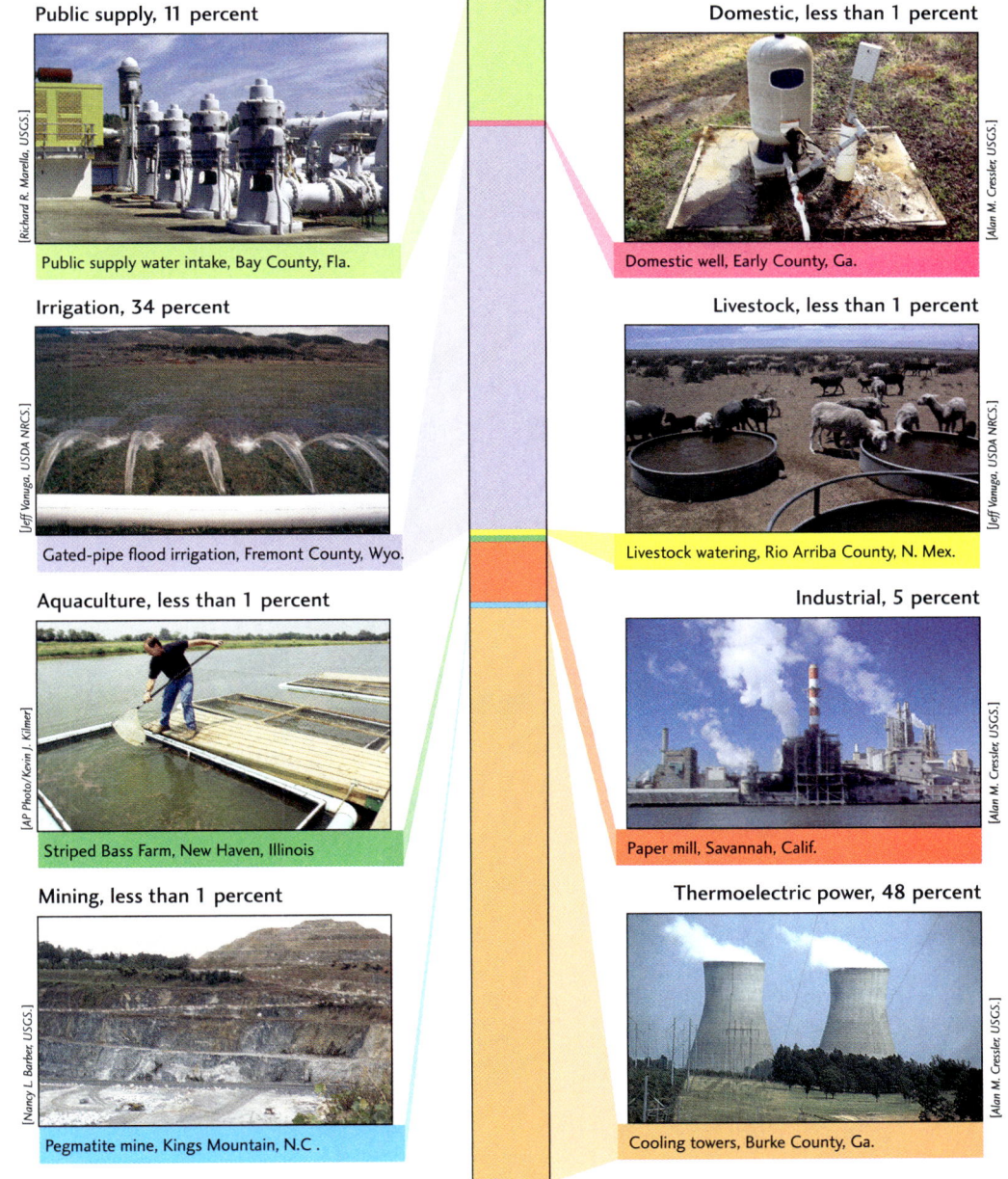

FIGURE 11.6 Uses of water by category in 2000. [Data from U.S. Geological Survey, USGS Circular 1268.]

- Although the states in the western United States receive one-fourth of the country's rainfall, their per capita water use (mostly for irrigation) is 10 times greater than that of the eastern states, and at much lower prices. In California, for example, which imports most of its water, 85 percent of the water is used for irrigation, 10 percent for municipalities and personal consumption, and 5 percent for industry. A 15 percent reduction in irrigation use would almost double the amounts of water available for use by cities and industries.

- Global climate change may lead to reduced rainfall in western states, worsening the problems there and making long-term solutions even more urgent.

Droughts and the resulting water shortages generate conflict among sectors of society over who should get water.

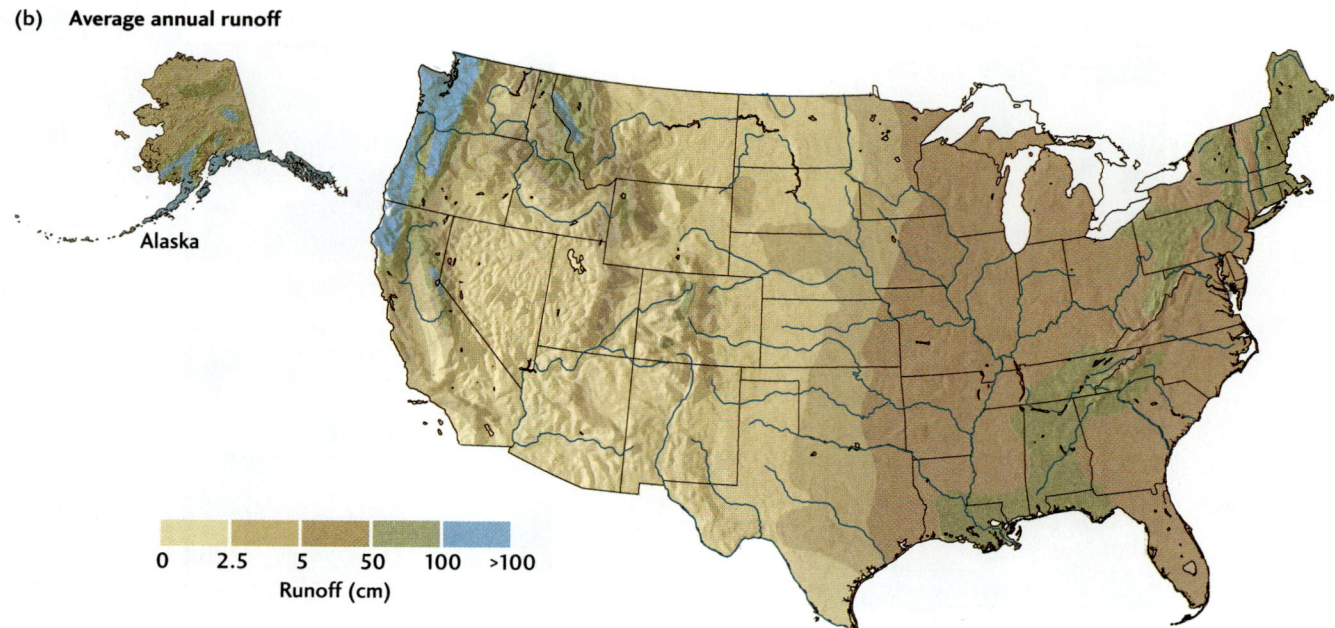

FIGURE 11.7 (a) Average annual precipitation in the United States. [Data from U.S. Department of Commerce, *Climatic Atlas of the United States*, 1968.] (b) Average annual runoff in the United States. [Data from USGS Professional Paper 1240-A, 1979.]

The Hydrology of Runoff

A dramatic example of how precipitation affects local stream and river runoff can be seen when flash flooding occurs after torrential rains. When levels of precipitation and runoff are measured over a large area (such as all the states drained by a major river) and over a long period (such as a year), the relationship between them is less direct, but still strong. The maps in **Figure 11.7** illustrate this relationship. When we compare them, we see that in areas of low precipitation—such as southern California, Arizona, and New Mexico—only a small fraction of precipitation ends up as runoff. In dry regions, much of the precipitation is lost by evaporation and infiltration. In humid areas, such as the southeastern United States, a much higher proportion of the precipitation runs off in rivers. A large river may carry large amounts of water from an area with high rainfall to an area with low rainfall. The Colorado River, for example, begins in an area of moderate rainfall in Colorado and then carries its water through arid western Arizona and southern California.

Rivers and streams carry most of the world's runoff. The millions of small and medium-sized streams carry about half the world's runoff; about 70 major rivers carry the other half. And the Amazon River of South America carries almost half of that. The Amazon carries about 10 times more water than the Mississippi, the largest river of North America (Table 11.1). These major rivers transport great volumes of water because they collect it from large networks of streams and rivers that cover very large areas. The Mississippi, for example, collects its water from a network of streams that covers about two-thirds of the United States (**Figure 11.8**).

Runoff collects and is stored in natural lakes as well as in artificial reservoirs created by the damming of rivers. Wetlands, such as swamps and marshes, also act as

TABLE 11.1 Water Flows of Some Major Rivers

River	Water Flow (m³/s)
Amazon, South America	175,000
La Plata, South America	79,300
Congo, Africa	39,600
Yangtze, Asia	21,800
Brahmaputra, Asia	19,800
Ganges, Asia	18,700
Mississippi, North America	17,500

FIGURE 11.8 The Mississippi River and its tributaries form the largest drainage network in the United States.

FIGURE 11.9 Like a natural lake or an artificial reservoir behind a dam, a wetland (such as a swamp or marsh) stores water during times of rapid runoff and slowly releases it during periods of little runoff.

reservoirs for runoff (**Figure 11.9**). If these reservoirs are large enough, they can absorb short-term inflows from major rainfall events, holding some of the water that would otherwise spill over riverbanks. During dry seasons or droughts, these reservoirs release water to streams or to water systems built for human use. These reservoirs help to control flooding by smoothing out seasonal or yearly variations in runoff and releasing steady flows of water downstream. For this reason, many governments have laws that regulate the artificial draining of wetlands for real estate development. Destruction of wetlands also threatens biological diversity because wetlands are breeding grounds for a great many types of plants and animals.

Runoff is carried in streams and rivers and collects in natural lakes, artificial reservoirs, and wetlands.

GROUNDWATER

Groundwater forms as raindrops and melting snow infiltrate soil and other unconsolidated surface materials and even sink into the cracks and crevices of bedrock. The enormous reservoir of groundwater stored beneath Earth's surface equals about 29 percent of all the fresh water stored in lakes and rivers, glaciers and polar ice, and the atmosphere. For thousands of years, people have drawn on this resource, either by digging shallow wells or by storing water that flows out onto the surface at natural springs. These springs are direct evidence of water moving below the surface (**Figure 11.10**).

Groundwater is formed as rain and melting snow infiltrate porous rocks and soils.

How Water Flows Through Soil and Rock

When water moves into and through the ground, what determines where and how fast it flows? With the exception of caves, there are no large open spaces for pools or rivers of water underground. The only spaces available for water are pores and cracks in soil and bedrock. Some pores, however small and few, are found in every kind of rock and soil, but large amounts of pore space are most often found in sandstones and limestones.

Recall from Chapter 5 that the amount of pore space in rock, soil, or sediment determines its *porosity*—the

FIGURE 11.11 Pores in rocks are normally filled partly or entirely with water. (Pores in oil- or gas-bearing sandstones and limestones are filled with oil or gas.)

Although a rock's porosity tells us how much water it can hold if all its pores are filled, it gives us no information about how rapidly water can flow through those pores. Water travels through a porous material by winding between grains and through cracks. The smaller the pore spaces and the more complex the path, the more slowly the water travels. The capacity of a solid to allow fluids to pass through it is its **permeability.** Generally, permeability increases as porosity increases.

> *The porosity of a material tells us how much water it can hold. Its permeability tells us how rapidly water can flow through it.*

Both porosity and permeability are important factors to geologists searching for groundwater supplies. In general, a good groundwater reservoir is a body of rock, sediment, or soil with both high porosity (so that it can hold large amounts of water) and high permeability (so that the water can be pumped from it easily). Well drillers in temperate climates, for example, know that they are most likely to find a good supply of water if they drill into porous sand or sandstone beds not far below the surface. A rock with high porosity but low permeability may contain a great deal of water, but because the water flows so slowly, it is hard to pump it out of the rock. Table 11.2 summarizes the porosity and permeability of various rock types.

FIGURE 11.10 Groundwater flows from a cliff in Vasey's Paradise, Marble Canyon, Grand Canyon National Park, Arizona, where hilly topography allows it to flow out onto the surface in a natural spring. [Larry Ulrich.]

percentage of its total volume that is taken up by pores. This pore space consists mainly of the spaces between grains and in cracks (**Figure 11.11**). It can vary from a small percentage of the total volume to as much as 50 percent where rock has been dissolved. In sedimentary rocks, porosity is typically 5 to 15 percent. Most metamorphic and igneous rocks have little pore space, except where fracturing has occurred.

TABLE 11.2 Porosity and Permeability of Various Rock Types

Rock Type	Porosity (Pore Space That May Hold Fluid)	Permeability (Ability to Allow Fluids to Pass Through)
Gravel	Very high	Very high
Coarse- to medium-grained sand	High	High
Fine-grained sand and silt	Moderate	Moderate to low
Sandstone, moderately cemented	Moderate to low	Low
Fractured shale or metamorphic rocks	Low	Very low
Unfractured shale	Very low	Very low

FIGURE 11.12 The groundwater table is the boundary between the unsaturated zone and the saturated zone. The saturated and unsaturated zones can be in unconsolidated material or in bedrock.

The Groundwater Table

As well drillers bore deeper into soil or rock, the samples they bring up become wetter. At shallow depths, the material is unsaturated—the pores contain some air and are not completely filled with water. This level is called the **unsaturated zone** (often termed the *vadose zone*). Below it is the **saturated zone** (often termed the *phreatic zone*), in which the pores are completely filled with water. The saturated and unsaturated zones can be in unconsolidated material or in bedrock. The boundary between the two zones is the **groundwater table,** usually shortened to *water table* (**Figure 11.12**). When a hole is drilled below the water table, water from the saturated zone flows into the hole and fills it to the level of the water table.

Groundwater moves under the force of gravity, so some of the water in the unsaturated zone may be on its way down to the water table. A fraction of that water, however, remains in the unsaturated zone, held in small pore spaces by surface tension—the attraction between the water molecules and the surfaces of the grains. Surface tension is what keeps the sand on a beach moist, even though gravity acts to move the water to spaces below.

If we were to drill wells at several sites and measure the elevations of the water levels in those wells, we could construct a map of the water table. A cross section of the landscape might look like the one shown in **Figure 11.13**. The water table follows the general shape of the surface topog-

> *The groundwater table is the boundary between the saturated zone, where water fills all available pore space, and the unsaturated zone, where pores are partly filled with air.*

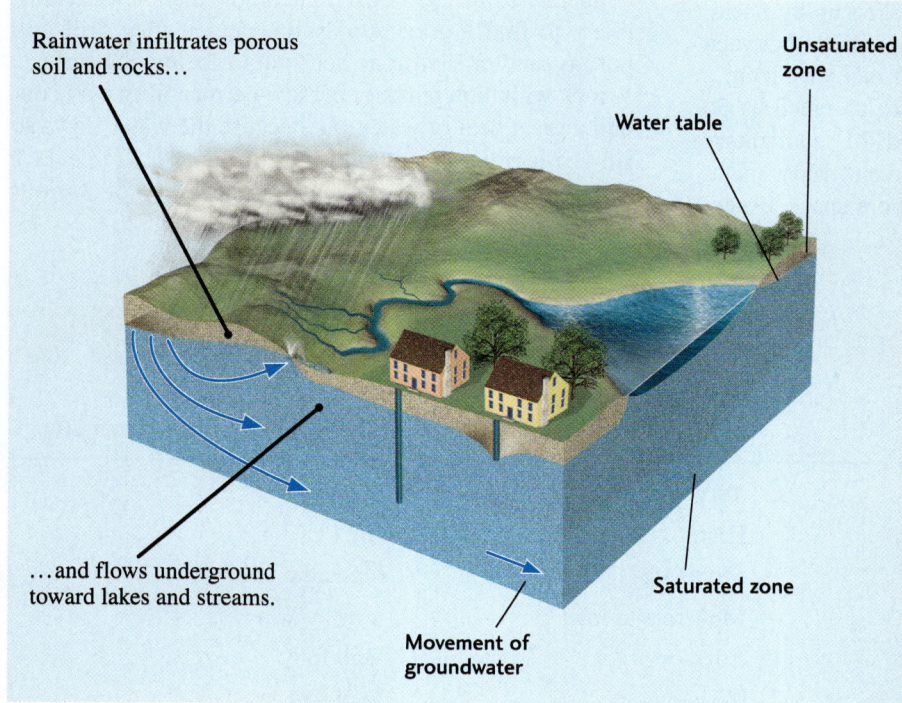

FIGURE 11.13 A cross section of a landscape illustrating the dynamics of a water table. Wells of different depths may tap different parts of the aquifer, where groundwater may be absent at shallower levels.

Groundwater flows downhill from areas where the water table elevation is high to areas where it is lower.

raphy, but its slopes are gentler. The water table is at the surface in riverbeds and lake beds and at springs. Under the influence of gravity, groundwater moves downhill from places where the water table elevation is high—under a hill, for example—to places where the water table elevation is low—such as a spring where it flows out onto the surface.

Water enters and leaves the saturated zone through recharge and discharge. **Recharge** is the infiltration of water into any subsurface formation. Rain and melting snow are the most common sources of recharge, but recharge may also take place through the bottom of a stream if the stream channel lies at an elevation above the water table. **Discharge,** the movement of groundwater to the surface, is the opposite of recharge. When a stream channel lies at an elevation below the water table, water discharges from the groundwater into the stream. Groundwater is also discharged through springs and by pumping from artificial wells.

Aquifers

Rock formations through which groundwater flows in sufficient quantity to supply wells are called **aquifers.** Groundwater may flow in unconfined or confined aquifers. In *unconfined aquifers,* the water travels through formations of more or less uniform permeability that extend to the surface in both discharge and recharge areas. The level of the reservoir in an unconfined aquifer is the same as the height of the water table (as in Figure 11.13).

Many permeable formations, however—typically sandstones—are bounded above and below by low-permeability beds, such as shales. These relatively impermeable formations are called **aquicludes.** Groundwater either cannot flow through them or flows through them very slowly. When aquicludes lie both over and under an aquifer, they form a *confined aquifer.*

The impermeable beds above a confined aquifer prevent rainwater from infiltrating the aquifer directly. Instead, a confined aquifer is recharged by precipitation over a *recharge area,* often a topographically higher upland characterized by outcropping rocks. Here there is no aquiclude preventing infiltration, so the rainwater travels down to and through the aquifer underground (**Figure 11.14**).

Water moving through a confined aquifer—known as an **artesian flow**—is under pressure. At any point in the aquifer, that pressure is equivalent to

Unconfined aquifers have relatively uniform permeability that extends to the surface. Confined aquifers are bounded by impermeable formations known as aquicludes.

FIGURE 11.14 A permeable formation situated between two aquicludes forms a confined aquifer, through which water flows under pressure.

FIGURE 11.15 Artesian wells are placed where the ground surface lies below the water table, so that water flows to the surface under its own pressure. [John Dominis/Time Life Pictures/Getty Images.]

the weight of all the water in the aquifer above that point. If we drill a well into a confined aquifer at a point where the elevation of the ground surface is lower than that of the water table in the recharge area, the water will flow out of the well spontaneously (Figure 11.15). Such wells are called *artesian wells,* and they are extremely desirable because no energy is required to pump the water to the surface. The water is brought up by its own pressure.

Artesian flows occur in confined aquifers and may be under sufficient pressure to drive water in wells to the surface without pumping.

In more complex geological environments, water tables may be more complicated. For example, if a relatively impermeable mudstone layer creates an aquiclude in an otherwise permeable sandstone formation, the aquiclude may form a shallow aquifer that lies above the main water table of a deeper aquifer (Figure 11.16). The water table in the shallow aquifer is called a *perched water table.* Many perched water tables are small, only a few meters thick and restricted in area, but some extend for hundreds of square kilometers.

Balancing Recharge and Discharge

When recharge and discharge are balanced, the groundwater reservoir in an aquifer and the elevation of the water table remain constant, even though water is continually flowing through the aquifer. For recharge to balance dis-

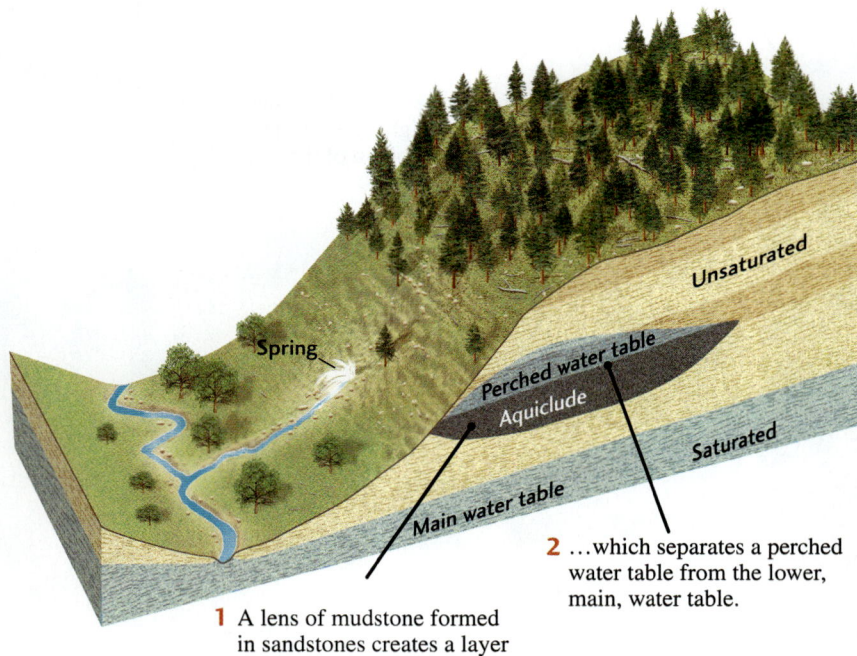

1 A lens of mudstone formed in sandstones creates a layer of very low permeability. This forms a local aquiclude,…

2 …which separates a perched water table from the lower, main, water table.

FIGURE 11.16 A perched water table forms in geologically complex situations—in this case, where a mudstone aquiclude is located above the main water table in a sandstone aquifer. The dynamics of the perched water table's recharge and discharge may be different from those of the main water table.

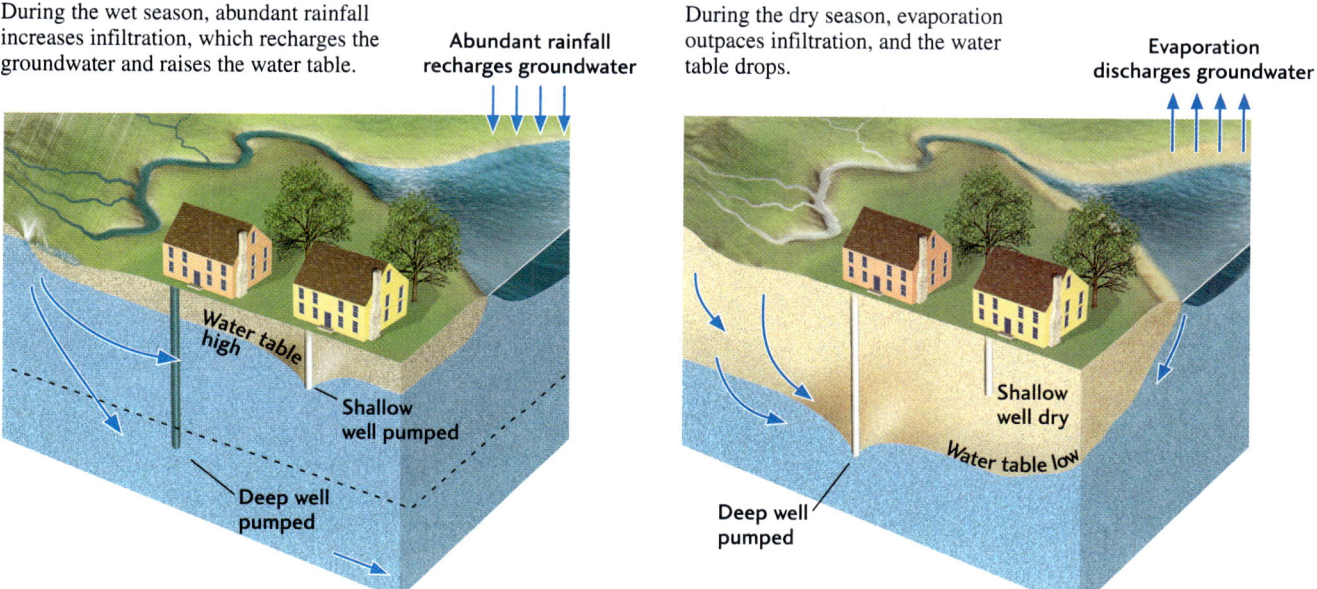

FIGURE 11.17 The water table fluctuates seasonally in response to the balance between recharge and discharge.

charge, rainfall must be frequent enough to compensate for runoff in streams and the outflow from springs and wells.

But recharge and discharge are usually not equal because recharge varies with rainfall from season to season. Typically, the water table drops in dry seasons and rises in wet seasons. A longer decrease in recharge, such as during a prolonged drought, will be followed by a longer-term imbalance and a greater lowering of the water table (Figure 11.17).

An increase in discharge, usually due to increased well pumping, can also produce a long-term imbalance and a lowering of the water table. Shallow wells may end up in the unsaturated zone and go dry. When a well pumps water out of an aquifer faster than recharge can replenish it, the water table in the aquifer is lowered in a cone-shaped area around the well, called a *cone of depression* (Figure 11.18). The water level in the well is lowered to the depressed level of the water table. If the cone of depression extends below the bottom of the well, that well goes dry. If the bottom of the well is above the base of the aquifer, extending the well deeper into the aquifer may allow more water to be withdrawn, even at continued high pumping rates. If the rate of pumping is maintained and the well is deepened so much that the entire aquifer is tapped, however, the cone of depression can reach the bottom of the aquifer and deplete it. The aquifer will recover only if the pumping rate is reduced enough to give it time to recharge.

> *Groundwater supplies remain constant only when recharge is equal to discharge.*

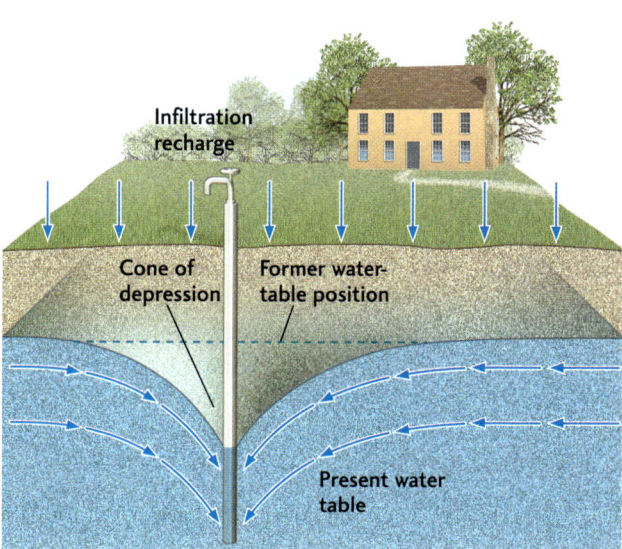

FIGURE 11.18 When discharge from a well exceeds recharge, the well draws down the water table into a cone-shaped depression. The water level in the well is lowered to the depressed level of the water table.

The Speed of Groundwater Flows

The speed at which water moves underground strongly affects the balance between discharge and recharge. Most

FIGURE 11.19 The rate of groundwater flow depends on several factors, including the slope of the water table and the distance between the recharge zone and the well.

The speed at which groundwater flows through an aquifer depends on the slope of the water table, the distance between the recharge area and the well where groundwater is being withdrawn, and the permeability of the aquifer.

groundwaters move slowly, a fact of nature that is responsible for our groundwater supplies. If groundwater moved as rapidly as streams, aquifers would run dry after a period without rain, just as many small streams do. The slow flow of groundwater also makes rapid recharge impossible if groundwater levels are lowered by excessive pumping.

Although all groundwaters flow through aquifers slowly, some flow more slowly than others. The rate of flow depends on several factors (**Figure 11.19**): it increases as the slope of the water table increases; it increases as the distance decreases between where the groundwater is recharged and the well where it is withdrawn; and it increases as the permeability of the aquifer increases.

HUMAN EFFECTS ON GROUNDWATER RESOURCES

Large parts of North America rely solely on groundwater for all their water needs. The demand for groundwater resources has grown as populations have increased and uses such as irrigation have expanded (see Figure 11.1). Many areas of the Great Plains and other parts of the Midwest rest on sandstone formations, most of which are confined aquifers that function like the one shown in Figure 11.14. Thousands of wells have been drilled into these aquifers, most of which transport water over hundreds of kilometers and thus constitute a major water resource. The aquifers are recharged from outcrops in the western high plains, some very close to the foothills of the Rocky Mountains. From there, the water runs downhill in an easterly direction. Extensive pumping withdraws water from such aquifers faster than the slow flow from distant recharge areas can fill them, so the reservoirs of groundwater they contain are being depleted. The depletion of the Ogallala aquifer, described in Chapter 1, is an example of the problems this region is facing.

Aquifers are depleted when the rate of pumping exceeds the rate of recharge.

A variety of innovative approaches are being used to enhance the sustainability of groundwater resources. Efforts to reduce excessive discharge have been supplemented by attempts to increase the recharge of aquifers artificially in some areas. On Long Island, New York, for example, the water authority drilled a large system of recharge wells to pump treated wastewater into the ground. The water authority also constructed large, shallow basins over natural recharge areas to increase infiltration by catching and diverting runoff, including stormwater and industrial waste drainage. The officials in charge of the program knew that urban development can decrease recharge by interfering with infiltration. As urbanization progresses, the impermeable materials used to pave large areas for streets, sidewalks, and parking lots increase runoff and prevent water from infiltrating the ground. Such decreases in natural infiltration may deprive aquifers of much of their recharge. One remedy is to catch and use stormwater runoff in a systematic program of artificial recharge, as the Long Island water authority did. The multiple efforts of the water authority helped rebuild the Long Island aquifer, though not to its original levels.

Orange County, near Los Angeles, California, receives only about 15 inches of rainfall per year, yet this water must supply a population of 2.5 million people. Groundwater pumped from beneath the western part of the county meets about 75 percent of its requirements. The water table is gradually dropping, however, threatening to diminish this supply. To help replenish the supply, the Orange County Water District operates 23 wells that inject treated wastewater mixed with groundwater from a second aquifer, located beneath the county's main aquifer. The recycled water meets drinking water standards with additional treatment, but most of the contaminants are filtered out by the aquifer's pore network.

EROSION BY GROUNDWATER

Every year, thousands of people visit caves, either on tours of popular attractions such as Mammoth Cave, Kentucky, or in adventurous explorations of little-known caves. Caves are produced by the dissolution of limestone—or, rarely, of other soluble rocks such as evaporites—by groundwater. Huge amounts of limestone have been dissolved to make some caves. Mammoth Cave, for example, has tens of kilometers of large and small interconnected chambers. The Big Room at Carlsbad Caverns, New Mexico, is more than 1200 m long, 200 m wide, and 100 m high (Figure 11.20).

Limestone is widespread in the upper parts of Earth's crust, but caves form only where these relatively soluble rocks are at or near the surface and enough carbon dioxide–rich or sulfur dioxide–rich water infiltrates the surface to dissolve extensive areas of limestone. The atmospheric carbon dioxide contained in rainwater creates a weak acid that enhances the dissolution of limestone. Water that infiltrates soil may pick up even more carbon dioxide from plant roots, microbes, and other soil-dwelling organisms that give off this gas. As this carbon dioxide–rich water moves down to the water table, through the unsaturated zone to the saturated zone, it creates openings as it dissolves carbonate minerals. These openings are enlarged as the limestone dissolves along joints and fractures, forming a network of rooms and passages. Such networks form extensively in the saturated zone, where—because the caves are filled with water—dissolution takes place over all surfaces, including floors, walls, and ceilings.

> *Caves are produced when acidic groundwater dissolves limestone or other highly soluble rocks.*

We can explore caves that were once below the water table but are now in the unsaturated zone because the water table dropped. In these caves, now air-filled, water saturated with calcium carbonate may seep through the ceiling. As each drop of water drips from the cave's ceiling, some of its dissolved carbon dioxide evaporates, escaping to the cave's atmosphere. Its evaporation makes the calcium carbonate in the groundwater solution less soluble, so each water droplet precipitates a small amount of calcium carbonate on the ceiling. These deposits accumulate, just as an icicle grows, in a long, narrow spike of carbonate called a *stalactite* suspended from the ceiling. When the drop of water falls to the cave floor, more carbon dioxide escapes, and another small amount of calcium carbonate is precipitated on the floor below the stalactite. These deposits also accumulate, forming a *stalagmite*. Eventually, a stalactite and a stalagmite may grow together to form a column (see Figure 11.20).

In some places, dissolution may thin the roof of a limestone cave so much that it collapses suddenly, producing a **sinkhole**—a small, steep depression in the land surface above the cave (Figure 11.21). Sinkholes are

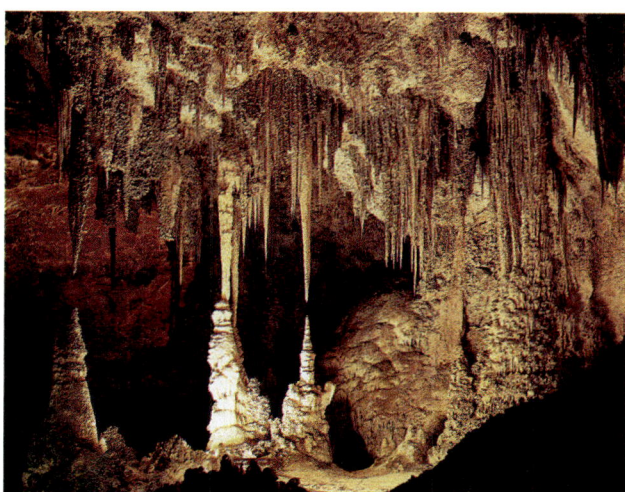

FIGURE 11.20 The Chinese Theater, in the Big Room of Carlsbad Caverns, New Mexico. Stalactites from the ceiling and stalagmites from the floor have joined to form a column. [David Muench.]

FIGURE 11.21 A large sinkhole formed by the collapse of a shallow underground cavern in Winter Park, Florida. Such collapses can occur so suddenly that moving cars are buried. [Leif Skoogfors/Woodfin Camp.]

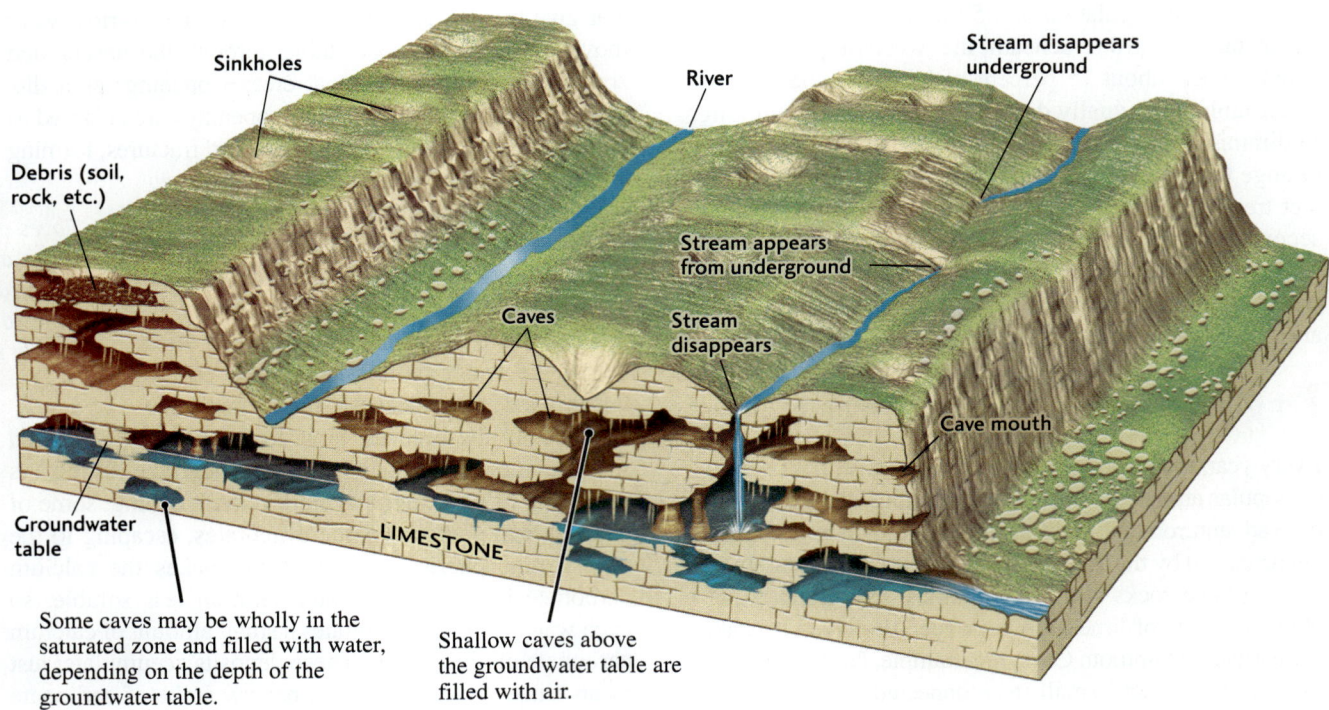

FIGURE 11.22 Some major features of karst topography are caves, sinkholes, and disappearing streams.

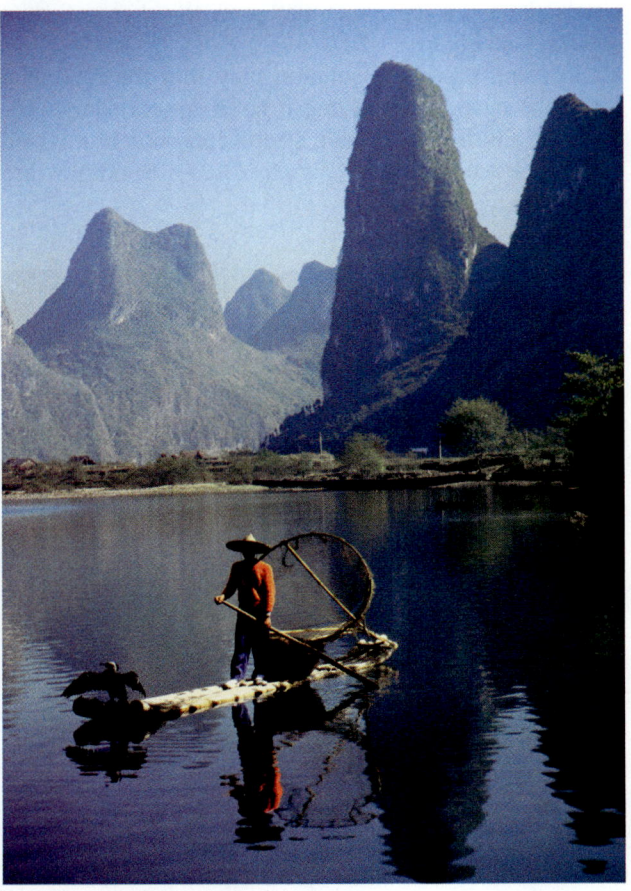

characteristic of a distinctive type of topography known as *karst*, named for a region in the northern part of Slovenia. **Karst topography** is an irregular, hilly type of terrain characterized by sinkholes, caves, and a lack of surface streams (Figure 11.22). Underground drainage channels replace the normal surface drainage system of small and large streams. The short, scarce streams often end in sinkholes, detouring underground and sometimes reappearing miles away.

Karst terrains often have environmental problems, including the potential for catastrophic cave-ins and surface subsidence. Karst topography is found in limestone terrains of Indiana, Kentucky, and Florida and on the Yucatán Peninsula of Mexico. It is also well developed on uplifted coral limestone terrains formed from tropical volcanic island arcs in the late Cenozoic era. The spectacular tower karst of China forms when cave networks collapse to form sinkholes, which then expand and merge to leave behind "towers" (Figure 11.23).

> *Karst topography is an irregular, hilly type of terrain where dissolution of rocks by groundwater has occurred.*

FIGURE 11.23 The tower karst of southeastern China is a spectacular terrain that features isolated hills with nearly vertical slopes. [Dennis Cox/Alamy.]

WATER QUALITY

Unlike people in many other parts of the world, North Americans are fortunate in that almost all of their public water supplies are free of bacterial contamination, and the vast majority are free enough of chemicals to drink safely. Yet as more rivers become polluted and more aquifers are contaminated by toxic wastes from surface dumps, North Americans are likely to see changes in water quality.

Sources of Contamination

The quality of groundwater is often threatened by a variety of contaminants. Generally, these contaminants are chemicals, though microorganisms can also have negative effects on human health under certain conditions.

Chemical Contaminants A number of human activities can contaminate groundwater (Figure 11.24). The disposal of chlorinated solvents—such as trichloroethylene (TCE), widely used as a cleaner in industrial processes—poses a formidable problem. These solvents persist in the environment because they are difficult to remove from contaminated waters. The burning of coal and the incineration of municipal and medical waste also emit mercury into the atmosphere, which then contaminates water supplies. Buried gasoline storage tanks can leak, and road salt inevitably drains into the soil and ultimately into aquifers. Rain can wash agricultural pesticides, herbicides, and fertilizers into the soil, from which they percolate downward into aquifers. In some agricultural areas where nitrate fertilizers are heavily used, groundwaters contain high concentrations of nitrate. In one recent study, 21 percent of the shallow wells sampled exceeded the maximum amounts of nitrate (10 ppm) allowed in drinking water in the United States. Such high nitrate levels pose a danger of "blue baby" syndrome (an inability to maintain healthy oxygen levels) to infants 6 months old and younger.

Radioactive Wastes There is no easy solution to the problem of groundwater contamination by radioactive waste. When radioactive wastes are buried underground, they may be leached by groundwater and find their way into aquifers. Storage tanks and burial sites at the atomic weaponry plants in Oak Ridge, Tennessee, and Hanford, Washington, have already leaked radioactive wastes into shallow groundwaters.

Microorganisms Microorganisms live in large numbers in groundwater at great depths (as much as several thousand meters), where they constitute a huge biomass. These microbes draw their energy from reactions involving chemicals and minerals in rocks. These reactions, aside from serving as a source of energy for the microbes, continue the weathering process underground. The chemicals released by the microbes during these reactions can cause deterioration of water quality.

> Groundwater can be contaminated by chemicals such as chlorinated solvents, gasoline, road salt, and fertilizers. Radioactive wastes and microorganisms may pose high local risks.

The most widespread causes of groundwater contamination by microbes, however, are leaky residential septic tanks and cesspools. These containers, widely used in neighborhoods that lack full sewer networks, are settling tanks buried at shallow depths in which bacteria decompose the solid wastes from household sewage. To prevent contamination of drinking water, cesspools should be replaced by septic tanks, which must be installed at sufficient distance from water wells in shallow aquifers.

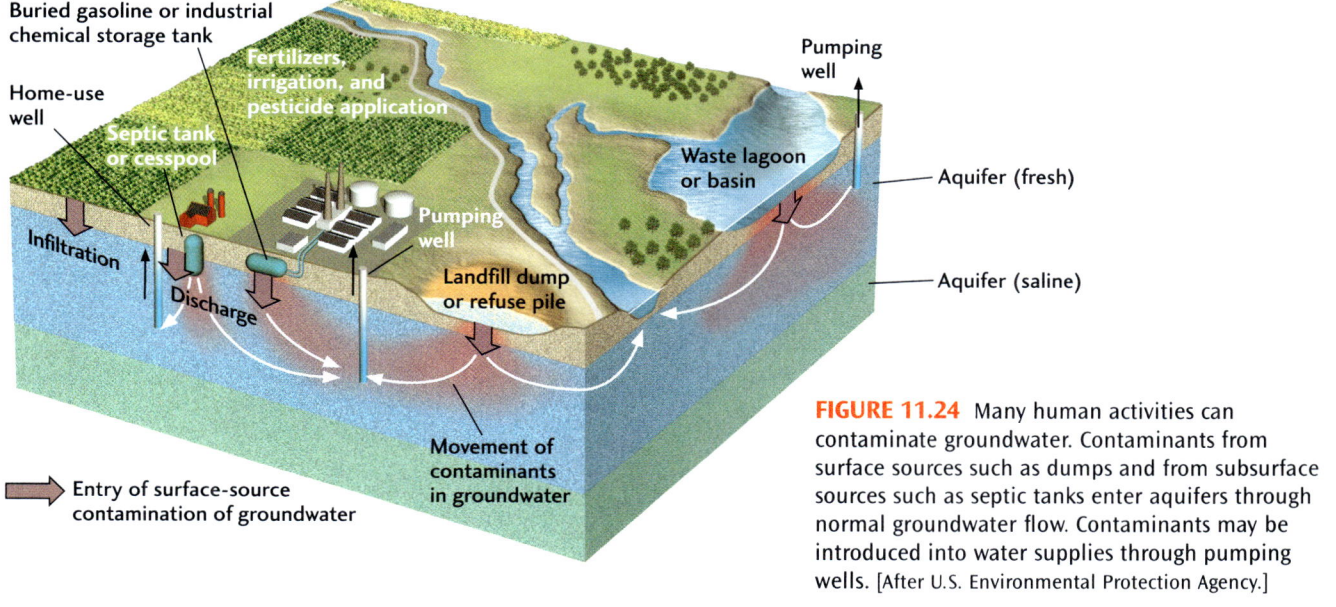

FIGURE 11.24 Many human activities can contaminate groundwater. Contaminants from surface sources such as dumps and from subsurface sources such as septic tanks enter aquifers through normal groundwater flow. Contaminants may be introduced into water supplies through pumping wells. [After U.S. Environmental Protection Agency.]

Reversing Contamination

Can we reverse the contamination of groundwater supplies? Yes, but the process is costly and very slow. The faster an aquifer recharges, the easier it is to decontaminate. If the recharge rate is rapid, fresh water moves into the aquifer as soon as we close off the sources of contamination, and in a relatively short time, the water quality is restored. Even a fast recovery, however, can take a few years.

The contamination of slowly recharging groundwater reservoirs is more difficult to reverse. The rate of groundwater movement may be so slow that contamination from a distant source takes a long time to appear. By the time it does, it is too late for rapid remediation. Even after the recharge area has been cleaned up, some contaminated deep reservoirs extending hundreds of kilometers from the recharge area may not be free of contaminants for many decades.

When public water supplies are polluted, we can pump the water and then treat it chemically to make it safe, but this is an expensive procedure. Alternatively, we can try to treat the water while it remains underground. In one moderately successful experimental procedure, contaminated water was funneled into a buried bunker full of iron filings that detoxified the water by reacting with the contaminants. The reactions produced new, nontoxic compounds that attached themselves to the iron filings.

Is the Water Drinkable?

Water that tastes agreeable and is not dangerous to human health is called **potable** water. The amounts of dissolved substances in potable waters are very small, usually measured by weight in parts per million (ppm). Potable groundwaters of good quality typically contain about 150 ppm total dissolved materials, because even the purest natural waters contain some dissolved substances derived from weathering. Only distilled water contains less than 1 ppm dissolved materials.

The many cases of groundwater contamination have led to the establishment of water quality standards based on medical studies. These studies have concentrated on the effects of ingesting average amounts of water containing various quantities of contaminants.

Once contaminated, groundwater is extremely difficult and expensive to decontaminate. Because groundwater flows very slowly, remediation processes take a long time.

FIGURE 11.25 Porosity and permeability, and therefore water content, generally decrease with increasing depth in Earth's crust.

Groundwater is almost always free of solid particles when it seeps into a well from a sand or sandstone aquifer. The complex passageways of the pore networks in the rock or sand act as a fine filter, removing small particles of clay and other solids and even straining out microbes and some large viruses. Limestone aquifers may have larger pores and so may filter water less efficiently.

Some groundwaters, although perfectly safe to drink, simply taste bad. Some have a disagreeable taste of "iron" or are slightly sour. Groundwaters passing through limestone dissolve carbonate minerals and carry away calcium, magnesium, and bicarbonate ions, making the water "hard." Hard water may taste fine, but it does not lather readily when used with soap. Water passing through waterlogged forests or swampy soils may contain dissolved organic compounds and hydrogen sulfide, which give the water a disagreeable smell similar to rotten eggs.

How do these differences in taste and quality arise in safe drinking waters? Some of the highest-quality, best-tasting public water supplies come from lakes and artificial surface reservoirs, many of which are simply collecting places for rainwater. Some groundwaters taste just as good; these tend to be waters that pass through rocks that weather only slightly. Sandstones made up largely of quartz, for example, contribute little in dissolved substances, and thus waters passing through them have a pleasant taste.

As we have seen, the contamination of groundwater in relatively shallow aquifers is a serious problem, and remediation is difficult. But are there deeper groundwaters that we can use?

Groundwater usually lacks solid particles, which are filtered out by pore networks, but it may contain dissolved substances that impart a disagreeable odor or taste.

WATER DEEP IN THE CRUST

Most crustal rocks below the groundwater table are saturated with water. Even in the deepest wells drilled for oil, some 8 or 9 km deep, geologists find water in permeable formations. At these depths, groundwaters move so slowly —probably less than a centimeter per year—that they have plenty of time to dissolve minerals from the rocks through which they pass. Thus, dissolved materials become more concentrated in these waters than in near-surface waters, making them unpotable. For example, groundwaters that pass through salt beds, which dissolve quickly, tend to contain large concentrations of sodium chloride.

At depths greater than 12 to 15 km, deep in the basement igneous and metamorphic rocks that underlie the sedimentary formations of the upper crust, porosities and permeabilities are very low. Although these rocks are saturated, they contain very little water (Figure 11.25). In some regions of the crust, such as along subduction zones, hot waters containing dissolved carbon dioxide play an important role in the chemical reactions of metamorphism. These waters help to dissolve some minerals and precipitate others. Even some mantle rocks are presumed to contain water, although in very minute quantities.

Most rocks deep in Earth's crust contain very little water because their porosities and permeabilities are very low.

Hydrothermal Waters

Natural hot springs are found in Yellowstone National Park; in Hot Springs, Arkansas; in Reykjavík, Iceland; and in many other places. Hot springs exist where **hydrothermal waters**—hot waters deep in the crust—migrate rapidly upward without losing much heat and emerge at the surface, sometimes at boiling temperatures.

Hydrothermal waters are loaded with chemical substances dissolved from rocks at high temperatures. As long as the water remains hot, the dissolved material stays in solution. As hydrothermal waters reach the surface, however, they cool quickly and precipitate various minerals, such as opal (a form of silica) and calcite or aragonite (forms of calcium carbonate). Crusts of calcium carbonate produced at some hot springs build up to form the rock travertine, which can form impressive deposits such as those seen at Mammoth Hot Springs in Yellowstone National Park (Figure 11.26). While still below the surface, hydrothermal waters deposit some of the world's richest metallic ores as they cool, as we saw in Chapter 4.

FIGURE 11.26 Travertine deposits at Mammoth Hot Springs, Yellowstone National Park, form large lobelike masses made of aragonite and calcite. [John Grotzinger.]

Hydrothermal waters form when groundwater is heated deep in the crust, where it accumulates high concentrations of dissolved minerals. When these waters are discharged at Earth's surface, they cool rapidly and precipitate those minerals.

Most hydrothermal waters of the continents come from surface waters that percolate downward to deeper regions of the crust (Figure 11.27). These surface waters originate primarily as **meteoric waters**—rain, snow, or other forms of water derived from the atmosphere (from the Greek *meteoron,* "phenomenon in the sky," which also gives us the word *meteorology*). These waters may be very old. It has been determined, for example, that the water at Hot Springs, Arkansas, derives from rain and snow that fell more than 4000 years ago and slowly infiltrated the ground.

Water that escapes from magma can also contribute to hydrothermal waters. In areas of igneous activity, sinking meteoric waters are heated as they encounter hot masses of rock. These hot waters then mix with water released from the nearby magma.

Hot Springs and Geysers

Hydrothermal waters may return to the surface as hot springs or geysers. Hot springs flow steadily; geysers erupt hot water and steam intermittently.

Geysers are probably connected to the surface by a system of very irregular and crooked fractures, recesses, and openings in rock, in contrast to the more regular and direct plumbing of hot springs. This irregular plumbing sequesters some water in recesses, thus helping to prevent the deeper waters from mixing with shallower waters and cooling. The bottom waters are heated by contact with hot rock. When they reach the boiling point, steam starts to ascend and heats the shallower waters, increasing the pressure and triggering an eruption. After the pressure is

Geysers form when hydrothermal waters boil in a restricted space, where pressure builds and triggers an eruption.

1 Meteoric water infiltrates the soil and filters down through permeable rocks.

2 As the water approaches the magma, it heats up and becomes less dense, thus setting up a circulation system that returns it to the surface.

3 Hot springs occur where heated groundwater is discharged at the surface.

4 The water in a geyser follows an irregular network of pores and cracks, which slows down the flow of water. Steam and boiling water are released to the surface under pressure, resulting in intermittent eruptions.

FIGURE 11.27 Circulation of water over a magma body produces geysers and hot springs.

released, the geyser becomes quiet as the fractures slowly refill with water.

In 1997, geologists reported the results of a novel technique for studying geysers. They lowered a miniature video camera to about 7 m below the surface of a geyser. They found that the geyser shaft was constricted at that point. Farther down, the shaft widened into a large chamber containing a wildly boiling mixture of steam, water, and what appeared to be carbon dioxide bubbles. These direct observations confirmed our ideas about how geysers work.

Geologists have turned to hydrothermal waters in the search for new and clean sources of energy. Northern California, Iceland, Italy, and New Zealand have already harnessed the steam produced by hot springs and geysers to drive electricity-generating turbines. Hydrothermal waters may soon be put to wider use for producing power, as we will see in Chapter 14.

Although hydrothermal waters provide human society with economically valuable ore deposits as well as electric power, these waters do not contribute to our drinking water supplies, primarily because they contain so much dissolved material.

SUMMARY

How does water move through the Earth system in the hydrologic cycle? The water movements of the hydrologic cycle maintain a constant balance among the major reservoirs of water on Earth. Evaporation from the oceans and the land transfers water to the atmosphere. Precipitation returns water from the atmosphere to the oceans and the land surface. Runoff returns part of the precipitation that falls on land to the ocean. The remainder infiltrates the ground and forms groundwater.

How does water move below the ground? Groundwater forms as precipitation infiltrates the ground and travels through porous and permeable formations called aquifers. The groundwater table is the boundary between the unsaturated and saturated zones of an aquifer. Groundwater moves downhill under the influence of gravity, eventually emerging at springs where the water table intersects the ground surface. Groundwater may flow through unconfined aquifers in formations of uniform permeability or in confined aquifers, which are bounded by aquicludes. Confined aquifers produce artesian flows and spontaneously flowing artesian wells. The rate of groundwater flow depends on several factors, including the slope of the water table, the permeability of the aquifer, and the distance of flow.

What factors govern human use of groundwater resources? As the human population grows, the demand for groundwater increases greatly, particularly where irrigation is widespread. As pumping discharge exceeds recharge, many aquifers, such as those of the Great Plains of North America, are being depleted, and there is no prospect of renewal for many years. Artificial recharge may help to renew some aquifers. The contamination of groundwater by industrial wastes, radioactive wastes, and sewage reduces the potability of some groundwater and limits our water resources.

What geologic processes are affected by groundwater? Erosion by groundwater in limestone terrains produces karst topography, with caves, sinkholes, and disappearing streams. The heating of meteoric waters deep in the crust forms hydrothermal waters, which may return to the surface as geysers and hot springs. At great depths in the crust, rocks contain extremely small quantities of water because their porosities are very low.

KEY TERMS AND CONCEPTS

aquiclude (p. 293)
aquifer (p. 293)
artesian flow (p. 293)
discharge (p. 293)
drought (p. 286)
groundwater (p. 283)
groundwater table (p. 292)
hydrologic cycle (p. 284)
hydrology (p. 282)
hydrothermal waters (p. 301)
infiltration (p. 284)
karst topography (p. 298)
meteoric waters (p. 302)
permeability (p. 291)
potable (p. 300)
precipitation (p. 284)
rain shadow (p. 285)
recharge (p. 293)
relative humidity (p. 285)
reservoir (p. 283)
runoff (p. 284)
saturated zone (p. 292)
sinkhole (p. 297)
unsaturated zone (p. 292)

EXERCISES

1. What are the main reservoirs of water at or near Earth's surface?

2. How do mountains form rain shadows?

3. What is an aquifer?

4. Why does water from an artesian well flow to Earth's surface without pumping?

5. What is the difference between the saturated and unsaturated zones of an aquifer?

6. How do aquicludes form a confined aquifer?

7. How do the slope of the water table and the permeability of an aquifer affect flow rate?

8. How does groundwater create karst topography?

9. What are the sources of water in hot springs?

10. What are some common contaminants in groundwater?

CHAPTER 12

SHAPING EARTH'S SURFACE: Streams, Coastlines, and Wind

Shipwrecks: Consequences of Coastal Geology

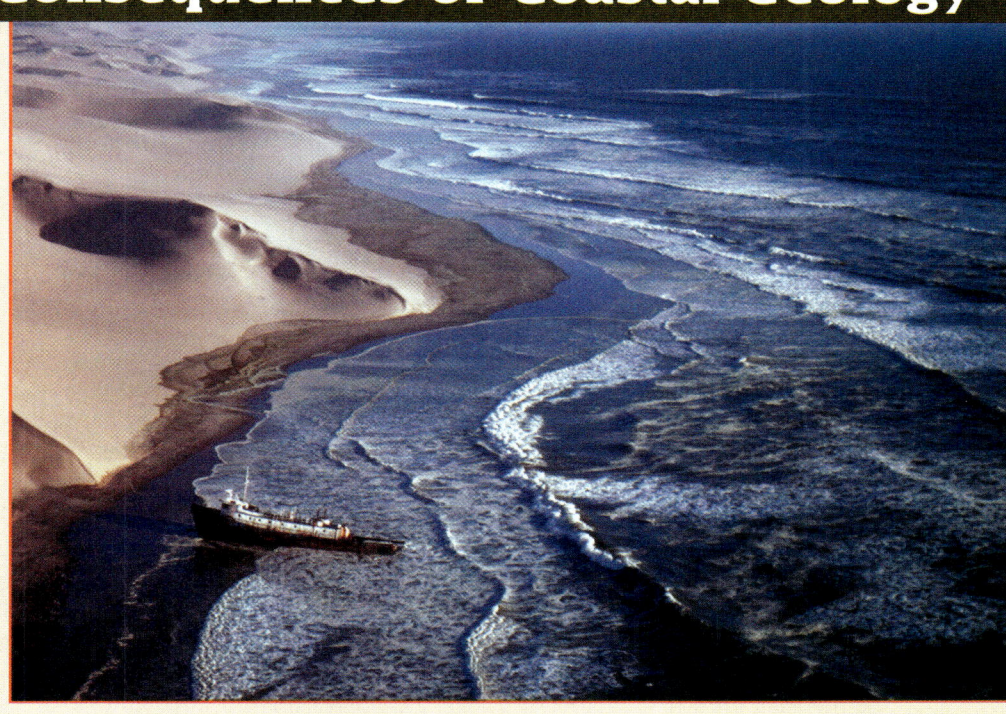

Beneath the world's oceans lie thousands of shipwrecks. Some of them tell a story of wartime battles on the high seas. Other vessels are undamaged and preserved in exquisite detail, offering a unique view into vanished civilizations. In every case, their very presence betrays human error, either in the creation or maintenance of ships or, more likely, in their operation. Most recorded shipwrecks take place within sight of land, along Earth's coastlines where land meets the sea. Simple navigation errors often conspire to produce errors in judgment, with disastrous consequences.

In reconstructing past events, it turns out that most shipwrecks involve strong currents that tend to push the ship toward points of land and the tendency of sand shoals to change their positions constantly. This latter problem has plagued mariners the most because even if it was known that there might be treacherous currents along a particular point of land, it was never certain where exactly its adjacent shoals would be. It turns out that sediment particles interact with currents to form sandbars, shoals, barrier islands, and a whole assortment of topographic features along coastlines, all of which are constantly changing. Come back the next year, or in 5 years, and the whole feature may have moved. In particularly active zones, sediments are so mobile that navigational charts are quickly outdated. Where the great rivers of the world, such as the Mississippi, Niger, and Amazon, pour their water—and sediments—into the oceans, the prospect of encountering unknown shoals is particularly great. The intersection of coastlines, rivers, and deserts—the subjects of this chapter—has profoundly influenced the history of human travel. ■

This shipwreck is one of hundreds that are strewn across the shores of the Skeleton Coast, Namibia. [Peter Johnson/CORBIS.]

CHAPTER OUTLINE

Earth's Surface Is Shaped by Currents 306

The Flow of Water and Wind 308

Sediment Transport and Deposition 310

Streams: Transport from Mountains to Oceans 313

Coastal Processes: Waves, Tides, and Hurricanes 325

The Shaping of Shorelines 331

Offshore: The Continental Shelf and Beyond 335

The Desert Environment 336

The most fabled shipwrecks have left people stranded along the desert coastlines of the world, particularly in Africa. Both the Sahara (in northern Africa) and Namib (in southern Africa) deserts have claimed the lives of many helpless men, women, and children marooned on desert beaches, often adjacent to rivers that only rarely flow with fresh water. There is no more treacherous coastline than that of Namibia, between Angola and South Africa. Mariners call it "the Skeleton Coast" and dread it. Its white sands are strewn with the skeletons of ships and travelers. Maps mark it merely as the Kaokoveld, which, freely translated from the local indigenous dialect, means "Coast of Loneliness."

As recently as 50 years ago, the Skeleton Coast had not been completely charted, and even today such charts are unreliable. The shoreline is in constant motion, moving westward, farther and farther out to sea. There is visible evidence of this movement in the wrecks of ships that are today high and dry in the sand, far from the water's edge. One vessel, stranded in 1909, is today well over half a mile inland.

In one epic drama of survival, the *Dunedin Star* ran aground on the Skeleton Coast on November 29, 1942, while carrying passengers and cargo from England to South Africa. In a first attempt at rescue, a second ship was wrecked even before reaching the *Dunedin Star*, stranding its own crew farther down the coast. Then an airplane was dispatched to deliver water, food, and medical supplies to the survivors, but when it landed, it became stuck in the loose sand, and its crew also became stranded. A second airplane was then sent to drop water and supplies to the survivors of the two shipwrecks and the airplane, but this second airplane crashed into the ocean, also stranding its crew, who survived and were able to swim to the coast. Finally, a convoy of trucks was sent to rescue all four groups of survivors, who just barely managed to make it back to civilization, almost a month after the wreck of the *Dunedin Star*.

Earth's surface is affected by a wide range of processes that regulate where water forms oceans, streams, and lakes, how the atmosphere circulates to form trade winds and deserts, and how our familiar landscapes are created. Why do deserts form? How do streams form, and why do they flow to the edges of continents? Why does the coastline have such a distinctive shape? This diverse list of apparently different topics has a common thread: moving currents of air and water. As these currents interact with Earth's surface, they sculpt it, they control where it rains, they provide navigational corridors into the deep interiors of continents, they transport vast quantities of gravel, sand, and silt, forming beaches and great fields of dunes both on land and under water, and they create navigational hazards at sea.

In this chapter, we'll discuss how water and air flow in currents and how these currents erode rock and carry sediments. Streams form as a result of interactions between the plate tectonic and climate systems, and we will see how they carve valleys and develop vast networks of channels. We will examine the processes that affect coastlines, including the effects of waves, tides, and damaging storms. Finally, we will describe desert environments, where wind plays its greatest role in shaping landscapes.

EARTH'S SURFACE IS SHAPED BY CURRENTS

Currents of water and air flow over all of Earth's surface. As they move, they help shape the land we live on, erode mountains, create the paths of rivers, form beaches, and control the locations of deserts.

Rivers and Streams

We use the word **stream** for any body of water, large or small, that flows over the land surface, and **river** for the major branches of a large stream system. Most streams run through well-defined troughs called **channels,** which allow water to flow over long distances. Streams are the lifelines of the continents. Their appearance directly records the interaction of climate and plate tectonic processes. Tectonic processes lift up the land, producing the steep topography and slopes of mountainous regions. Climate determines where rain and snow will fall. Rainwater runs downhill, eroding the rocks and soils of the mountains, forming channels and carving out valleys as it gathers into streams. Streams carry back to the sea the bulk of the precipitation that falls on land and much of the sediment produced by erosion of the land surface. Worldwide, streams carry about 25 billion tons of siliciclastic sediment and an additional 2 billion to 4 billion tons of dissolved matter each year.

All flowing bodies of water are called streams, and rivers are the major branches of a stream system. Most streams have channels that allow water to flow over long distances.

FIGURE 12.1 "Three rivers junction" in Pittsburgh, Pennsylvania, where the Allegheny and Monongahela rivers join to form the Ohio River. [Jupiter Images/Comstock Premium/Alamy.]

Streams have always been important to human civilization. A body of water flows through almost every town and city in most parts of the world. These streams have served as commercial waterways for barges and steamers and as water resources for resident populations and industries (Figure 12.1). Before cars and airplanes, people traveled across the oceans to coastal ports where they could trade goods and resupply their ships with food and water. Commonly these coastal ports were located on deltas, where great rivers met the ocean, allowing further access to the interiors of continents. The Nile River, for example, was vital to the agricultural economy of ancient Egypt and remains important to Egypt today. Living near a river also entails risks, however. When rivers flood, they destroy lives and property, sometimes on a huge scale.

Coastlines and Ocean Currents

Rivers meet oceans along Earth's **coastlines,** the broad zones where land meets the sea. Coastlines can give rise to a great diversity of landscapes, depending on what kinds of currents are active in the ocean and how tectonically active the coastal region is (Figure 12.2). Along the coast of North Carolina, for example, long stretches of sandy beaches extend for miles (Figure 12.2a). Here, tectonic activity is limited, and it is the currents produced by breaking waves that mold the coastline. The Oregon coastline, on the other hand, is dominated by rocky cliffs, showing that even though the effect of waves is considerable, it is tectonic uplift along faults that shapes that landscape (Figure 12.2b).

The major surface processes operating along coastlines are waves and tides. Together, they erode even the most resistant rocky shores. Waves and tides create currents, which transport sediments produced by erosion of the land and deposit them on beaches and in shallow waters along the shore. Coastlines are where hurricanes and other tropical storms affect the continents most strongly, causing devastation not just because of their strong winds, but also because of coastal flooding.

The landscapes of coastlines are shaped by tectonic processes and by currents driven by tides and waves.

Deserts: The Role of Wind

In deserts, rainfall is infrequent, water is scarce, and strong winds can howl for days on end. Some coastlines, like that of Namibia, are flanked by broad deserts. Other deserts cover broad regions of the interiors of continents. The locations of deserts are determined by the global flow of air, and

(a)

FIGURE 12.2 (a) Waves break along the shoreline and transport sand to form a beach at Cape Hatteras National Seashore, North Carolina. [Stephen Saks/Lonely Planet.]

(b)

(b) Waves pound a rocky shoreline at Cape Arago State Park, Coos County, Oregon. [Steve Terrill/CORBIS.]

wind plays a significant role in shaping their landscapes. Wind is much like water in its ability to erode, transport, and deposit sediments, and it is capable of moving enormous quantities of sand, silt, and dust over large regions of continents and oceans. The ancient Greeks called the god of winds Aeolus, and geologists today use the term **eolian** for the geologic processes powered by wind.

The locations of deserts are determined by global wind patterns.

Recently, concern over the expansion of Earth's deserts has increased. The population of southern Spain, for example, increasingly wonders whether the Sahara has jumped the Mediterranean Sea and is now encroaching on southern Europe. The process of desertification, in which land is degraded by decreases in rainfall resulting from various factors such as climate variations and human activities, has become a major focus of scientists trying to understand Earth's climate system.

THE FLOW OF WATER AND WIND

All currents, regardless of whether we are talking about water or air currents, share some basic characteristics of flowing fluids. We can illustrate two kinds of fluid flow by using lines of motion called *streamlines* (Figure 12.3). In **laminar flow,** the simplest kind of fluid movement, straight or gently curved streamlines run parallel to one another without mixing or crossing between layers. The slow movement of thick syrup over a pancake, with strands of unmixed melted butter flowing in parallel but separate paths, is a laminar flow. **Turbulent flow** has a more complex pattern of movement, in which streamlines mix, cross, and form swirls and eddies. Fast-moving river waters typically show this kind of motion. Turbulence—the degree to which there are irregularities and eddies in the flow—may be low or high.

In laminar flow, streamlines run parallel to each other. In turbulent flow, streamlines mix, cross, and form swirls and eddies.

Whether a flow is laminar or turbulent depends primarily on three factors:

1. Its velocity (rate of movement)
2. Its depth
3. Its viscosity

Viscosity is a measure of a fluid's resistance to flow. The more viscous (the thicker) a fluid is, the more it resists flow. The higher the viscosity, the greater the tendency for laminar flow. The viscosity of most fluids,

FIGURE 12.3 The two basic patterns of fluid flow: laminar flow and turbulent flow. The photograph shows the transition from laminar to turbulent flow in water along a flat plate, revealed by the injection of a dye. [ONERA.]

including water, decreases as the temperature increases. Given enough heat, a fluid's viscosity may decrease sufficiently to change a laminar flow into a turbulent one.

Water has low viscosity in the range of temperatures at Earth's surface. For this reason alone, most streams in nature tend toward turbulent flow. In addition, the rapid movement of water in most streams makes them turbulent. In nature, we are likely to see laminar flows of water only in thin sheets of rain runoff flowing slowly down nearly level slopes. In cities, we may see small laminar flows in street gutters. Because most streams and rivers are broad and deep and flow quickly, their flows are almost always turbulent.

A stream may show turbulent flow over much of its width and laminar flow along its edges, where the water is shallower and is moving more slowly. The flow velocity is highest near the center of a stream. We commonly refer to a rapid flow of water as a strong current.

Like flows of water in streams, air flows are nearly always turbulent. The extremely low density and viscosity of air make it turbulent even at the velocity of a light breeze.

Wind is a natural flow of air that is parallel to the surface of the rotating planet. Winds obey all the laws of fluid flow that apply to water in streams. Unlike flows of water in stream channels, however, winds are generally unconfined by solid boundaries, except for the ground surface and narrow valleys. Air flows are free to spread out in all directions, including upward into the atmosphere. Furthermore, the much lower density of wind makes it less powerful than water currents, even though wind speeds are often much greater than those in currents of water.

Winds vary in speed and direction from day to day, but over the long term they tend to come mainly from one direction because Earth's atmosphere flows in global prevailing wind belts (**Figure 12.4**). In the temperate latitudes, which include most of North America, the prevailing winds come from the west and so are referred to as the *westerlies*. Temperate climates are located at latitudes

> *Whether a flow is laminar or turbulent depends on its velocity, its depth, and its viscosity.*

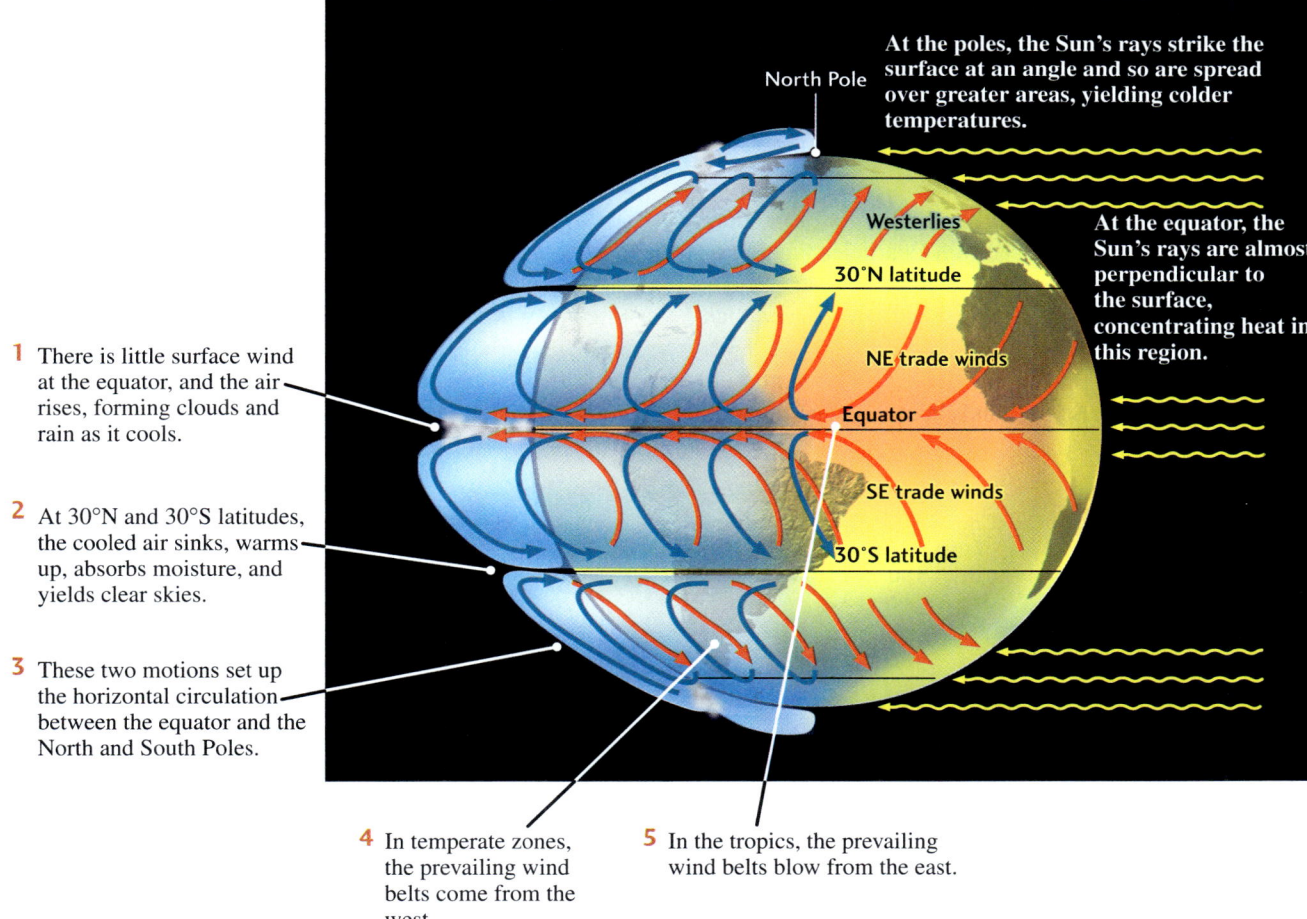

1 There is little surface wind at the equator, and the air rises, forming clouds and rain as it cools.

2 At 30°N and 30°S latitudes, the cooled air sinks, warms up, absorbs moisture, and yields clear skies.

3 These two motions set up the horizontal circulation between the equator and the North and South Poles.

4 In temperate zones, the prevailing wind belts come from the west.

5 In the tropics, the prevailing wind belts blow from the east.

At the poles, the Sun's rays strike the surface at an angle and so are spread over greater areas, yielding colder temperatures.

At the equator, the Sun's rays are almost perpendicular to the surface, concentrating heat in this region.

FIGURE 12.4 Earth's atmosphere circulates in prevailing wind belts created by variable heating of Earth's surface by the Sun.

between 30° and 60° N and 30° and 60° S. In the tropics, which are between 30° S and 30° N of the equator, the *trade winds* (named for an archaic use of the word *trade* to mean "track" or "course") blow from the east.

These prevailing wind belts arise because the Sun warms a given amount of land surface most intensely at the equator, where the Sun's rays are almost perpendicular to Earth's surface. The Sun heats the land less intensely at high latitudes and at the poles because there its rays strike Earth's surface at an angle (see Figure 12.4). Hot air, which is less dense than cold air, rises at the equator and flows toward the poles, gradually sinking and releasing water vapor as it cools. The sinking air reaches ground level at about 30° S and 30° N, then flows back along Earth's surface toward the equator to form the trade winds. This cool, dry air warms and absorbs moisture as it sinks, producing clear skies and arid climates. It overlies many of the world's deserts, such as the Sahara.

> *Global prevailing wind belts result from variable heating of Earth's surface by the Sun.*

SEDIMENT TRANSPORT AND DEPOSITION

> *Turbulent flows can generally transport larger sediment particles than laminar flows. Particles suspended in a current are its suspended load, whereas particles that slide and roll across its bed are its bed load.*

Currents of water and air vary in their ability to erode and carry sand grains and other sediments. Laminar flows of water can lift and carry only the smallest, lightest clay-sized particles. Turbulent flows of water, depending on their speed, can move particles ranging in size from clay to pebbles and cobbles. Winds cannot normally carry particles larger than sand grains. As turbulence lifts particles into a flow, the flow carries them downcurrent. Turbulence also rolls and slides larger particles. A current's **suspended load** includes all the material temporarily or permanently suspended in the flow. Its **bed load** is the material the current carries along Earth's surface by sliding and rolling (Figure 12.5). The *bed* is the layer of loose sediment that interacts with the current.

Suspension, Saltation, and Settling

A current's ability to carry sediment depends on a balance between turbulence, which lifts particles, and the competing downward pull of gravity, which makes them settle out of the current and become part of the bed. Small grains of silt and clay are easily lifted into the current and settle slowly, so they tend to stay in suspension. Larger particles, such as medium- and coarse-grained sand, settle more quickly. Most larger particles therefore stay suspended in the current only a short time before they settle.

Sand grains in a current typically move by **saltation**—an intermittent jumping motion along the bed. The grains are sucked up into the flow by turbulent eddies, move with the current for a short distance, and then fall back to the bed (see Figure 12.5). If you were to stand in a rapidly flowing sandy stream of water, you might see a cloud of saltating sand grains moving around your ankles. The bigger the grain, the longer it will tend to remain on the bed before it is picked up. Once a large grain is in the current, it will settle quickly. The smaller the grain, the more frequently it will be picked up, the higher it will "jump," and the longer it will take to settle.

> *During saltation, particles are lifted into a flowing current, move downcurrent a short distance, then land on the bed.*

Thus, turbulent flows transport sediment by suspension (clays), saltation (sands), and rolling and sliding along the bed (sand and gravel). To study how

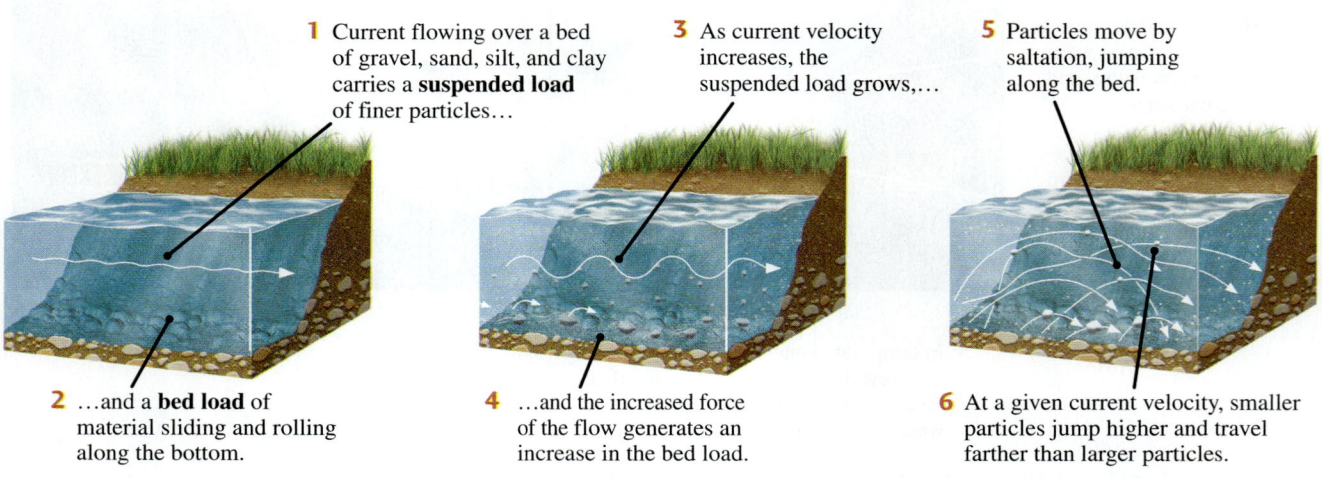

1. Current flowing over a bed of gravel, sand, silt, and clay carries a **suspended load** of finer particles…

2. …and a **bed load** of material sliding and rolling along the bottom.

3. As current velocity increases, the suspended load grows,…

4. …and the increased force of the flow generates an increase in the bed load.

5. Particles move by saltation, jumping along the bed.

6. At a given current velocity, smaller particles jump higher and travel farther than larger particles.

FIGURE 12.5 A current flowing over a bed of sand, silt, and clay transports particles in two ways: by saltation and in suspension.

Air has a staggering capacity to hold dust in suspension. Large dust storms may carry more than 100 million tons of dust and deposit it in layers several meters thick. Fine-grained particles from the Sahara have been found as far away as England and have been traced across the Atlantic Ocean to Florida. Scientists on oceanographic research vessels have measured airborne dust far out to sea, and today it can be observed directly from space (**Figure 12.6**).

Dust includes microscopic rock and mineral fragments of all kinds, especially silicates, as might be expected from their abundance as rock-forming minerals. Two of the most important sources of silicate minerals in dust are clays from soils on dry plains and volcanic ash from eruptions.

Windblown dust has complex effects on climate. Mineral dust in the atmosphere scatters the incoming visible (light) radiation from the Sun and absorbs the outgoing infrared (heat) radiation from Earth.

Flows of air can hold large amounts of dust in suspension. Large dust storms may deposit suspended particles in layers several meters thick.

Sediment Bed Forms: Dunes and Ripples

When a current transports sand grains by saltation, they tend to form dunes and ripples (see Chapter 6). **Dunes** are mounds or ridges of sand up to many meters high that form in flows of both wind and water over a sandy bed (**Figure 12.7**). As a dune grows, the whole mound starts to migrate

FIGURE 12.6 Satellite photograph of a dust storm originating in the Namib Desert in September 2002. Dust and sand are being transported from right (east) to left (west) by strong winds blowing out to sea. These sediments can be transported for hundreds to thousands of kilometers across the ocean. [NASA.]

a particular river carries sediments, geologists and engineers measure the relationship between particle size and the force the flow exerts on the particles in the suspended and bed loads. This relationship shows them how much sediment a particular flow can move and how rapidly it can move it. This information allows them to design dams and bridges or to estimate how quickly artificial reservoirs behind dams will fill with sediments.

As we saw in Chapter 6, the greater a current's velocity, the larger the sediment particles it can transport. Thus, when the current slows, the largest particles settle out first. Geologists can infer the velocities of ancient currents from the sizes of particles in sedimentary rocks.

Windblown Dust

Most North Americans are familiar with rainstorms or snowstorms: high winds accompanied by heavy precipitation. We may have less experience of dry storms, during which high winds blowing for days on end carry enormous amounts of sand, silt, and dust.

FIGURE 12.7 Sand dunes in the Namib Desert, in southwestern Africa, are among the tallest in the world. [John Grotzinger.]

downcurrent by the combined movements of a host of individual grains as sand is eroded from the windward side and deposited on the leeward (downcurrent) side. Sand grains constantly saltate to the top of the low-angled windward slope and then fall over into the wind shadow on the leeward slope, as shown in **Figure 12.8**. These grains gradually build up a steep, unstable accumulation on the upper part of the leeward slope. Periodically, this accumulation gives way and spontaneously slips or cascades down this *slip face*, as it is called, to a new slope at a lower angle.

Ripples are very small dunes—with heights ranging from less than a centimeter to several centimeters—whose long dimension is formed at right angles to the current (**Figure 12.9**). Although underwater ripples and dunes are harder to observe than those produced on land by air currents, they form in the same way and are just as common. The steady downcurrent transfer of grains across their ridges causes ripples and dunes to migrate downcurrent. The speed of this migration is much slower than the movement of individual grains and very much slower than the current.

> Transport of individual sand particles by saltation forms dunes and ripples. Ripples usually form first and grow into dunes as current velocity increases.

1. A ripple or dune advances by the movements of individual grains of sand. The whole form moves forward slowly as sand erodes from the windward slope and is deposited on the leeward slope.

2. Particles of sand arriving on the windward slope of the dune move by saltation over the crest,...

3. ...where the wind velocity decreases and the sand deposited slips down the leeward slope.

4. This process acts like a conveyor belt that moves the dune forward.

5. The dune stops growing vertically when it reaches a height at which the wind is so fast that it blows the sand grains off the dune as quickly as they are brought up.

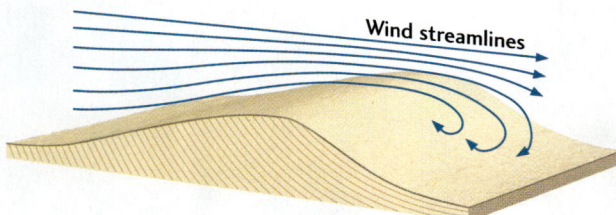

FIGURE 12.8 Dunes form when a current transports sand particles by saltation.

FIGURE 12.9 Wind ripples in sand at Stovepipe Wells, Death Valley, California. [Tom Bean.]

FIGURE 12.10 The form of a sediment bed changes with increasing current velocity. [After D. A. Simmons and E. V. Richardson, "Forms of Bed Roughness in Alluvial Channels." *American Society of Civil Engineers Proceedings* 87 (1961): 87–105.]

The shapes of ripples and dunes and their migration speeds change as the velocity of the current increases. At the lowest current velocities, few grains are saltating, and the sediment bed is flat. At slightly higher velocities, the number of saltating grains increases. A rippled bed forms, and the ripples migrate downstream (**Figure 12.10**). As the velocity increases further, the ripples grow larger and migrate faster until, at a certain point, dunes replace the ripples. Both ripples and dunes have a cross-bedded structure, and as the current flows over their tops, it can actually reverse and flow backward along their leeward side. As the dunes grow larger, small ripples form on them. These ripples tend to climb over the backs of the dunes because they migrate more quickly than the dunes.

STREAMS: Transport from Mountains to Oceans

In this section we'll look at how moving currents of water shape the continents and transport sediments to the oceans. These currents modify the surfaces of continents to form mountain valleys and major rivers. The sediments produced by weathering and erosion and transported in streams are ultimately delivered to the edges of continents, where deltas are formed in low-lying regions.

Streamflow

The flow of a stream may appear steady when you view it from a bridge for a few minutes or canoe along it for a few hours, but its volume and velocity at a single place may change appreciably from month to month and season to season. Streams are constantly changing, shifting from low waters to floods over a few hours or days and reshaping their valleys over longer periods (**Figure 12.11**). The flow and dimensions of a stream also change as it moves downslope, from narrow valleys in its upland headwaters to broader floodplains in its middle and lower courses. Most of these longer-term changes are adjustments in the normal (nonflood) volume and velocity of flow as well as the depth and width of the channel.

From their headwaters to their termination points in lakes and oceans, all streams react to changes in climate (such as changes in precipitation) and tectonic processes (uplift or subsidence of Earth's crust). Near their headwaters, streams gather into ever larger streams, eventually forming a single large strand, as in the case of the Mississippi River. Precipitation in the headwaters may be felt far downriver, where a river's volume may exceed the volume of the channel and then spill over its banks to create a flood.

As streams move across the land surface—in some places over bedrock, in others over unconsolidated sediments—they erode these materials and create valleys. Identifying and mapping stream valleys were essential tasks for Meriwether Lewis and William Clark when they first explored and mapped the western United States 200 years ago. As they traveled upstream and the river branched, they had to choose which branch was the larger of the two. They used two observations about streams to help them make this choice: the width of the stream valley and the depth of the stream channel. Was the valley wide enough, and the channel deep enough, for their boats? Narrow valleys and shallow channels would mean that the branch led into a much shorter, and therefore less desirable, route; wider valleys and deeper channels, on the other hand, promised a longer passage up the main branch of the river.

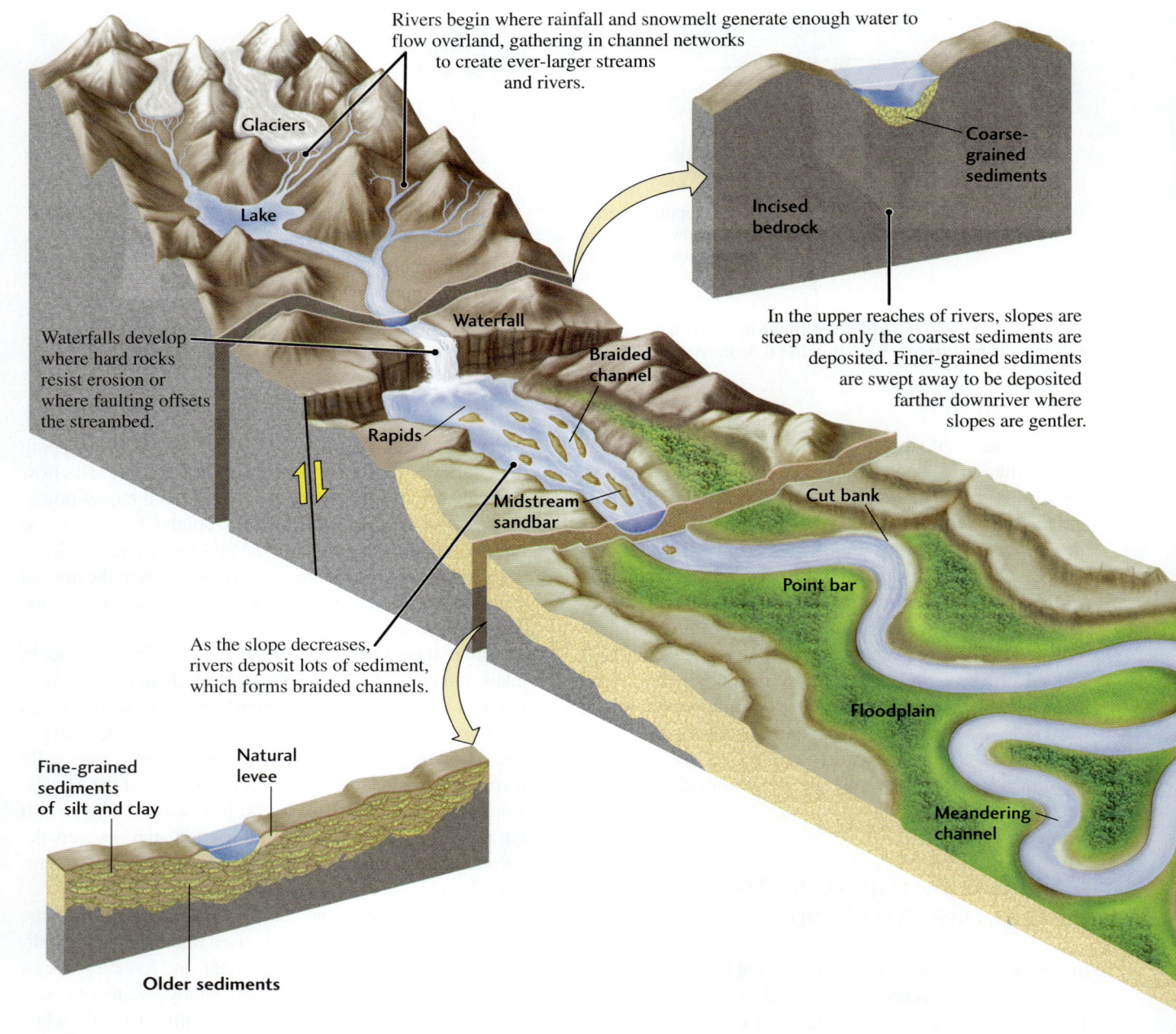

FIGURE 12.11 Streams transport water and sediments from the mountains to the oceans.

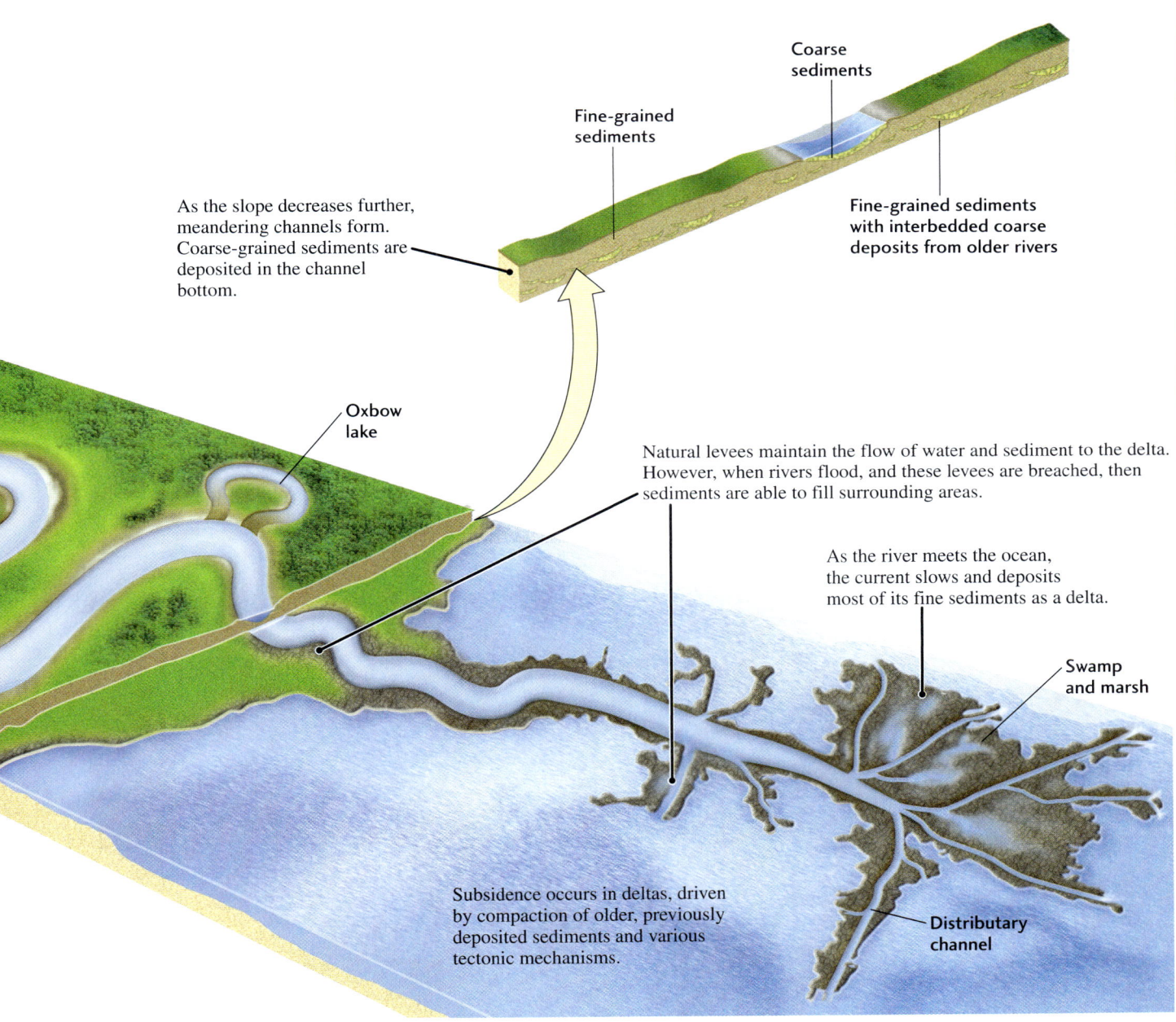

Stream Valleys

A **stream valley** encompasses the entire area between the tops of the slopes on both sides of the stream. The cross-sectional profile of many stream valleys is V-shaped, but many other valleys have a broad, low profile like that shown in **Figure 12.12**. At the bottom of the valley is the channel, the trough through which the water runs. The channel carries all the water during normal, nonflood times. At low water levels, the stream may run only along the bottom of the channel. At high water levels, the stream occupies most of the channel. In broad valleys, a **floodplain**—a flat area about level with the top of the channel—lies on either side of the channel. It is the part of the valley that is flooded when the river spills over its banks, carrying with it silt and sand from the channel.

In high mountains, stream valleys are narrow and have steep walls, and the channel may occupy most or all of the valley bottom. A small floodplain may be visible only at low water levels. In such valleys, the stream is actively cutting into the bedrock, a process that is characteristic of newly uplifted highlands in tectonically active areas. In lowlands, where tectonic uplift has long since ceased, the stream shapes its valley by eroding sediment particles and transporting them downstream. With a long time to operate, these processes produce gentle slopes and floodplains many kilometers wide (**Figure 12.13**).

Stream valleys are formed when streams break apart and erode rock by abrasion. The sand and pebbles the stream carries create a sandblasting action that wears away even the hardest rock. In some streambeds, pebbles and cobbles rotating inside swirling eddies grind deep *potholes* into the bedrock (**Figure 12.14**). At low water,

FIGURE 12.12 This drawing shows a river flowing in a channel that moves over a broad, flat floodplain in a wide valley. Floodplains may be narrow or absent in steep valleys.

FIGURE 12.13 Aerial view of the meandering Adelaide River, in Australia. Its appearance is typical of meandering streams in lowland environments. [Peter Bowater/Photo Researchers.]

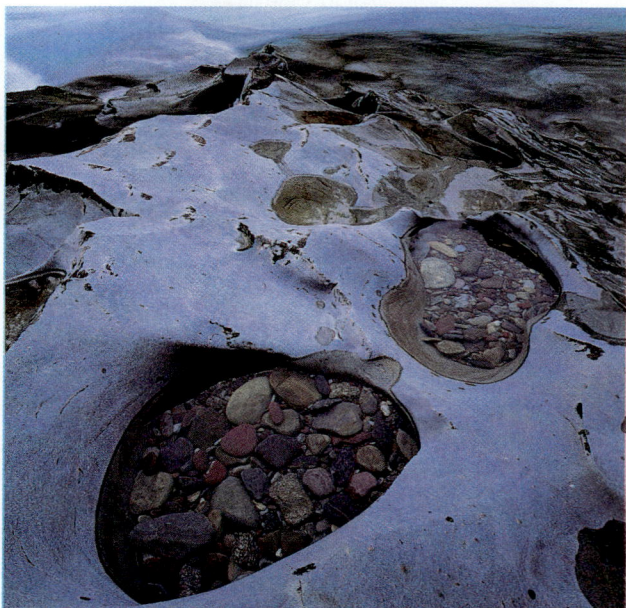

FIGURE 12.14 Potholes in river rock along McDonald Creek, Glacier National Park, Montana. The pebbles rotate inside the potholes, grinding deep holes in the bedrock. [Carr Clifton/Minden Pictures.]

A stream valley contains a channel through which the water runs. Normally, the channel carries all the water, but during floods it spills out of the channel onto the adjacent floodplain.

pebbles and sand can be seen lying quietly at the bottom of exposed potholes.

Stream channels begin where rainwater, draining off the surface of the land, flows so fast that it abrades the soil and bedrock and carves into it (**Figure 12.15**). Once a small gully forms, it will capture more of the flow of water over the land surface, and thus the tendency of the stream to cut downward will increase. As the gully progressively deepens, the rate of downcutting increases as more water is captured.

Rock erosion is particularly strong at rapids and waterfalls. *Rapids* are places in a stream where the flow is extremely fast because the slope of the streambed suddenly steepens, typically at rocky ledges. The speed and turbulence of the water quickly break rocks into smaller pieces that are carried away by the strong current.

Streams carrying coarse sediments are capable of eroding solid rock. This erosion is greatest at rapids and waterfalls, where flows are fastest and most turbulent.

The tremendous impact of huge volumes of plunging water and tumbling boulders quickly erodes streambeds below waterfalls. Waterfalls also erode the underlying rock of the cliffs that form the falls. As erosion undercuts these cliffs, the upper streambed collapses, and the falls recede upstream (**Figure 12.16**).

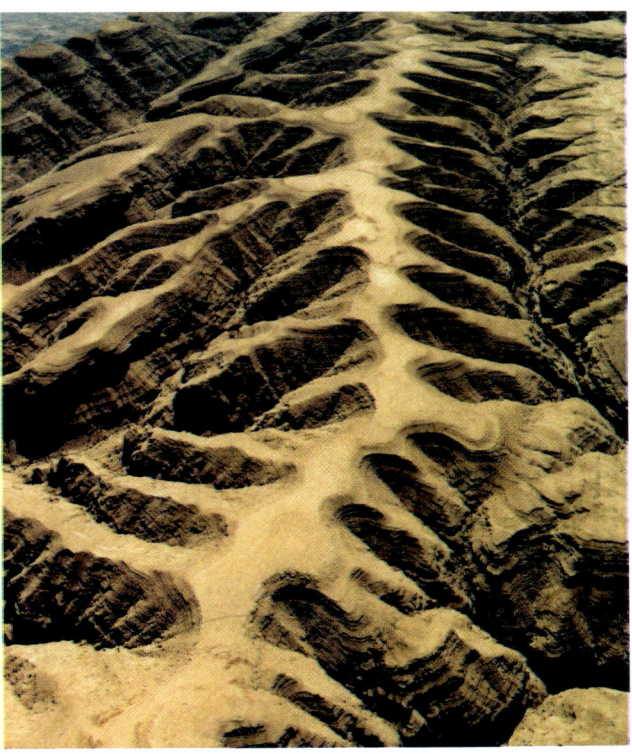

FIGURE 12.15 Streams create gullies when the action of water flowing across Earth's surface causes erosion of soil or bedrock. The smallest gullies converge to form larger stream channels, and farther downslope these become river channels. These gullies were formed in the desert of Oman by occasional rainstorms that inundate the surface with rapidly flowing water that erodes the bedrock. [Petroleum Development Oman.]

FIGURE 12.16 This waterfall on the Iguaçú River, Brazil, is retreating upstream as falling water and sediment pound the cliff's base and undercut it. From the center to the upper left of the photo, one can see steep walls, the remnants of the waterfall's retreat upstream. [Donald Nausbaum.]

Stream Channel Patterns

As a stream channel makes its way along the bottom of a valley, it may run straight in some stretches and take a snaking, irregular path in others, sometimes splitting into multiple channels. The channel may flow along the center of the floodplain or hug one edge of the valley.

Meanders On a great many floodplains, stream channels follow curves and bends called **meanders** (see Figure 12.13). Meanders are the normal pattern for streams flowing through plains or lowlands, where their channels typically cut through unconsolidated sediments—fine sand, silt, or mud—or easily eroded bedrock. Meanders are less pronounced, but still common, in streams flowing down steeper slopes over harder bedrock. In such terrain, meandering stretches may alternate with long, relatively straight ones.

A stream that has cut deeply into the curves and bends of its channel may produce incised meanders (**Figure 12.17**). Other streams may meander on somewhat wider

FIGURE 12.17 This section of the San Juan River, Utah, is a good example of an incised meander belt, a deeply eroded, meandering, V-shaped valley with virtually no floodplain. [Tom Bean.]

1. Meanders shift from side to side in a snaking motion.
2. The current is faster at outside banks, which are eroded,…
3. …and sediments get deposited at inside banks where the current is slower, forming point bars.
4. As the erosion and deposition process continues, the bends grow closer and the point bars bigger.
5. During a major flood, when velocity and water volume increase, the river takes a new, shorter course, cutting across the loop.
6. The abandoned loop remains as an oxbow lake.

FIGURE 12.18 (*top*) Meanders on floodplains migrate over a period of many years. [Peter Kresan.] (*bottom*) Oxbow lake in Blackfoot River valley, Montana. [James Steinberg/Photo Researchers.]

floodplains bounded by steep, rocky valley walls. Meanders on a floodplain migrate over periods of many years as the stream erodes the outside banks of bends, where the current is strongest (Figure 12.18). As the outside banks are eroded, sediments are deposited to form curved sandbars called **point bars** along the inside banks, where the current is slower. In this way, meanders slowly shift position from side to side, as well as downstream, in a snaking motion something like that of a long rope being snapped.

Meanders migrate over time as a stream cuts into the outsides of bends and deposits sediments on the insides of bends as point bars.

As meanders migrate, sometimes unevenly, the bends may grow closer and closer together, until finally the river bypasses one of them, often during a major flood. The river takes a new, shorter course like that shown in Figure 12.18. In its abandoned path, it leaves behind an *oxbow lake*—a crescent-shaped, water-filled loop.

Braided Streams Some streams have many channels instead of a single one. A **braided stream** is one whose channel divides into an interlacing network of channels, which then rejoin in a pattern resembling braids of hair (Figure 12.19). Braided streams are found in many settings, from broad lowland valleys to wide, sediment-filled rift valleys adjacent to mountain ranges. Braids tend to form in streams with large variations in volume of flow combined with a high sediment load and banks that are easily eroded. They are well developed, for example, in the sediment-choked streams formed at the edges of melting glaciers.

Braided streams have interlacing networks of channels.

FIGURE 12.19 This stretch of the Chitina River, Alaska, is a braided stream, with many channels that divide and then rejoin [Tom Bean.]

Alluvial Fans Sometimes streams leave narrow mountain valleys and enter broad, relatively flat valleys at lower elevations. Along sharply defined mountain fronts, typically at steep fault scarps, streams drop large amounts of sediment in cone- or fan-shaped accumulations called **alluvial fans** (Figure 12.20). This deposition results from the sudden decrease in current velocity that occurs as the channel widens abruptly. Coarse materials, from boulders to sand, dominate on the steep upper slopes of the fan. Lower down, finer sands, silts, and muds are deposited.

Alluvial fans are deposits of sediment that form where current velocity decreases as streams emerge from narrow mountain valleys.

FIGURE 12.20 An alluvial fan (Tucki Wash) in Death Valley, California. Alluvial fans are large cone- or fan-shaped accumulations of sediment deposited when stream velocity slows, as at a mountain front. [Marli Miller.]

The Stream Floodplain

We can measure the size of a stream's flow by its **discharge**—the volume of water that passes a given point in a given time as it flows through a channel of a certain width and depth. Floods occur when discharge increases such that more water flows into the stream than can flow out, so that the excess water spills over the banks. Streams flood regularly, some at infrequent intervals, others almost every year. Some floods are large, with very high water levels lasting for days. At the other extreme are minor floods that barely break out from the channel before they recede. Small floods are more frequent, occurring every 2 or 3 years on average. Large floods are generally less frequent, usually occurring every 10, 20, or 30 years.

A stream channel migrating over the floor of a valley creates a floodplain. As floodwaters spread out over the floodplain, the velocity of the water slows, and the current loses its ability to carry sediment. The floodwater velocity drops most quickly along the immediate borders of the channel. As a result, the current deposits large amounts of coarse sediment, typically sand and gravel, along a narrow strip at the edge of the channel. Successive floods build up **natural levees,** ridges of coarse material that confine the stream within its banks between floods, even when water levels are high (**Figure 12.21**). Where natural levees have reached a height of several meters and the stream almost fills the channel, the floodplain level is below the stream level. You can walk the streets of an old river town built on a floodplain, such as Vicksburg, Mississippi, and look up at the levee, knowing that the river waters are rushing by above your head.

Floods form natural levees by depositing coarse sediments at the margins of stream channels.

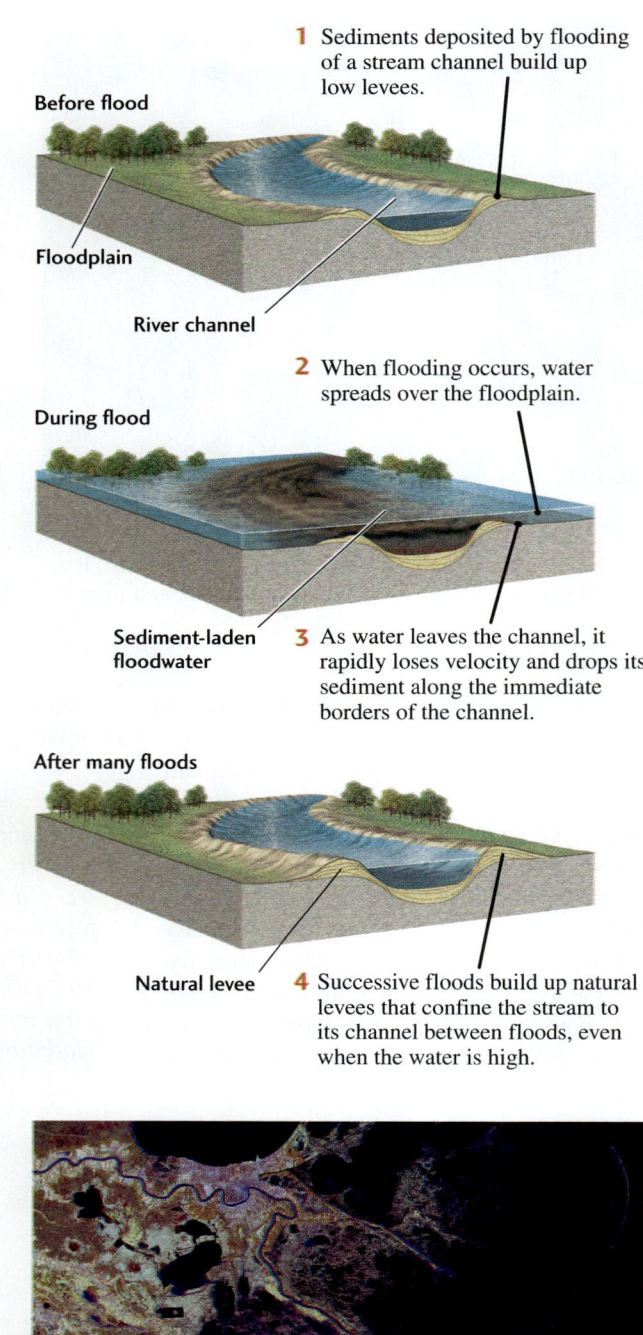

FIGURE 12.21 Floods form natural levees along the banks of a river. The false color LANDSAT image shows natural levees along the main channel of the Mississippi River near South Pass, Louisiana. [USGS National Wetlands Research Center.]

The Development of Cities on Floodplains

Floodplains are natural sites for urban settlements because they combine easy transportation along a river with access to fertile agricultural lands. Such sites, however, remain subject to the floods that formed the floodplains.

About 4000 years ago, cities began to dot floodplains in Egypt along the Nile, in the ancient land of Mesopotamia along the Tigris and Euphrates rivers, and in Asia along the Indus River of India and the Yangtze and Huang Ho of China. Later, many of the capital cities of Europe were built on floodplains: Rome on the Tiber, London on the Thames, Paris on the Seine. Floodplain cities in North America include St. Louis on the Mississippi, Cincinnati on the Ohio, and Montreal on the St. Lawrence.

worst flood in U.S. history (behind the flooding of New Orleans by Hurricane Katrina in 2005). Some geologists believe that the construction of levees and dikes to confine the Mississippi contributed to these record floods because the river can no longer erode its banks and widen its channel to accommodate some of the additional water flowing through it during times of high discharge.

What are cities and towns in this position to do? Some have urged a halt to all construction and development on the lowest parts of the floodplains. Some have called for the elimination of federally subsidized disaster funds for rebuilding in such areas. Harrisburg, Pennsylvania, hit hard by a flood in 1972, turned some of its devastated riverfront area into a park. Still other towns have moved off the floodplain to a completely new location, as described in Chapter 1. Yet the benefits of living on floodplains continue to attract people to those sites.

> Floodplains are natural sites for urban development because they provide river transportation as well as fertile soil for agriculture. But cities on floodplains are also prone to destructive floods.

Drainage Networks

Every topographic rise between two streams, whether it measures a few meters or a thousand, forms a **divide**—a ridge of high ground along which all rainfall runs off down one side or the other. A **drainage basin** is an area of land, bounded by divides, that funnels all its water into the network of streams draining that area (**Figure 12.23**). Drainage basins occur at many scales, from a ravine surrounding a small

> A drainage basin is an area bounded by divides that contains a network of streams draining that area.

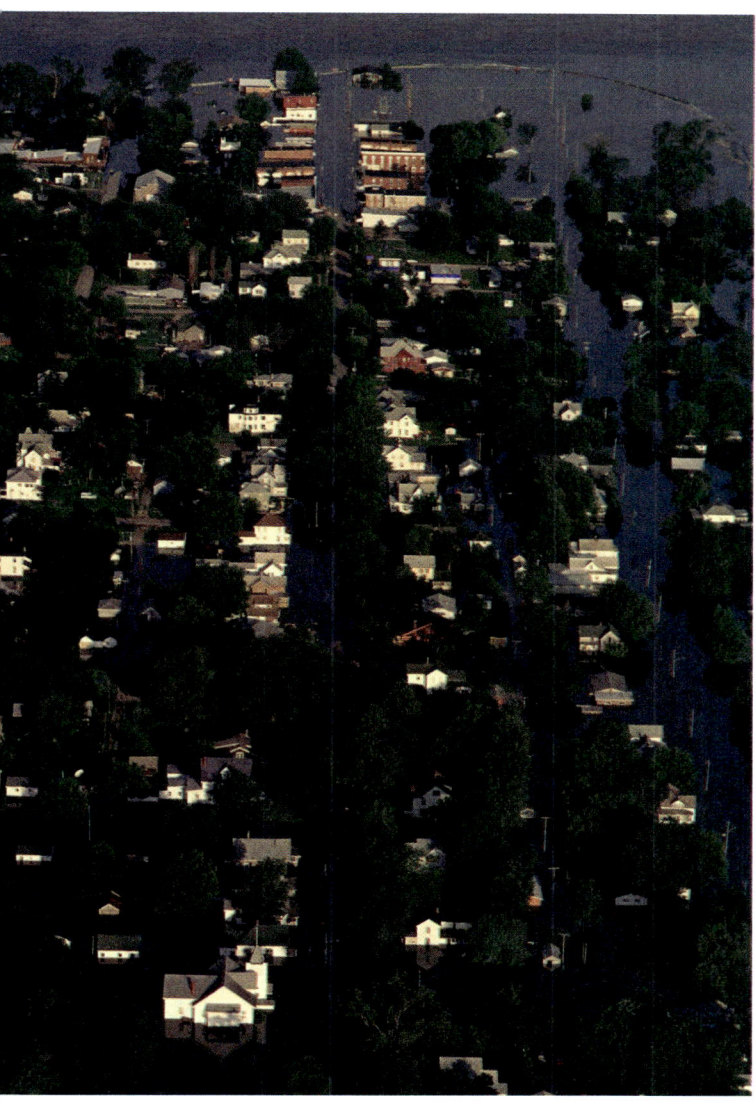

FIGURE 12.22 Residential neighborhood along the Mississippi River is completely inundated during the 1993 flood. [Jupiter Images/Creatas/Alamy.]

Floods periodically destroyed sections of these ancient and modern cities on the lower parts of the floodplains, but each time the inhabitants rebuilt them. Today, most large cities are protected by artificial levees that strengthen and heighten the river's natural levees. Extensive systems of dams can help control flooding that would affect these cities, but they cannot eliminate the risk entirely. In 1973, for example, the Mississippi went on a rampage with a flood that continued for 77 consecutive days at St. Louis. The river reached a record 4.03 m above flood stage (the height at which the river first overflows the channel banks). In 1993, the Mississippi and its tributaries broke loose again, shattering the old record in a disastrous flood (**Figure 12.22**) that has been officially designated the second

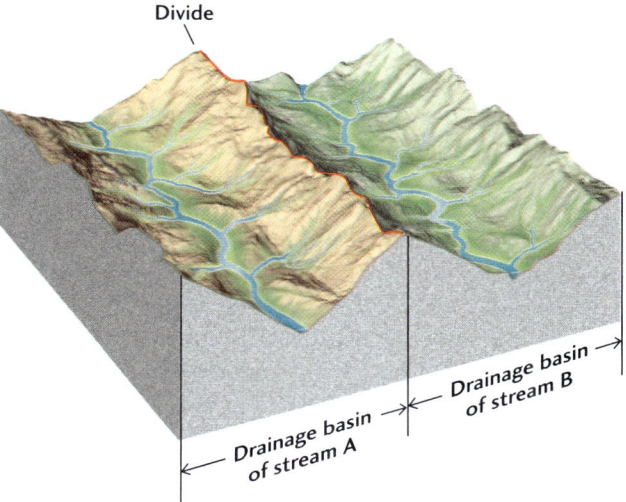

FIGURE 12.23 Drainage basins are separated by divides, which occur at many scales, from ridges to gentle uplands to mountain ranges.

FIGURE 12.24 The natural drainage basin of the Colorado River covers about 630,000 km², constituting a large part of the southwestern United States. The basin is surrounded by divides that separate it from the neighboring drainage basins. [After U.S. Geological Survey.]

stream to a great region drained by a major river and its tributaries (**Figure 12.24**).

A continent has several major drainage basins separated by major divides. In North America, the continental divide formed by the Rocky Mountains separates all waters flowing into the Pacific Ocean from all those entering the Atlantic. Lewis and Clark followed the Missouri River upstream to its headwaters at the continental divide in western Montana. When they crossed over the divide, they descended from the headwaters of the Columbia River, which they followed to the Pacific Ocean.

A map showing the courses of all the large and small streams in a drainage basin reveals a pattern of connections called a *drainage network*. If you followed a stream from its mouth to its headwaters, you would see that it steadily divides into smaller and smaller tributaries, forming a drainage network that shows a characteristic branching pattern (**Figure 12.25**). Branching is a general property of many kinds of networks in which material is collected and distributed. The network of the human circulatory system, for example, distributes blood to the body through a branching system of arteries and collects it through a corresponding system of veins.

> The streams in a drainage basin are joined together in a drainage network with a characteristic branching pattern

1 **Dendritic drainage** is characterized by branches similar to the limbs of a tree.

2 **Rectangular drainage,** developed on a strongly jointed rocky terrain, tends to follow the joint pattern.

3 **Trellis drainage** develops in valley and ridge terrain, where rocks of varying resistance to erosion are folded into anticlines and synclines.

4 **Radial drainage** patterns develop on a single large peak, such as a large dormant volcano.

FIGURE 12.25 Some typical drainage network patterns.

Deltas: The Mouths of Rivers

Sooner or later, all rivers end as they flow into a lake or an ocean, mix with the surrounding water, and—no longer able to travel downslope—gradually lose their forward momentum. The largest rivers, such as the Amazon and the Mississippi, can maintain some current many kilometers out to sea. Where smaller rivers enter a turbulent, wave-swept sea, the current disappears almost immediately beyond the river's mouth.

Delta Sedimentation Patterns As its current gradually dies out, a river progressively loses its power to transport sediments. The coarsest material, typically sand, is dropped first, right at the mouths of most rivers. Finer sands are dropped farther out, followed by silt and, still farther out, by clay. As the floor of the lake or ocean slopes to deeper water away from the shore, the deposited materials build up a platform called a **delta.** (We owe the name *delta* to the Greek historian Herodotus, who traveled through Egypt about 450 B.C. The roughly triangular shape of the sediment deposit at the mouth of the Nile prompted him to name it after the Greek letter Δ, delta.)

As a river approaches its delta, it reverses its upstream-branching drainage pattern. Instead of collecting more water from tributaries, it discharges water into **distributaries**—smaller streams that receive water and sediments from the main channel, branch off *downstream,* and thus distribute the water and sediments into many channels. Materials deposited on top of the delta, typically sand, make up horizontal *topset beds.* Downstream, on the outer front of the delta, fine-grained sand and silt are deposited to form gently inclined *foreset beds,* which resemble large-scale cross-beds. Spread out on the seafloor seaward of the foreset beds are thin, horizontal *bottomset beds* of mud, which are eventually buried as the delta continues to grow. **Figure 12.26** shows how these structures form in a typical large marine delta.

> Where a river enters a lake or ocean, a delta is formed as the current slows and drops its sediments.

The Growth of Deltas As a delta builds outward into the ocean, the mouth of its river advances seaward, leaving new land in its wake. Much of this land is just a few meters above sea level. The delta of the Mississippi River, like many other major river deltas, has been growing for millions of years. About 150 million years ago, it started out around what is now the junction of the Ohio and the Mississippi rivers, at the southern tip of Illinois. It has advanced about 1600 km since then, creating almost the entire states of Louisiana and Mississippi as well as major parts of adjacent states. **Figure 12.27** shows the growth of the Mississippi delta over the past 6000 years as well as the direction its growth is likely to take in the future.

> Deltas can build new areas of land very quickly.

FIGURE 12.26 A typical large marine delta, many kilometers in extent, in which the fine-grained foreset beds are deposited at a very low angle, typically only 4° to 5° or less. Sandbars form at the mouths of the distributaries, where the current velocity suddenly decreases. The delta builds forward by the advance of the sandbars and the topset, foreset, and bottomset beds. Between distributary channels, shallow bays fill with fine-grained sediment and become salt marshes. This general structure is found on the Mississippi delta.

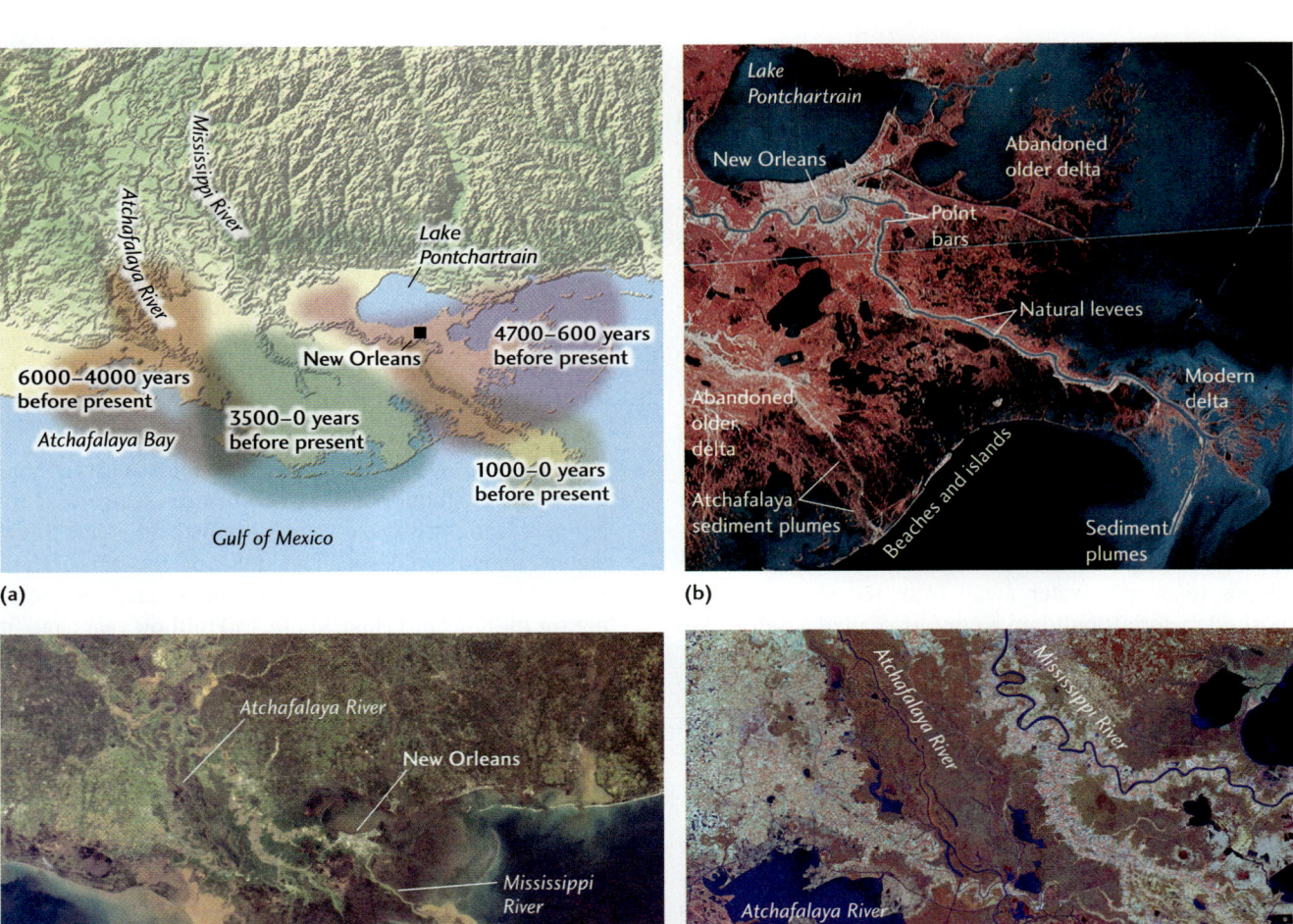

FIGURE 12.27 The Mississippi delta. Over the past 6000 years, the river has built its delta first in one direction and then in another as water flow shifted from one major distributary to another. (a) The modern delta was preceded by deltas deposited to the east and west. (b) The infrared-sensitive film used to record this satellite image of the Mississippi delta causes vegetation to appear red, relatively clear water to appear dark blue, and water with suspended sediment to appear light blue. At the upper left are New Orleans and Lake Pontchartrain. Well-defined natural levees and point bars can be seen at the center. At the lower left are beaches and islands that formed as waves and currents transported river-deposited sand from the delta. [From G. T. Moore, "Mississippi River Delta from Landsat2," *Bulletin of the American Association of Petroleum Geologists* (1979).] (c) Satellite photograph of the Mississippi delta. [NASA.] (d) This image shows the discharge of sediment into the Gulf of Mexico from the Mississippi River delta and the Atchafalaya River. A major flood could divert the main flow of the Mississippi into the Atchafalaya, causing a new delta to form. Construction of artificial levees by the Army Corps of Engineers has prevented this so far. [USGS National Wetlands Research Center.]

COASTAL PROCESSES: Waves, Tides, and Hurricanes

The landscapes of coastlines, the broad regions where land and streams meet the oceans, present striking contrasts (**Figure 12.28**). On the coast of North Carolina, for example, long, straight, sandy beaches stretch for miles along low coastal plains (Figure 12.28a). In most of New England, by contrast, rocky cliffs bound elevated shores, and the few beaches that exist are made of gravel (Figure 12.28b). Rocky coastlines also occur along the south coast of Australia, where wave erosion generates spectacular landforms (Figure 28c). Many of the seaward edges of islands in the tropics, such as those in the Caribbean Sea, are coral reefs, the delight of divers (Figure 12.28d).

The major geologic forces operating at the **shoreline** —the line where the water surface intersects the land— are ocean currents created by waves and tides. Together, they erode even the most resistant rocky shores. Waves and tides transport sediments produced by erosion of the land and deposit them on beaches and in shallow waters along the shore.

Coastal processes that affect shorelines result from interactions within the climate system and the solar system. Tides are caused by gravitational interactions between Earth and the Sun and Moon, and coastal surf and storms —such as hurricanes—result from interactions between the atmosphere and the hydrosphere.

How can we monitor these processes and gauge their effects? Ordinary waves and tides are responsible for the slow, steady change of the shape of coastlines and the

(a)

(b)

(c)

(d)

FIGURE 12.28 Coastlines exhibit a variety of geologic forms. (a) Long, straight, sandy beach coastline, Pea Island, North Carolina. [Courtesy of Bill Birkemeier/U.S. Army Corp of Engineers.] (b) Rocky coastline, Mount Desert Island, Maine. This formerly glaciated coastline has been uplifted since the end of the last ice age, about 10,000 years ago. [Neil Rabinowitz/CORBIS.] (c) The Twelve Apostles, a group of stacks at Port Campbell, Australia, developed in horizontal beds of sedimentary rock. These remnants of shore erosion are left as the shoreline retreats under the action of waves. [Kevin Schafer.] (d) Coral reef along the Florida coastline. [Hays Cummins, Miami University.]

movement of sediments up and down the coast. These patterns may be modified or accelerated when storms arrive at the coast. What is present today may be gone tomorrow, and vice versa.

As we learned earlier in this chapter, currents are the key to understanding geologic processes at Earth's surface, and coastal processes are no exception. Let's look at the various types of currents that shape our shorelines.

Currents Created by Waves

Centuries of observation have taught us that waves are constantly changing. In quiet weather, waves with calm troughs between them roll regularly into shore. In the high winds of a storm, however, waves move in a confusion of shapes and sizes. To understand the dynamics of shorelines, and to make sensible decisions about coastal development, we need to understand how waves work.

The wind blowing over the surface of the ocean creates waves by transferring its energy of motion from air to water. As a gentle breeze of 5 to 20 km/hour starts to blow over a calm sea surface, ripples—little waves less than a centimeter high—take shape. As the speed of the wind increases to about 30 km/hour, the ripples grow to full-sized waves. Stronger winds create larger waves and blow off their tops to make whitecaps. The height of waves depends on three factors:

- The wind speed
- The length of time over which the wind blows
- The distance the wind travels over water

Storms blow up large, irregular waves that radiate outward from the storm center, like the ripples moving outward from a pebble dropped into a pond. As the waves travel out from the storm center in ever-widening circles, they become more regular, changing into low, broad, rounded waves called *swell*, which can travel hundreds of kilometers. Several storms at different distances from a shoreline, each producing its own pattern of swell, account for the often irregular intervals between waves approaching the shore.

> The height of waves depends on wind speed, how long the wind blows, and the distance the wind travels over water.

The Surf Zone Swell becomes higher as it approaches the shore, where it assumes the familiar sharp-crested wave shape (**Figure 12.29**). These waves are called *breakers* because, as they come closer to shore, they break and form surf—a foamy, bubbly surface. The *surf zone* is the belt along which breaking waves collapse as they approach the shore. Breaking waves pound the shore, eroding and carrying away sand, weathering and breaking up solid rock, and destroying structures built close to the shoreline.

The transformation from swell to breakers starts where water depth decreases to less than one-half the distance between waves. At that point, wave motion just above the bottom becomes restricted. The restricted motion of the water molecules slows the wave, which causes it to steepen (**Figure 12.30**a). As the wave rolls toward the shore, it becomes so steep that the water can no longer support itself, and the wave breaks with a crash in the surf zone. Gently sloping bottoms cause waves to break farther from shore, while steeply sloping bottoms make waves break closer to shore. Where rocky shores are bordered by deep water, the waves break directly on the rocks with a force amounting to tons per square meter,

FIGURE 12.29 Wave motion is influenced by water depth and shape of the shoreline. [Galen Rowell/CORBIS.]

Coastal Processes: Waves, Tides, and Hurricanes 327

FIGURE 12.30 Breaking waves create longshore drift and longshore currents. (a) Swell becomes breaking waves as it approaches the shoreline. (b) When breaking waves approach the shoreline at an angle, their swash and backwash move sand and water in a zigzag pattern.

throwing water high into the air (see Figure 12.2b). It is not surprising that concrete seawalls built to protect structures along the shore quickly start to crack and must be repaired constantly.

After breaking at the surf zone, the waves, now reduced in height, continue to move in, breaking again right at the shoreline. They run up onto the sloping front of the beach, forming an uprush of water called *swash*. The water then runs back down again as *backwash*. Swash can carry sand and even large pebbles and cobbles if the waves are high enough. The backwash carries the particles seaward again.

This back-and-forth motion of the water near the shore is strong enough to carry sand grains and even gravel. Wave action in water as deep as about 20 m can move fine sand. Large waves caused by intense storms can scour the bottom at much greater depths, down to 50 m or more. At shallower water depths, these big storms transport sediments in an offshore direction, often depleting beaches of their fine sand.

Wave Motion and Longshore Currents Far from shore, the lines of swell are parallel to one another, but as the waves approach a beach over a shallowing bottom, they gradually bend to a direction more parallel to the shore. When a wave encounters the shallowing bottom of the shoreline, the front of the wave slows. Then the next part of the wave meets the bottom and it, too, slows. Thus, in a continuous transition along the wave crest, the wave line bends toward the shore as it slows. As a result, most waves still approach the shore at a shallow angle.

As the waves break on the shore, the swash moves up the beach slope at a slight angle, and the backwash runs down the

> *Surf is created when waves arrive at the shoreline, where water depth decreases, and break. Water from breaking waves flows onto the beach as swash, then flows back toward the sea as backwash.*

> *Because most waves approach the beach at a shallow angle, the swash and backwash of breaking waves transport sand down the beach in a zigzag motion known as longshore drift.*

slope in the opposite direction. The combination of these two motions moves the water a short way down the beach (Figure 12.30b). Sand grains carried by swash and backwash are thus moved along the beach in a zigzag motion known as *longshore drift*.

Waves approaching the shoreline at an angle can also cause a **longshore current,** a shallow-water current that flows parallel to the shore. The movement of swash and backwash in and out from the shore at an angle creates a zigzag path of water molecules that transports sediments along the shallow bottom in the same direction as the longshore drift. Much of the transport of sand along many beaches results from longshore currents. Longshore currents are prime determiners of the shape and extent of sandbars and other depositional shoreline features. At the same time, because of their ability to erode loose sand, longshore currents may remove large amounts of sand from a beach. Longshore drift and longshore currents working together are potent processes in the transport of sand on beaches and in very shallow waters. In slightly deeper waters (less than 50 m), longshore currents—especially those running during large storms—strongly affect the bottom.

> Waves breaking at an angle to the shore can produce longshore currents in shallow water that flow parallel to the shore.

Currents Created by Tides

Tides are caused by the gravitational attraction between Earth and the Moon. The strength of this attraction varies across Earth's surface. On the side of Earth closest to the Moon, the ocean water experiences a greater gravitational attraction than the average for the whole of the solid Earth. This pull produces a bulge in the water. On the side of Earth farthest from the Moon, the solid Earth, being closer to the Moon than the water, is pulled toward the Moon more than the water is, and the water therefore appears to be pulled away from Earth as another bulge. Thus, two bulges of water occur on Earth's oceans: one on the side nearest the Moon, and the other on the side farthest from the Moon (Figure 12.31). As Earth rotates, these bulges stay approximately aligned: one always faces the Moon; the other is always directly opposite the Moon. These bulges of water passing over the rotating Earth are the high tides.

Although tides occur regularly everywhere, the difference between high and low tides varies in different parts of the ocean. As Earth rotates, the tidal bulges of water move along the surface of the ocean, encountering obstacles, such as continents and islands, that hinder the flow of water. In the middle of the Pacific Ocean—in Hawaii, for example, where there is little to obstruct the flow of the tides—the difference between low and high tides is only 0.5 m. Near Seattle, where the shape of the shoreline along Puget Sound is very irregular and the tidal flow must move through narrow passageways, the difference between the two tides is about 3 m. Extraordinary tides occur in a few places, such as the Bay of Fundy in eastern Canada, where the tidal range can be more than 12 m. Many people living along the coast need to know when tides will occur, so governments publish tide tables showing predicted tide heights and times.

Tides moving near shorelines cause currents that can reach speeds of a few kilometers per hour. As the tide rises, water flows in toward the shore as a *flood tide,* moving through narrow passages into inlets and bays, into shallow coastal marshes, and up small streams. As the tide passes the high stage and starts to fall, water flows out as an *ebb tide,* and low-lying coastal areas are exposed. Tidal currents meander across *tidal flats,* the muddy or sandy areas that lie above low tide but are flooded at high tide (Figure 12.32). Where obstacles restrict tidal flow and increase tidal range, such as in the Bay of Fundy, current velocities become very high. Large sand

> The gravitational attraction between Earth and the Moon produces bulges of ocean water that create high tides as Earth rotates.

> The difference between high and low tides depends on the presence of obstacles that hinder tidal flow.

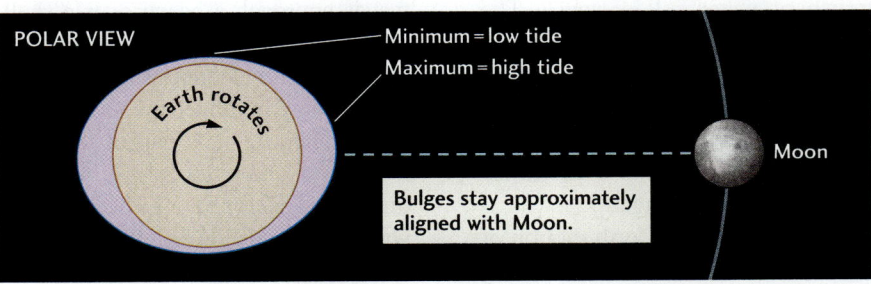

FIGURE 12.31 Tides are caused by the gravitational attraction of Earth and the Moon. The Moon's gravitational pull causes two bulges of water on Earth's oceans, one on the side of Earth nearest the Moon and the other on the side farthest from the Moon. As Earth rotates, these bulges remain aligned with the Moon and pass over Earth's surface, creating the high tides.

Coastal Processes: Waves, Tides, and Hurricanes

FIGURE 12.32 Tidal flats, such as this one at Mont-Saint-Michel, France, may be extensive areas covering many square kilometers, but most often are narrow strips seaward of the beach. When a very high tide advances on a broad tidal flat, it may move so rapidly that areas are flooded faster than a person can run. The beachcomber is well advised to learn the local tides before wandering. [Thierry Prat/CORBIS Sygma.]

ridges many meters high may be formed in these tidal channels.

Hurricanes and Coastal Storm Surges

Few things in nature can compare with the destructive force of a hurricane. Regarded as the greatest storms on Earth, hurricanes can pound shorelines and coastal regions with sustained winds, lasting many hours, of 260 km/hour (160 miles/hour) or more, combined with intense rainfall and a storm surge caused by the storm's extremely low atmospheric pressure. Earth scientists have estimated that during its life cycle, a hurricane can expend as much energy as 10,000 nuclear bombs!

Hurricanes form over tropical parts of Earth's oceans, between 8° and 20° latitude, in areas of high humidity, light winds, and warm sea surface temperatures (typically 26°C or greater). These conditions usually occur in the summer and early fall in the tropical North Atlantic and North Pacific oceans. For this reason, hurricane "season" in the Northern Hemisphere runs from June through November (**Figure 12.33**).

Most hurricanes that affect the Atlantic Ocean and Gulf of Mexico originate just off the coast of West Africa

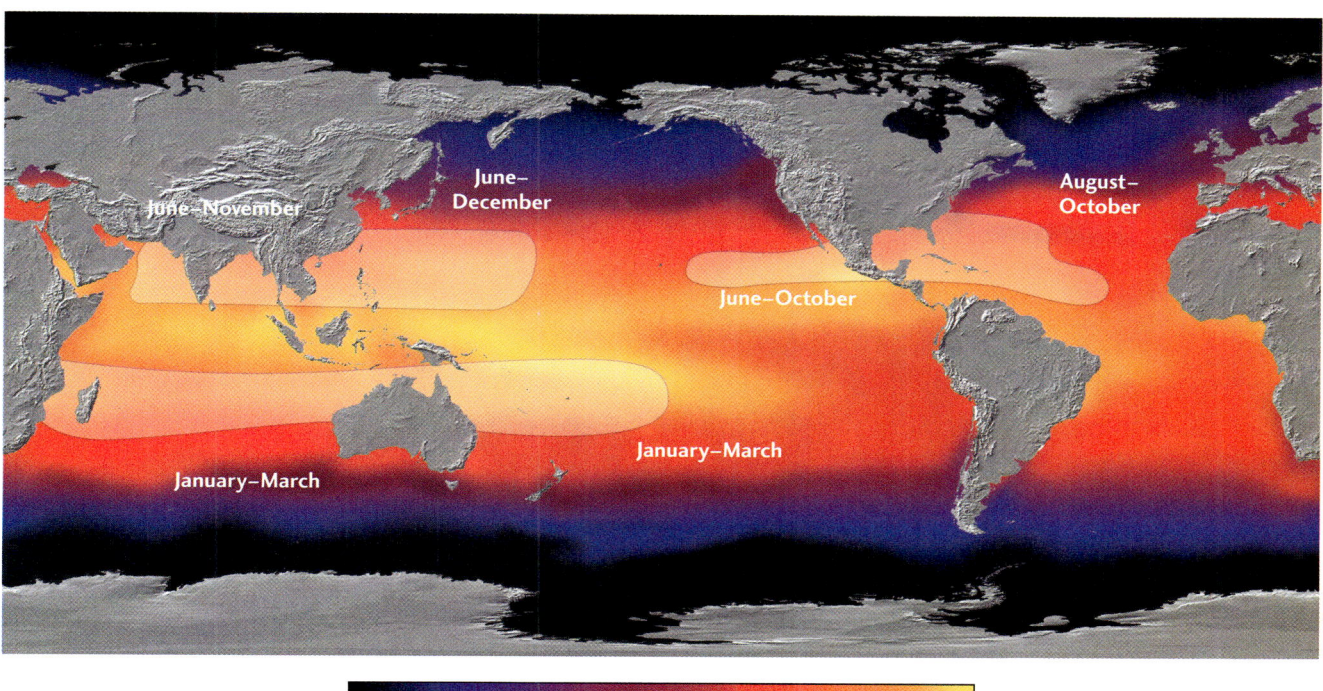

FIGURE 12.33 Hurricanes arise in summer and early fall when ocean temperatures are warmest. The light-shaded areas indicate places where hurricanes are most common. The times of year when they are most abundant are also shown. [NASA/GSFC.]

FIGURE 12.34 Hurricane Katrina on August 28, 2005, a few hours before it struck New Orleans. In the Northern Hemisphere, winds circulate in a counterclockwise direction around the "eye" of the hurricane, which is the location of lowest surface pressure. Katrina was one of the most powerful hurricanes on record, with maximum sustained winds of up to 280 km/hour (175 miles/hour) and gusts up to 360 km/hour (225 miles/hour). [NASA/Jeff Schmaltz, MODIS Land Rapid Response Team.]

and intensify as they move westward across the open ocean. Once sustained wind speeds reach 37 km/hour (23 miles/hour), the storm system is called a *tropical depression*. In the Northern Hemisphere, the increasing winds begin to circulate in a counterclockwise pattern around the storm's area of lowest pressure, which ultimately becomes the "eye" of the hurricane (Figure 12.34). When the wind speeds reach 119 km/hour (74 miles/hour), the storm is classified as a hurricane.

As a hurricane intensifies, a dome of seawater—known as a **storm surge**—rises above the level of the surrounding ocean surface. Large swells, high surf, and wind-driven waves ride atop this dome as it floods coastal land areas, causing extensive damage to structures and the shoreline environment (Figure 12.35). Any landmass in the path of a storm surge will be affected to a greater or lesser extent, depending on a number of factors. The stronger the storm and the shallower the offshore waters, the higher the storm surge.

The storm surge is the most deadly of a hurricane's associated hazards, as underscored by Hurricane Katrina in 2005. The magnitude of a hurricane is usually described in terms of its wind speed, but coastal flooding causes many more deaths than high wind.

Hurricanes are storms that form over tropical parts of the oceans and have sustained wind speeds of at least 119 km/hour.

A storm surge is a dome of water formed during a hurricane that can flood coastal regions. A storm surge can be the most damaging force in a hurricane.

FIGURE 12.35 Hurricane storm surges along coastlines may result in the complete destruction of residential buildings, which pile up as lines of debris well inland of the shoreline. The damage seen here was caused by Hurricane Katrina in 2005. [U.S. Navy/Getty Images.]

FIGURE 12.36 A tide terrace exposed at low tide. This shallow depression between an outer ridge (a sandbar at high tide) and the upper beach is rippled by the tidal flow in many places. [James Valentine.]

Boats ripped from their moorings, utility poles, and other debris floating atop a storm surge often demolish those buildings not destroyed by extreme winds. Even without the weight of floating debris, a storm surge can severely erode beaches and highways and undermine bridges. Because much of the United States' densely populated Atlantic and Gulf Coast shorelines lie less than 3 m above mean sea level, the danger from storm surges is tremendous.

THE SHAPING OF SHORELINES

At shorelines, we can observe the constant motion of ocean waters and their effects on the shore. Current environmental problems such as coastal erosion and pollution of shallow waters have made the geology of shorelines and shallow seas a critical area of research. Waves, longshore currents, and tidal currents interact with the geologic structures and plate tectonic processes of the coast to shape shorelines into a multitude of forms (see Figure 12.28). We can see these factors at work in the most popular of shorelines: beaches.

Beaches

A **beach** is a shoreline made up of sand and pebbles. Beaches may change shape from day to day, week to week, season to season, and year to year. Waves and tides sometimes broaden and extend a beach by depositing sand and sometimes narrow it by carrying sand away.

Many beaches are straight stretches of sand that range from 1 km to more than 100 km long; others are smaller crescents of sand between rocky headlands. Belts of dunes border the landward edge of many beaches; bluffs or cliffs of sediment or rock border others. A beach may have a *tide terrace*—a flat, shallow area between the upper beach and an outer bar of sand—on its seaward side (**Figure 12.36**).

> A beach is a shoreline made up of sand and pebbles. Its shape is often changed by the action of waves and tides.

The Structure of a Beach Figure 12.37 shows the major parts of a beach. These parts may not all be present at all times on any particular beach. Farthest out is the *offshore,* bounded by the surf zone, where the bottom begins to become shallow enough for waves to break. The *foreshore* includes the surf zone; the tide terrace; and, right at the

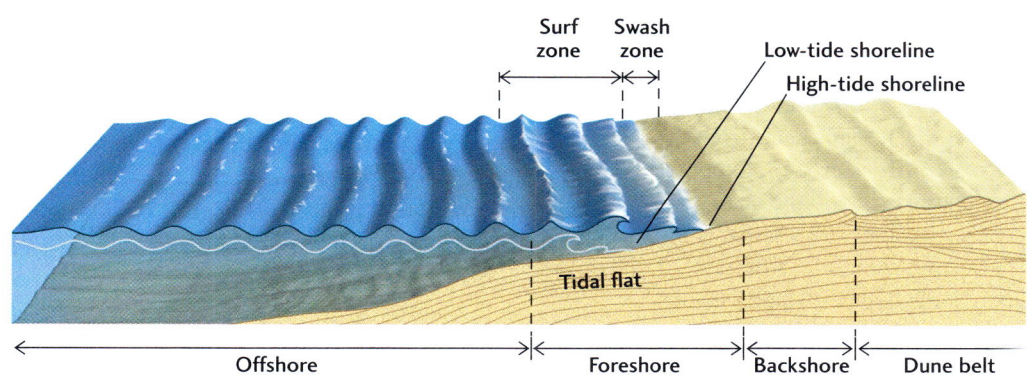

FIGURE 12.37 A profile of a beach, showing its major parts.

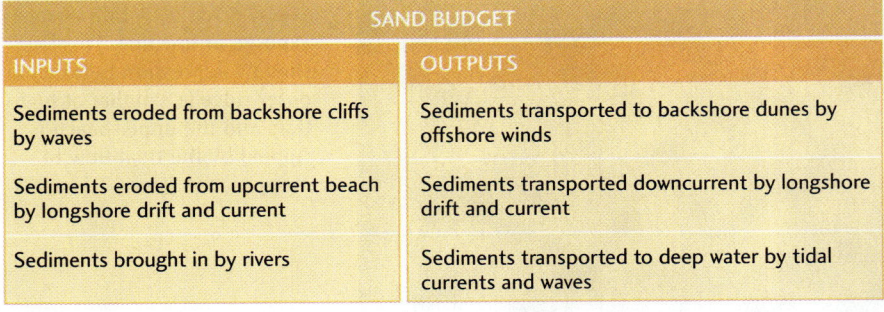

FIGURE 12.38 The sand budget of a beach is a balance between inputs and outputs of sand by erosion, sedimentation, and transport.

A beach can be subdivided into three parts: offshore, foreshore, and backshore.

shore, the swash zone, a slope dominated by the swash and backwash of the waves. The *backshore* extends from the swash zone up to the highest level of the beach.

The Sand Budget of a Beach A beach is a scene of constant movement. Each wave moves sand back and forth with swash and backwash. Both longshore drift and longshore currents move sand down the beach. At the end of a beach, and to some extent along it, sand is removed and deposited in deep water. In the backshore or along sea cliffs, sand and pebbles are freed by erosion and replenish the beach. The wind that blows over the beach transports sand, sometimes offshore into the water and sometimes onshore onto the land.

All these processes together maintain a balance between the addition and removal of sand, resulting in a beach that may appear to be stable but is actually exchanging its material with the environments on all sides. Figure 12.38 illustrates the sand budget of a beach—the inputs and outputs caused by erosion, sedimentation, and transport. At any point along a beach, it gains sand from the inputs: material along the backshore is eroded, longshore drift and longshore currents bring sand to the beach, and rivers that enter the ocean along the shore bring in sediment. The beach loses sand from the outputs: winds carry sand to backshore dunes, longshore drift and currents carry it downcurrent, and deep-water currents and waves transport it during storms.

If the total input balances the total output, the beach will keep the same general form. If input and output are not balanced, the beach grows or shrinks. Temporary imbalances are natural over weeks, months, or years. A series of large storms, for example, might move large amounts of sand from the beach to deeper waters on the far side of the surf zone, narrowing the beach. Then, weeks of mild weather and low waves might move the sand onto the shore and rebuild a wide beach. Without this constant shifting of sands, beaches might be unable to recover from the effects of trash, litter, and other kinds of pollution. Within a year or two, even oil from spills is transported or buried out of sight, although the tarry residue may later be uncovered in spots.

The form of a beach depends on a balance between inputs and outputs of sand. If that balance is upset, the beach may grow or shrink.

Erosion and Deposition at Shorelines

The topography of the shoreline, like that of the land interior, is a product of tectonic forces elevating or depressing Earth's crust, erosion wearing it down, and sedimentation filling in the low spots. Thus, several factors are directly at work:

- Uplift of the coastal region, which leads to erosion at the coast

- Subsidence of the coastal region, which leads to sediment buildup at the coast

- The nature of the rocks or sediments at the shoreline

- Changes in sea level, which affect the drowning or emergence of a shoreline

- The average and storm wave heights

- The heights of the tides, which affect both erosion and sedimentation

Erosional Coastal Forms Erosion is active along tectonically uplifted rocky coasts. Along these coasts, prominent cliffs and headlands jut into the sea, alternating with narrow inlets and irregular bays with small beaches. Waves crash against the rocky shorelines, undercutting cliffs and causing huge blocks of rock to fall into the water, where they are gradually eroded away. As the sea cliffs retreat, isolated remnants called *stacks* may be left standing in the sea, far from the shore (see Figure 12.28c). Erosion by waves also planes the rocky surface beneath the surf zone and creates a **wave-cut terrace,** which is sometimes visible at low tide (Figure 12.39). Wave erosion over long periods may straighten shorelines as headlands retreat faster than recesses and bays.

Where relatively soft sediments or sedimentary rocks make up the coastal region, the slopes are gentler and the heights of shoreline bluffs are lower. Waves erode these softer materials efficiently, and erosion of bluffs on such shores may be extraordinarily rapid. The high sea cliffs of soft glacial materials along the Cape Cod National Seashore in Massachusetts, for instance, are retreating about a meter each year. Since Henry David Thoreau walked the entire length of the beach below those cliffs in the mid-nineteenth century and wrote of his travels in *Cape Cod,* about 6 km² of coastal land have been eaten away by the ocean, equivalent to about 150 m of beach retreat.

In recent decades, more than 70 percent of the total length of the world's sand beaches has retreated at a rate of at least 10 cm/year, and 20 percent of the total length has retreated at a rate of more than 1 m/year. Much of this loss can be traced to the damming of rivers, which decreases the sediment supply to the shoreline.

Depositional Coastal Forms Sediment builds up in areas where subsidence depresses Earth's crust along a coastline. Such coastlines are characterized by long, wide beaches and wide, low-lying coastal plains of sedimentary rock. Shoreline forms along these coastlines include sandbars, low-lying sandy islands, and extensive tidal flats. Long beaches grow longer as longshore currents carry sand to the downcurrent end of the beach. There it builds up, first forming a submerged sandbar, then rising above the surface and extending the beach by a narrow addition called a **spit.**

Offshore, long sandbars may build up into **barrier islands** that form a barricade between open ocean waves and the main shoreline. Barrier islands are common, especially along low-lying coasts composed of sediments that are easily eroded and transported or of poorly cemented sedimentary rocks where longshore currents are strong. Some of the most prominent barrier islands are found along the coast of New Jersey; at Cape Hatteras, North Carolina; and along the Texas coast of the Gulf of Mexico, where one—Padre Island—is 130 km long. As the sandbars build up above the waves, vegetation takes hold, stabilizing the islands and helping them resist wave erosion during storms.

> *Wave action is strong enough to erode rocky coastlines. Softer sediments and sedimentary rocks can be more easily eroded.*

FIGURE 12.39 Multiple wave-cut terraces on the California coastline. [Photo by Dan Muhs/USGS. Daniel R. Muhs, Kathleen R. Simmons, George L. Kennedy, and Thomas K. Rockwell, "The Last Interglacial Period on the Pacific Coast of North America: Timing and Paleoclimate." *Geological Society of America Bulletin* (May 2002): 569–592.]

Barrier islands are separated from the coast by tidal flats or shallow lagoons. Like beaches on the main shore, barrier islands are affected by natural changes in climate or in wave and current patterns, as well as by real estate development. Disruption or devegetation can lead to increased erosion, and barrier islands may even disappear beneath the sea surface. Barrier islands may also grow larger and more stable if sedimentation increases.

Over hundreds of years, these shorelines may undergo significant changes. Hurricanes and other intense storms, such as the "storm of the century" that hit the Eastern Seaboard of the United States in March 1993, may form new inlets, elongate spits, or breach existing spits and barrier islands. Such changes have been documented by aerial photographs taken at various time intervals. The shoreline of Chatham, Massachusetts, at the elbow of Cape Cod, has changed enough in the past 160 years or so that a lighthouse has had to be moved. **Figure 12.40** illustrates the many changes that have taken place in the configuration of the barrier islands to the north and to the long spit of Monomoy Island, as well as several breaches of the barrier islands. Many homes in Chatham are now at risk, but there is little that the residents or the state can do to prevent these beach processes from taking their natural course.

> Spits, sandbars, and barrier islands may form along coastlines where sediments are easily eroded and transported.

(a) Beach near Chatham Light

The 1987 breach in the barrier spit, shown at the right below, closed again before this photo was taken.

(b)

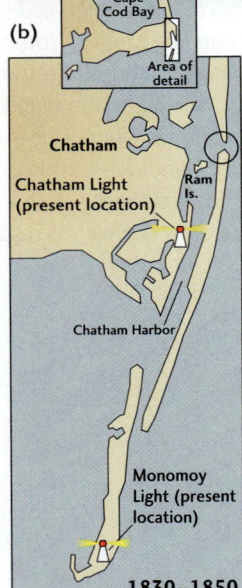

1830–1850
Circle shows approximate location of 1846 breach in barrier spit. Ram Island later disappears.

1870–1890
Beach south of the inlet breaks up and migrates southwest toward the mainland and Monomoy.

1910–1930
The southern beach has disappeared, and its remnants soon will connect Monomoy to the mainland.

1950–1970
The northern beach steadily grows with cliff sediment; Monomoy breaks from the mainland.

1987
140-year cycle begins again with Jan. 2 breach in the barrier spit across from Chatham Light (circle).

FIGURE 12.40 Migrating barrier islands at Chatham, Massachusetts, at the southern tip of Cape Cod. (a) Aerial view of Monomoy Point. This spit has advanced into deep water to the south (*foreground*) from barrier islands along the main body of the Cape to the north (*background*). [Steve Dunwell/The Image Bank.] (b) Changes in the shoreline at Chatham over the past 160 years. [After Cindy Daniels, *Boston Globe* (February 23, 1987).]

OFFSHORE: The Continental Shelf and Beyond

Offshore, beyond the shoreline, lies the continental shelf. At its edge is a continental slope that descends more or less steeply into the depths of the ocean. At the foot of the slope is a continental rise, a gently sloping apron of sediment extending to the abyssal plain at the bottom of the ocean basin (**Figure 12.41**).

Continental shelves consist of essentially flat-lying sediments several kilometers thick. These sediments accumulate mostly as a result of transport by waves and tidal currents. The continental shelf is one of the most economically valuable parts of the ocean. Georges Bank off New England and the Grand Banks of Newfoundland, for example, are among the world's most productive fishing grounds. Recently, oil-drilling platforms have been used to extract huge quantities of oil and gas from the continental shelf, especially off the Gulf Coast of Louisiana and Texas.

Seaward of the continental shelf, the seafloor drops off abruptly. Here, the waters are too deep for the seafloor to be affected by waves and tidal currents. As a consequence, sediments that have been carried across the shallow continental shelf by waves and tides come to rest on the continental slope. Deposits of sand, silt, and mud on both the continental slope and the continental rise indicate active transport of sediments into deeper waters. These sediments are transported by **turbidity currents**—turbulent flows of muddy water down the slope. Because of its suspended load of mud, the turbid water is denser than the overlying clear water and flows beneath it.

Turbidity currents start when the sediment draped over the edge of the continental shelf slumps onto the continental slope (see Figure 12.41). Such a sudden submarine landslide, which can occur spontaneously or be triggered by an earthquake, throws mud into suspension, creating a dense, turbid layer of water near the bottom. This turbid layer starts to flow, accelerating down the slope. As the turbidity current reaches the foot of the slope, it slows. Some of the coarser sandy sediment starts to settle, often forming a *submarine fan*—a deposit something like an alluvial fan on land. Stronger turbidity currents may continue across the continental rise, cutting channels into submarine fans. Where these currents reach the abyssal plain, they spread out and come to rest in beds of sand, silt, and mud called *turbidites*.

Some of the world's largest oil and gas fields are located in very deep water where turbidites have accumulated, forming an ideal environment for trapping these fossil fuels. Their exploitation is a tricky and

> The continental shelf is formed by thick accumulations of sediments transported offshore by waves and tides.

> Turbidity currents transport sediments from the continental shelf down the continental slope to the abyssal plain.

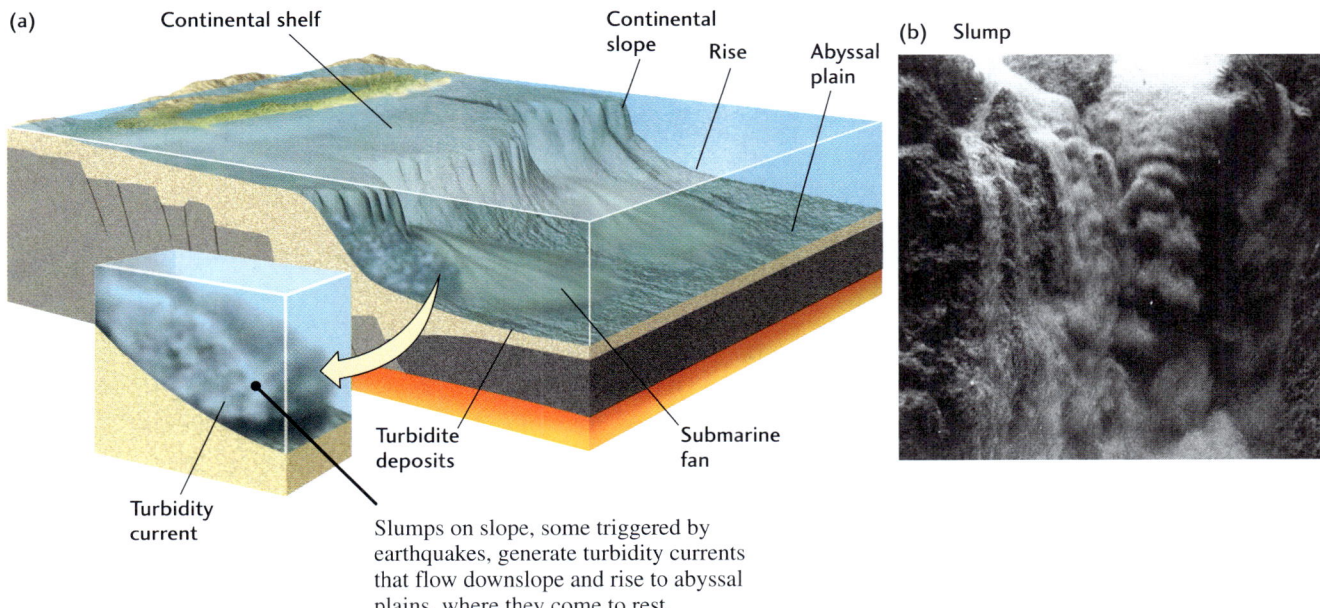

FIGURE 12.41 Turbidity currents transport sediments from the continental shelf into deeper waters. (a) How a turbidity current forms. These currents can erode and transport large quantities of sand down the continental slope. (b) A slump at the edge of the continental shelf. [U.S. Navy.]

expensive business, as deep-sea wells may cost hundreds of millions of dollars to drill.

So far we have considered environments and processes dominated by flows of water. Let's turn our attention to deserts, where water is scarce and wind has a major role in sculpting the landscape.

THE DESERT ENVIRONMENT

The deserts of the world are among the most hostile environments for humans. Yet many of us are fascinated by these hot, dry, apparently lifeless zones, full of bare rocks and sand dunes. Desert environments amount to one-fifth of Earth's land area, about 27.5 million square kilometers. Of all Earth's environments, the desert is where wind is best able to do its work of erosion, transportation, and sedimentation. Today's deserts may have been wet regions in the past but may have dried out as a result of long-term climate change. The dry climate of deserts creates harsh yet fragile conditions, where human impacts last for decades.

Where Deserts Are Found

Rainfall is the major factor determining the locations of the world's great deserts (**Figure 12.42**). The Sahara and Kalahari deserts of Africa and the Great Australian Desert get extremely low amounts of rainfall, normally less than 25 mm/year and in some places less than 5 mm/year. These deserts are found between 30° N and 30° S latitudes, where prevailing wind patterns cause dry air to sink to ground level and atmospheric pressures are high (see Figure 12.4). Because the relative humidity is extremely low, clouds are rare, and the chance for precipitation is very small. The Sun beats down week after week.

Deserts also exist at midlatitudes—between 30° N and 50° N and between 30° S and 50° S—in regions where rainfall is low because moisture-laden winds either are blocked by mountain ranges or must travel great distances from the ocean, their source of moisture. The Great Basin and Mojave deserts of the western United States, for example, lie in rain shadows created by the western coastal mountains (see Figure 11.2). The Gobi and other deserts of central Asia are so deep in the interior of the continent that the winds reaching them have precipitated all their ocean-derived moisture long before they arrive there.

Another kind of desert is found in polar regions. There is little precipitation in these cold, dry areas because the frigid air can hold only small amounts of moisture. The dry valley region of southern Victoria Land in Antarctica is so dry and cold that its environment resembles that of Mars.

Deserts develop at latitudes where global wind patterns limit rainfall. They also form where mountains create rain shadows, in the interior of continents, and in polar regions.

FIGURE 12.42 Major desert areas of the world (exclusive of polar deserts) in relation to prevailing wind patterns and major mountain ranges. Notice that sand dunes make up only a small proportion of the total desert area. [After K. W. Glennie, *Desert Sedimentary Environments.* New York: Elsevier, 1970.]

The Role of Climate Change Changes in a region's climate may transform semiarid lands into deserts, a process called **desertification.** Climate changes that we do not fully understand may decrease precipitation for decades or even centuries. After such a dry period, a region may return to a milder, wetter climate. Over the past 10,000 years, climates of the Sahara appear to have oscillated between drier and wetter conditions. We have evidence from orbiting satellites above the Sahara that an extensive system of river channels existed a few thousand years ago (**Figure 12.43**). Now dry and buried by more recent sand deposits, these ancient drainage systems carried abundant running water across the northern Sahara during wetter periods.

The Sahara Desert may now be expanding northward. The Desert Watch project, led by the European Space Agency, reports that over 300,000 km² of Europe's Mediterranean coast (an area almost as large as the state of New York), with a population of 16 million, has been enduring the longest drought in recorded history. During 2005, fires raged along the southern Spanish coast, and temperatures set new record highs for weeks on end. Was this merely a long, hot summer, or are these the initial symptoms of desertification, made worse by overpopulation and overdevelopment within the fragile ecosystems of dry landscapes? Evidence to support the latter scenario is building.

The Role of Humans Climate change in the Sahara and other deserts occurs naturally, but human activities are responsible for some desertification. The growth of human populations and increased agriculture and animal grazing may result in the expansion of deserts. When population growth and periods of drought coincide, the results in semiarid regions can be disastrous. In Spain, the greatest urban and agricultural expansion is taking place in the driest region. Former farmlands have been stripped of vegetation due to overfarming (up to four crops per year), which depletes water and strips soils. A tourism boom and its accompanying development are literally paving over the dry lands and desiccating the countryside that

> *Long-term climate change can create deserts where milder, wetter climates once were present. This process is called desertification.*

> *Human activities, such as depletion of water sources, overfarming, and urbanization, can result in desertification.*

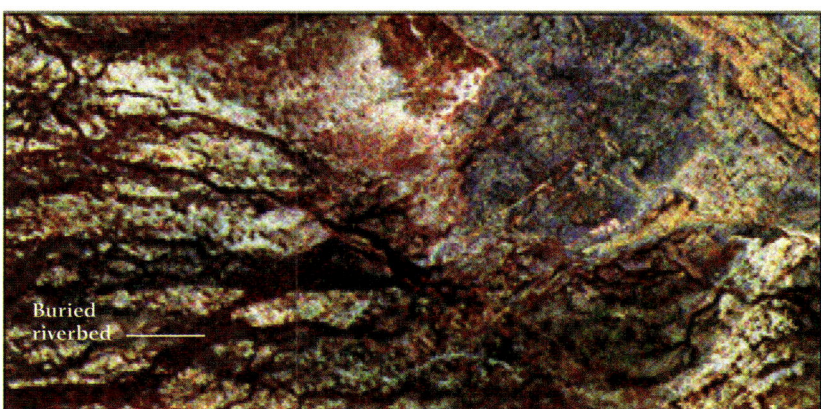

FIGURE 12.43 Remote sensing techniques that look at the surface of Earth from orbiting satellites see only evidence for sand in the Sahara (*top*). However, remote sensing techniques that penetrate a few meters below the surface see a dense network of buried streambeds (*bottom*). [NASA/JPL Imaging Radar Team.]

is left. In 2004, more than 350,000 new homes were built on Spain's Mediterranean coast, many with backyard swimming pools and nearby golf courses requiring large amounts of water. In isolation, any one of these human activities might not have a negative effect. Together, however, they add up to desertification.

Desert Weathering

As unique as they are, the same geologic processes operate in deserts as elsewhere. Weathering and transportation work in the same way in deserts as they do everywhere, but with a different balance. In deserts, physical weathering predominates over chemical weathering. Chemical weathering of feldspars and other silicates to clay minerals proceeds slowly because the water required for those reactions is lacking. The little clay that does form is usually blown away by strong winds before it can accumulate. Slow chemical weathering and rapid wind transport combine to prevent the buildup of any significant thickness of soil, even where sparse vegetation binds some of the particles. Thus, desert soils are thin and patchy. Sand, gravel, rock fragments of many sizes, and bare bedrock are characteristic of much of the desert surface.

Physical weathering predominates over chemical weathering in deserts because of the lack of water.

FIGURE 12.44 Petroglyphs scratched in desert varnish by Native Americans at Newspaper Rock, Canyonlands, Utah. The scratches are several hundred years old but appear fresh on the varnish, which has accumulated over thousands of years. [Peter Kresan.]

The Colors of the Desert The rusty, orange-brown colors of many weathered surfaces in the desert come from the ferric iron oxide minerals hematite and limonite. These minerals are produced by the slow chemical weathering of iron silicate minerals such as pyroxene. The iron oxides, even when present only in small amounts, stain the surfaces of sands, gravels, and clays.

Desert varnish is a distinctive dark brown, sometimes shiny coating found on many rock surfaces in the desert. It is a mixture of clay minerals with smaller amounts of manganese and iron oxides. Desert varnish probably forms when silicate minerals weather in the presence of dew to form iron and manganese oxides that adhere to exposed rock surfaces. The process is so slow that Native American inscriptions scratched in desert varnish hundreds of years ago still appear fresh, with a stark contrast between the dark varnish and the light unweathered rock beneath (**Figure 12.44**). Desert varnish requires thousands of years to form, and some particularly ancient varnishes in North America are of Miocene age. However, recognizing varnish as such on ancient sandstones is difficult.

Desert varnish is a shiny, dark coating on rock surfaces that forms primarily by accumulation of iron and manganese oxides.

Desert Streams Wind plays a larger role in erosion in the desert than it does elsewhere, but it cannot compete with the erosive power of streams. Even though it rains so seldom that most streams flow only intermittently, streams do most of the erosional work in the desert when they do flow.

Even the driest desert gets occasional rain. In sandy and gravelly areas of deserts, rainfall infiltrates the soil and permeable bedrock. There, some of it evaporates. A smaller amount eventually reaches the groundwater table far below—in some places, as much as hundreds of meters below the surface. Desert oases form where the groundwater table comes close enough to the surface that roots of palms and other plants can reach it.

When rain occurs in heavy cloudbursts, so much water falls in such a short time that infiltration cannot keep pace, and the bulk of the water runs off into streams. Unhindered by vegetation, the runoff is rapid and may cause flash floods along valley floors that have been dry for years. Thus, a large proportion of streamflows in the desert consist of floods. When floods occur, they have great erosive power because most of the loose sediment is not held in place by vegetation. Streams may become so choked with sediment that they look more like fast-moving mudflows than rivers. The abrasiveness of this sediment load moving rapidly at flood velocities makes such streams efficient eroders of bedrock valleys.

Most desert streams flow only intermittently, but a large proportion of streamflows in deserts are floods.

Desert Sediments and Sedimentation

Deserts are composed of a diverse set of depositional environments. These environments may change dramatically when rain suddenly forms raging rivers and widespread

(a)

(b)

FIGURE 12.45 (a) A desert valley during a summer thunderstorm at Saguaro National Park, Arizona. (b) The same valley a day after the flooding. The coarse sediment deposited by such sudden desert floods may cover the entire valley floor. [Peter Kresan.]

lakes. Prolonged dry periods intervene, during which sediments are blown into sand dunes.

Alluvial Deposits As sediment-laden flash floods dry up, they leave distinctive deposits on the floors of desert valleys. In many cases, a flat fill of coarse sediment covers the entire valley floor, and the ordinary differentiation of the stream into channel, natural levees, and floodplain is absent (**Figure 12.45**). The sediments of many other desert valleys clearly show the intermixing of stream-deposited channel and floodplain sediments with eolian sediments. This pattern creates deposits containing extensive layers of eolian sandstone separated by stream flood sediments and ancient river sandstones.

Large alluvial fans are prominent features at mountain fronts in deserts because desert streams deposit much of their sediment load on the fans (see Figure 12.20). The rapid infiltration of stream water into the permeable sediments that make up the fan deprives the stream of the water required to carry the sediment load any farther downstream.

Alluvial deposits in deserts are generally spread across the valley floor.

Eolian Sediments By far the most dramatic sedimentary accumulations in deserts are sand dunes. Dune fields range in size from a few square kilometers to the "seas of sand" found in Namibia (see Figure 12.7). These sand seas—or *ergs*—may cover as much as 500,000 km², twice the area of the state of Nevada. Although film and television portrayals might lead one to think that deserts are mostly sand, actually only one-fifth of the world's desert area is covered by sand (see Figure 12.42). The other four-fifths are rocky or covered with desert pavement—a coarse, gravelly surface that forms as finer-grained particles are eroded away. Sand covers only a little more than one-tenth of the Sahara Desert, and sand dunes are far less common in the deserts of the southwestern United States.

Eolian sediments, typified by vast dune fields, are common in deserts.

Evaporite Sediments **Playa lakes** are permanent or temporary lakes that occur in arid mountain valleys or basins where water pools after rainstorms (**Figure 12.46**). As the

FIGURE 12.46 A desert playa lake in Death Valley, California. [David Muench.]

Playa lakes form where water pools after rainstorms and dissolved minerals are concentrated and precipitated as it evaporates.

lake water evaporates, dissolved minerals are concentrated and gradually precipitated. Playa lakes are sources of evaporite minerals such as sodium carbonate, borax (sodium borate), and other unusual salts. The water in playa lakes may be deadly to drink because of the high concentrations of dissolved minerals. Desert streams carry large amounts of dissolved salts, and these salts accumulate in playa lakes when the streams redissolve evaporite minerals deposited by evaporation from earlier runoff. If evaporation is complete, the lakes become *playas,* flat beds of clay that are sometimes encrusted with precipitated salts.

SUMMARY >>

How do water and air behave as fluids, and how do they transport and deposit sediments? A fluid can move in either laminar or turbulent flow, depending on its velocity, viscosity, and depth. The turbulence that characterizes most water and wind currents is responsible for transporting sediments by suspension (clays), saltation (sands), and rolling and sliding along the bed (sand and gravel). The tendency for particles to be carried in suspension is countered by the gravitational force that pulls them to settle to the bottom.

Where do winds form and how do they flow? Earth is encircled by belts of prevailing winds that develop because the Sun warms Earth most intensely at the equator, which causes the air to rise there and flow toward the poles. As the air moves toward the poles, it gradually cools and begins to sink. This cool, dense air then flows back along Earth's surface to the equator.

How do winds deposit sand dunes and dust? Dunes migrate downwind as sand grains saltate up their gentler windward slopes and fall over onto their steeper downwind slip faces. As the velocity of dust-laden winds decreases, the dust settles in thick layers.

How do stream valleys and their channels and floodplains evolve? As a stream flows, it carves a valley with steep to gently sloping walls and a floodplain on either side of its channel. The channel may be straight, meandering, or braided. During normal, nonflood periods, the channel carries the flow of water and sediments within its banks. When stream discharge increases to flood stage, the sediment-laden water overflows the banks of the channel and inundates the floodplain. The velocity of the floodwater decreases as it spreads over the floodplain. The water drops sediments, which build up natural levees and floodplain deposits.

How do drainage networks work as collection systems and deltas as distribution systems for water and sediment? Rivers and their tributaries constitute an upstream-branching drainage network that collects water and sediments from a drainage basin. Each drainage basin is separated from other drainage basins by a divide. Where it meets the ocean, the river drops its sediments to form a delta, branching downstream into distributaries that spread sediments across the delta.

What processes shape shorelines? Waves and tides, interacting with plate tectonic processes, shape the landscapes of coastlines, which vary from beaches and tidal flats to uplifted rocky coasts. Winds blowing over the ocean generate swell; as those waves approach the shore, they are transformed into breakers in the surf zone. Waves approaching the shoreline at an angle result in longshore drift and longshore currents, which transport sand along beaches. Tides, which are generated by the gravitational pull of the Moon on ocean water, can generate currents that transport sediments to tidal flats and form bars in tidal channels.

How do hurricanes affect coastal areas? Hurricanes are intense storms with extremely high winds and very low atmospheric pressures. The low pressure results in the formation of a dome of seawater, known as a storm surge. As the storm surge moves onshore, it floods low-lying areas, often causing more extensive damage than the storm's high winds.

What forces shape offshore environments? Waves and tides affect the continental shelf, but the continental slope is shaped primarily by turbidity currents. These deep-water currents carry large loads of sediment from the continental shelf down the continental slope to the continental rise and abyssal plain, where they form deposits known as turbidites.

How do wind and water combine to shape the desert environment and its landscape? Deserts occur in subtropical regions of constant high atmospheric pressure, in the rain shadows of mountain ranges, and in the interiors of continents. Physical weathering predominates over chemical weathering because of the lack of water. Most desert soils are thin, and bare rock surfaces are common. Wind plays a larger role in shaping the landscape in deserts than it does elsewhere, but streams are responsible for most erosion and sedimentation in deserts even though they flow only intermittently.

KEY TERMS AND CONCEPTS >>

alluvial fan (p. 319)
barrier island (p. 333)
beach (p. 331)
bed load (p. 310)
braided stream (p. 319)
channel (p. 306)

coastline (p. 307)
delta (p. 323)
desertification (p. 337)
discharge (p. 320)
distributary (p. 323)
divide (p. 321)

drainage basin (p. 321)
dune (p. 311)
eolian (p. 308)
floodplain (p. 316)
laminar flow (p. 308)
longshore current (p. 328)
meander (p. 318)
natural levee (p. 320)
playa lake (p. 339)
point bar (p. 319)
ripple (p. 312)
river (p. 306)
saltation (p. 310)
shoreline (p. 325)
spit (p. 333)
storm surge (p. 330)
stream (p. 306)
stream valley (p. 316)
suspended load (p. 310)
tides (p. 328)
turbidity current (p. 335)
turbulent flow (p. 308)
wave-cut terrace (p. 333)

EXERCISES >>

1. What is the difference between the way wind transports dust and the way it transports sand?

2. What is a natural levee, and how is it formed?

3. What is a delta distributary?

4. How are ocean waves formed?

5. How does the Moon interact with Earth to create tidal currents?

6. What is a storm surge?

7. Where do turbidity currents form?

8. What is desertification?

9. What are the geologic processes that form playa lakes?

10. Why might the flow of a very small, shallow stream be laminar in winter and turbulent in summer?

11. In some places, engineers have artificially straightened a meandering stream. If such a straightened stream were then left free to adjust its course naturally, what changes would you expect?

12. In a 100-year period, the southern tip of a long, narrow, north-south beach has become extended about 200 m to the south by natural processes. What shoreline processes could have caused this extension?

13. There are large areas of sand dunes on Mars. From this fact alone, what can you infer about conditions on the Martian surface?

14. How does desert weathering differ from or resemble weathering in more humid climates?

EARTHQUAKES AND EARTH STRUCTURE

CHAPTER 13

The Great Alaska Earthquake

At 5:36 P.M. on March 27, 1964, the crust beneath Prince William Sound began to rupture. The slip started on a small patch of the fault surface and accelerated outward like a crack through glass. The rupture ran rampant down the Alaska coastline for over 700 km, to beyond Kodiak Island, at speeds reaching 2.7 km/s (6000 miles/hour). By the time it stopped 6 minutes later, the Pacific Plate had slipped under the North American Plate by as much as 20 m along a fault surface the size of Iowa. Seismic waves propagated away from the rupture over Earth's surface and through its deep interior, causing the planet to ring like a bell for many days.

The thrusting of the North American Plate up into the shallow sea displaced several hundred billion tons of water, which moved away from the fault in a huge tsunami. The water waves, slower than the seismic waves but more deadly, passed through the bays and inlets of the mountainous coast like an undulating monster, gaining height as they approached the shore. The district geologist of Valdez, Ralph Migliaccio, filed the following report:

Within seconds of the initial tremors, it was apparent to eyewitnesses that something violent was occurring in the area of the Valdez waterfront.... Men, women, and children were seen staggering around the dock, looking for something to hold onto.... Almost immediately a large wave rose up, smashing everything in its path.... Several people stated the wave was 30 to 40 feet high, or more.... Approximately 10 minutes after the initial wave receded, a second wave crossed the waterfront carrying large amounts of wreckage.... There followed a lull of approximately 5 or 6 hours during which time search parties were able to search the waterfront area for possible survivors. There were none.

The city of Anchorage, 100 kilometers west of the epicenter, was shielded from the tsunami, but it experienced considerable damage, especially in a low-lying region of soft sediments that became liquefied by the shaking, causing a massive landslide. Robert Atwood, who lived in the Turnagain Heights residential section, described his experiences during the landslide:

I had just started to practice playing the trumpet when the earthquake occurred.... I headed for the door.... Tall trees were falling in our yard. Pieces of the ground in jigsaw-puzzle shapes moved up and down, tilted at all angles.... My neighbor's house collapsed and slid into the chasm.

The tsunami propagated across the Pacific Ocean, striking coastlines thousands of kilometers away. In Crescent City, California, a series of large waves inundated the harbor, beginning $4\frac{1}{2}$ hours after the rupture. After the first two waves had struck, seven people returned to a seaside tavern to recover their valuables. Since the ocean seemed normal, they remained to have a drink and were caught by the third wave, which killed five of them. ■

The Turnagain Heights subdivision near Anchorage, Alaska, after the landslide caused by the great 1964 Alaska earthquake. [NOAA/ Central Library.]

CHAPTER 13: Earthquakes and Earth Structure

CHAPTER OUTLINE

- What Is an Earthquake? 344
- Measuring Earthquakes Using Seismic Waves 349
- Exploring Earth's Interior Using Seismic Waves 356
- Earthquakes and Patterns of Faulting 363
- Earthquake Destructiveness 366
- Can Earthquakes Be Predicted? 375

Earthquakes rival all other natural disasters in the threat they pose to human life and property. Our fragile "built environment" is necessarily anchored in Earth's active crust, making it vulnerable to seismic movements and their secondary effects, such as landslides and tsunamis.

To cope with all-too-frequent earthquake disasters, we have long sought to improve our ability to predict where and when earthquakes might occur and our understanding of what happens when they do. Science has shown that most seismic activity can be understood in terms of the basic machinery of plate tectonics. Consequently, efforts to reduce earthquake risk have become increasingly tied to the quest for a more fundamental understanding of our active planet.

This chapter will examine what happens during an earthquake, how scientists use seismic waves to locate and measure earthquakes, and what can be done to reduce earthquake risk. We will also see how seismic waves have been used to image Earth's layered structure—its crust, mantle, and core—and how they are now being used to image the motions of mantle convection.

We have already discussed why earthquakes are concentrated on the faults that form the boundaries between shifting tectonic plates (see Chapter 3), as well as the basic types of faulting produced in brittle crust by the forces acting at those boundaries (see Chapter 7). Here we will see that the pattern of faulting in many earthquake-prone areas, such as California, can be much more complex than the idealized boundaries of simple plate tectonics.

These complexities contribute to the difficulty of earthquake prediction, which remains an unsolved scientific problem. As we will see, however, there are several important ways in which geological knowledge can be used to reduce earthquake risk.

WHAT IS AN EARTHQUAKE?

We saw in Chapter 7 how plate movements generate forces at the boundaries between tectonic plates. These forces act on crustal rocks in ways that can be described by the concepts of stress, strain, and strength. *Stress* is the local force per unit area that causes rocks to deform. *Strain* is the relative amount of deformation, expressed as the percentage of distortion (for example, compression of a rock by 1 percent of its length). Rocks *fail*—that is, they lose cohesion and break into two or more parts—when they are stressed beyond a critical value, called their *strength*.

An **earthquake** occurs when brittle rocks being stressed suddenly fail along a geologic fault. Most large earthquakes are caused by ruptures of preexisting faults, where past earthquakes have already weakened the rocks on the fault surface. The two blocks of rock on either side of the fault slip suddenly, releasing energy in the form of *seismic waves*, which are manifested as ground vibrations. When the fault slips, the stress is reduced, dropping to a level below the rock strength. After the earthquake, the stress begins to increase again, and the cycle is repeated (Figure 13.1). Earthquakes occur most often at plate boundaries, where stresses are concentrated and straining of the crust is intense.

Earthquakes occur when rocks on a fault surface are stressed beyond their strength and the fault slips.

The Elastic Rebound Theory

On a fine April morning in 1906, the citizens of northern California were awakened by the roar and violent shaking caused by the breaking of the San Andreas fault. This rupture began just west of the Golden Gate and extended over 400 km, from the mission town of San Juan Bautista all the way to Cape Mendocino (Figure 13.2), creating the most destructive earthquake the United States has yet experienced. The ensuing fires destroyed the city of San Francisco; by the time the flames died away, nearly 3000 of its inhabitants were dead (Figure 13.3).

The earthquake on the San Andreas fault that devastated San Francisco in 1906 was studied more than any earthquake had been up to that time. In 1910, a geologist who investigated the catastrophe, Henry Fielding Reid of Johns Hopkins University, advanced the **elastic rebound theory** to explain why earthquakes recur on active faults in Earth's crust.

To visualize what happens in an earthquake, imagine the following experiment carried out across a strike-slip fault between two crustal blocks. Suppose that surveyors had painted straight lines on the ground, running perpendicular to a fault and extending from one block to the other, as in Figure 13.1. The two blocks are being pushed in opposite directions by plate movements. The weight of the overlying rock presses them together, however, so friction locks them in place along the fault. They do not move, just as a car does not move when the emergency brake is engaged. Instead of slipping along the fault as stress builds up, the blocks are strained elastically near the

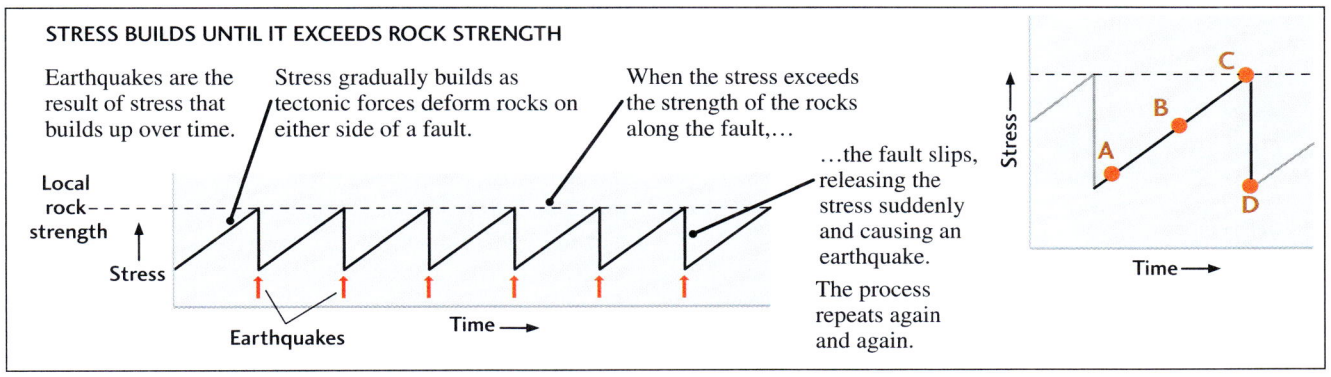

FIGURE 13.1 The elastic rebound theory explains the earthquake cycle.
[Photo by G. K. Gilbert/USGS.]

FIGURE 13.2 Map of California, showing the segments of the San Andreas fault that ruptured in 1680, 1857, and 1906. [Southern California Earthquake Center.]

fault, as shown by the bent lines in Figure 13.1. By *elastically*, we mean that the blocks would spring back and return to their undeformed, stress-free shape if the fault were suddenly to unlock.

As the slow plate movements continue to push the blocks in opposite directions, the strain in the rocks—evidenced by the bending of the survey lines—continues to build up over decades, centuries, or even millennia. At some point, the strength of the rocks is exceeded. The frictional bond that locks the fault can no longer hold, and somewhere along the fault, it breaks. The blocks slip suddenly, and the rupture extends over a section of the fault.

The bottom drawing in Figure 13.1 shows how the two blocks have rebounded—sprung back to their undeformed state—after the earthquake. The imaginary bent lines have straightened, and the two blocks have been displaced. The distance of the displacement is called the **fault slip.** After the fault has slipped, it locks up again. The steady movement of the blocks on either side of the fault causes the stresses to begin to rise again, and the cycle repeats.

The energy that is slowly built up by elastic strain as two blocks are pushed in opposite directions is like the elastic energy stored in a rubber band when it is slowly stretched. The sudden release of energy when a fault slips is like the violent backlash, or *rebound*, that occurs when the rubber band breaks. In the same way, elastic energy accumulates and is stored for many decades in rocks under stress. The energy is released at the moment the fault breaks, and some of it is radiated as seismic waves in the few minutes of an earthquake.

According to the elastic rebound theory, stressed rocks are strained elastically, so that when a fault breaks, the rocks on either side of the fault return to their undeformed, stress-free shape.

The elastic rebound theory implies that there should be a periodic buildup and release of elastic energy at faults. However, real earthquakes rarely exhibit this simple be-

FIGURE 13.3 This photograph, taken from a balloon by George Lawrence 5 weeks after the great earthquake of April 18, 1906, shows the devastation of San Francisco caused by the quake and subsequent fire. The view is looking over Nob Hill toward the business district. [CORBIS.]

FIGURE 13.4 Irregularities in the earthquake cycle can be caused by variations in rock strength and stress accumulation. [H. Kanamori and E. E. Brosky, *Physics Today* (June 2001): 34–39.]

havior. For instance, all of the strain accumulated since the last earthquake may not be released in the next—that is, the rebound may be incomplete—or the stress may not build up at a steady rate (Figure 13.4). This irregularity is one reason why earthquakes are so difficult to predict.

Fault Rupture During Earthquakes

The point at which fault slipping begins is the **focus** of the earthquake (Figure 13.5). The **epicenter** is the geographic

FIGURE 13.5 During an earthquake, fault slipping begins at the focus and spreads out along the fault surface. Panel C is just before the earthquake; panel D is just after, as in

Figure 13.1. Panels 1–4 are snapshots of the fault rupture corresponding to the numbered points on the graph.

point on Earth's surface directly above the focus. For example, you might hear in a news report: "The U.S. Geological Survey reports that the epicenter of last night's destructive earthquake in California was located 6 kilometers east of Los Angeles City Hall. The depth of the focus was 14 kilometers."

For most earthquakes occurring in continental crust, focal depths range from about 2 to 20 km. Continental earthquakes below 20 km are rare, because under the high temperatures and pressures found at greater depths, the crust behaves as a ductile rather than a brittle material (just as hot wax flows when stressed, whereas cold wax breaks; see Chapter 7). In subduction zones, however, where cold oceanic lithosphere plunges into the mantle, earthquakes can originate at depths as great as 690 km.

The fault rupture does not happen all at once. It begins at the focus and expands outward along the fault surface, typically at 2 to 3 km/s (see Figure 13.5). The rupture stops where the stresses become insufficient to continue breaking the fault (where the rocks are stronger) or where the rupture enters ductile material in which it can no longer propagate as a fracture. As we will see later in this chapter, the magnitude of an earthquake is related to the total area of fault rupture. Most earthquakes are very small, with rupture dimensions much less than the depth of the focus, so that the rupture never breaks the ground surface. In large, destructive earthquakes, however, surface breaks are common. The great 1906 San Francisco earthquake, for example, caused surface displacements averaging about 4 m along the 400-km section of the San Andreas fault that ruptured in that event (see inset of Figure 13.1 and Figure 13.2). Fault ruptures in the largest earthquakes can extend for more than 1000 km, and the fault slip can be as large as 20 m. Generally, the longer the fault rupture, the greater the fault slip.

As we have seen, the sudden slipping of the blocks at the time of the earthquake reduces the stress on the fault and releases much of the stored elastic energy. Most of this stored energy is converted to frictional heat in the fault zone, but part of it is released as seismic waves that travel outward from the rupture, much as waves ripple outward from the spot where a stone is dropped into a still pond. The focus of the earthquake generates the first seismic waves, although slipping parts of the fault continue to generate waves until the rupture stops. In a large event such as the 1906 San Francisco earthquake, the propagating fracture continues to produce waves for many tens of seconds. These waves radiate all along the fault rupture and can cause damage far from the epicenter.

When a fault slips, stored elastic energy is released as seismic waves, which travel outward from the earthquake focus.

Foreshocks and Aftershocks

Almost all large earthquakes trigger smaller earthquakes called **aftershocks.** Aftershocks follow the triggering event, or *mainshock,* in sequences, and their foci are distributed in and around the rupture plane of the mainshock (Figure 13.6). Although fault slipping during the mainshock decreases the stress along most of the rupture surface, it can increase the stress on parts of the fault surface that did not slip, or where the slip was incomplete, as well as in surrounding regions. Aftershocks happen where that stress exceeds the rock strength.

Aftershock sequences exemplify the complexities of earthquakes that cannot be described by simple elastic rebound theory. The number and sizes of aftershocks depend on the magnitude of the mainshock, and their frequencies decrease with time after the mainshock. The aftershock sequence of a magnitude 5 earthquake might last for only a few weeks, whereas that of a magnitude 7 earthquake could persist for several years. The size of the largest aftershock is usually about one magnitude unit smaller than the mainshock. According to this rough rule of thumb, a magnitude 7 earthquake might have an aftershock as large as magnitude 6 (and sometimes even larger). In populated regions, the shaking from large aftershocks

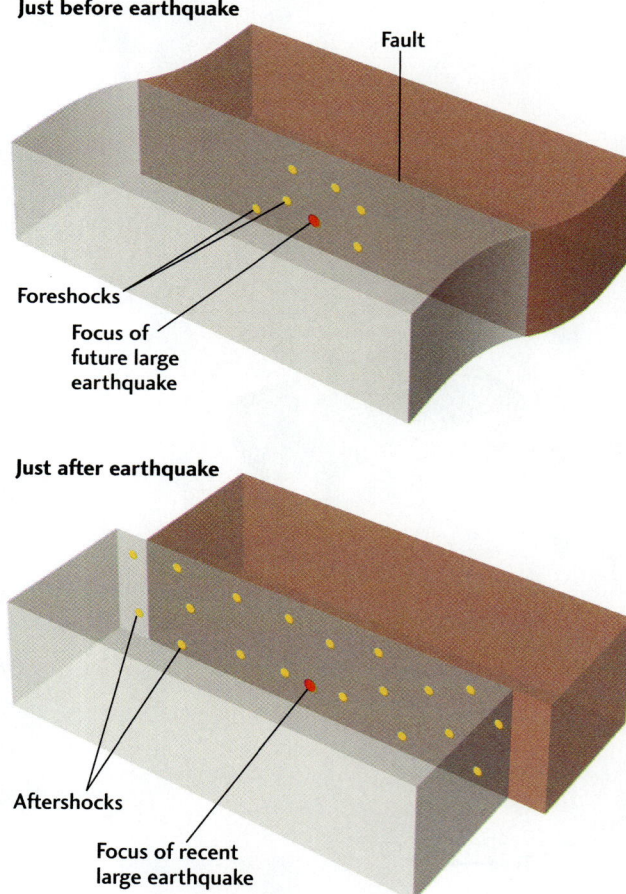

FIGURE 13.6 Aftershocks are smaller shocks that follow a large earthquake (the mainshock). Foreshocks occur near, but before, the mainshock.

can be very dangerous, compounding the damage caused by the mainshock.

A **foreshock** is a small earthquake that occurs near, but before, a mainshock (see Figure 13.6). One or more foreshocks have preceded many large earthquakes, so scientists have tried to use them to predict when and where large earthquakes might happen. Unfortunately, it is usually very hard to distinguish foreshocks from other small earthquakes that occur randomly and frequently on active faults, so this prediction method has only rarely proved successful.

> *Most large earthquakes are followed by smaller earthquakes called aftershocks. They may also be preceded by one or more foreshocks.*

MEASURING EARTHQUAKES USING SEISMIC WAVES

The **seismograph,** an instrument that records the seismic waves generated by earthquakes, is to the Earth scientist what the telescope is to the astronomer: a tool for peering into inaccessible regions (**Figure 13.7**). The ideal seismograph would be a device affixed to a stationary frame not attached to Earth. When the ground shook, the seismograph would measure the changing distance between the frame, which did not move, and the vibrating ground, which did. As yet, we have no way to position a seismograph that is not attached to Earth—although modern space technology, such as the Global Positioning System, is beginning to remove this limitation. So we compromise. We attach a dense mass, such as a piece of steel, to Earth so loosely that the ground can vibrate up and down or side to side without causing much motion of the mass.

This loose attachment can be achieved by suspending the mass from a spring (Figure 13.7a). When seismic waves move the ground up and down, the mass tends to remain stationary because of its inertia (an object at rest tends to stay at rest), but the mass and the ground move relative to each other because the spring can compress or stretch. In this way, the vertical displacement of the ground caused by seismic waves can be recorded by a pen on chart paper or, more likely these days, digitally on a computer; such a record is called a *seismogram.*

FIGURE 13.7 A seismograph can be designed to record (a) vertical or (b) horizontal motion. Because of its inertia and its loose coupling to Earth through (a) a spring or (b) a hinge, the dense mass does not keep up with the motion of the ground. A typical observatory has seismographs set up to measure three components of ground motion: vertical, horizontal east-west, and horizontal north-south.

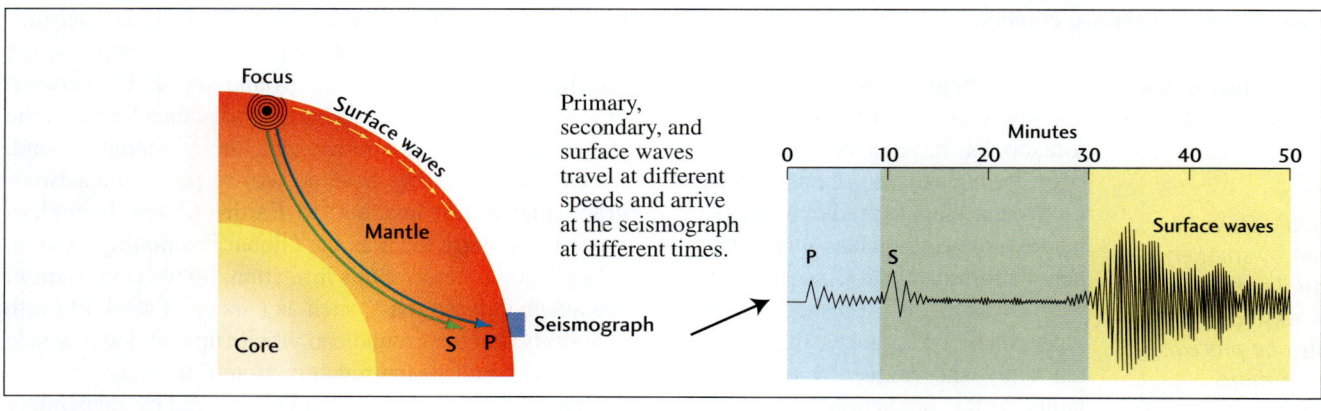

FIGURE 13.8 Seismographs record P-wave, S-wave, and surface-wave motions.

Loose attachment of the mass can also be achieved with a hinge. A seismograph that has its mass suspended on hinges like a swinging gate (Figure 13.7b) can record the horizontal displacement of the ground. Modern seismographs can detect ground oscillations of less than a billionth of a meter—an astounding feat, considering that such small displacements are of atomic size!

Seismic Waves

Install a seismograph anywhere, and within a few hours it will record the passage of seismic waves generated by an earthquake somewhere on Earth. The waves travel from the earthquake focus through Earth and arrive at the seismograph in three distinct groups. The first waves to arrive are called primary waves, or **P waves.** The secondary waves, or **S waves,** follow. Both P waves and S waves travel through Earth's interior. Finally come the **surface waves,** which travel around Earth's surface (**Figure 13.8**).

P waves in rock are similar to sound waves in air, except that P waves travel through the solid rock of Earth's crust at about 6 km/s, which is about 20 times faster than sound waves travel through air. Like sound waves, P waves are *compressional waves,* so called because they travel through solid, liquid, or gaseous materials as a succession of compressions and expansions. P waves can be thought of as push-pull waves: they push or pull particles of matter in the direction of their path of travel.

S waves travel through solid rock at a little more than half the velocity of P waves. They are called *shear waves* because they displace material at right angles to their path of travel. Shear waves cannot travel through liquids or gases.

> P waves are compressional waves that can travel through solid, liquid, or gaseous materials. S waves are shear waves that can travel only through solid materials.

The velocities at which P and S waves travel are higher when the resistance to their motion is greater. It takes more force to compress solids than to shear them, so P waves always travel faster than S waves through a solid, which is why the P waves from an earthquake arrive at a seismograph before the S waves. This physical principle also explains why S waves cannot travel through air, water, or Earth's liquid outer core: gases and liquids put up no resistance to shear.

Surface waves are confined to Earth's surface and outer layers, like waves on the ocean. Their velocity is slightly less than that of S waves. One type of surface wave sets up a rolling motion in the ground; another type shakes the ground sideways (see Figure 13.8).

People have felt seismic waves and witnessed their destructiveness throughout history, but not until the close of the nineteenth century were scientists able to devise seismographs to record them accurately. Seismic waves enable seismologists (scientists who study these waves and their sources) to locate earthquakes and determine the nature of the faulting that produces them, and they provide our most important means of probing Earth's deep interior.

Locating the Focus

Locating a quake's focus is like deducing the distance of a lightning strike on the basis of the time interval between the flash of light and the sound of thunder—the greater the distance to the lightning bolt, the longer the time interval. Light travels faster than sound, so the lightning flash may be likened to the P waves of an earthquake and the thunder to the slower S waves.

The time interval between the arrival of P waves and S waves depends on the distance the waves have traveled from the focus: the longer the time interval, the longer the distance the waves have traveled. Seismologists have calibrated how long it takes seismic waves to travel a certain distance using the waves from earthquakes and underground nuclear explosions at known distances from seismographs. To determine the approximate distance to a quake's focus, they read from a seismogram the amount of time that elapsed between the arrival of the first P waves and the later arrival of the S waves. Then they use a table or a graph like the one shown in **Figure 13.9** to determine the distance from the seismograph station to the focus. If they can determine the distances from three or more seismographic stations, they can locate the focus. They can also deduce the time of the quake at the focus, because the arrival time of the P waves at each station is known, and it is possible to determine from a graph or table how long those waves took to reach the station. Today, this entire process is carried out repeatedly by a computer until the data from a large number of seismographs agree on where the epicenter is, the depth of the focus, and the time when the earthquake began.

> Seismologists can determine the location of an earthquake's focus by using the time interval between the arrival of P waves and S waves at three or more seismographic stations.

Measuring the Size of an Earthquake

Locating earthquakes is only one step on the way to understanding them. We must also determine their sizes, or magnitudes. Other things (such as distance from the focus and regional geology) being equal, an earthquake's magnitude is the main factor that determines the intensity of the seismic waves and thus the earthquake's potential destructiveness.

Richter Magnitude In 1935, Charles Richter, a California seismologist, devised a simple procedure that assigned to each earthquake a numerical size, now called the

FIGURE 13.9 Readings from three or more seismographic stations can be used to determine the location of an earthquake's epicenter.

FIGURE 13.10 The maximum amplitude of ground motion, corrected by the P-S wave interval, is used to assign a Richter magnitude to an earthquake. [California Institute of Technology.]

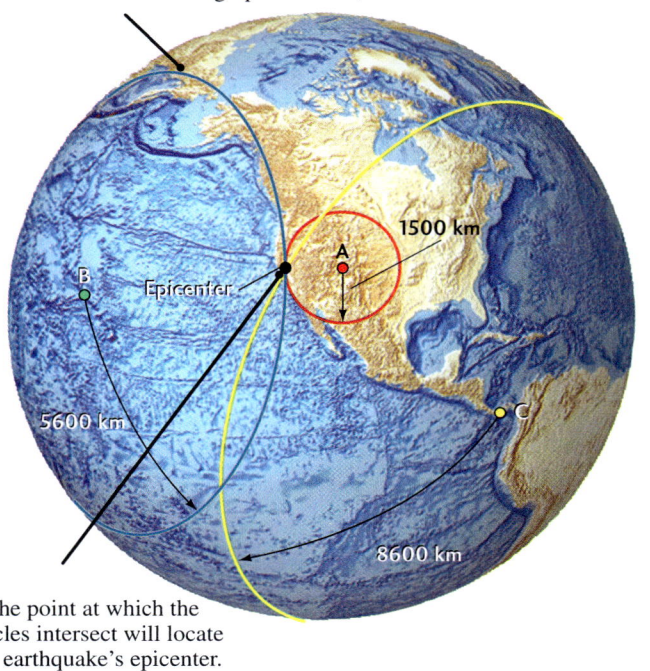

4 If the geologist then draws a circle with a radius calculated from the travel time curves around each seismographic station,…

5 …the point at which the circles intersect will locate the earthquake's epicenter.

Richter magnitude (**Figure 13.10**). Richter studied astronomy as a young man and learned how astronomers use a logarithmic scale to measure the brightness of stars over a huge range of values. Adapting this idea to earthquakes, Richter took as his measure of earthquake size the logarithm of the largest ground motion registered by a standard seismograph at a standard distance, thus defining a **magnitude scale**.

On Richter's magnitude scale, two earthquakes at the same distance from a seismograph differ by one magnitude unit if the size of their ground motions differs by a factor of 10. The ground motion of an earthquake of magnitude 3, therefore, is 10 times that of an earthquake of magnitude 2. Similarly, a magnitude 6 earthquake produces ground motions that are 100 times greater than those of a magnitude 4 earthquake. The energy released as seismic waves increases even more with earthquake magnitude, by a factor of about 32 for each Richter unit. A magnitude 8 earthquake releases 32 × 32, or about 1000, times the energy of a magnitude 6 earthquake.

> *The Richter magnitude scale estimates the size of an earthquake using the logarithm of the largest ground motion registered by a seismograph.*

Seismic waves gradually weaken as they move away from the focus, so to make his procedure work for any seismograph, Richter had to find a way to correct the measurement of ground motion for the distance between the seismograph and the focus. He devised simple graphical methods that allowed seismologists anywhere to quickly come up with nearly the same value for the magnitude of an earthquake no matter what distance their instruments were from the focus (see Figure 13.10).

Moment Magnitude Although the "Richter scale" has become a household term, seismologists prefer a measure of earthquake size that is more directly related to the physical properties of the faulting that causes the earthquake. The "moment" of an earthquake is defined as a number proportional to the product of the area of faulting and the average fault slip. The corresponding *moment magnitude* increases by about one unit for every 10-fold increase in the area of faulting. Although Richter's method and the moment method produce roughly the same numerical values, the moment magnitude can be measured more accurately from seismograms, and it can also be determined directly from field measurements of the fault.

> *The moment magnitude estimates the size of an earthquake from the area of faulting and the fault slip.*

Large earthquakes occur much less often than small ones. This observation can be expressed by a very simple relationship between earthquake frequency and magnitude (**Figure 13.11**). Worldwide, approximately 1,000,000 earthquakes with magnitudes greater than 2 take place each year, and this number decreases by a factor of 10 for each magnitude unit. Hence, there are about 100,000 earthquakes with magnitudes greater than 3, about 1000 with magnitudes greater than 5, and about 10 with magnitudes greater than 7.

According to these statistics, there should be, on average, about 1 earthquake with a magnitude greater than 8 per year and 1 earthquake with a magnitude greater than 9 every 10 years. In fact, the very largest earthquakes, such as the ones that occurred on thrust faults in the subduction zones off Sumatra in 2004 (moment magnitude 9.2), Alaska in 1964 (moment magnitude 9.1), and Chile in 1960 (moment magnitude 9.5), are not quite this common because there is an upper limit to how big fault ruptures can be.

> *Large earthquakes occur less often than small ones. The number of quakes per year decreases by a factor of 10 with each magnitude unit.*

Shaking Intensity Earthquake magnitude does not by itself describe the earthquake hazard, because the shaking that causes destruction generally weakens with distance from the fault rupture. A magnitude 8 earthquake in a remote area far from the nearest city might cause no human or economic losses, whereas a magnitude 6 quake

FIGURE 13.11 Relationship between moment magnitude, energy release, and number of earthquakes per year worldwide. Several other large sources of sudden energy release are included for comparison. [IRIS Consortium, http://www.iris.edu.]

immediately beneath a city is likely to cause serious damage.

In the late nineteenth century, before Richter invented his magnitude scale, seismologists and earthquake engineers developed methods for estimating the intensity of shaking directly from an earthquake's destructive effects. Table 13.1 shows the scale that remains in most common use today, called the *modified Mercalli intensity scale* after Giuseppe Mercalli, the Italian scientist who proposed it in 1902. This **intensity scale** assigns a value,

TABLE 13.1 Modified Mercalli Intensity Scale

Intensity Level	Description
I	Not felt.
II	Felt only by a few people at rest. Suspended objects may swing.
III	Felt noticeably indoors. Many people do not recognize it as an earthquake. Parked cars may rock slightly.
IV	Felt indoors by many, outdoors by few. Dishes, windows, doors rattle. Parked cars rock noticeably.
V	Felt by most; many awakened. Some dishes, windows broken. Unstable objects overturned.
VI	Felt by all. Some heavy furniture moves. Damage slight.
VII	Slight to moderate damage in well-built structures; considerable damage in poorly built structures; some chimneys broken.
VIII	Considerable damage in well-built structures. Damage great in poorly built structures. Fall of chimneys, factory stacks, columns, monuments, walls.
IX	Damage great in well-built structures, with partial collapse. Buildings shifted off foundations.
X	Some well-built wooden structures destroyed; most masonry and frame structures destroyed. Rails bent.
XI	Few if any masonry structures remain standing. Bridges destroyed. Rails bent greatly.
XII	Damage total. Lines of sight and level are distorted. Objects thrown into the air.

FIGURE 13.12 Modified Mercalli intensities measured for the New Madrid earthquake of December 16, 1811, a magnitude 7.7 event near the juncture of Missouri, Arkansas, and Tennessee. Regions near the fault rupture show intensities greater than IX, and intensities as high as VI were observed 200 km from the epicenter (see Table 13.1). [Carl W. Stover and Jerry L. Cossman, USGS Professional Paper 1527, 1993.]

given as a Roman numeral from I to XII, to the intensity of the shaking at a particular site. For example, a location where an earthquake was barely felt by a few people is assigned an intensity of II, whereas one where it was felt by nearly everyone is given an intensity of V. Numbers at the upper end of the scale describe increasing amounts of damage. The narrative attached to the highest value, XII, is tersely apocalyptic: "Damage total. Lines of sight and level are distorted. Objects thrown into the air." Not a place you would want to be!

By making observations at many sites and interviewing many people who experienced an earthquake, or even by examining historical records, seismologists can make maps showing contours of equal intensity. **Figure 13.12** shows an intensity map for the New Madrid earthquake of December 16, 1811, a magnitude 7.7 event near the southern tip of Missouri, which was felt as far away as Boston. Although earthquake intensities are generally highest near the fault rupture, they also depend on the local geology. For example, when sites at equal distances from the rupture are compared, the shaking tends to be more intense on soft sediments (especially water-saturated sediments near shorelines) than on hard basement rocks. Intensity maps thus provide engineers with crucial data for designing structures to withstand earthquake shaking.

Determining Fault Mechanisms

The pattern of ground shaking also depends on the orientation of the fault rupture and the direction of slipping, which together specify the **fault mechanism** of an earthquake. The fault mechanism tells us whether the rupture was on a *normal, reverse,* or *strike-slip* fault. If the rupture was on a strike-slip fault, the fault mechanism also tells us whether the motion was *right-lateral* or *left-lateral* (see Figure 7.7 for the definition of these terms). We can then use this information to infer the regional pattern of tectonic forces (**Figure 13.13**).

For shallow ruptures that break the surface, we can sometimes determine the fault mechanism from field observations of the fault scarp. As we have seen, however, most ruptures are too deep to break the surface, so we must deduce the fault mechanism from the motions measured on seismographs.

For large earthquakes at any depth, this turns out to be easy to do, because there are enough seismographs around

FIGURE 13.13 The three main types of fault mechanisms that initiate earthquakes and the stresses that cause them. (a) A fault before movement takes place. (b) Normal faulting due to tensional stress. (c) Reverse faulting due to compressive stress. (d) Strike-slip faulting due to shearing stress (in this case, left-lateral).

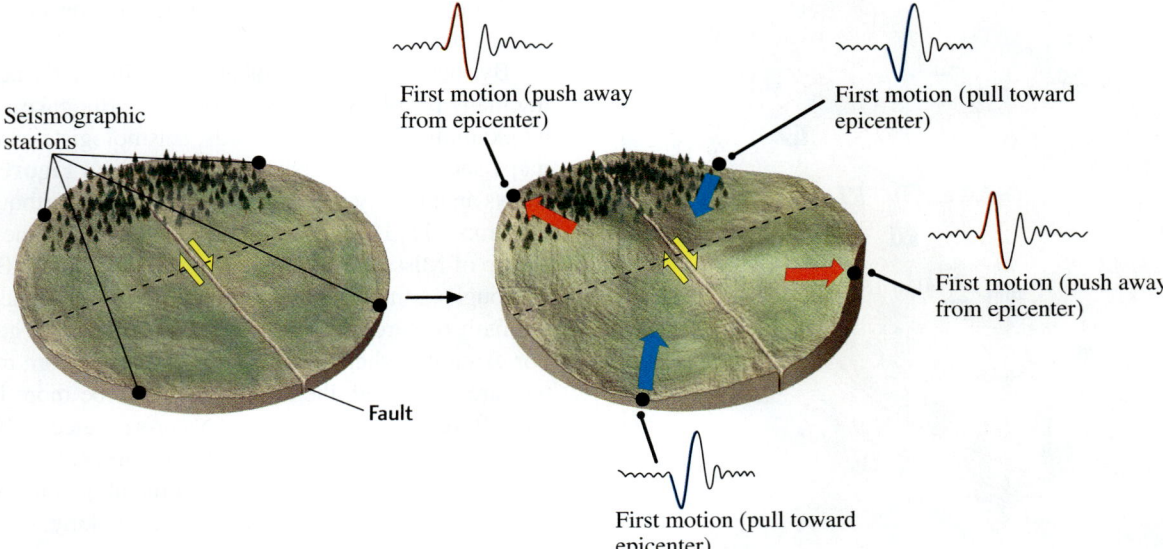

FIGURE 13.14 The motion of the first P wave arriving at a seismographic station is used to determine the orientation of the fault surface and the direction of fault slip. The case shown here is the rupture of a right-lateral strike-slip fault. Note that the alternating pattern of pushes and pulls would remain the same if the plane perpendicular to the fault ruptured with left-lateral displacement. Seismologists can usually choose between the two possibilities using additional information, such as field mapping of the fault scarp or the alignment of aftershocks along the fault surface.

the world to surround any earthquake's focus. In some directions from the focus, the very first movement of the ground recorded by a seismograph—a P wave—is a *push away* from the focus, causing upward motion on a vertical seismograph. In other directions, the initial ground motion is a *pull toward* the focus, causing downward motion on a vertical seismograph. In other words, the slipping on a fault looks like a push if you view it from one direction but like a pull if you view it from another. The locations of pushes and pulls can be plotted and divided into four sections based on the positions of the seismographic stations, as shown in **Figure 13.14**. One of the two boundaries between those sections is the fault orientation; the other is a plane perpendicular to the fault. The slip direction is determined from the arrangement of pushes and pulls. In this manner, without surface evidence, seismologists can deduce whether the horizontal crustal forces that triggered an earthquake were tensional, compressive, or shearing forces.

> *The fault mechanism of an earthquake—the orientation of the fault rupture and the direction of slipping—can tell seismologists what type of fault has ruptured: normal, reverse, or strike-slip.*

EXPLORING EARTH'S INTERIOR USING SEISMIC WAVES

The same seismic waves that cause terrible shaking during earthquakes can illuminate Earth's deepest regions, allowing us to construct models of its internal layering, make three-dimensional images of crustal geology and mantle convection, and understand the workings of its central core.

Paths of Seismic Waves Through Earth

If Earth were made of a single material with constant properties from the surface to the center, P and S waves would travel from the focus of an earthquake to a distant seismographic station along straight lines through Earth's interior (just as the Sun's rays travel in straight lines through the atmosphere). When the first global networks of seismographs were installed about a century ago, however, seismologists discovered that the structure of Earth's interior was much more complicated (see Chapter 2). Observations of long-distance seismic waves showed that the paths of P and S waves are *refracted* (bent) upward through the mantle and can be *reflected* by sharp boundaries, such as Earth's solid surface and the boundary between its mantle and core. These types of wave paths are illustrated in **Figure 13.15**.

From the travel times and the amount of upward refraction of seismic waves, seismologists were able to conclude that P waves travel much faster through rock at great depths than they do through rock at Earth's surface. This was hardly surprising, because rock subjected to great pressures in Earth's interior is squeezed into tighter crystal

> *The travel times and paths of P and S waves can be used to construct models of Earth's internal layers.*

FIGURE 13.15 (a) P and S waves are reflected upward from the core-mantle boundary and can reflect off Earth's solid surface. An S wave that has reflected once off the surface is labeled SS. (b) Seismograms recorded at various distances (in angular degrees) from an earthquake in the Aleutian Islands, Alaska. The colored lines identify the arrival times of the P and S waves (in blue), the surface waves (in yellow), and the S waves reflected from Earth's solid surface (the SS waves, in green).

structures. The atoms in these tighter structures are more resistant to further compression, which causes P waves to travel through them more quickly.

The Core Shadow Seismologists were very surprised, however, by what they found at progressively greater distances from an earthquake focus (**Figure 13.16**). After the P waves and S waves had traveled beyond about 11,600 km, they suddenly disappeared! (Like airplane pilots and ship captains, seismologists prefer to measure distances traveled on Earth's surface in angular degrees, from 0° at the earthquake focus to 180° at a point on the opposite side of

The pattern of P-wave paths through Earth's interior is shown by solid blue lines. The dashed lines show the progress of the waves through the interior at 2-minute intervals. Distances are measured in angular degrees from the earthquake focus. The P-wave shadow zone extends from 105° to 142°. P waves cannot reach the surface within this zone because of the way they are bent when they enter and leave the core.

S-wave paths through Earth's interior are shown by solid green lines. The larger S-wave shadow zone extends from 105° to 180°. Although S waves strike the core, they cannot travel through its fluid outer region and therefore never emerge beyond 105° from the focus.

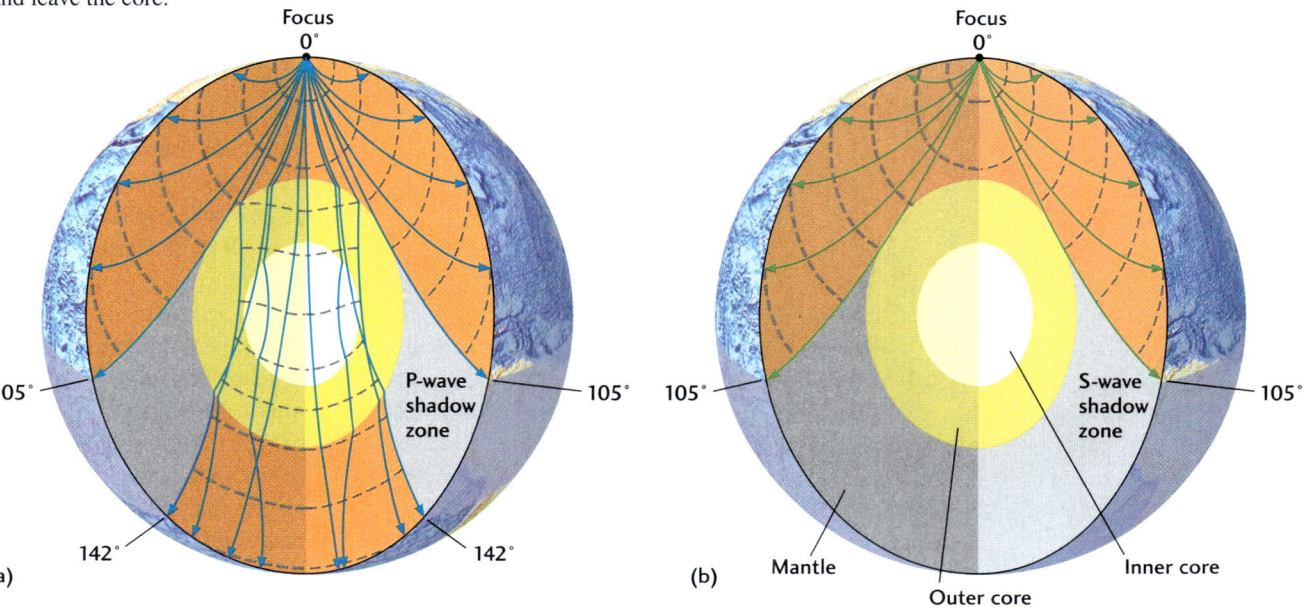

FIGURE 13.16 Earth's core creates P-wave and S-wave shadow zones.

Earth. Each degree measures 111 km at the surface, so 11,600 km corresponds to an angular distance of 105°, as shown in Figure 13.16.) When they looked at seismograms recorded beyond this distance, they did not see the distinct P- and S-wave arrivals that were so clear on seismograms recorded at shorter distances. Then, beyond about 15,800 km from the focus (142°), the P waves suddenly reappeared, but they were much delayed compared with their expected travel times. The S waves never reappeared.

In 1906, the British seismologist R. D. Oldham put these observations together to provide the first evidence that Earth has a liquid outer core. S waves cannot travel through the liquid outer core because fluids have no resistance to shear; thus, there is an S-wave **shadow zone** beyond 105° from the earthquake focus (see Figure 13.16b). The propagation of P waves is more complicated (see Figure 13.16a). At 105°, their paths just miss the core, whereas waves that would have traveled to greater distances encounter the core-mantle boundary. At the core-mantle boundary, P-wave velocity drops by almost a factor of two. Therefore, the waves are refracted downward into the core and emerge at greater distances after the delay caused by their detour through the core. This refraction effect forms a P-wave shadow zone at angular distances between 105° and 142°.

> The absence of shear waves on seismograms recorded at large distances from earthquake foci revealed the liquid state of Earth's core.

Reflections at Internal Boundaries The core-mantle boundary turned out to be very sharp, so when seismologists looked at records of seismic waves made at angular distances of less than 105° from an earthquake focus, they found waves that must have been reflected from this boundary. In 1914, a German seismologist, Beno Gutenberg, used the travel times of these reflected waves to determine an accurate depth for the core-mantle boundary: just under 2900 km. Twenty-two years later, Danish seismologist Inge Lehmann (**Figure 13.17**) discovered Earth's inner core, a solid sphere at Earth's center with only about one-third the core radius. Waves reflected from the inner core–outer core boundary showed that the depth of this boundary is about 5150 km.

> Seismologists have used seismic waves reflected by sharp boundaries between Earth's layers to estimate the depths of those boundaries.

Seismic Exploration of Near-Surface Layering

The probing of Earth's crust with seismic waves has a number of practical applications. Seismic waves generated by artificial sources, such as dynamite explosions, are reflected by geologic structures at shallow depths in the crust. Recording these reflections has proved to be the most successful method for finding deeply buried oil and gas reservoirs (see Figure 1.11). This type of seismic exploration is now a multibillion-dollar industry. Reflected seismic waves are also used to measure the depth of water tables and the thickness of glaciers. At sea, compressional waves can be generated by mechanical sources similar to loudspeakers, and oceanographic ships routinely use the underwater sound they produce to measure the depth of the ocean and the thickness of sediments on the seafloor.

Layering and Composition of Earth's Interior

Seismograms of earthquakes around the world have allowed seismologists to devise a model of Earth's interior layering (**Figure 13.18**), which we will explore by taking an imaginary downward journey through Earth's interior, from its outer crust to its inner core.

The Crust Earth's outermost layer, the crust, is thin (about 7 km) under oceans, thicker (about 40 km) under continents, and thickest (as much as 70 km) under high mountains. P waves move through crustal rocks at about 6 to 7 km/s. By measuring the velocities of seismic waves passing through samples of various materials typical of the crust and mantle in the laboratory, we can compile a library of seismic wave velocities. Rough values for P-wave velocities in igneous rocks, for example, are as follows:

FIGURE 13.17 Danish seismologist Inge Lehmann discovered Earth's inner core in 1936. [Courtesy of Beverley Bolt.]

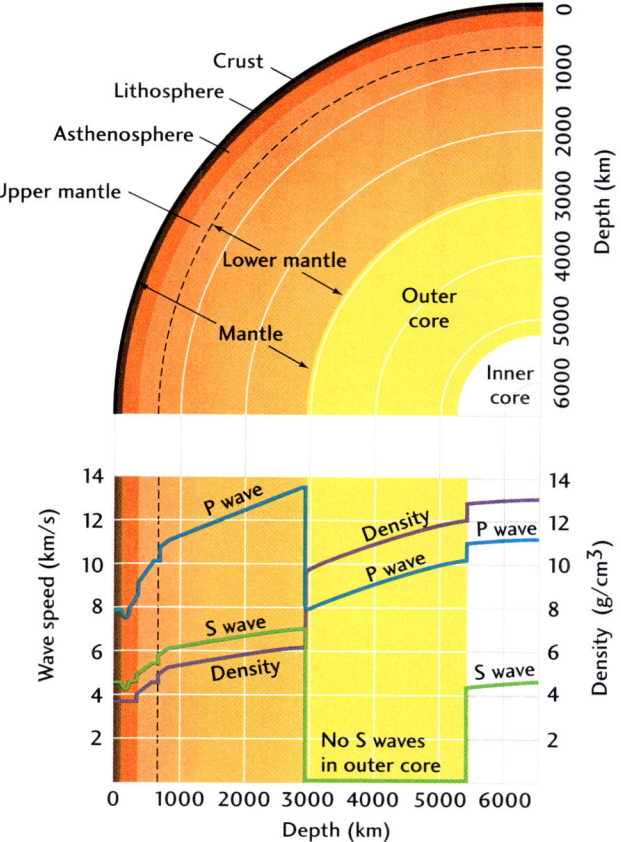

FIGURE 13.18 Earth's layering as revealed by seismology. The lower diagram shows changes in P-wave and S-wave velocities and rock density with depth. The upper diagram is a cross section through Earth on the same depth scale, showing how these changes are related to the major layers.

- Felsic rocks typical of the upper continental crust (granite): 6 km/s
- Mafic rocks typical of oceanic crust or the lower continental crust (gabbro): 7 km/s
- Ultramafic rocks typical of the upper mantle (peridotite): 8 km/s

P-wave velocities vary because they depend on a rock's density and its resistance to compression and shear, all of which change with chemical composition and crystal structure. In general, higher densities correspond to higher P-wave velocities; typical densities for granite, gabbro, and peridotite are 2.6 g/cm^3, 2.9 g/cm^3, and 3.3 g/cm^3, respectively.

We know from correlations of P-wave velocity with rock type that the upper part of the continental crust is made up mostly of low-density granitic rocks and that no granite exists on the floor of the deep sea. The crust there consists entirely of basalt and gabbro overlain by sediments. The velocity of P waves increases abruptly to 8 km/s at the Mohorovičić discontinuity, or Moho, which marks the base of the crust (see Chapter 2).

Exploring Earth's Interior Using Seismic Waves

The Mantle The P-wave velocity of 8 km/s indicates that the mantle below the Moho is made mostly of peridotite, a dense, ultramafic rock composed primarily of olivine and pyroxene. These minerals have less silica and more magnesium and iron than those in typical crustal rocks (see Chapter 5).

S-wave velocities have been used to explore the layering of the mantle. The layering of the upper mantle (**Figure 13.19**) is primarily caused by the effects of increasing pressure and temperature on peridotite. The pressure in Earth's interior increases steadily with depth. The increase of temperature with depth, which defines a curve known as the

> *Correlations of seismic velocities with rock types in the laboratory have allowed seismologists to identify rock types deep in Earth's crust and mantle.*

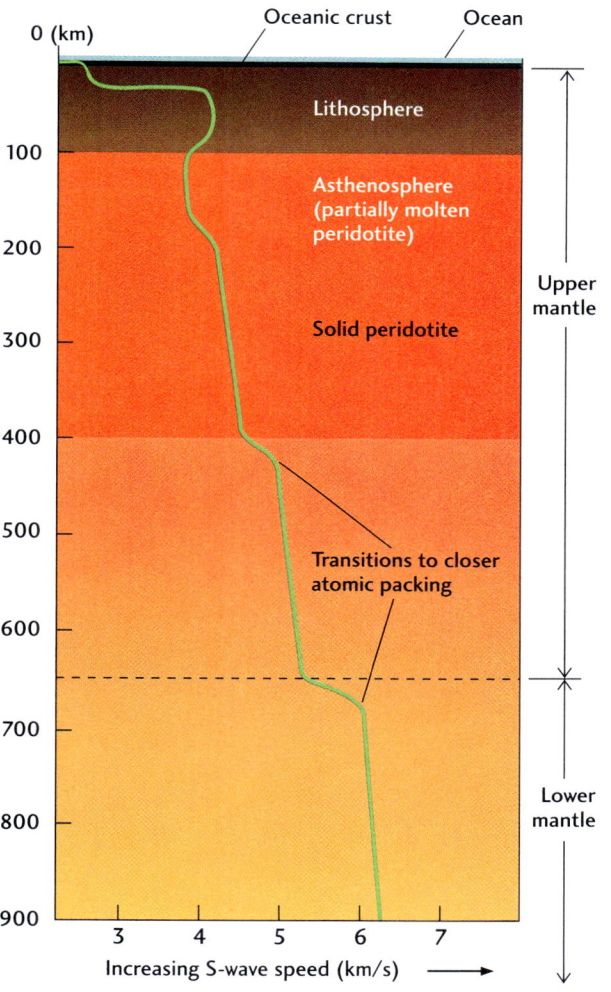

FIGURE 13.19 The structure of the mantle beneath an old ocean basin, showing S-wave velocities to a depth of 900 km. Changes in S-wave velocity mark the strong lithosphere, the weak asthenosphere, and two zones at which increasing pressure forces a rearrangement of atoms into denser and more compact crystal structures. [After D. P. McKenzie, "The Earth's Mantle." *Scientific American* (September 1983): 66.]

geotherm, is more complicated. In Figure 13.20, we compare the geotherm (in yellow) with the temperatures at which Earth materials begin to melt (in red).

The mantle just below the Moho is relatively cold. Like the crust, it is part of the lithosphere, the rigid layer that forms the tectonic plates (see Chapter 2). On average, the thickness of the lithosphere is about 100 km, but it is highly variable geographically, ranging from almost no thickness at all near spreading centers, where new oceanic lithosphere is forming from hot, rising mantle material, to over 200 km beneath the cold, stable continental cratons.

The temperature rises rapidly with depth in the lithosphere, and near 100 km it gets hot enough that some minerals in the peridotite begin to melt. Although the amount of melting is small (in most places, less than 1 percent), it is sufficient to cause an abrupt drop in the rigidity of the rock, which decreases the S-wave velocity, forming a **low-velocity zone.** Because partial melting allows peridotite to flow more easily, geologists identify the low-velocity zone with the top part of the asthenosphere—the weak, ductile layer on which the lithospheric plates slide. This idea fits nicely with evidence that the asthenosphere is the source of most basaltic magma (see Chapter 5).

In the asthenosphere, the melting temperature of rock rises because of the increasing pressure, whereas mantle convection causes the geotherm to flatten (for the same reason that stirring your bathwater evens out the temperatures). Therefore, the geotherm drops below the melting curve as we go deeper into the mantle, and the S-wave velocities begin to increase again.

About 400 km below the surface, both the S-wave and P-wave velocities increase by about 10 percent within a narrow zone less than 20 km thick. This jump in seismic velocities can be explained by a **phase change** in olivine, the major mineral constituent of the upper mantle, which transforms its ordinary crystal structure into a denser, more closely packed structure at high pressures. This hypothesis has been confirmed by laboratory experiments. When olivine is subjected to high pressures in the laboratory, the atoms that form its crystal structure shift into a more compact arrangement at the temperatures and pressures corresponding to depths of about 400 km. Moreover, the jumps in the P- and S-wave velocities measured in the laboratory match the increase observed for seismic waves at this depth.

In the region from 420 to 650 km below the surface, mantle properties change slowly as depth increases. Near 660 km, however, the seismic velocities abruptly increase again, indicating a second major phase change to an even more closely packed crystal structure. Laboratory experiments on olivine have also confirmed the existence of another major mineralogical phase change at this depth.

Below the transition at a depth of 660 km, the seismic wave velocities increase gradually and do not show any more unusual features until close to the core-mantle boundary (see Figure 13.18). This relatively homogeneous region, more than 2000 km thick, is called the **lower mantle.**

The Core-Mantle Boundary
At the **core-mantle boundary,** about 2890 km below the surface, we encounter the most extreme change in properties found

> Below the lithosphere, seismic velocities decrease because of the partial melting of peridotite, which contributes to the ductile properties of the asthenosphere.

> At two depths in the mantle, 410 km and 660 km below the surface, seismic velocities jump, indicating phase changes in the crystal structures of mantle minerals.

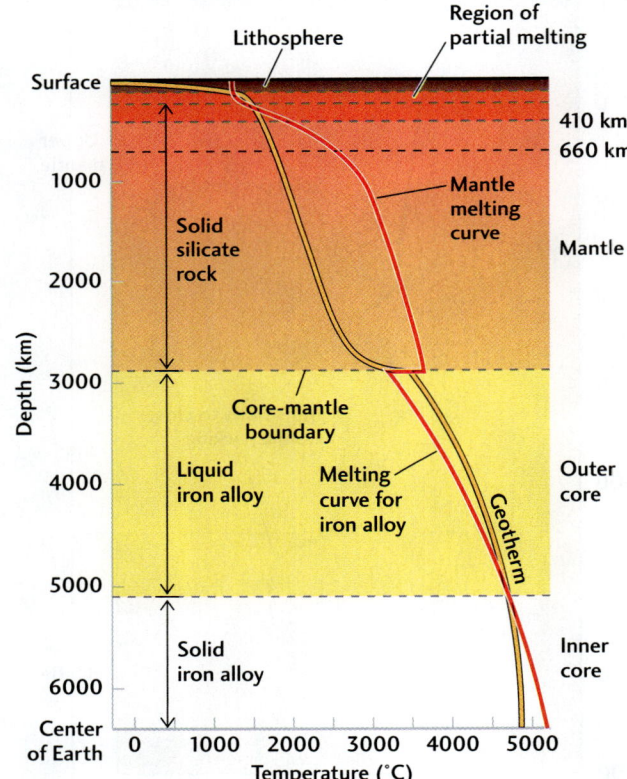

FIGURE 13.20 An estimate of Earth's geotherm, which describes the increase in temperature with depth (yellow line). The geotherm first rises above the melting curve, the temperature at which rock begins to melt (red line) in the upper mantle, forming the partially molten low-velocity zone. It does so again in the outer core, where the iron-nickel alloy is in a liquid state. The geotherm falls below the melting curve throughout most of the mantle and in the solid inner core.

anywhere in Earth's interior. From the way seismic waves reflect from this boundary, seismologists can tell that it is a very sharp interface. Here the material changes abruptly from a solid silicate rock to a liquid iron alloy. Because of the complete loss of rigidity, the S-wave velocity drops from about 7.5 km/s to zero and the P-wave velocity drops from more than 13 km/s to about 8 km/s, causing the core shadow zones. Density, on the other hand, increases by about 4.5 g/cm^3 (see Figure 13.18). This large density jump, which is even greater than the increase in density at Earth's solid surface, keeps the core-mantle boundary very flat (you could probably skateboard on it!) and prevents any large-scale mixing of the mantle and core.

The core-mantle boundary appears to be a very active place. Heat conducted out of the core increases the temperatures at the base of the mantle by as much as 1000°C (see Figure 13.20). Indeed, seismic waves that pass near the base of the mantle show peculiar complications, suggesting a region of exceptional geologic activity. Some geologists believe this hot region to be the source of mantle plumes that rise all the way to Earth's surface, creating volcanic hot spots such as Hawaii and Yellowstone.

The lowest boundary layer of the mantle, a region about 300 km thick, may be the ultimate graveyard of some lithospheric material subducted at the surface, such as the dense, iron-rich parts of the oceanic crust. It is possible that this region experiences an upside-down version of the tectonics we see at Earth's surface. For example, accumulations of heavy, iron-rich material might form chemically distinct "anticontinents" that are constantly pushed to and fro across the core-mantle boundary by convection currents. Seismologists are teaming with other geologists who study mantle and core convection to learn more about whatever geologic processes might be active in this strange place.

The reflection of seismic waves at a depth of 2890 km reveals a sharp core-mantle boundary, where density increases dramatically and seismic wave velocities drop.

The Core Many lines of evidence support the hypothesis that Earth's core is made up of iron and nickel. These metals are abundant in the universe (see Chapter 2); in addition, they are dense enough to explain the mass of the core (about one-third of Earth's total mass) and to be consistent with the theory that the core formed by gravitational differentiation (see Chapter 9).

Laboratory measurements at appropriately high pressures and temperatures have led to a slight revision of this hypothesis. A pure iron-nickel alloy turns out to be about 10 percent too dense to match the data for the outer core. Therefore, it has been proposed that the core includes minor amounts of some lighter element. Oxygen and sulfur are leading candidates, although the precise composition remains the subject of research and debate.

Seismology tells us that the core below the mantle is a fluid, but it is not fluid to the very center of Earth. As Lehmann first discovered, P waves that penetrate to depths of 5150 km suddenly speed up, indicating the presence of an inner core, a metallic sphere two-thirds the size of the Moon. Seismologists have recently shown that the inner core does transmit shear waves, confirming early speculations that it is solid. In fact, some calculations suggest that the inner core spins at a slightly faster rate than the mantle, acting like a "planet within a planet."

The very center of the planet is not a place you would want to be. The pressures are immense, over 4 million times the atmospheric pressure at Earth's surface. And it's also very hot. In Figure 13.20, you can see that the geotherm rises above the melting curve in the outer core but falls below it in the inner core, which explains why the former is liquid and the latter is solid.

A jump in P-wave velocity marks the boundary between the fluid outer core and the solid inner core at a depth of 5150 km.

The Three-Dimensional Structure of the Mantle

So far, we have investigated how the properties of Earth's materials vary with depth. Such a one-dimensional description would suffice if our planet were perfectly symmetrical, but of course it is not. At the surface, we can see *lateral variations* (geologic differences) in Earth's structure associated with oceans and continents and with the basic features of plate tectonics: mid-ocean ridges at spreading centers, deep-sea trenches at subduction zones, and mountain belts lifted up by continent-continent collisions.

Below the crust, we can expect that convection will cause changes in temperature from one part of the mantle to another. Downwelling currents, such as those associated with subducted lithospheric plates, will be relatively cold, whereas upwelling currents, such as those associated with mantle plumes, will be relatively hot. Computer models tell us that lateral variations in temperature due to mantle convection should be on the order of several hundred degrees. From laboratory experiments on rocks, we know that such temperature differences should cause small variations in seismic wave velocities from place to place.

Seismologists have been able to make three-dimensional images of these small lateral variations by adapting a medical technique known as computerized axial tomography. CAT scanners construct three-dimensional images of bones and organs by measuring small differences in X rays that sweep the body in many directions.

362 CHAPTER 13 Earthquakes and Earth Structure

Seismic tomography uses the travel times of seismic waves recorded at many seismographic stations to create three-dimensional images of Earth's interior.

Focus

Mantle
Liquid outer core
Solid inner core

Seismographic stations

Regions with faster S-wave speeds (blue and purple) indicate relatively colder, denser rock. Regions with slower S-wave speeds (red and yellow) indicate relatively hotter, less dense rock.

A tomographic section through Earth reveals hot rocks, such as a mantle plume rising from Earth's core beneath South Africa,…

North America
Africa
Cross section

…and colder rocks, such as the descending remnants of the Farallon Plate under the North American Plate.

Global maps at four different depths

70 km

Near Earth's surface, hot rocks in the asthenosphere slow S waves, as revealed by the warm colors (red and yellow) along oceanic spreading centers.

200 km

Moving deeper, we see the cold lithosphere of stable continental cratons (blue and purple regions) and warmer asthenosphere beneath ocean basins (red regions).

500 km

Deeper in the mantle, the features no longer match the continental positions.

2800 km (near core-mantle boundary)

Near the core-mantle boundary, the S-wave patterns reveal colder regions around the Pacific that may be the "graveyards" of sinking lithospheric slabs.

FIGURE 13.21 A three-dimensional model of Earth's mantle created by seismic tomography. [S-wave velocities courtesy of G. Ekström and A. Dziewonski, Harvard University. Cross section from M. Gurnis, *Scientific American* (March 2001): 40. Maps courtesy of L. Chen and T. Jordan, University of Southern California.]

Similarly, we can use the velocities of seismic waves from earthquakes, as recorded on thousands of seismographs all over the world, to sweep Earth's interior in many different directions and construct a three-dimensional image of what's inside, a technique called **seismic tomography.**

In the 1990s, researchers at Harvard University constructed the tomographic model of the mantle shown in **Figure 13.21**. The model is displayed as a cross section of Earth and as a series of global maps at depths ranging from just below the crust down to the core-mantle boundary. Near Earth's surface, you can clearly see the structure of plate tectonics. The low S-wave velocities caused by the upwelling of hot asthenosphere along the mid-ocean ridges are visible in warm colors; the high S-wave velocities caused by cold lithosphere in old ocean basins and beneath the continental cratons are visible in cool colors.

At greater depths, the features become more variable and less coherent with surface tectonic features, reflecting what is probably a complex pattern of mantle convection. Some large-scale features stand out particularly well. You will notice that, just above the core-mantle boundary, there is a red region of relatively low S-wave velocities beneath the central Pacific Ocean, surrounded by a broad blue ring of higher S-wave velocities. Seismologists have speculated that the high velocities represent a "graveyard" of oceanic lithosphere subducted beneath the Pacific's volcanic arcs—the Ring of Fire—during the last 100 million years or so.

> *Seismic tomography is a technique that maps small variations in the velocities of seismic waves to create three-dimensional images of Earth's interior.*

Seismic tomography has shown that cold lithospheric slabs plunge all the way through the mantle, demonstrating that the plate tectonic system involves convection extending almost as deep as the core-mantle boundary (see Chapter 3). The cross section through the mantle in Figure 13.21 clearly reveals material from the once-large Farallon Plate, which has been almost completely subducted under North America (see Chapter 9). The obliquely sinking slab material (in blue) appears to have penetrated the entire mantle. The image also indicates sinking colder rock beneath Indonesia. In addition, a large yellow blob of hotter rock, thought to be a "superplume," can be seen rising at an angle from the core-mantle boundary to a position beneath southern Africa. This hot, buoyant mass pushing up the cooler material above it may explain the uplifted, mile-high plateaus of South Africa.

Seismologists thus use the waves produced by earthquakes to understand the forces in Earth's interior that are responsible for the plate tectonic processes observed at Earth's surface. Of course, it is those processes that generate earthquakes in the first place. By mapping patterns of earthquake faulting, we can learn more about the forces acting within and between the lithospheric plates.

EARTHQUAKES AND PATTERNS OF FAULTING

As we have seen, seismologists are using networks of sensitive seismographs to locate earthquakes around the world, measure their magnitudes, and deduce their fault mechanisms. The **seismicity map** in **Figure 13.22** shows the epicenters of earthquakes recorded in a 27-year period.

The Big Picture: Earthquakes and Plate Tectonics

The most obvious features of the global seismicity map are the belts of seismic activity that mark the major plate boundaries. The fault mechanisms observed for earthquakes in these belts are consistent with the types of faulting along plate boundaries that we discussed in Chapter 7.

Divergent Boundaries The narrow belts of shallow earthquakes that run through ocean basins coincide with mid-ocean ridge crests and their offsets on transform faults. The first P-wave motions recorded from these quakes indicate that they are caused by normal faulting. The faults strike parallel to the ridge and dip toward the mid-ocean rift valley. Normal faulting implies that tensional forces are at work as the plates are pulled apart during seafloor spreading, which explains why rift valleys develop at the ridge crests. Earthquakes also have normal fault mechanisms in zones where continental crust is being pulled apart, such as in the East African rift valleys and in the Basin and Range province of western North America.

Transform-Fault Boundaries Earthquake activity is even greater along the transform-fault boundaries that offset mid-ocean ridge segments. These earthquakes show strike-slip fault mechanisms, just as one would expect where plates slide past each other in opposite directions. Moreover, for earthquakes along these transform faults, the slip direction indicated by the fault mechanisms is left-lateral where the ridge crest steps right and right-lateral where it steps left. These directions are the opposite of what would be needed to create the offsets of the ridge crest but are consistent with the direction of slip predicted by seafloor spreading. In the mid-1960s, seismologists used this property of transform faults to support the hypothesis of seafloor spreading (see Figure 13.22). Slip directions on transform faults that run through continental crust, such as California's San Andreas

> *Tensional forces at divergent plate boundaries cause normal faulting. Strike-slip faulting occurs along transform faults, including those that offset mid-ocean ridge segments.*

FIGURE 13.22 Global map of seismic activity, showing the concentration of earthquakes along the boundaries between tectonic plates. The inset figures illustrate the types of faulting observed at different types of plate boundaries. [Map based on data from Harvard CMT catalog; plot by M. Boettcher and T. Jordan.]

fault and New Zealand's Alpine fault (both right-lateral), also agree with the predictions of plate tectonics.

Convergent Boundaries The world's largest earthquakes occur at convergent plate boundaries. The three greatest earthquakes of the last hundred years were of this type: the 1964 Alaska earthquake, described in the opening story of this chapter (moment magnitude 9.1); the 2004 earthquake in the subduction zone on the west side of the Indonesian island of Sumatra (moment magnitude 9.2); and the largest of all, the 1960 earthquake in the subduction zone west of Chile (moment magnitude 9.5).

The fault mechanisms of these great earthquakes show that they were caused by horizontal compression along huge thrust faults, called *megathrusts,* that form the boundaries where one plate subducts beneath another (see Figure 13.22). Large ruptures on megathrusts can displace the seafloor, generating devastating tsunamis. Both the 1960 Chile and 1964 Alaska earthquakes sent tsunamis across the Pacific Ocean, causing damage and killing people thousands of kilometers from their source. The Sumatra earthquake sent a huge tsunami across the Indian Ocean, drowning more than 300,000 people living on coastlines from Thailand to Africa.

Earth's deepest earthquakes also occur at convergent boundaries. Almost all earthquakes originating below 100 km rupture the descending plate in a subduction zone. The fault mechanisms of these deep earthquakes show a variety of orientations, but they are consistent with the deformation expected within the descending plate as gravity pulls it back into the convecting mantle. The deepest earthquakes, located at depths of 600 to 700 km, take place in the oldest—and therefore coldest—descending plates, such as those beneath South America, Japan, and the island arcs of the western Pacific Ocean.

> *Earth's largest earthquakes occur along megathrusts under compressive stress at convergent boundaries. Ruptures of descending plates in subduction zones cause Earth's deepest earthquakes.*

Intraplate Earthquakes Although most earthquakes occur at plate boundaries, a small percentage of global seismic activity originates within plate interiors (see Figure 13.22). The foci of these *intraplate earthquakes* are relatively shallow, and most occur on continents. Among these earthquakes are some of the most famous in American history: a sequence of three large events near New Madrid, Missouri, in 1811–1812; the Charleston, South Carolina, earthquake of 1886; and the Cape Ann earthquake, near Boston, Massachusetts, in 1755. Many of these earthquakes occur on old faults that were once parts of ancient plate boundaries. These faults no longer form plate boundaries but remain zones of crustal weakness that concentrate and release intraplate stresses.

One of the deadliest intraplate earthquakes (magnitude 7.6) occurred near Bhuj in the state of Gujarat, in western India, in 2001. It is estimated that some 20,000 lives were lost. The epicenter was 1000 km south of the boundary between the Indian Plate and the Eurasian Plate. Indian geologists believe that the compressive stresses responsible for the Bhuj earthquake originated in the continuing northward collision of India with Eurasia. Apparently, strong crustal forces can still develop and cause faulting within a lithospheric plate far from modern plate boundaries—in this case, triggering a previously unknown thrust fault at a depth of about 20 km.

> *Intraplate earthquakes often occur on old faults that remain zones of crustal weakness.*

Regional Fault Systems

Although the fault mechanisms of most major earthquakes conform to the predictions of plate tectonic theory, a plate boundary can rarely be described as a single fault, particularly when the boundary involves continental crust. Rather, the zone of deformation between two moving plates usually comprises a network of interacting faults—a *fault system*. The fault system in Southern California provides an interesting example (**Figure 13.23**).

The "master fault" of this system is our old nemesis, the San Andreas—a right-lateral strike-slip fault that runs northwestward through California from the Salton Sea near the Mexican border to Cape Mendocino, where it goes offshore in the northern part of the state (see Figure 13.2). However, there are a number of subsidiary faults on either side of the San Andreas that generate large earthquakes. In fact, most of the damaging earthquakes in Southern California during the last century have occurred on those subsidiary faults.

Why is the San Andreas fault system so complex? Part of the explanation has to do with the geometry of the San Andreas fault itself. Notice in Figure 13.23a that although the strike of the San Andreas is approximately in the direction of the relative movement of the Pacific and North American plates, as expected for a transform-fault boundary, the fault is not straight but curves westward just north of Los Angeles. This leftward "Big Bend" causes thrust faulting south of the San Andreas (compare with Figure 7.14, part c). This thrust faulting was responsible for two recent deadly earthquakes, the San Fernando earthquake of 1971 (magnitude 6.6, 65 people killed) and the Northridge earthquake of 1994 (magnitude 6.9, 58 people killed). Over the past several million years, thrust faulting associated with the Big Bend has raised the San Gabriel Mountains to elevations of 1800 to 3000 m.

Another complication is the plate extension taking place east of California in the Basin and Range province, which spans the state of Nevada and much of Utah and

(a) Southern California fault traces

In the Carrizo Plain of central California, the San Andreas fault is parallel to the relative movement of the Pacific and North American plates and shows right-lateral strike-slip faulting.

San Andreas Fault
San Gabriel Mountains
North American Plate
Pacific Plate
Los Angeles
Movement of Pacific Plate relative to movement of North American Plate

The westward "Big Bend" causes the Pacific Plate to compress the North American Plate, causing thrust faulting in the Los Angeles region south of the fault. This convergence raises the San Gabriel Mountains.

(b) Southern California earthquakes (July 1970–June 1995)

Northridge 1994
Magnitude 6.9

San Fernando 1971
Magnitude 6.6

Landers 1992
Magnitude 7.3

Key:
5+
<5

July 1970–June 1995

FIGURE 13.23 (a) A map of the fault system of Southern California (yellow lines), showing the surface traces of the San Andreas fault and its subsidiary faults. (b) Locations of earthquakes in Southern California during the period July 1970–June 1995. [Southern California Earthquake Center.]

Arizona (see Chapters 7 and 9). This broad zone of extensional deformation is connected with the San Andreas fault system through a series of faults that run along the eastern side of the Sierra Nevada and through the Mojave Desert. Faults of this system were responsible for the 1992 Landers earthquake (magnitude 7.3) and the 1999 Hector Mine earthquake (magnitude 7.1), as well as the 1872 Owens Valley earthquake (magnitude 7.6).

EARTHQUAKE DESTRUCTIVENESS

Over the last century, earthquakes worldwide have caused an average of 13,000 deaths per year and hundreds of billions of dollars in economic losses. Two California earthquakes—the 1989 Loma Prieta earthquake (magnitude 7.1),

FIGURE 13.24 Sixteen people died in the Northridge Meadows apartment building in Los Angeles during the 1994 Northridge earthquake. The victims lived on the first floor and were crushed when the upper levels collapsed. Many more buildings would have collapsed if the newer buildings in the area had not been constructed according to stringent building codes. [Nick Ut, Files/AP Photo.]

which occurred 80 km south of San Francisco, and the 1994 Northridge earthquake in Los Angeles—were among the costliest disasters in U.S. history. Damage amounted to more than $10 billion in the Loma Prieta earthquake and $40 billion in the Northridge quake because of their proximity to urban centers. About 60 people died in each event, but the death toll would have been many times higher if stringent building codes had not been in place (Figure 13.24).

Destructive earthquakes are even more frequent in Japan than in California. The recorded history of destructive earthquakes in Japan, going back 2000 years, has left an indelible impression on the Japanese people. Perhaps that is why Japan is the best prepared of any nation in the world to deal with earthquakes. It has impressive public education campaigns, building codes, and warning systems. Despite this preparedness, more than 5600 people were killed in a devastating (magnitude 6.9) earthquake that struck Kobe on January 16, 1995 (Figure 13.25). The large numbers of casualties and structural failures (50,000 buildings destroyed) resulted partly from the less stringent building codes that were in effect before 1980, when much of the city was built, and from the location of the earthquake rupture so close to the city.

How Earthquakes Cause Damage

Earthquakes can damage our environment in several ways. The damaging events proceed as chain reactions in which the primary effects of faulting and ground shaking induce secondary effects, such as landslides and tsunamis, as well

FIGURE 13.25 This elevated expressway, in Kobe, Japan, was overturned during an earthquake in 1995. [Tom Wagner/CORBIS/SABA.]

as destructive processes within the built environment, particularly fires.

Faulting and Shaking The *primary hazards* of earthquakes are the ruptures in the ground surface that occur when faults break the surface, the permanent subsidence and uplift of the ground surface caused by faulting, and the ground shaking caused by seismic waves radiated during the quake. Seismic waves can shake structures so hard that they collapse. The ground accelerations near the epicenter of a large earthquake can approach and even exceed the acceleration of gravity, so an object lying on the ground surface can literally be thrown into the air. Very few structures built by human hands can survive such severe shaking, and those that do are severely damaged. The collapse of buildings and other structures is the leading cause of casualties and economic damage during earthquakes, including the high death tolls in the Tangshan, China, earthquake of 1976 (more than 240,000 killed); the Spitak, Armenia, earthquake of 1988 (25,000 killed); the Izmit, Turkey, earthquake of 1999 (15,600 killed); the Bhuj, India, earthquake of 2001 (20,000 killed); the Bam, Iran, earthquake of 2003 (30,000 killed); and the Pakistan earthquake of 2005 (86,000 killed).

Landslides and Other Types of Ground Failure The primary hazards of faulting and ground shaking generate *secondary hazards* such as landslides. When seismic waves shake water-saturated soils, those soils can behave like a liquid and become unstable. The ground simply flows away, taking buildings, bridges, and everything else with it. Such soil *liquefaction* destroyed the residential area of Turnagain Heights near Anchorage, Alaska, in the 1964 earthquake (see the photo at the opening of this chapter); the Nimitz Freeway near San Francisco in the 1989 Loma Prieta earthquake; and areas of Kobe in the 1995 earthquake.

In some instances, landslides cause more damage than the ground shaking itself. A great earthquake in China's Kansu Province in 1920 triggered an extensive flow of debris that covered a region larger than 100 km^2 and resulted in roughly 200,000 deaths. A landslide caused by the 1970 earthquake in Peru killed more than 66,000 people (see Figure 1.32). Many of the 844 people killed in the 2001 earthquake in El Salvador were buried by a muddy landslide loosened from a slope in the suburbs of the country's capital, San Salvador.

Tsunamis A large earthquake that occurs beneath the ocean can generate a destructive sea wave, sometimes called a tidal wave but more accurately named a *tsunami* (Japanese for *harbor wave*), since it has nothing to do with tides. Tsunamis are by far the deadliest and most destructive hazards associated with the world's largest earthquakes: the megathrust events that occur in subduction zones. When a megathrust ruptures, it can push the seafloor landward of the deep-sea trench upward by as much as 10 m, displacing a large mass of the overlying ocean

FIGURE 13.26 Megathrust earthquakes may generate tsunamis that can propagate across ocean basins. [NOAA, Pacific Marine Environmental Laboratory.]

FIGURE 13.27 The tsunami caused by the 2004 Sumatra earthquake struck without warning on a beach in Phuket, Thailand. [Courtesy of David Rydevik.]

water. This disturbance flows outward in waves that travel across the ocean at speeds of up to 800 km/hour (about as fast as a commercial jetliner). In the deep sea, a tsunami is hardly noticeable, but when it approaches shallow coastal waters, the waves slow down and pile up, inundating the shoreline in walls of water that can reach heights of tens of meters (Figure 13.26).

The destructive power of a great tsunami was brought home by the terrifying video images captured on December 26, 2004, as the tsunami unleashed by the magnitude 9.2 Sumatra earthquake swept over low-lying coastal areas of the Indian Ocean from Indonesia and Thailand to Sri Lanka, India, and the east coast of Africa (Figure 13.27). Within 15 minutes, the first wave ran up the Sumatran coastline. Few eyewitnesses survived there, but geologic investigations after the tsunami indicated that the maximum wave height on the beaches of the west-facing coast was about 15 m (Figure 13.28), and the run-up attained heights of 25 to 35 m, reaching inland up to 2 km and wiping out most built structures, vegetation, and human life in its path.

FIGURE 13.28 This small headland near Banda Aceh on the west coast of Sumatra was previously covered by dense jungle to the waterline, but was stripped clean to a height of about 15 m by the 2004 tsunami. [José Borrero, University of Southern California/ Tsunami Research Group.]

It is believed that more than 200,000 people perished along the Sumatran coastline, though no one will ever be sure because many bodies were washed out to sea.

The tsunami moved faster in the deep sea to the west and more slowly in the shallow waters to the east, so that the waves arrived at the heavily populated coastlines of Sri Lanka and Thailand at nearly the same time, some 2 hours after the earthquake. In Sri Lanka and Thailand, more than 40,000 and 15,000 people were killed, respectively. The tsunami struck India at 3 hours (15,000 deaths), the Maldive Islands at 4 hours (108 deaths), and Somalia on the African coast at 9 hours (298 deaths) after the quake. Enough time was available to forewarn the coastal residents of the impending tsunami, but no system for this purpose existed in the Indian Ocean, so the waves appeared without any warning.

Disturbances of the seafloor caused by landslides or volcanic eruptions can also produce tsunamis. The 1883 explosion of Krakatau, a volcano in Indonesia, generated a tsunami that reached 40 m in height and drowned 36,000 people on nearby coasts.

Fires The secondary hazards of earthquakes also include destructive processes that stem from the nature of the built environment itself, such as the fires ignited by ruptured gas lines or downed electrical power lines. Damage to water mains in an earthquake can make fire fighting all but impossible—a circum-

> *The primary effects of earthquakes, faulting and ground shaking, induce secondary effects, such as landslides and tsunamis, as well as destructive processes within the built environment, such as fires.*

FIGURE 13.29 Seismic hazard map for the United States. The region of highest hazard lies along the San Andreas fault system in California, with a branch extending into eastern California and western Nevada. High hazard levels are also found along the coast of the Pacific Northwest. In the central and eastern United States, the areas of highest hazard are near New Madrid, Missouri, and Charleston, South Carolina; in eastern Tennessee; and in portions of the Northeast. [U.S. Geological Survey, http://geohazards.cr.usgs.gov/eq/.]

stance that contributed to the burning of San Francisco after the 1906 earthquake (see Figure 13.3). Most of the 140,000 fatalities in the 1923 Kanto earthquake, one of Japan's greatest disasters, resulted from fires in the cities of Tokyo and Yokohama.

Reducing Earthquake Risk

In Chapter 1, we discussed why it is important to distinguish between hazard and risk when considering how to prepare for natural disasters. Here we illustrate those issues with earthquakes.

Seismic hazard describes the *intensity* of seismic shaking and ground disruption that can be expected over the long term at some specified location. It depends on the proximity of the site to active faults that might generate earthquakes and can be expressed in the form of a seismic hazard map. Figure 13.29 displays the national seismic hazard map produced by the U.S. Geological Survey.

In contrast, **seismic risk** describes the *damage* that can be expected over the long term in a specified region, such as a county or state, usually measured in terms of average dollar loss per year. A region's risk depends not only on its seismic hazard, but also on its *exposure* to seismic damage (its population and number of buildings and other infrastructure) and its *fragility* (the vulnerability of its built structures to seismic shaking). Because so many geologic and economic variables must be considered, estimating seismic risk is a complex job. The results of the first comprehensive study of seismic risk in the United States, published by the Federal Emergency Management Agency in 2001, are presented in Figure 13.30.

The differences between seismic hazard and seismic risk can be appreciated by comparing the two types of national maps. For instance, although the seismic hazard levels in Alaska and California are both high (see Figure 13.29), California's exposure is much greater, yielding a much larger total risk (see Figure 13.30). California leads the nation in seismic risk, with about 75 percent of the national total; in fact, a single county, Los Angeles, accounts for 25 percent. Nonetheless, the problem is truly national: 46 million people in metropolitan areas outside of California face substantial earthquake risks. Those areas include Hilo, Honolulu, Anchorage, Seattle, Tacoma, Portland, Salt Lake City, Reno, Las Vegas, Albuquerque, Charleston, Memphis, Atlanta, St. Louis, New York, Boston, and Philadelphia.

Not much can be done about seismic hazard because we have no way to prevent or control earthquakes. However, there are many important steps that society can take to reduce seismic risk if the hazard is properly characterized.

Hazard Characterization The first step is to follow the advice of the old proverb, "Know thy enemy." We still have much to learn about the sizes and frequencies of ruptures on active faults. For example, it is only in the past decade that we have come to appreciate that an earthquake in the Cascadia subduction zone, which stretches from northern California through Oregon and Washington to British Columbia, could produce a tsunami as large as the one that devastated the Indian Ocean region in

> *Seismic hazard describes the intensity of seismic shaking and ground disruption expected at a specified location. Seismic risk describes the damage that can be expected at that location.*

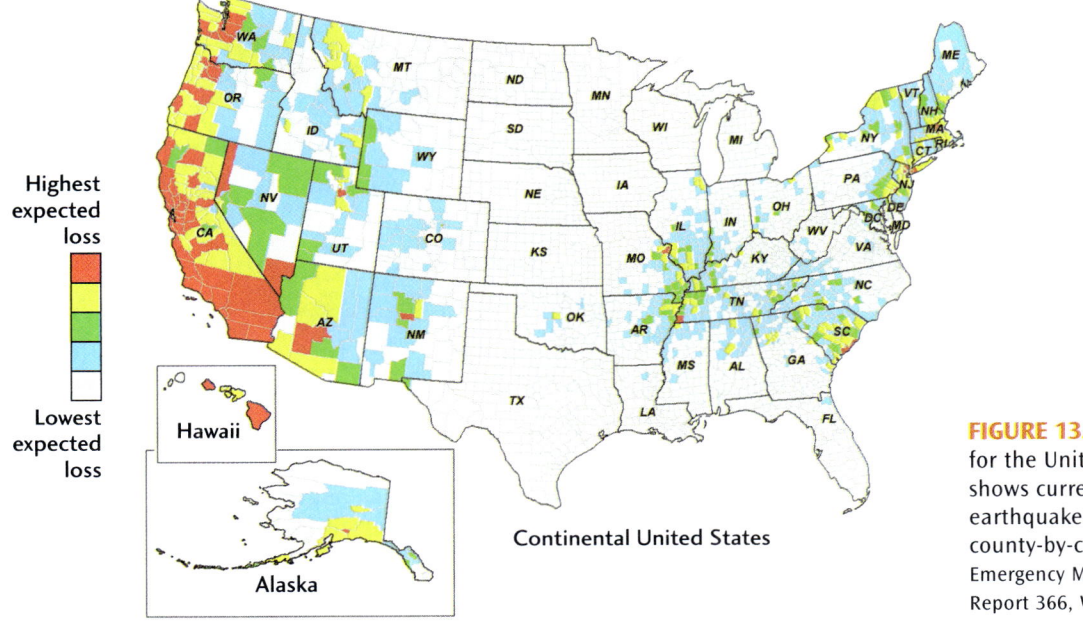

FIGURE 13.30 Seismic risk map for the United States. The map shows current annualized earthquake losses (AEL) on a county-by-county basis. [Federal Emergency Management Agency, Report 366, Washington, D.C., 2001.]

FIGURE 13.31 World seismic hazard map. [K. M. Shedlock et al., *Seismological Research Letters* 71(2000): 679–686.]

2004. These dangers became apparent when geologists found evidence of a magnitude 9 earthquake that occurred in 1700, before any written historical accounts of the area existed. This monstrous rupture caused major ground subsidence along the Cascadia coastline and left a record of flooded, dead coastal forests. A tsunami at least 5 m high hit Japan, where historical records pin down its exact date (January 26, 1700). The average time between magnitude 9 earthquakes in this subduction zone is thought to be about 500 to 600 years.

Although we have a good understanding of seismic hazards in some parts of the world—the United States and Japan, in particular—we know much less about others. During the 1990s, the United Nations sponsored an effort to map seismic hazards worldwide as part of the International Decade of Natural Disaster Reduction. This effort resulted in the first global seismic hazard map, shown in **Figure 13.31**. The map is based primarily on historical earthquakes, so it may underestimate the hazard in some regions where the historical record is short. Much more needs to be done to characterize seismic hazards on a global scale.

Land-Use Policies The exposure of buildings and other structures to earthquake hazard can be reduced by policies that restrict land uses in high-hazard areas. This approach works well where the hazard is localized, as in the case of known faults. Erecting buildings on active faults, as was done in the case of the residential developments pictured in **Figure 13.32**, is clearly unwise, because few buildings

FIGURE 13.32 Housing tracts constructed within the San Andreas fault zone, on the San Francisco Peninsula, before the state passed legislation restricting this practice. The white line indicates the approximate fault trace, along which the ground ruptured and slipped about 2 m during the earthquake of 1906. [R. E. Wallace/USGS.]

can withstand the deformation to which they might be subjected during a quake. In the 1971 San Fernando earthquake, a fault ruptured under a densely populated area of Los Angeles, destroying almost 100 structures. The state of California responded in 1972 with a law that restricts the construction of new buildings across an active fault. If an existing residence is on or very near a fault, real estate agents are required to disclose that information to potential buyers. A notable omission is that the act does not cover publicly owned or industrial facilities.

Earthquake Engineering Although land-use policies help to reduce the risk from localized hazards such as ground ruptures and soil liquefaction, they are less effective where seismic shaking is distributed across large regions. The risks from seismic shaking can best be reduced by good engineering and construction. Standards for the design and construction of new buildings are regulated by building codes enacted by state and local governments. A **building code** specifies the forces a structure must be able to withstand, based on the maximum intensity of shaking expected. In the aftermath of an earthquake, engineers study buildings that were damaged and recommend modifications to building codes that could reduce future damage from similar earthquakes.

As a result of this procedure, U.S. building codes have been largely successful in preventing loss of life during earthquakes. For example, from 1983 to 2004, 131 people died in nine severe earthquakes in the western United States, whereas more than 460,000 people were killed by earthquakes worldwide. Nevertheless, more can be done. Damage to structures could be reduced by using specialized construction materials and advanced engineering methods, such as putting entire buildings on movable supports to isolate them from the shaking.

Emergency Response Public authorities must plan ahead and be prepared with emergency supplies, rescue teams, evacuation procedures, fire-fighting plans, and other steps to minimize the consequences of a severe earthquake. Once an earthquake happens, networks of seismographs can transmit signals automatically to central processing facilities. In a fraction of a minute, computers can pinpoint the earthquake focus, measure its magnitude, and determine its fault mechanism. If equipped with strong-motion sensors that accurately record the most violent shaking, these automated systems can also deliver accurate maps in nearly real time showing where the ground shaking was strong enough to cause significant damage. Such information can help emergency managers and other officials deploy equipment and personnel as quickly as possible to save people trapped in rubble and to reduce further property losses from fires and other secondary effects. Bulletins about the magnitude and area of the shaking can also be channeled through the mass media to reduce public confusion during disasters and allay the fears aroused by minor tremors.

Earthquake Early Warning With the technology just described, it is possible to issue a warning tens of seconds before the arrival of destructive seismic waves. This "earthquake early warning" capability is possible because radio waves travel much faster than seismic waves. The most damaging vibrations from a nearby earthquake travel at a velocity of about 3.3 km/s, whereas radio signals travel at the speed of light (about 300,000 km/s).

A system recently deployed in Mexico City has demonstrated how this technology can work. The earthquakes that induce the strongest ground motions occur some 300 km from the city along the subduction zone where the Cocos Plate is being thrust under the North American Plate. In 1985, a magnitude 8.1 earthquake on that megathrust killed 10,000 people in Mexico City (see Figure 1.34). Computers attached to a special network of seismographs in the megathrust zone can now determine the location and time of an earthquake within seconds, and satellite radio links can transmit the information almost instantaneously, providing a 50- to 80-second warning before the most destructive waves reach the city. Even such an extremely short warning would allow schoolchildren to duck under their desks, utilities to shut off the flow of electricity and gas, hospitals to turn on emergency power, nuclear reactors to prepare to minimize damage, and so forth. Earthquake early warning systems have been deployed in Japan and Taiwan, as well as in Mexico, but they have not yet been developed for the United States.

Tsunami Warning Because tsunami waves travel 10 times more slowly than seismic waves, there is enough time to warn people on distant shorelines of an impending disaster. The Pacific Tsunami Warning Center, based in Hawaii, rapidly locates oceanic earthquakes using their fast-moving seismic waves, estimates their potential for triggering a tsunami, and quickly notifies the countries that may be in danger. An advance warning may be broadcast as much as a few hours before the arrival of the tsunami, allowing time for the evacuation of coastal populations. Unfortunately, no such system had been installed in the Indian Ocean, so the great 2004 tsunami struck with essentially no warning. Realizing how many tens of thousands of lives could have been saved, nations around the world are now cooperating to develop tsunami warning systems for the Indian Ocean and to upgrade the existing systems in the Pacific and Atlantic oceans.

The most difficult situation arises when a tsunami arrives so quickly that there is no time to warn nearby

FIGURE 13.33 Tsunami barrier in the town of Taro, Japan. [Courtesy of Taro, Japan.]

communities. This happened in the deadly 1998 Papua New Guinea tsunami, which killed as many as 3000 people in coastal villages near the epicenter of the quake that caused it. Such communities could be protected by building barrier walls to block inundation by the sea, but this type of construction is expensive and has been tried only in Japan (Figure 13.33). In this case, the best warning system is a very simple one: if you feel a strong earthquake, move quickly away from the coastal lowlands to higher ground!

Seven Steps to Earthquake Safety

As the last exclamation emphasizes, individuals living in seismically active areas need to prepare for earthquakes and know how to respond when one strikes. Here are seven steps to earthquake safety recommended by the Southern California Earthquake Center, which you can use to protect yourself and your family (Figure 13.34).

Before an earthquake:

1. Identify potential hazards in your home and begin to fix them. As buildings are becoming better designed to withstand seismic shaking, more of the damage and injuries that occur are resulting from falling objects. You should secure items in your home that are heavy enough to cause damage or injury if they fall or valuable enough to be a significant loss if they break.

2. Create your disaster plan. With your family or housemates, plan now what you will do before, during, and after an earthquake. The plan should include safe spots you can go to during the shaking, such as under sturdy desks and tables; a safe spot outside your home where you can meet after the shaking stops; and contact phone numbers, including someone outside the area who can be called to relay information in case local communications are disrupted.

3. Create your disaster kit. Stock your disaster kit with essential items. Your personal kit should include medications, a first-aid kit, a whistle, sturdy shoes, high-energy snacks, a flashlight with extra batteries, and personal hygiene supplies. Your home kit should include a fire extinguisher, wrenches to turn off gas and water mains, a portable radio, drinking water, food supplies, and extra clothing.

4. Identify your building's potential weaknesses and begin to fix them. Consult a building inspector or contractor to identify potential safety problems. Common problems include inadequate foundations, unbraced cripple walls, weak first stories, unreinforced masonry, and vulnerable pipes.

FIGURE 13.34 Seven steps to earthquake safety. For further information, see *Putting Down Roots in Earthquake Country*, Southern California Earthquake Center, 32 pp., 2004. This pamphlet is available online at http://www.scec.org.

During an earthquake:

5. *Drop, cover, and hold on.* During an earthquake or severe aftershock, drop to the floor, take cover under a sturdy desk or table, and hold on to it so that it doesn't move away from you. Wait there until the shaking stops. Stay away from danger zones, such as those near the exterior walls of buildings, near windows, and under architectural façades.

After an earthquake:

6. *After the shaking stops, check for damage and injuries needing immediate attention.* Take care of your own situation first; get to a safe location and remember your disaster plan. If you are trapped, protect your mouth, nose, and eyes from dust; signal for help using a cell phone or whistle or by knocking loudly on a solid part of the building three times every few minutes (rescuers will be listening for such knocks). Check for injuries and treat people needing assistance. Check for fires, gas leaks, damaged electrical systems, and spills. Stay away from damaged structures.

7. *When safe, follow your disaster plan.* Be in communication by turning on your radio and listening for advisories; check your phones, call your out-of-area contact to report your status, and then stay off the phone except for emergencies. Check your food and water supplies and check on your neighbors.

CAN EARTHQUAKES BE PREDICTED?

If we could predict earthquakes reliably, communities could be prepared, people could be evacuated from dangerous locations, and many aspects of the impending disaster might be averted. How well can we predict earthquakes?

Predicting an earthquake means specifying its time, location, and size. By combining information from plate tectonic theory and from detailed geologic mapping of regional fault systems, geologists can reliably predict which faults are likely to produce earthquakes over the long term. However, specifying precisely *when* a particular fault will rupture in a large earthquake has turned out to be very difficult.

Long-Term Forecasting

Ask a seismologist to predict the time of the next great earthquake at a particular location and the response is likely to be, "The longer the time since the last big shock, the sooner the next one will be." The **recurrence interval**—the average time between large earthquakes—can be estimated in several ways. According to the elastic rebound theory, the recurrence interval is the time required to accumulate the strain that will be released by fault slipping in a future earthquake. The recurrence interval can be calculated from the fault slip rate and the size of the expected fault slip (as estimated from the displacements observed in past earthquakes). On the San Andreas fault, for example, the slip rate is about 30 mm/year, so it takes about 150 years to accumulate enough strain to cause a fault slip of 4 m (about the amount expected for a magnitude 7.5 earthquake).

Geologists can also estimate the intervals between large earthquakes up to several thousand years in the past by finding and dating soil layers that were offset by fault displacements (**Figure 13.35**). Although these methods usually give similar results, the uncertainty of the predictions turns out to be large—as much as 50 percent of the average recurrence interval. In Southern California, the recurrence interval for the San Andreas fault is estimated to be 100 to 300 years. One part of this fault experienced a great earthquake in 1857, whereas another part (the southernmost) appears to have remained locked since a large earthquake that occurred around 1680 (see Figure 13.2). Therefore, an earthquake can be expected at any time—tomorrow, or decades from now.

Because of its large uncertainties (decades to centuries), this method of earthquake prediction is called *long-term forecasting* to distinguish it from what most people would really want—a *short-term prediction* of a large rupture on a specific fault accurate to within days or even hours of the actual event.

> *The earthquake recurrence interval on a fault can be estimated by dividing the expected fault slip by the fault slip rate.*

FIGURE 13.35 Geologist Gordon Seitz (left) examines layers of rock and peat that have been distorted by prehistoric earthquakes in a trench crossing a major strand of the San Andreas fault system in Southern California. By dating the peat using the carbon-14 method, geologists can reconstruct the history of large earthquakes on this fault. Such information helps scientists to forecast future events. [Courtesy of Tom Rockwell, San Diego State University.]

Short-Term Prediction

There have been a few successful short-term earthquake predictions. In 1975, an earthquake was predicted only hours before it occurred near Haicheng, in northeastern China. Chinese seismologists used what they considered to be *precursors* to make their predictions: swarms of tiny earthquakes and a rapid deformation of the ground several hours before the mainshock. Almost a million people, prepared in advance by a public education campaign, evacuated their homes and factories in the hours before the shock. Although many towns and villages were destroyed and a few hundred people were killed, there is no doubt that tens of thousands of lives were saved. The very next year, however, an unpredicted earthquake struck the Chinese city of Tangshan, killing more than 240,000. Obvious precursors such as those seen in Haicheng have not been repeated in subsequent large events.

Although many schemes have been proposed, we have not yet found a reliable method of predicting earthquakes from minutes to weeks ahead of time. Although we cannot say that short-term earthquake prediction is impossible, seismologists are pessimistic that short-term prediction will be feasible in the near future.

Intermediate-Term Prediction

Seismologists are more optimistic that the uncertainties in long-term forecasting can be reduced by studying the behavior of regional fault systems. The key is to generalize the elastic rebound theory. The simple version of the theory depicted in Figure 13.1 describes how the tectonic stress steadily building on an isolated fault segment is released in a periodic sequence of fault ruptures. However, as we have seen in the case of Southern California (see Figure 13.23), faults are rarely isolated. Instead, they are connected together in complex networks. A rupture on one fault segment changes the stresses throughout the surrounding region. Depending on the geometry of the fault network, these changes can either increase or decrease the likelihood of earthquakes on nearby fault segments. In other words, when and where earthquakes happen in one part of a fault system influences when and where they happen elsewhere in the system.

If Earth scientists could understand how variations in stress raise or lower the frequency of small seismic events, they might be able to predict earthquakes over intervals as short as a few years, or maybe even a few months, although still with substantial uncertainties. Monitoring of such events on networks of seismographs could then provide a regional "stress gauge." Someday you might hear a news report that says, "The National Earthquake Prediction Evaluation Council estimates that, during the next year, there is a 50 percent probability of a magnitude 7 or larger earthquake on the southern segment of the San Andreas fault." Much current research aims to establish a scientific basis for this type of *intermediate-term prediction*.

The ability to issue such predictions would raise difficult questions. How should society respond to a threat that is neither imminent nor long term? An intermediate-term prediction would give the probability of an event only on time scales of months to years—not precise enough to evacuate areas that might be damaged by an impending large earthquake. What effect would credible predictions have on property values and other investments in the threatened region? False alarms would be common. How should communities deal with this uncertainty? These are questions more suited to policy makers than to scientists.

> *Intermediate-term earthquake prediction might be possible in the future if we can better understand how variations in stress raise or lower the frequency of seismic events in a regional fault system.*

SUMMARY >>

What is an earthquake? An earthquake is a shaking of the ground that occurs when brittle rocks being stressed by tectonic forces break suddenly. When they break, the elastic energy built up over years of slow deformation is released rapidly, and some of it is radiated as seismic waves.

Where do earthquakes occur? Most earthquakes originate at plate boundaries, where stresses are concentrated and straining of the crust is intense. The foci of most continental earthquakes are above 20 km; below that depth, the crust behaves as a ductile material. In subduction zones, however, where cold oceanic lithosphere plunges back into the mantle, earthquakes can occur at depths as great as 690 km.

What are the three types of seismic waves? Two types of waves travel through Earth's interior: P (primary) waves, which are transmitted by all forms of matter and move fastest, and S (secondary) waves, which are transmitted only by solids and move at a little more than half the velocity of P waves. P waves are compressional waves that travel as a succession of compressions and expansions. S waves are shear waves that displace material at right angles to their path of travel. Surface waves are confined to Earth's surface and outer layers, like waves on the ocean. They move more slowly than S waves.

What is earthquake magnitude and how is it measured? Earthquake magnitude is a measure of the size of an earthquake. The Richter magnitude is determined from the amplitude of the ground motions recorded by seismographs. Seismologists prefer to use the moment magnitude because it is more directly related to the physical properties of the faulting that causes the earthquake: it varies as the logarithm of the area of faulting and the average fault slip. The magnitude of an earthquake is also proportional to the logarithm of the seismic energy released during the rupture.

How frequently do earthquakes occur? About 1,000,000 earthquakes with magnitudes greater than 2 take place each year. This number decreases by a factor of 10 for each magnitude unit. Hence, there are about 100,000 earthquakes with magnitudes greater than 3, about 1000 with magnitudes greater than 5, and about 10 with magnitudes greater than 7. The largest earthquakes, with magnitudes of 9 to 9.5, are rare and are confined to subduction zones.

What do seismic waves reveal about the layering of Earth's crust and mantle? Correlations of seismic wave velocities with rock types have made it possible to use seismic waves to explore the composition of Earth's interior. These explorations revealed that the continental crust is made up mostly of low-density granitic rocks, and that the deep seafloor is composed of basalt and gabbro. An increase in P-wave velocity marks the Moho, below which lies the mantle, composed mostly of peridotite. At the top of the asthenosphere, the temperature is high enough to partially melt peridotite, forming a zone where S-wave velocities decrease with depth. Below about 200 km, seismic wave velocities increase with depth down to Earth's core. However, at two depths, 410 km and 660 km below the surface, they show jumps caused by phase changes in mantle minerals. Below 660 km lies the lower mantle, a layer 2000 km thick in which seismic wave velocities increase steadily.

What do seismic waves tell us about the layering of Earth's core? Seismic waves reflected from the core-mantle boundary locate this sharp boundary at a depth of 2890 km. The failure of S waves to penetrate below the core-mantle boundary indicates that the outer core is fluid. A jump in P-wave velocity marks the boundary between the fluid outer core and the solid inner core at a depth of 5150 km. Several lines of evidence show that the core is composed mostly of iron and nickel, with minor amounts of some lighter element, such as oxygen or sulfur.

What has seismic tomography revealed about structures in the mantle? Seismologists use seismic tomography to create three-dimensional images of Earth's interior. These images reveal geographic variations in Earth's crust as well as many features of mantle convection, such as the sinking of lithospheric slabs into the lower mantle and the rising of superplumes from deep within the mantle.

What governs the type of faulting that occurs in an earthquake? The fault mechanism of an earthquake is determined by the type of plate boundary at which it occurs. Normal faulting under tensional stress occurs at divergent boundaries, thrust faulting under compressive stress occurs at convergent boundaries, and strike-slip faulting occurs along transform-fault boundaries.

What causes the destructiveness of earthquakes? Faulting and ground shaking during an earthquake can damage or destroy buildings and other structures. They can also trigger secondary effects, such as landslides and fires. Earthquakes on the seafloor can excite tsunamis, which sometimes cause widespread destruction when they reach shallow coastal waters.

What can be done to reduce the risks of earthquakes? Construction in earthquake zones can be regulated so that buildings and other structures will be strong enough to withstand the expected intensity of ground shaking, and land-use policies can prevent new building in high-hazard areas. Systems using seismograph networks and radio technology are being developed to provide early warnings of earthquakes and tsunamis. Public authorities can plan ahead, be prepared, and put early warning systems in place. People living in earthquake-prone areas can be informed about what to do when an earthquake occurs.

Can scientists predict earthquakes? Scientists can characterize the level of hazard in a region, but they cannot consistently predict earthquakes with the accuracy that would be needed to alert a population hours to weeks in advance. The best hope of making such predictions in the future may lie in a better understanding of how variations in stress raise or lower the frequency of seismic events in a regional fault system.

▌KEY TERMS AND CONCEPTS ≫

aftershock (p. 348)
building code (p. 373)
core-mantle boundary (p. 360)
earthquake (p. 344)
elastic rebound theory (p. 344)
epicenter (p. 347)
fault mechanism (p. 355)
fault slip (p. 346)
focus (p. 347)
foreshock (p. 349)
geotherm (p. 360)
intensity scale (p. 354)
lower mantle (p. 360)
low-velocity zone (p. 360)
magnitude scale (p. 353)
P wave (p. 351)
phase change (p. 360)
recurrence interval (p. 375)
S wave (p. 351)
seismic hazard (p. 371)
seismic risk (p. 371)
seismic tomography (p. 363)
seismicity map (p. 363)
seismograph (p. 349)
shadow zone (p. 358)
surface wave (p. 351)

EXERCISES

1. Seismographic stations report the following S-wave–P-wave time differences for an earthquake: Dallas, S-P = 3 minutes; Los Angeles, S-P = 2 minutes; San Francisco, S-P = 2 minutes. Use a map of the United States (or better yet, a globe) and the travel time curves in Figure 13.9 to obtain a rough location for the epicenter of the earthquake.

2. Describe two scales for measuring the size of an earthquake. Which is the more appropriate scale for measuring the amount of faulting that caused the earthquake? Which is more appropriate for measuring the amount of shaking experienced by a particular observer?

3. How much more energy is released by a magnitude 7.5 earthquake than by a magnitude 6.5 earthquake?

4. In Southern California, a magnitude 5 earthquake occurs about once per year. Approximately how many magnitude 4 earthquakes would you expect each year? How many magnitude 2 earthquakes?

5. What evidence suggests that the asthenosphere is partially molten?

6. What evidence indicates that Earth's outer core is molten and composed mostly of iron and nickel?

7. How can features of mantle convection, such as rising and descending convection currents, be imaged by seismic tomography?

8. How does the distribution of earthquake foci correlate with plate boundaries?

9. What are the fault mechanisms of earthquakes at the three types of plate boundaries?

10. In Figure 13.1, the right-lateral fault offsets the fence line to the right. In Figure 13.22, the mid-ocean ridge crest is also offset to the right. Why, then, is the transform fault in Figure 13.22 left-lateral?

11. The belts of shallow-focus earthquakes at divergent plate boundaries, shown by the blue dots in Figure 13.22, are wider and more diffuse in the continents than in the oceans. Why? (In answering this question, you might want to review Chapter 7.)

12. Why are earthquakes with focal depths greater than 20 km infrequent in continental lithosphere?

13. Why do the largest earthquakes occur on subduction megathrusts and not, say, on continental strike-slip faults?

14. Destructive earthquakes occasionally occur within plates, far from plate boundaries. Why?

15. At a location along the boundary fault between the Nazca Plate and the South American Plate, the relative plate movement is 80 mm/year. The last great earthquake, in 1880, showed a fault slip of 12 m. When should local residents begin to worry about another great earthquake?

16. Taking into account the possibility of false alarms, mass hysteria, economic depression, and other possible negative consequences of earthquake prediction, do you think the objective of predicting earthquakes should have a high priority?

HUMAN IMPACT ON EARTH'S ENVIRONMENT

CHAPTER 14

The Carbon Crisis

On a hot afternoon in late-summer Los Angeles, the carbon crisis is easy to observe. The freeways are crowded with returning vacationers, and smog gives the atmosphere a gray-brown tinge. Southern California Edison is urging everyone to keep the air conditioning low because the generating capacity of the state's power grid is stretched thin. The hills are brown tinderboxes because of the drought that grips the Southwest, and disputes over dwindling water supplies in the Lower Colorado River Basin are growing more contentious. War in the Middle East threatens as the price of oil hovers around $100 per barrel. The radio reports that Arctic sea ice has reached an all-time low and may melt entirely in the late summers 50 years from now. The magnificent glaciers of Kilimanjaro, the Andes, Switzerland, and Tibet are not expected to last even that long. Ecosystems are becoming more isolated, and some are collapsing, banishing their species to shrinking terrains, where many will face extinction.

> *The real voyage of discovery consists not in seeking new lands, but in seeing with new eyes.*
> — Marcel Proust

These problems all stem from interactions between human civilization and its environment, and all of them are connected through the processes that regulate the cycling of carbon within the Earth system. The crux of the problem can be stated simply enough: our economy depends on the burning of fossil fuels—a nonrenewable resource—which produces carbon dioxide—a powerful greenhouse gas.

How shall we confront this carbon crisis? From our perspective as Earth scientists, an old environmental adage, "Think globally, act locally," makes sense. The carbon crisis must be understood on a global scale, and the measures for dealing with it must be worldwide in scope. A plausible interim goal for the next half-century would be to stabilize our total carbon emissions at current rates, rather than allowing them to accelerate. As we will see, even this stopgap measure will require the implementation of new technologies on a Herculean scale.

Our leaders are not likely to have the stomach for such actions without a strong social consensus to move beyond "business as usual." This consensus will have to be forged within our local communities by educated people who understand the crisis and believe they can work together to improve our common lot. Cooperation among communities will have to trump competition among nations.

At the outset of our journey together, we offered to give you new eyes to view our planet and its problems. Now we urge you to keep them wide open as you chart the future course of Spaceship Earth. ■

—*T.H.J. and J.P.G., Los Angeles, September 2007*

Burnoff from an oil refinery during a power blackout that left nearly half of Los Angeles in the dark on September 12, 2005. [David McNew/Getty Images.]

CHAPTER OUTLINE

The Carbon Economy 380

Fossil-Fuel Resources 382

Alternative Energy Resources 390

The Carbon Cycle 395

Global Climate Change 400

Effects of Climate Change 403

Earth System Engineering and Management 409

The human habitat is a thin interface where Earth meets sky, where the global geosystems—the climate system, the plate tectonic system, and the geodynamo—interact to provide a life-sustaining environment. We have increased our standard of living by discovering clever ways to exploit this environment: to grow food, extract minerals, build structures, transport materials, and manufacture goods of all kinds. One result has been an explosion in the human population.

Roughly 100 million people were living on the planet 10,000 years ago (Figure 14.1). The population grew slowly, taking about 5000 years to double. The first doubling, to 200 million, was achieved early in the Bronze Age, when humans first learned how to mine ores and refine them into metals such as copper and tin (of which bronze is an alloy). The second doubling, to 400 million, was not achieved until the late Middle Ages. But once industrialization began, the global population really took off, climbing to 1 billion in 1804, 2 billion in 1927, and 4 billion in 1974. By the mid-twentieth century, the doubling time for the human population had dropped to only 47 years—less than a human lifetime. About 6.8 billion people are living today (2007), and our numbers are expected to reach 8 billion by 2028.

As the population has exploded, our appetites for energy and other resources have become voracious. Because the natural resources of Spaceship Earth are necessarily limited, the progress of civilization cannot be taken for granted. Environmental conditions and overall prosperity are not improving in some parts of the world, and the prospects for detrimental changes to the global environment loom large. Balancing our use of resources against detrimental effects on our environment, and thus on the ability of our environment to sustain a prosperous society, raises new challenges for Earth scientists and for all of human society.

In this final chapter, we survey the energy resources that power our civilization and examine how energy production and other human activities affect the global environment. We focus on two pressing issues: the depletion of petroleum as an energy resource and the potential for anthropogenic climate change.

These two issues are connected because they both involve the flow of carbon from one component of the Earth system to another. Petroleum depletion results from the movement of carbon-based fossil fuels from the lithosphere into the energy system of civilization, and the threat of climate change results from the release of carbon dioxide, a greenhouse gas, from that energy system into the atmosphere (Figure 14.2). Humans have greatly modified the flow of carbon among the components of Spaceship Earth, and we must understand what the consequences might be and how we might adapt our energy system to minimize the adverse effects of our activities.

THE CARBON ECONOMY

Petroleum and other fossil fuels are mixtures of combustible compounds that are rich in hydrogen and carbon. We mine these *hydrocarbons* from reservoirs in the lithosphere. We refine them into fuels such as diesel fuel and gasoline, which we burn to produce energy for transportation and other useful work.

Figure 14.3 describes the energy system of the United States in the year 2001. Consumption of energy from all

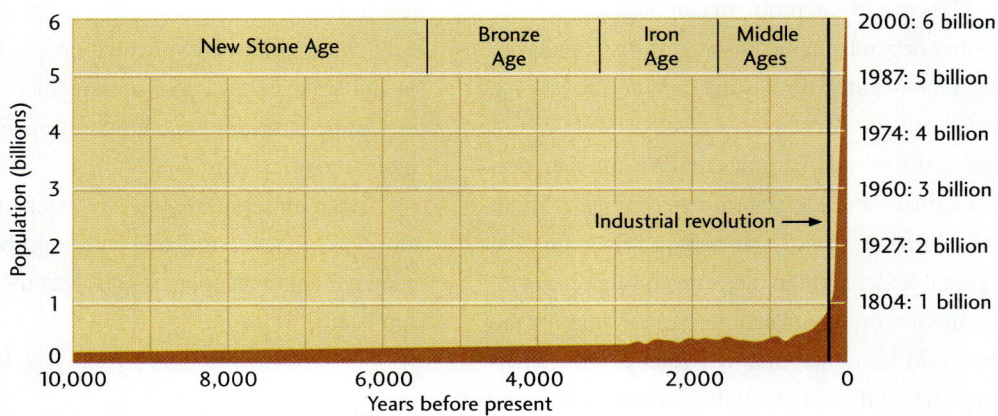

FIGURE 14.1 Human population growth over the last 10,000 years. The global population is expected to reach 8 billion by 2028.

The Carbon Economy

sources totaled 97 quadrillion Btu (quads). Nonrenewable fossil fuels—crude oil, natural gas, and coal—provided 86 percent of that total, and renewable biomass accounted for another 3 percent. Given that the nation depends on these carbon-based resources for almost 90 percent of its energy, we can fairly call the civilization fed by this energy system a **carbon economy.** You will notice that the flow of energy through the system is not particularly efficient: about 40 percent of the energy was used to do useful work, while 60 percent was wasted.

You can also see that the U.S. energy system released about 1.7 gigatons of carbon into the atmosphere each year, primarily as carbon dioxide (1 gigaton = 1 billion tons = 10^{12} kg). Worldwide, the total amount of carbon released by human energy production was about 6.8 gigatons during 2001, and the annual rate has since increased to over 8 gigatons per year. That's a lot of carbon!

Carbon plays a major role in the Earth system. Carbon compounds are continually moving between the atmosphere, biosphere, hydrosphere, cryosphere, and lithosphere in a natural *carbon cycle.* In the prehuman world, the exchange of carbon between the lithosphere

FIGURE 14.2 Coal-fired power plants such as this one in San Juan County, New Mexico, are converting fossil carbon previously buried in Earth's crust into atmospheric carbon dioxide. [Larry Lee Photography/CORBIS.]

FIGURE 14.3 The energy system of the United States consumes primarily non-renewable fossil fuels to produce useful energy and emits carbon dioxide, a greenhouse gas. The figures shown here are for the year 2001. From 2001 to 2006, total energy consumption in the United States grew by approximately 4 percent, from 97 quads to 101 quads, with only small changes in the relative consumption of different energy types. [Adapted from a special study by Lawrence Livermore National Laboratory.]

> *Our almost complete dependence on carbon-based fuels has disrupted the natural carbon cycle by pumping huge amounts of carbon from the lithosphere directly into the atmosphere. The result has been an increase in atmospheric concentrations of carbon dioxide.*

and the other components of the Earth system was regulated by the slow rates at which geologic processes buried and unearthed organic matter. This natural carbon cycle has been disrupted by the rise of the carbon economy, which is now pumping huge amounts of carbon from the lithosphere directly into the atmosphere.

This massive inflow of carbon has increased atmospheric concentrations of carbon dioxide from preindustrial levels of about 280 parts per million (ppm) to more than 380 ppm today (see Figure 10.13). If the burning of fossil fuels continues unabated, the amount of carbon dioxide in the atmosphere will double by mid-century. Because CO_2 is a greenhouse gas, that increase is likely to lead to enhancement of the greenhouse effect and global climate warming.

Our current reliance on the carbon economy poses some difficult questions: How long will our fossil-fuel resources last? To what extent will the increase in atmospheric carbon dioxide concentrations caused by fossil-fuel combustion adversely affect the global climate? How quickly will we need to replace fossil fuels with alternative energy sources? How can we use our resources to achieve sustainable development—to meet the needs of the present without compromising the ability of future generations to meet their own needs? As we rewrite the operating manual for Spaceship Earth to address these problems, we will be constrained by the availability of fossil fuels and alternative sources of energy, which we will now consider in more detail.

FOSSIL-FUEL RESOURCES

In Chapter 1, we provided an overview of renewable and nonrenewable natural resources. We distinguished *reserves*—the confirmed supplies of materials that are exploitable under current economic conditions—from *resources*—the total amounts of those materials, which are not precisely known but "guesstimated" based on past experience or theoretical models.

Figure 14.4 gives a rough estimate of the world's remaining nonrenewable energy resources of all types: about 360,000 quads. The largest resource is uranium (240,000 quads), which can be used to produce nuclear energy, followed by coal (67,500 quads).

Calculations based on these numbers can be deceptive, however. For example, simply dividing total resources by current (2007) annual consumption, which is about 460 quads (see Figure 1.10), might lead one to conclude (mistakenly) that many hundreds of years' worth of re-

FIGURE 14.4 A rough estimate of total remaining nonrenewable world energy resources amounts to about 360,000 quads. Amounts are given in conventional units of weight (metric tons) or volume (barrels) and by energy content (quads). [World Energy Council.]

sources remain before we have to worry about depletion of energy supplies. The economics of this issue are much more complicated, however, because some energy sources will give out before others, the various sources of energy are not readily interchangeable, and the environmental costs of converting some of them into useful forms of energy may be too great.

We can better understand what energy resources will be available in the future by considering the geologic circumstances in which these resources are found and the problems related to their recovery and use.

How Oil and Gas Form

Economically valuable deposits of our most important energy resources, crude oil and natural gas, develop under special environmental and geologic conditions. Both come from the organic debris of former life: plants and microorganisms (such as bacteria and algae) that have been buried, transformed, and preserved in sediments.

Oil and gas begin to form in sedimentary basins where the production of organic matter is high and the supply of oxygen in the sediments is inadequate to decompose all the organic matter they contain. Many offshore sedimentary basins on continental shelves satisfy both these conditions. In such environments, and to a lesser degree in some river deltas and inland seas, the rate of sedimentation is high, and organic matter is buried and protected from decomposition.

During millions of years of burial, chemical reactions triggered by the elevated temperatures and pressures found deep in the sediments slowly transform some of the organic material in these *source beds* into combustible hydrocarbons. The simplest hydrocarbon is methane gas (CH_4), the compound we call **natural gas.** Raw petroleum, or **crude oil,** includes a diverse class of liquids composed of more complex hydrocarbons.

Crude oil forms at a limited range of pressures and temperatures, known as the **oil window,** usually located at depths between about 2 and 5 km. The temperatures above the oil window are too low to transform organic material into hydrocarbons, whereas the temperatures below the oil window are so high (greater than 150°C) that the hydrocarbons that form are broken down into methane, producing only natural gas.

As burial progresses, compaction of the source beds forces crude oil and natural gas into adjacent beds of permeable rock (such as sandstones or porous limestones), which act as *hydrocarbon reservoirs.* The relatively low density of oil and gas causes them to rise, so that they float atop the water that almost always occupies the pores of permeable rock formations.

The conditions that favor large-scale accumulation of oil and natural gas are combinations of geologic structures and rock types that create an impermeable barrier to upward migration—an **oil trap** (**Figure 14.5**). Some oil traps are created by structural deformation and are called *structural traps.* One type of structural trap is formed by an anticline in which an impermeable layer of shale overlies a permeable sandstone formation (Figure 14.5a). The oil and gas accumulate at the crest of the anticline—the gas highest, the oil next—both floating on the groundwater that saturates the sandstone. Similarly, an angular unconformity or displacement at a fault may place a dipping permeable limestone formation opposite an impermeable shale, creating a structural trap (Figure 14.5b).

> *Crude oil and natural gas form in sediments that are rich in organic material and poor in oxygen. Elevated temperatures and pressures transform that organic material into combustible hydrocarbons.*

> *Oil and gas accumulate where geologic structures create impermeable barriers to their upward migration, forming oil traps.*

FIGURE 14.5 Oil and gas accumulate in traps formed by geologic structures. Four types of traps are illustrated here.

FIGURE 14.6 Estimated world oil reserves at the end of 2004 by region. [*British Petroleum Statistical Review of World Energy 2005*, June 2005.]

Other oil traps are created by the original pattern of sedimentation, as when a dipping permeable sandstone formation thins out against an impermeable shale (Figure 14.5c). These structures are called *stratigraphic traps*. Oil can also be trapped against an impermeable mass of salt in a *salt dome trap* (Figure 14.5d).

Distribution of Oil Reserves

In their search for petroleum resources, geologists have mapped thousands of oil traps throughout the world. Only a fraction of them have proven to contain economically valuable amounts of oil or gas, because traps alone are not enough to create a hydrocarbon reservoir. A trap will contain oil only if source beds were present, if the necessary chemical reactions took place, and if the oil could migrate into the trap and stay there without being disturbed by subsequent heating or deformation. Although oil and gas are not rare, most of the large, easy-to-find deposits have already been located, and the discovery of new resources is becoming more difficult.

The worldwide reserves of oil are estimated to be about 1.2 trillion barrels (1 barrel = 42 gallons); these reserves are broken down by region in **Figure 14.6**. The oil fields of the Middle East—including Iran, Kuwait, Saudi Arabia, Iraq, and the Baku region of Azerbaijan—contain nearly two-thirds of the world's total. Here sediments rich in organic material have been folded and faulted by the closure of the ancient Tethys Ocean, forming a nearly

FIGURE 14.7 The Ghawar oil field, in Saudi Arabia, is the world's largest, with a total reserve of at least 88 billion barrels. [Robert Azzi/Woodfin Camp Associates.]

> Worldwide reserves of oil are clustered in the Middle East, where closure of the Tethys Ocean created a nearly ideal environment for oil accumulation.

ideal environment for oil accumulation. The extensive reservoirs discovered in this vast convergence zone include the enormous Ghawar field in Saudi Arabia, the world's largest. Ghawar began production in 1948 and is expected to produce at least 88 billion barrels of oil over its lifetime (**Figure 14.7**).

Most of the oil reserves in the Western Hemisphere are located in the highly productive Gulf Coast–Caribbean area, which includes the Louisiana-Texas region, Mexico, Colombia, and Venezuela. Saudi Arabia ranks first among oil-producing nations, with 22 percent of the world's reserves; the United States ranks tenth, with 2.5 percent.

Oil Production and Consumption

In 2004, the amount of oil pumped out of the ground worldwide was about 30 billion barrels, an increase of 3.4 percent from the previous year. The United States produced about 2 billion barrels, more than any other nation except Saudi Arabia and Russia, but it consumed almost four times more—7.5 billion barrels, one-fourth of the total world production. This gap between production and consumption is growing at 5 percent per year, and it must be filled by importing oil, at an annual cost of hundreds of billions of dollars. This imbalance, $174 billion in 2004, contributes more than any other factor to the massive U.S. foreign trade deficit.

Thirty-one U.S. states have commercial oil reserves, and small, noncommercial resources can be found in most of the others. The United States, however, is a "mature" oil producer, in the sense that most of the petroleum resources within its borders have already been exploited; production reached a maximum in 1970 and is now in decline. The history of its oil production follows a bell-shaped curve (**Figure 14.8**). The high point of the curve is **Hubbert's peak,** named for petroleum geologist M. King Hubbert. In 1956, Hubbert used a simple mathematical relationship between the production rate and the rate of discovery of new reserves to predict that U.S. oil production, which was growing rapidly at the time, would actually begin to decline sometime in the early 1970s. His arguments were roundly dismissed as overly pessimistic, but history has proved him right.

> Hubbert's peak, the high point in oil production, can be predicted by a mathematical relationship between the production rate and the rate of discovery of new reserves.

When Will We Run Out of Oil?

At the current production rate, the world will consume all of the known oil reserves in just 40 years. Does that mean we will run out of oil before mid-century? No, because oil resources are much greater than oil reserves.

In fact, we will never really "run out" of oil. As the resources diminish, prices will eventually rise to such high levels that buyers cannot afford to waste oil by burning it as a fuel. Its main use will then be as a feedstock for the petrochemical industry, which uses it as a raw material to produce plastics, fertilizers, and a host of other products. The petrochemical industry is already a very big business,

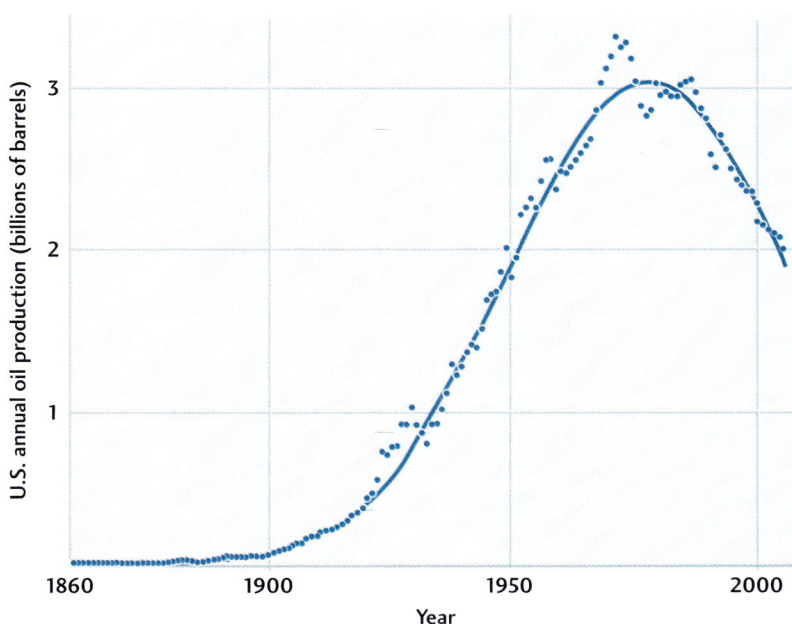

FIGURE 14.8 U.S. annual oil production from 1860 to 2000. Production peaked in 1970 and subsequently declined. [Modified from K. Deffeyes, *Hubbert's Peak*. Princeton, N.J.: Princeton University Press, 2001.]

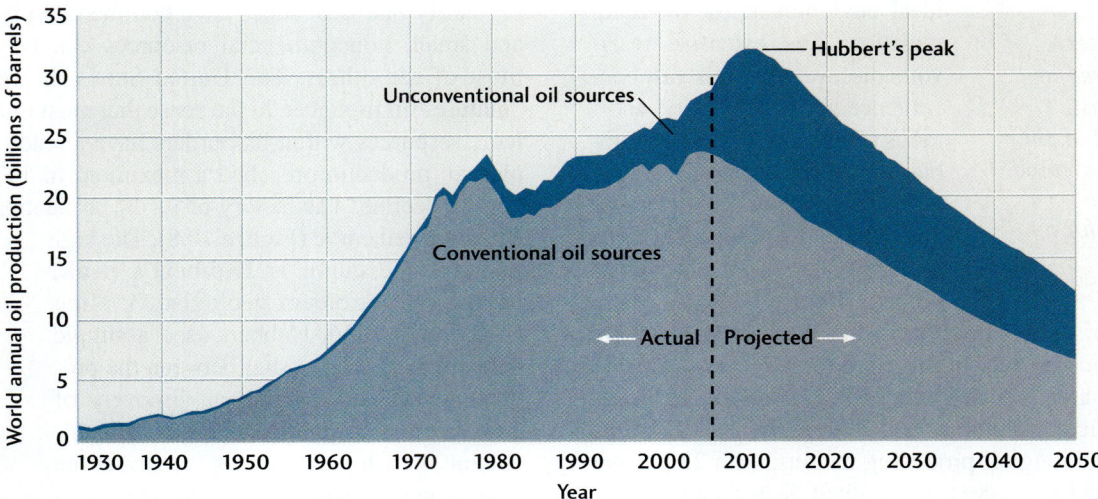

FIGURE 14.9 A pessimistic projection of world oil production, showing Hubbert's peak occurring around 2011.

[Colin J. Campbell, "The General Depletion Picture," *Association for the Study of Peak Oil and Gas Newsletter*, 2007.]

consuming 7 percent of global oil production. As oil geologist Ken Deffeyes has noted, future generations will probably look back on the Petroleum Age with a certain amount of disbelief: "They burned it? All those lovely organic molecules, and they just burned it?"

The key issue is not when oil will run out, but when oil production will stop rising and begin to decline. This milestone—Hubbert's peak for world oil production—is the real tipping point; once it is reached, the gap between supply and demand will grow rapidly, driving oil prices sky-high.

So how close are we to Hubbert's peak? The answer to this question is the subject of considerable debate. Oil optimists believe there is enough undiscovered oil to satisfy world demand for several decades into the future. The U.S. Geological Survey (USGS), the federal agency responsible for estimating energy resources, sits squarely in the optimist camp. It issued a worldwide assessment in 2000 that pegged total oil resources at more than twice the proven reserves, boosting by 20 percent the estimate it had made 6 years earlier. This resource value is the one used in Figure 14.4.

Oil pessimists, on the other hand, believe that we are fast approaching Hubbert's peak. Their views are supported by the same type of analysis Hubbert used to predict the 1970 peak in U.S. oil production. In particular, the pessimists note that the rate of oil discovery, which determines the growth of reserves, is declining too rapidly to be consistent with the USGS's optimistic scenario. One projection, by petroleum geologist Colin Campbell, shows global oil production declining shortly after 2010 (**Figure 14.9**). If this projection is correct, the estimate in Figure 14.4 may be too high by a factor of two.

A respected economist and spokesman for the optimists, Morris Adelman of the Massachusetts Institute of Technology (MIT), challenges pessimistic estimates of remaining oil supplies: "Nobody knows how much hydrocarbon exists or what percentage of that will be recoverable. The tendency to deplete a resource is counteracted by increases in knowledge." He believes that with improved oil exploration and production technology, oil supplies will continue to increase. Whether the oil optimists or the oil pessimists are correct is a central issue for Earth scientists and policy makers.

Oil Production and the Environment

In addition to the global effects of carbon dioxide emissions, our dependence on fossil fuels has other detrimental effects on the environment. Oil production, in particular, raises a number of regional and local issues. There are large resources of oil and gas, for example, under the coastal plain of northern Alaska and under the submerged continental shelves of North America. Many oil producers argue that these areas will eventually have to be drilled to satisfy the world's growing energy needs. A skeptical public, however, is not convinced that drilling can be done without serious threat to the environment. They remember 1969, when the ocean and beaches around Santa Barbara, California, were damaged by oil accidentally released from an offshore drilling platform. In 1979, a well that was being drilled in the Gulf of Mexico off the Yucatán coast "blew out," spilling as much as 100,000 barrels of oil per day for

Oil optimists and oil pessimists disagree on how soon world oil production will reach Hubbert's peak.

FIGURE 14.10 The oil spilled from the tanker *Exxon Valdez* in Prince William Sound, Alaska, had many detrimental effects on wildlife. [UPI/CORBIS-Bettmann.]

many weeks before it could be capped. The grounding of the tanker *Exxon Valdez* off the coast of Alaska in 1989, which released 240,000 barrels of crude oil into pristine coastal waters, was covered widely by television and newspapers and heightened public awareness of the severe ecological damage that can result from an oil spill (**Figure 14.10**). Despite such incidents and the difficulty of guaranteeing the safety of a well or a tanker, proponents of oil development believe that careful design of equipment and safety procedures can greatly reduce the chances of a serious accident.

The United States is currently embroiled in a strident political debate about whether to allow drilling for oil and natural gas in Alaska's Arctic National Wildlife Refuge (ANWR). The total ANWR resource has not been fully evaluated, but it could be as much as 20 billion to 30 billion barrels of oil. The USGS estimates that if oil prices were high enough, 4 billion to 12 billion barrels of this oil could be produced economically using current technologies. There is no doubt that these resources would contribute to the national economy. But oil and gas production would require the building of roads, pipelines, and housing in a very delicate ecological environment that is a particularly important breeding area for caribou, musk-oxen, snow geese, and other wildlife. Policy makers must weigh economic benefits against possible long-term environmental losses in making this decision.

> *The drilling and transportation of oil can pose serious threats to local and regional environments.*

Natural Gas

The world's resources of natural gas are comparable to its crude oil resources (see Figure 14.4) and may exceed them in the decades ahead. Estimates of natural gas resources have been rising in recent years. Exploration for this relatively clean-burning fuel has increased, and geologic traps have been identified in new settings, such as very deep formations, overthrust belts, coal beds, tight (less permeable) sandstones, and shales. The world's resources of natural gas are less depleted than its oil resources because natural gas is a relative newcomer on the energy scene.

The combustion of methane produces only carbon dioxide and water. Natural gas therefore burns much more cleanly than coal or oil, which also produce ash, sulfur dioxide (the major cause of acid rain, see Chapter 1), and other pollutants. Moreover, it emits 30 percent less carbon dioxide per unit of energy than oil and more than 40 percent less than coal. In addition, it is easily transported across continents through pipelines. Getting it from source to market across oceans has been more difficult. The construction of tankers and ports that can handle liquefied natural gas (LNG) is beginning to solve this problem, although the potential dangers (e.g., the risks of large explosions) have made LNG facilities controversial in the communities where they would be located.

> *Natural gas, a resource as abundant as oil, is a premium fuel because it burns more cleanly and produces less carbon dioxide than coal or oil.*

For all these reasons, natural gas is a premium fuel. It accounts for about 24 percent of all fossil-fuel consumption in the United States each year, most of it for industry and commerce (55 percent), followed by residential use (24 percent) and the generation of electric power (21 percent). More than half of U.S. homes, and a great majority of commercial and industrial buildings, are connected to a network of underground pipelines that draw gas from fields in the United States, Canada, and Mexico. At current rates of use, natural gas reserves in the United States will last about 10 years, and natural gas resources are likely to provide a substantial fraction of the nation's energy budget for several decades longer.

Coal

The abundant plant fossils found in **coal beds** are evidence that coal forms from large accumulations of plant material of the type that exist in wetlands. As the luxuriant plant growth of a wetland dies, it falls to the waterlogged soil. Rapid burial by falling leaves and immersion in water protect the dead twigs, branches, and leaves from complete decay because the bacteria that decompose vegetative matter are cut off from the oxygen they need. The vegetation accumulates and gradually turns into *peat,* a porous brown mass of organic matter in which twigs, roots, and other plant parts can still be recognized (**Figure 14.11**). The accumulation of peat in oxygen-poor environments can be seen in modern swamps and peat bogs. When dried, peat burns readily because it is 50 percent carbon.

Over time, with continued burial, the peat is compressed and heated. Chemical transformations increase the peat's already high carbon content, and it becomes *lignite,* a very soft, brownish black, coal-like material containing about 70 percent carbon. The higher temperatures and structural deformation that accompany greater depths of burial may metamorphose the lignite into *subbituminous* and *bituminous coal,* or soft coal, and ultimately into *anthracite,* or hard coal. The higher the grade of metamorphism, the harder and more vitreous the coal, and the higher its carbon content, which increases its energy content. Anthracite is more than 90 percent carbon.

Coal Resources There are huge resources of coal in sedimentary rocks. The global energy system has used only about 2.5 percent of the world's minable coal. According to some estimates, the amount of coal remaining is about 3.1 trillion metric tons, which is capable of producing 67,500 quads of energy (see Figure 14.4). About 85 percent of the world's coal resource is concentrated in the former Soviet Union, China, and the United States, which has extensive deposits in many states (**Figure 14.12**); these areas are also the largest coal producers. Domestic coal resources in the United States would last for a few hundred years at the nation's current rates of use—about a billion tons a year. Coal has supplied an increasing proportion of U.S. energy needs since 1975, when the price of oil began to rise; it currently accounts for about 22 percent of energy consumption in the nation (see Figure 1.6).

Environmental Costs of Coal The extraction and use of coal present serious problems that make it a less desirable fuel than oil or natural gas. When burned, coal releases, on average, 25 percent more carbon dioxide per

> Coal is formed when wetland vegetation accumulates in an oxygen-poor environment and undergoes burial, compression, and metamorphism.

FIGURE 14.11 The formation of coal beds begins with the deposition of vegetation in oxygen-poor environments.

1 Surface accumulation of leaf litter, twigs, branches, and other fragments of vegetation is buried as swamp deposits, partly decayed, and compressed to peat.

2 Shallow burial transforms peat to lignite.

3 Further burial under hundreds to thousands of meters of sediment transforms lignite to soft (bituminous) coal.

4 Continued burial and structural deformation, plus heat, metamorphose soft coal to hard (anthracite) coal.

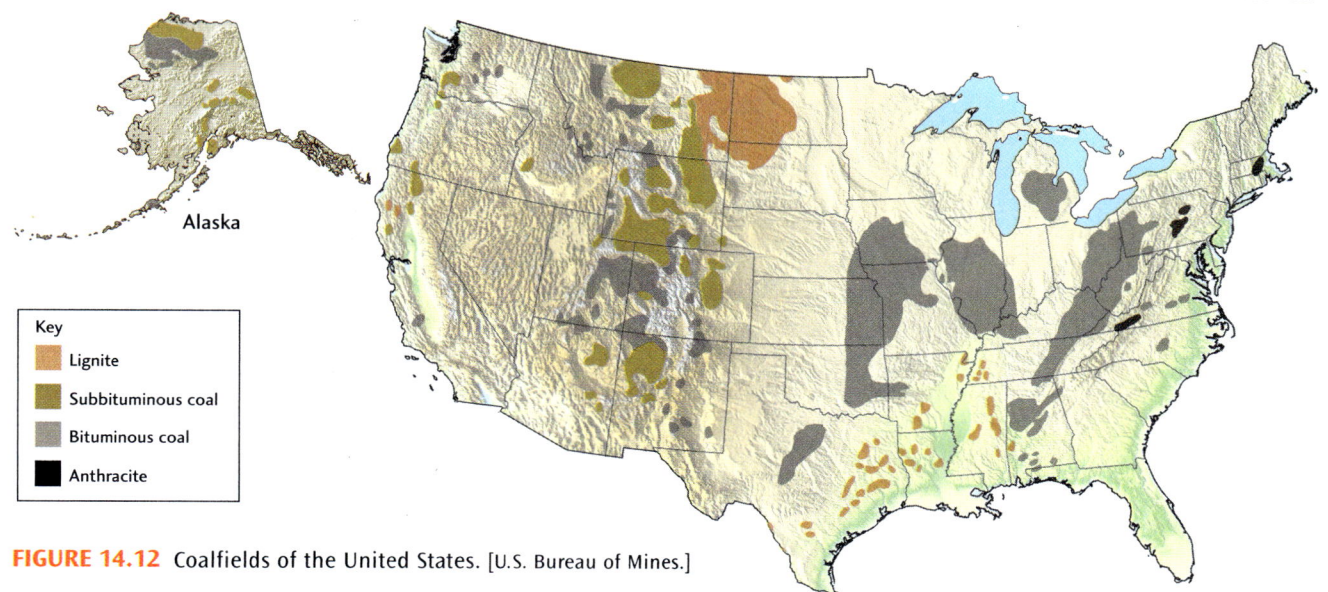

FIGURE 14.12 Coalfields of the United States. [U.S. Bureau of Mines.]

Coal produces more carbon dioxide than other fossil fuels, and its combustion releases sulfur oxides into the atmosphere, which cause acid rain.

unit of energy than oil and 70 percent more than natural gas. Most coal also contains appreciable amounts of sulfur, which is released into the atmosphere as noxious sulfur oxides when the coal is burned. Acid rain, which forms when these gases combine with rainwater, has become a severe problem in Canada, Scandinavia, the northeastern United States, and eastern Europe (see Chapter 1).

Coal ash is the inorganic residue that remains after coal is burned. It contains metal impurities from the coal, some of which are toxic. Ash can amount to several tons for every 100 tons of coal burned, so it poses a significant disposal problem. It can also escape from smokestacks, creating a health risk to people downwind.

Underground coal mining is a dangerous profession; more than 4000 miners are killed each year in China alone. Many more suffer from black lung, a debilitating inflammation of the lungs caused by the inhalation of coal particles. Strip mining, the removal of soil and surface sediments to expose coal beds, can ravage the countryside if the land is not restored (**Figure 14.13**).

United States government regulations now require industries that burn coal to adopt technologies for "clean"

FIGURE 14.13 *(left)* Coal strip mine, Buskin, Indiana. *(right)* Same strip mine after restoration. [Photos courtesy of Vigo Coal Company.]

Although the extraction and burning of coal presents a number of risks to human life and the environment, its use is likely to increase because it is so abundant relative to other fossil fuels.

coal combustion, which have reduced emissions of sulfur and toxic chemicals such as mercury. Federal laws also mandate the restoration of land disrupted by strip mining and the reduction of danger to miners. But these measures are expensive and add to the cost of coal. This drawback, however, is unlikely to prevent the increased use of coal, which is so much more abundant than oil. Many countries have no other fuel resources, and some countries will not be able to afford to import oil, which will increase in price as the supply diminishes.

Unconventional Fossil-Fuel Resources

Extensive deposits of hydrocarbons occur in two other forms: source beds that are rich in organic material but never reached the oil window, such as *oil shales,* and formations that once contained oil but have since "dried out" to form *heavy oil* or a tarlike substance called *natural bitumen* (not to be confused with bituminous coal). One deposit of the latter type, the **tar sands** of Alberta, Canada, is estimated to contain a hydrocarbon reserve equivalent to 180 billion barrels of oil and a total resource perhaps 10 times that amount. More than 400 million barrels of oil are now extracted from the Alberta tar sands each year, and Canadian production is projected to increase fivefold by 2030, providing as much as 5 percent of world demand.

Development of the tar sands raises important environmental concerns, however. It takes 2 tons of mined sand to produce 1 barrel of oil, leaving lots of waste sand, which is polluting. Moreover, production of oil from tar sands is an energy-intensive process that sucks up about two-thirds of the energy they ultimately render—a very low energy efficiency—and emits six times more carbon dioxide than conventional oil production.

A second unconventional source of fossil fuel is **oil shale,** a fine-grained, clay-rich sedimentary rock containing relatively large amounts of organic matter. In the 1970s, oil companies began trying to commercialize the extensive oil shales of western Colorado and eastern Utah, but those efforts were largely abandoned by the 1980s as oil prices fell, concerns over environmental damage increased, and technical problems persisted. As is the case with tar sands, the efficiency of producing energy from oil shale is very low, and the environmental costs per unit of energy are very high. For example, shale oil and combustible gas can be extracted from these rocks, but the process requires huge amounts of water, a scarce resource in the western United States. Nevertheless, renewed interest in energy production from domestic oil shale has been sparked by higher oil prices and is being encouraged by national energy policies.

The production of fossil fuels from unconventional sources, such as tar sands and oil shales, is limited by high energy costs and negative environmental effects.

ALTERNATIVE ENERGY RESOURCES

If crude oil and natural gas continue to be the major resources used to satisfy the world's voracious appetite for energy, the great bulk of the world's supplies will be exhausted within a century. Coal will eventually become the predominant fossil fuel in many countries. It may be reassuring to know that if energy demand grows at modest annual rates—say, 3 percent per year—coal and other fossil fuels can meet the world's energy needs for about 100 years or longer. The environmental costs of coal may be unacceptable, however. In particular, the threat of climate change could force a shift away from fossil fuels before they are depleted.

These estimates do not consider the possibility that we may begin to meet our energy needs in nontraditional ways: through increased efficiency in the use of fossil fuels and through the development and use of alternative energy sources. To the extent that alternative sources can be used, we can reduce the pressure on our fossil-fuel resources as well as their negative environmental effects.

Nuclear Energy

Although the first large-scale use of the radioactive isotope uranium-235 to produce energy was in the atomic bomb in 1944, the nuclear physicists who first observed the vast energy released when its nucleus split spontaneously (a phenomenon called *fission*) foresaw the possibility of peaceful applications of this new energy source. After World War II, their predictions were fulfilled as countries around the world built nuclear reactors to produce **nuclear energy.** In these reactors, the fission of uranium-235 releases heat that is used to make steam, which then drives turbines to create electricity. A typical commercial reactor produces about 1000 megawatts of electrical power (1 megawatt = 1 million watts). In the United States, some 110 nuclear reactors now provide about 22 percent of the nation's electrical energy. More than 400 nuclear reactors in 25 countries account for about 6 percent of global energy production. If its full potential were to be realized, nuclear energy could meet the world's needs for electrical energy for hundreds of years.

The geologic problems associated with nuclear energy include the assessment of uranium reserves and the underground disposal of nuclear wastes.

Nuclear energy has the potential to meet the world's electrical energy needs for hundreds of years.

Uranium Reserves Uranium constitutes only 0.00016 percent of the average continental crustal rock. Moreover, only a small portion of this element is uranium-235; its other, much more abundant isotopes (for example, uranium-238) are not radioactive enough to be used as fuel in nuclear reactors. However, uranium is still the world's largest minable energy resource by far, with a potential energy-generating capacity of at least 240,000 quads (see Figure 14.4). It is typically found as small quantities of uraninite, a uranium oxide mineral (also called pitchblende), in veins in granites and other felsic igneous rocks. Uranium in igneous rocks near Earth's surface may oxidize and dissolve, be transported in groundwater, and later be reprecipitated as uraninite in sedimentary rocks.

Uranium is the world's largest minable energy resource.

Hazards of Nuclear Energy The biggest drawbacks of nuclear energy are concerns about the safety of nuclear reactors, the risk of environmental contamination, and the potential use of radioactive fuels for making nuclear weapons.

Two accidents have raised questions about the safety of nuclear energy. The first was at the Three Mile Island reactor in Pennsylvania in 1979. A reactor was destroyed, and radioactive debris was released, although it was confined within the containment building. No one was harmed, but most experts agree that it was a close call. Much more serious was the destruction of a nuclear reactor in the town of Chernobyl in Ukraine in 1986. The reactor went out of control because of poor design and human error. Radioactive debris spilled into the atmosphere and was carried by winds over Scandinavia and western Europe. Contamination of buildings and soil has made hundreds of square miles of land surrounding Chernobyl uninhabitable. Food supplies in many countries were contaminated by the radioactive fallout and had to be destroyed. Deaths from cancer caused by exposure to the fallout may be in the thousands.

The uranium consumed in nuclear reactors leaves behind dangerous radioactive wastes. A system of safe long-term waste disposal is not yet available, and reactor wastes are being held in temporary storage at reactor sites. In a few years, the limits of the space available for temporary storage in the United States will be reached. Although many scientists believe that geological containment—the burial of nuclear wastes in deep, stable, impermeable rock formations—is a workable solution, there is not yet a generally approved plan for the safe storage of the most dangerous wastes for the hundreds of thousands of years required before they cease to be radioactive. France and Sweden have built underground nuclear waste repositories. A major facility is also being developed in the United States—the Yucca Mountain Nuclear Waste Repository at the Nevada Test Site (**Figure 14.14**)—but it is embroiled in litigation as the state of Nevada battles the federal government to keep it from being built. No state wants to be the dumping ground for a long-term environmental hazard.

In the United States, these unresolved problems, plus the danger that nuclear fuels could be used to build nuclear weapons, have essentially halted the installation of new nuclear power plants, and new installations have been slowed in other countries as well.

Concerns about the safety and security of nuclear reactors and the unresolved problem of nuclear waste disposal may limit the use of nuclear energy.

FIGURE 14.14 Aerial view of the north entrance to the Yucca Mountain Nuclear Waste Repository being developed at the Nevada Test Site, north of Las Vegas. Yucca Mountain is the high ridge to the right of the entrance. [U.S. Department of Energy.]

Solar Energy

Solar energy enthusiasts remind us that "every month Earth receives more energy from sunshine than is stored in all of Earth's nonrenewable energy resources!" Light from the Sun can be converted into heat and electricity, or it can be used to liberate hydrogen from water, which can then be used as a clean fuel. **Solar energy** is nondepletable—the Sun will continue to shine for at least the next several billion years. The bad news is that the technology currently available for large-scale conversion of solar energy into useful forms, such as heat and electricity, is inefficient and expensive; the good news is that it is improving.

In the near term, the only use of solar energy that is likely to be economical will be to heat and provide hot water for homes and industrial and agricultural processes. Solar generation of electricity is not yet cost-effective, though solar power plants are being built as demonstration projects (**Figure 14.15**). The efficiency of converting sunlight into electricity is still low, and the costs of building and installing solar systems are still high. Moreover, commercial-scale solar electric power plants can present significant environmental problems. A solar plant with a 100-megawatt generating capacity (about 10 percent of the capacity of a typical nuclear reactor) located in a desert region would require at least a square mile of land and might alter the local climate by significantly changing the balance of solar radiation in the area.

In 2004, solar energy supplied only 0.06 percent of U.S. energy consumption. Solar energy enthusiasts believe that this amount could increase to some 20 quads per year in a decade or so, an amount equivalent to about half the oil the nation uses today. A more realistic figure is probably less than 10 quads per year. Even this more conservative scenario would yield important social benefits: conservation of other energy resources, diversification of energy supplies so that consumers are not overly dependent on a single source, and reduction of fuel imports. With adequate research and development, solar energy can probably become economically competitive and a major source of energy within a few decades.

Efforts to increase the use of solar energy are focused on improving the efficiency of converting sunlight to electricity.

Hydroelectric Energy

Hydroelectric energy is derived from water moving under the force of gravity that is made to drive a turbine that generates electricity. Waterfalls or artificial reservoirs behind dams usually provide the water. Hydroelectric energy is renewable because it depends on rainfall, and it is also relatively clean, risk-free, and cheap.

The Three Gorges Dam, on the Yangtze River in China (**Figure 14.16**), will be the world's largest hydroelectric facility when it is completed in 2009. It will be able to generate 22,500 megawatts, close to 5 percent of China's total electricity demand. The project has been controversial, however, because the damming of the Yangtze has caused flooding that is displacing over a million people.

In the United States, hydroelectric dams deliver about 3 quads annually, or about 3 percent of the nation's annual energy consumption. The U.S. Department of Energy has identified more than 5000 sites where new hydroelectric dams could be built and operated economically, boosting the nation's hydroelectric capacity by as much as 40 percent. Such expansion would probably be resisted, however, because the dams would drown farmlands and wilderness areas under artificial reservoirs but add only a few percentage points to U.S. energy production. This is one energy issue on which most people agree that the environmental costs would exceed the economic gains. For this reason, energy experts expect that almost no new hydroelectric generating capacity will be added through 2020 and that its share of the nation's electricity production will decline to about 6 percent from about 10 percent today.

Use of hydroelectric power is expected to decline because of the environmental costs associated with building hydroelectric dams and artificial reservoirs.

FIGURE 14.15 Aerial view of the Gut Erlasse Solar Park, a 12-megawatt solar electric power plant located amid Bavarian cropland near the town of Arnstein, Germany. The panels of solar cells tilt and rotate to face the Sun throughout the day. [Daniel Karmann/European Pressphoto Agency.]

FIGURE 14.16 The Three Gorges Dam on China's Yangtze River. When it is completed in 2009, 32 generators will produce 22,500 megawatts of hydroelectric power. The dam is about 2335 m (7660 feet) long and 185 m (616 feet) high. [AP photo/Xinhua Photo, Xia Lin.]

Wind Power

Wind power is produced by using windmills to drive electric generators. Its use is growing in many places as windmill designs improve and costs are brought down to where they are competitive with conventional energy sources (**Figure 14.17**). The U.S. Department of Energy estimates that winds sufficient for power generation blow across 6 percent of the land area of the continental United States, and that they have the potential to supply more than one and a half times the nation's current electricity consumption. But that's a lot of area—550,000 km^2—and harvesting this energy would require millions of 30-m windmills.

Geothermal Energy

In regions where geothermal gradients are steep, Earth's internal heat can sometimes be tapped to drive electric generators and heat homes. **Geothermal energy** depends on the heating of water as it passes through a region of hot rock (a *heat reservoir*) that may be hundreds or thousands of meters beneath Earth's surface. Hot water or steam is brought to the surface through boreholes drilled for the purpose. Usually, the water is naturally occurring groundwater that seeps downward along fractures in rock. Less typically, the water is artificially introduced by pumping from the surface.

At least 46 countries now use some form of geothermal energy. By far the most abundant source is naturally occurring groundwater that has been heated to the relatively low temperatures of 80° to 180°C and is used for residential, commercial, and industrial heating. Warm underground water drawn from a heat reservoir in the Paris sedimentary basin now heats more than 20,000 apartments in France. Iceland, as we saw in Chapter 5, sits on the Mid-Atlantic Ridge, where hot mantle material wells up as the North American and Eurasian plates separate. Reykjavík, the capital of Iceland, is almost entirely heated by geothermal energy derived from this volcanic heat.

FIGURE 14.17 These rows of windmills in Altamont, California, are producing electric power. [Glen Allison/Photodisc/Getty Images.]

FIGURE 14.18 The Geysers, one of the world's largest supplies of natural steam. Its geothermal energy is converted into electricity for San Francisco, 120 km to the south. [Pacific Gas and Electric.]

In regions where geothermal gradients are steep, Earth's internal heat can be tapped as a source of energy.

Geothermal energy sources with temperatures above 180°C are useful for generating electricity. They are present primarily in regions of recent volcanism as hot, dry rock, natural hot water, or natural steam. The latter two sources are limited to those few areas where surface water seeps down through underground faults and fractures to reach deep rocks heated by recent magmatic activity. Naturally occurring water heated above the boiling point and naturally occurring steam are highly prized resources. The world's largest facility for producing electricity from natural steam, located at The Geysers, 120 km north of San Francisco, generates more than 600 megawatts of electricity (**Figure 14.18**). Some 70 geothermal electricity-generating plants operate in California, Utah, Nevada, and Hawaii, producing 2800 megawatts of power—enough to supply about a million people.

According to one Icelandic estimate, as much as 40 quads of electricity could be generated each year from accessible geothermal energy sources, but so far only a tiny fraction of that amount, about 0.15 quad per year, is actually being generated. Another 0.12 quad of geothermal energy is used for direct heating.

Like most of the other energy sources we have examined, geothermal energy presents some environmental problems. Regional ground subsidence can occur if hot groundwater is withdrawn without being replaced. In addition, geothermally heated waters can contain salts and toxic materials dissolved from the hot rock.

Biofuels

Before the coal-fired industrial revolution of the mid-nineteenth century, the burning of wood and other biomass derived from plants and animals (e.g., whale oil, dried buffalo dung) satisfied most of society's energy needs (see Figure 1.6). Even today, the energy derived from biomass exceeds the total derived from all other renewable resources.

Biomass is an attractive alternative to fossil fuels because, at least in principle, it is *carbon-neutral;* that is, the CO_2 produced by the combustion of biomass is eventually recycled by plant photosynthesis to produce new biomass, allowing us to harvest energy from sunlight without increasing CO_2 concentrations in the atmosphere. In particular, liquid **biofuels** derived from biomass, such as *ethanol* (ethyl alcohol: C_2H_6O), could replace gasoline as our main automobile fuel.

The use of biofuels in transportation is hardly new. The first four-stroke internal combustion engine, invented by Nikolaus Otto in 1876, ran on ethanol, and the diesel engine, patented by Rudolf Diesel in 1898, ran on vegetable oil. Henry Ford's Model T car, first produced in 1903, was designed to operate completely on ethanol. But soon thereafter, petroleum from the new reserves discovered in Pennsylvania and Texas became widely available, and cars and trucks were converted almost entirely to gasoline and diesel fuel (see Chapter 1).

Ethanol remains the most popular biofuel and can be mixed with gasoline to run most car engines built today. It is produced mainly from corn in the United States and from sugarcane in Brazil. For the last 30 years, the Brazilian government has been pushing to replace imported oil with domestic ethanol; today, more than 30 percent of Brazil's automobile fuels come from sugarcane, saving the country about $50 billion in oil imports.

The biofuel ethanol, which is produced from plant biomass, is an alternative to gasoline that can reduce the net anthropogenic emission of carbon dioxide.

FIGURE 14.19 Switchgrass, a perennial plant native to the Great Plains, is an efficient source of ethanol, the most popular biofuel. Here, geneticist Michael Casler harvests switchgrass seed as part of a breeding program to increase the plant's ethanol yield. [Wolfgang Hoffmann.]

The United States has initiated a crash program to develop more efficient methods for producing biofuels, with the goal of reducing oil imports by at least 75 percent before 2025. A promising biomass crop is switchgrass, a perennial plant native to the Great Plains (**Figure 14.19**). Switchgrass has the potential to produce up to 1000 gallons of ethanol per acre per year, compared with 665 gallons for sugarcane and 400 gallons for corn.

Can biofuels really be carbon-neutral? If the energy it takes to fertilize the plants, transform them into biofuels, and deliver the biofuels to market comes primarily from fossil fuels, the answer is no. The widespread use of biofuels for transportation would no doubt reduce the pumping of carbon from the lithosphere to the atmosphere, but experts are still arguing about the magnitude of that reduction.

This question raises a more general one: what will be the net effect of our various energy-producing strategies on carbon dioxide concentrations in the atmosphere? To answer this question, we need to understand how carbon moves through the Earth system. That movement is described by the carbon cycle, the topic we consider next.

THE CARBON CYCLE

Because of the high rate of fossil-fuel burning that sustains our carbon economy, the CO_2 concentration of the atmosphere is now increasing at the unprecedented rate of 0.4 percent per year (see Figure 10.13), faster than at any time in recent geologic history.

Yet the situation could be worse. In the 1980s, human civilization emitted about 5.4 gigatons of carbon each year by fossil-fuel burning and other industrial activities, and an additional 1.7 gigatons by the burning of forests and other changes in land use. If all of that carbon had stayed in the air, the atmospheric CO_2 increase would have been closer to 0.9 percent per year, more than twice the observed rate. Instead, 3.8 gigatons of carbon were removed from the atmosphere each year by natural processes. Where did all that carbon go?

We will answer this question by examining the **carbon cycle**—the continual movement of carbon between the atmosphere and its other principal reservoirs: the lithosphere, hydrosphere, and biosphere. We will then be able to consider an even deeper set of issues: What are the interactions between the carbon cycle and the climate system? Is the net feedback between these systems positive or negative—that is, will changes in the carbon cycle enhance or reduce global warming? Finally, how might humans intervene in the carbon cycle to stabilize the climate system and prevent global warming?

> The carbon cycle describes the movement of carbon among its principal geochemical reservoirs: the atmosphere, the lithosphere, the hydrosphere, and the biosphere.

Geochemical Cycles and How They Work

In Chapter 11, we saw that the hydrologic cycle involves the transfer of a chemical substance—water—between reservoirs. Similarly, when discussing the carbon cycle, we view the major components of the Earth system—atmosphere, hydrosphere, cryosphere, lithosphere, and biosphere—as **geochemical reservoirs** for the storage of carbon and other terrestrial chemicals, and we consider the processes that transport those chemicals among reservoirs.

Geochemical Fluxes Geochemical cycles trace the flow, or *flux*, of chemicals from one reservoir to another. These cycles include the evaporation and precipitation of water, the airborne transfer of dust from the ground surface to the oceans, volcanic discharges, and gas exchange between the sea surface and the atmosphere (**Figure 14.20**). On time scales relevant to the climate system, the major fluxes into and out of a reservoir are usually in approximate balance. Sedimentation in the oceans, for example, balances the influx of sediments from rivers. As seafloor sediments are buried, they become part of the oceanic crust. There they stay until they move into the mantle through subduction or become part of the continental crust through accretion. Over the long term, the uplifting of continental regions by plate convergence and other orogenic processes exposes crustal rocks to weathering and erosion, maintaining the balance of fluxes among the reservoirs.

FIGURE 14.20 A number of processes result in fluxes of chemical substances between components of the climate system.

The biosphere is a unique reservoir because each individual organism is constantly interacting with its environment, which often includes three of the other major reservoirs: the atmosphere, hydrosphere, and lithosphere. The most important fluxes into and out of the biosphere are the inflow and outflow of atmospheric gases by respiration, the inflow of nutrients from the lithosphere and hydrosphere, and the outflow of chemicals through the death and decay of organisms. When the biosphere plays a key role in the flux of a chemical, geologists refer to the cycle that describes that flux as a *biogeochemical cycle*. The carbon cycle, which depends critically on the pumping of carbon into and out of the atmosphere by living organisms, is clearly a biogeochemical cycle.

> Geochemical cycles describe the flux of a chemical among its geochemical reservoirs. Biogeochemical cycles are cycles in which the biosphere plays a key role.

Residence Times Reservoirs gain chemicals from inflows and lose chemicals from outflows. If inflow equals outflow, the amount of the chemical in the reservoir stays the same, even though the chemical is constantly entering and leaving it. On average, a molecule of the chemical spends a certain amount of time, called the **residence time,** in a reservoir.

We can visualize a chemical's residence time in the ocean, for example, as the average time that elapses between the entry of a molecule into the ocean and its removal through sedimentation or some other process. For example, the residence time of sodium in the ocean is extremely long—about 48 million years—because sodium is highly soluble in seawater (that is, the capacity of the reservoir to store sodium is high) and because rivers contain relatively small amounts of sodium (its inflow into the reservoir is low). In contrast, iron has a residence time in the ocean of only about 100 years because its solubility in seawater is very low and the inflow from rivers is relatively high.

The residence times of chemicals in the atmosphere are usually shorter than those in the ocean because the atmosphere is a smaller reservoir than the ocean and because fluxes into and out of the atmosphere can be relatively high. Sulfur dioxide, for example, has a residence time in the atmosphere of hours to weeks, and oxygen, which makes up about 21 percent of the atmosphere, has a residence time of 6000 years. An exception is the most common atmospheric gas, nitrogen, which constitutes about 78 percent of the atmosphere; because it is abundant and stable, its residence time in the atmosphere is almost 400 million years. A molecule of nitrogen that entered the atmosphere in the late Paleozoic era, about 300 million years ago, is still likely to be there!

Some chemicals are relatively inert, but most participate in chemical reactions. In many cases, these reactions govern a chemical's residence time in a reservoir. Calcium ions, for example, are removed from seawater when they react with bicarbonate ions to form calcium carbonate, which can precipitate out of the seawater as limestone sediments. The dissolving of more CO_2 in the oceans makes them more acidic ("ocean acidification"), which decreases their bicarbonate concentration and thus increases the solubility of calcium, allowing less limestone to form. As this example illustrates, chemical reactions can couple the geochemical cycle of one element (calcium) with that of another (carbon). For that reason, geo-

FIGURE 14.21 The calcium cycle, showing fluxes into and out of the ocean. Fluxes are given in gigatons (Gt; 10^{12} kg) per year. The inflow of calcium to the ocean approximately balances the outflow.

chemical cycles can rarely be considered in isolation but must be treated as interacting geosystems.

An Example: The Calcium Cycle Before we examine the carbon cycle in more detail, let's take a look at the calcium cycle, which provides a simpler illustration of the concepts involved in geochemical cycles (Figure 14.21). The ocean contains about 560,000 gigatons of calcium, dissolved in a total ocean mass of about 1.4×10^9 gigatons. Calcium steadily enters this reservoir in large quantities through the rivers of the world, which transport dissolved and suspended calcium derived from the weathering of minerals such as calcite, gypsum, plagioclase feldspars, and other calcium-rich silicates. A much smaller amount enters the ocean via transport by windblown dust.

If the ocean received this continuous inflow of calcium liberated by weathering without there being any way to remove it, the ocean would quickly become supersaturated with calcium. The flux that removes most of the excess calcium from the ocean is the precipitation of calcium carbonate. A smaller amount of calcium is precipitated as gypsum in evaporites.

The amount of calcium the ocean can hold is much larger than the inflow and outflow of calcium, so calcium has a fairly long residence time in the ocean. By dividing the total annual influx (0.9 gigaton/year) by the ocean's calcium capacity (560,000 gigatons), we obtain a residence time of about 600,000 years.

The Cycling of Carbon

Carbon cycles among four main reservoirs of the Earth system: the atmosphere; the oceans, including marine organisms; the land surface, including terrestrial plants and soils; and the deeper lithosphere (Figure 14.22).

The greatest carbon flux, 120 gigatons per year, is the cycling of CO_2 between the biosphere and the atmosphere by land plants and animals through photosynthesis, respiration, and decay. Plants take in this entire amount during photosynthesis and respire about half of it back into the atmosphere. The other half is incorporated into plant tissues—leaves, wood, and roots. Animals eat the plants, and microorganisms decompose them; both processes result in the breakdown of plant tissues and the respiration of CO_2. Much of the organic carbon, about three times the total plant mass, is stored in soils. A significant fraction (about 4 gigatons/year) is directly oxidized through forest fires and other combustion of plant material. A smaller fraction (0.4 gigaton/year) is dissolved in surface waters and transported by rivers to the ocean, where it is respired back into the atmosphere by marine organisms.

The greatest flux in the carbon cycle is between the biosphere and the atmosphere.

The exchange of CO_2 directly across the interface between the oceans and the atmosphere amounts to a carbon flux of about 90 gigatons per year. This flux depends on many factors, including air and sea temperatures and the composition of the seawater, but it is particularly sensitive to wind velocity, which increases the transfer of CO_2

FIGURE 14.22 The carbon cycle describes the fluxes of carbon between the atmosphere and its other principal reservoirs. Fluxes are given in units of gigatons per year. [IPCC, *Climate Change 2001: The Scientific Basis.* Cambridge University Press, 2001.]

and other gases by stirring up the surface water. Carbon dioxide dissolved in seawater escapes from solution and enters the atmosphere by evaporating from sea spray, while atmospheric CO_2 enters the ocean by dissolving in sea spray and rain or directly across the sea surface.

The weathering of carbonate rocks removes about 0.2 gigaton of carbon per year from the lithosphere and an equal amount from the atmosphere. This inorganic carbon is dissolved in surface waters, primarily as bicarbonate ions, and is transported by rivers to the oceans. Here shell-forming marine organisms reverse the weathering reaction, precipitating 0.2 gigaton of calcium carbonate and releasing an equal amount of carbon into the atmosphere as CO_2.

Other geologic processes are also involved in the carbon cycle. Volcanism releases minor amounts of CO_2 into the atmosphere, and the weathering of silicate rocks removes CO_2 from the atmosphere (through the reaction of silicates with carbon dioxide to produce calcium carbonate and silica). The net flux of carbon by these processes is relatively small (less than 0.1 gigaton/year), so they are usually neglected in considerations of short-term effects on climate. Over the long term, however, their effects can be substantial. According to one controversial hypothesis, mentioned in Chapter 10, the uplifting of the Himalayas and the Tibetan Plateau, which began about 40 million years ago, increased weathering rates enough to reduce the concentration of CO_2 in the atmosphere, and this reduction may have contributed to the subsequent climate cooling that led to the Pleistocene glaciations.

Human Effects on the Carbon Cycle

With this background, let's return to the fate of anthropogenic carbon emissions. We have good estimates of the extent to which human activities have perturbed the carbon cycle during the late twentieth century. Over the decade 1980–1989, which has been exceptionally well studied, the average mass of carbon emitted into the atmosphere by human activities was 7.1 gigatons per year: 5.3 gigatons per year from fossil-fuel burning; 1.7 gigatons per year from anthropogenic land-use changes, primarily deforestation and agriculture; and a tiny contribution (0.1 gigaton/year) from cement production, which also releases CO_2.

Figure 14.23 shows what happened to this carbon. Only 46 percent of the total (3.3 gigatons/year) remained in the atmosphere as CO_2. The rest was absorbed in nearly equal amounts by the oceans and the land surface. On land, the uptake was primarily by plant growth at temperate latitudes in the Northern Hemisphere, caused in part by the "fertilization effect" of increasing CO_2 (plants love the stuff, and when provided with more of it, they grow more) and in part by the rapid growth of new forests on land previously used for agriculture.

In the 1990s, anthropogenic carbon emissions rose to 6.3 gigatons per year from fossil-fuel burning and

Geologic processes regulate the carbon cycle over the long term.

1 Human activities release a total of 7.1 Gt of carbon into the atmosphere each year.

2 New plant growth and air-sea gas exchange remove 3.8 Gt/yr,…

3 …yielding a net atmospheric increase of 3.3 Gt/yr.

ATMOSPHERE

Fossil-fuel burning (5.3 Gt/yr)

Land-use change: deforestation, agriculture (1.7 Gt/yr)

Land uptake by new plant growth (1.9 Gt/yr)

Ocean uptake by air-sea gas exchange (1.9 Gt/yr)

Terrestrial biosphere — LAND

Cement production (0.1 Gt/yr)

OCEAN

Fossil organic carbon | Rock carbonates

GEOLOGIC RESERVOIRS

FIGURE 14.23 Humans add CO_2 to the atmosphere by burning fossil fuels; by releasing carbon through deforestation, agriculture, and other land-use changes; and by producing cement. Some of this carbon is absorbed by the oceans and by increases in plant growth on land. The remainder stays in the atmosphere, increasing the concentration of CO_2. The fluxes shown in this figure, given in units of gigatons per year, are for the 1980s. Since then, the amount of carbon produced by fossil-fuel burning has increased to more than 8 gigatons per year. [IPCC, *Climate Change 2001: The Scientific Basis.* Cambridge University Press, 2001.]

The oceans absorb much of the excess carbon dioxide emitted by human activities, but continued increases in those emissions may decrease their capacity to do so.

2.2 gigatons per year from land-use changes, an increase of 20 percent from the 1980s. The amount added to the atmosphere each year as CO_2, however, remained equal to the 1980s value of 3.3 gigatons. The land and oceans are doing more than their fair share of absorbing our extra carbon dioxide!

How will these numbers change as atmospheric CO_2 concentrations continue to rise? Scientists believe that the percentage of human carbon emissions taken up by the oceans will decrease as more CO_2 is dissolved in seawater, causing it to become more acidic. This "ocean acidification" will also increase the solubility of calcium in seawater, making it more difficult for key marine organisms—such as corals and some plankton—to form their external calcium carbonate skeletons (**Figure 14.24**). If the present trends continue, ocean

FIGURE 14.24 Shell-forming marine organisms, such as these corals in the Great Barrier Reef of Australia, are threatened by ocean acidification. [© 2004 Richard Ling, www.rling.com.]

acidification could begin to kill off certain types of marine organisms within the next few decades. In addition, the enhanced greenhouse effect resulting from rising atmospheric CO_2 concentrations will warm the oceans, which will further decrease their capacity to absorb carbon from the atmosphere—an example of positive feedback in the climate system (see Chapter 10). The overall magnitude of these effects is not yet certain, but it has become clear that the changes in the carbon cycle caused by human activities are likely to amplify greenhouse warming.

Now that we have described the carbon cycle, we are better prepared to examine the problem of global climate change in more detail. We introduced the problem in Chapter 1 and examined the evidence for recent anthropogenic warming in Chapter 10. Here we will focus on two key questions: how will the modifications of the carbon cycle brought about by our carbon economy change the climate system, and what might be the effects of those changes on our environment?

GLOBAL CLIMATE CHANGE

We have seen in earlier chapters that the history of the Earth system is punctuated by extreme events—short periods of rapid global change. Geologists have ranked these changes according to their effects on the fossil record, and they have used major transitions in life-forms to delineate the eras, periods, and epochs of the geologic time scale (see Figure 8.13).

The time scale's most recent division was set at about 11,000 years ago, when the Pleistocene epoch, a period characterized by a cold climate with sporadic ice ages, gave way to a warm, stable period—the Holocene epoch (see Chapter 10). Average global temperatures jumped by 7°C over an interval that may have been as short as a few decades, sea level rose, ecosystems migrated poleward with the receding ice, and many large mammals went extinct. At about the same time, our own species, *Homo sapiens*, began to alter the environment. In particular, humans developed new social structures and technologies that allowed them to grow crops, raise livestock, and build cities—the novel landscaping processes of civilization.

Holocene Climate Stability

Some scientists think that if human civilization had not come along, Earth's climate might by now be plunging into another ice age, driven by decreasing amounts of solar energy (due to the Milankovitch cycles described in Chapter 10) and accompanied by decreasing atmospheric concentrations of greenhouse gases. According to paleoclimatologist William Ruddiman, the expansion of civilization began to release significant amounts of carbon dioxide and methane from the biosphere to the atmosphere as early as 8000 years ago, primarily through deforestation and the rise of agriculture. Ruddiman hypothesizes that this extra source of greenhouse gases has been responsible for extending the warm interglacial period beyond its natural limit.

Whatever the reasons, the Holocene warm period has been unusually long and stable when compared with the previous interglacial periods of the Pleistocene epoch (see Chapter 10). Measurements from ice cores indicate that, from the end of the Pleistocene until about 200 years ago,

FIGURE 14.25 Atmospheric concentrations of carbon dioxide, methane, and nitrous oxide over the last 10,000 years (large panels) and since 1750 (inset panels). These measurements, compiled by the Intergovernmental Panel on Climate Change, were derived from ice cores (symbols with different colors represent different studies) and atmospheric samples (red lines). [IPCC, *Climate Change 2007: The Physical Science Basis.* Cambridge University Press, 2007.]

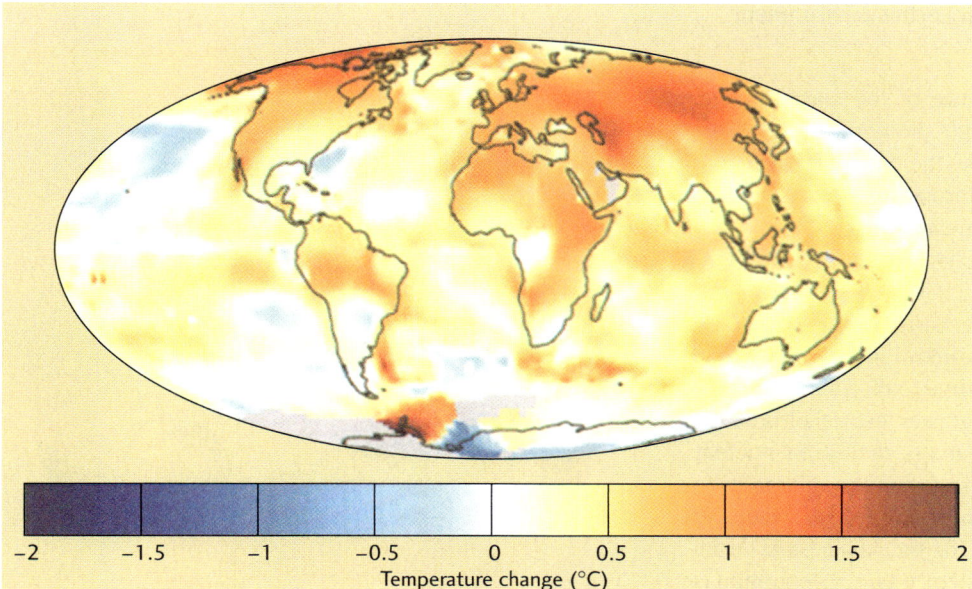

FIGURE 14.26 Map of the differences between average surface temperatures measured during the period 1995–2004 and those measured at the same location during the baseline period 1940–1980. [NASA Goddard Institute for Space Studies.]

The Holocene epoch has been an unusually long and stable interglacial period. Anthropogenic emissions of greenhouse gases may have prevented the onset of another ice age.

the average global temperature fluctuated by only a few degrees, and the atmospheric concentrations of the major greenhouse gases stayed within a rather narrow range. The average CO_2 concentration, for example, fluctuated only between 260 and 280 ppm—less than a 10 percent variation over this entire period.

Industrialization and Global Warming

Atmospheric concentrations of greenhouse gases began to increase rapidly with the industrialization of society and the explosion of the human population about 200 years ago. **Figure 14.25** shows the atmospheric concentrations of three greenhouse gases—carbon dioxide, methane, and nitrous oxide (N_2O, "laughing gas")—over the past 10,000 years. In all three cases, there is a remarkable correspondence with the history of the human population as you can see by comparing this figure with Figure 14.1: the concentrations remained relatively constant throughout the Holocene but shot upward after the industrial revolution.

The global atmospheric concentration of methane, for example, has increased by almost 150 percent from its preindustrial value, and the concentration of carbon dioxide has risen 36 percent. In both cases, the observed increases can be explained by human activities, predominantly agriculture and fossil-fuel use. Methane's greenhouse effect is weaker than that of carbon dioxide, however, so even though its relative concentration has gone up more, its contribution to greenhouse warming is only about 30 percent as large. The postindustrial increase in nitrous oxide, primarily from agriculture, has been 18 percent; its contribution to the greenhouse effect is a small fraction of carbon dioxide's.

The average global temperature increased by about 0.6°C during the twentieth century (see Figure 10.13), with most of the warming occurring in the latter half of the century. **Figure 14.26** is a map of the average temperature differences between two time periods, 1995–2004 and 1940–1980. The average global increase was 0.42°C, but the changes have not been geographically uniform: they have been greater than this average over the land surface and less over the oceans. In particular, you can see that the rise observed in large regions of the northern continents exceeds 1°C.

Predictions of Future Global Warming

The warming trend observed in the twentieth century is continuing, but how much hotter will the planet get, and how will this global warming affect local climates? Answering such questions requires a sequence of uncertain predictions.

The first step is to forecast rates of anthropogenic greenhouse gas emissions. To do this, we need to make some predictions about the future of the carbon economy, as well as the steps human society might take to curb its emissions. However, uncertainties about the extent of our fossil-fuel and alternative energy resources, as well as the unpredictability of economic and political conditions, make these forecasts difficult.

The second step is to calculate how much of the carbon pumped into the atmosphere will remain there as carbon dioxide and how much will be absorbed by the oceans and the land surface. To do this, we need to devise a model

Atmospheric concentrations of greenhouse gases began to increase rapidly with the industrialization of society and the explosion of the human population.

Predictions of future climate warming depend on several factors: future rates of anthropogenic greenhouse gas emissions, the responses of the carbon cycle to rising carbon emissions, and the effects of changing atmospheric concentrations of greenhouse gases on temperatures at Earth's surface.

that predicts the future of the carbon cycle. That is another tough problem, because we don't quite know how the carbon cycle might respond to rising anthropogenic emissions.

The Intergovernmental Panel on Climate Change (IPCC; see Chapter 1) has modeled increases in atmospheric CO_2 concentrations under three different scenarios (**Figure 14.27**a): continued reliance on fossil fuels as our major energy source, sometimes called the "business-as-usual" scenario; greater use of cleaner alternative energy sources, including nuclear energy and renewable resources; and an even more rapid conversion from fossil fuels to cleaner alternatives. In the business-as-usual scenario, the atmospheric CO_2 concentration is predicted to exceed 900 ppm by 2100—more than three times the preindustrial value—whereas the rapid-conversion scenario yields "only" twice the preindustrial value.

The last step is to calculate how changes in greenhouse gas concentrations and other factors, such as land-

FIGURE 14.28 Numerical climate models are used to predict future global climate change. This climate model, developed with support from the U.S. Department of Energy, portrays interactions among the atmosphere, hydrosphere, cryosphere, and land surface. [Warren Washington and Gary Strand/National Center for Atmospheric Research.]

use changes, will change temperatures at Earth's surface. To do this, we need a **climate model,** a sophisticated computer model that is capable of predicting the circulation patterns of air and water as they are driven by solar energy, including small disturbances such as storms in the atmosphere and eddies in the oceans. Such models represent the properties of the atmosphere and oceans on three-dimensional grids comprising millions of geographic points (**Figure 14.28**).

The predictions of temperature increases by the late twenty-first century made with one such climate model, developed by the United Kingdom's Hadley Centre for Climate Change, are mapped in **Figure 14.29**. These predictions were based on the business-as-usual scenario. The geographic pattern of temperature changes displays some similarities with the observed pattern of twentieth-century warming in Figure 14.26. In particular, the temperate and polar regions of the Northern Hemisphere show the most warming, and the average temperature change for the continents is higher than for the oceans.

The IPCC has made predictions of global warming under its three scenarios by averaging the results of many climate models (Figure 14.27b). Taking into account the political, economic, and scientific uncertainties, the range of likely global temperature increases during the twenty-first century is 1° to 6°C. However, the lower values

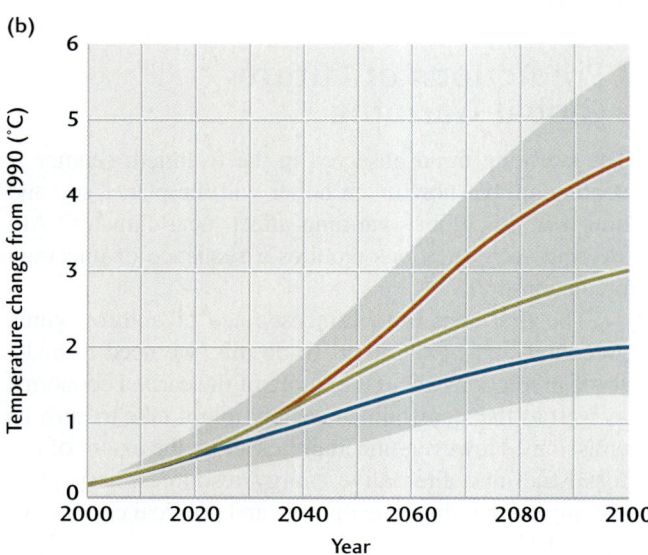

FIGURE 14.27 Projections of (a) atmospheric carbon dioxide concentrations and (b) average surface temperatures for the next 100 years based on three economic scenarios. [IPCC, *Climate Change 2001: The Scientific Basis.* Cambridge University Press, 2001.]

According to current climate models, global temperatures are likely to increase by 1° to 6°C during the twenty-first century.

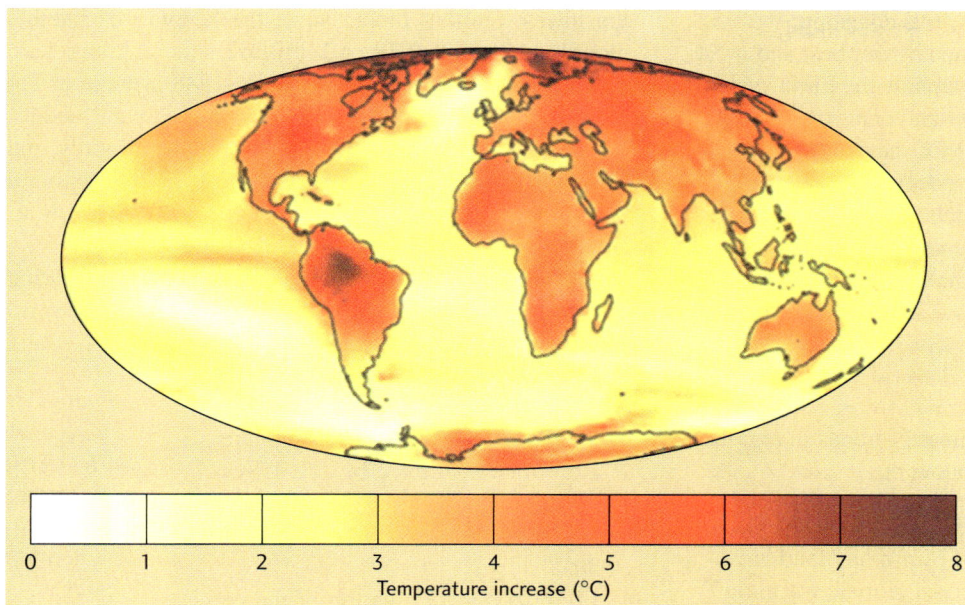

FIGURE 14.29 Map showing average surface temperatures predicted for 2070–2100, expressed as differences from average surface temperatures measured at the same location during the baseline period 1960–1990. This prediction was made using a numerical climate model and "business-as-usual" projections of carbon dioxide and other greenhouse gas emissions (see Figure 14.27a). [Data from Hadley Centre for Climate Change; map courtesy of Robert A. Rohde.]

in this range can be achieved only through reductions in fossil-fuel burning and the introduction of clean and resource-efficient energy technologies.

EFFECTS OF CLIMATE CHANGE

It seems clear that human emissions of greenhouse gases will cause further global warming and may result in major changes in the climate system. It is also clear that these changes have the potential to affect civilization in both positive and negative ways. Some regional climates may improve—now may be the time to buy land in Canada—while others may deteriorate—the authors' home city of Los Angeles, for example, is likely to become hotter and dryer. Some potential effects of climate change are listed in Table 14.1.

Predicting the effects of climate change is even more difficult than predicting the climate changes themselves. Moreover, the question of how to respond to these possible threats has political and economic dimensions that extend far beyond Earth science, so it does not have strictly

TABLE 14.1 Potential Effects of Climate Change on Various Systems

Systems	Potential Effects
Forests and other terrestrial ecosystems	Migration of vegetation; reduction in ecosystem ranges; altered ecosystem composition
Species diversity	Loss of diversity; migration of species; invasion of new species
Coastal wetlands	Inundation of wetlands; migration of wetland vegetation
Aquatic ecosystems	Loss of habitat; migration to new habitats; invasion of new species
Coastal resources	Inundation of coastal structures; increased risk of flooding
Water resources	Changes in water supplies; changes in patterns of drought and flooding; changes in water quality
Agriculture	Changes in crop yields; shifts in relative productivity among regions
Human health	Shifts in ranges of infectious disease organisms; changes in patterns of heat-stress and cold-weather afflictions
Energy	Increase in cooling demand; decrease in heating demand; changes in hydroelectric energy resources

Source: Office of Technology Assessment, U.S. Congress.

scientific answers. Nevertheless, the decisions we will make as a society must be informed by our best and most realistic scientific predictions about how the Earth system will change over the next decades and centuries.

The Earth scientists of the IPCC have developed an extensive catalogue of the observed and predicted effects of global warming, a few of which we described briefly in Chapter 1. Here we will describe some of the large-scale effects of particular concern.

Changes in Regional Weather Patterns

The IPCC has documented a number of trends in regional weather patterns, including the following:

- During the past century, the frequency of heavy precipitation events has increased over many land areas in a manner consistent with observed climate warming and increases in atmospheric water vapor concentrations. Significantly increased precipitation has been observed in eastern parts of North and South America, northern Europe, and northern and central Asia.

- Drying has been observed in the Sahel, the Mediterranean, southern Africa, and parts of southern Asia. More intense and longer droughts have been observed over wider areas since the 1970s, particularly in the tropics and subtropics.

- Widespread changes in temperature extremes have been observed over the last 50 years. Cold days, cold nights, and frost have become less frequent, while hot days, hot nights, and heat waves have become more frequent.

- Intense hurricane activity has increased in the North Atlantic since about 1970. This increase is correlated with increases in tropical sea surface temperatures, although there is no clear trend in the annual number of hurricanes.

Global warming is expected to produce heavy precipitation in some regions and drought in others, more frequent heat waves, and more intense hurricane activity.

As we noted in our comparison of Figures 14.26 and 14.29, climate warming in the twenty-first century is likely to have geographic patterns similar to those observed over the past several decades. Therefore, these trends in regional weather patterns will probably continue. Business for the Weather Channel should be good!

Changes in the Cryosphere

Nowhere are the effects of global warming more evident than in the cryosphere. The amount of sea ice in the Arctic Ocean is decreasing, and the downward trend seems to be accelerating. The sea ice cover in September 2007 was the lowest for that month since the keeping of satellite records began in 1978: 4.1 million square kilometers, down by 45 percent from the 1978–1988 average of 7.4 million square kilometers. According to climate models, the north polar ice cap will continue to shrink rapidly, and much of the Arctic Ocean will become ice-free within a few decades (**Figure 14.30**). The shrinkage of sea ice is already severely disrupting Arctic ecosystems.

Mountain glaciers at lower latitudes retreated during the twentieth-century warming (**Figure 14.31**). Careful fieldwork by Lonnie Thompson (see Chapter 10) and other geologists has demonstrated that rates of glacial retreat and snow cover loss are increasing in both hemispheres. According to a study by the USGS, Glacier National Park in northern Montana will lose the last of its glaciers by 2030!

Temperatures at the top of the permafrost layer in the Arctic have gone up by 3°C since the 1980s, and the melting of permafrost is destabilizing structures such as the Trans-Alaska oil

Climate warming is most evident in the cryosphere: reduction of Arctic sea ice and permafrost, retreat of mountain glaciers, and breakoffs from Antarctic ice shelves.

FIGURE 14.30 Global warming is melting the north polar ice cap. This map of the Arctic compares the average extent of the polar ice cap at the end of the summer during the period 1972–1990 with its projected extent in 2030. The change is expected to disrupt Arctic ecosystems. One benefit for humans, however, could be the opening of the Northwest Passage and other shorter sea routes between the Atlantic and Pacific oceans within the twenty-first century. [U.S. Navy.]

FIGURE 14.31 Photographs of Qori Kalis Glacier in Peru from the same vantage point (*left*) in July 1978 and (*right*) in July 2004. Between 1998 and 2001, the ice front retreated an average of 155 meters per year, an alarming 32 times faster than the average annual retreat from 1963 to 1978. [Lonnie G. Thompson, Byrd Polar Research Center, Ohio State University.]

pipeline (Figure 14.32). The maximum area covered by seasonally frozen ground has decreased by about 7 percent in the Northern Hemisphere since 1900, with a decrease in spring of up to 15 percent.

Using high-resolution radar satellite mapping, geologists have observed that several Antarctic glaciers have retreated more than 30 km in just 3 years. Over the past 20 years or so, enormous pieces of ice have snapped off Antarctic glaciers. In March 2000, an iceberg slightly smaller than 10,000 km^2 (larger than the state of Delaware) calved from the Ross Ice Shelf (see Figure 10.6b). More recently, in February and March of 2002, a portion of the Larsen Ice Shelf larger than Rhode Island (about 3250 km^2) shattered and separated from the eastern side

FIGURE 14.32 Permafrost melting could destabilize structures at high latitudes, such as the Trans-Alaska oil pipeline. Its 1300-km (800-mile) route from Prudhoe Bay to Valdez includes 675 km of permafrost. Where this pipeline crosses permafrost, it is perched on specially designed vertical supports. Because thawing of the permafrost would cause the supports to become unstable, they are outfitted with heat pumps designed to keep the ground around them frozen. [Galen Rowell/CORBIS.]

FIGURE 14.33 Collapse of the Larsen Ice Shelf. This satellite image was taken on March 7, 2002, toward the end of a 2-month period during which a huge piece of the ice shelf separated from land and splintered into thousands of icebergs. The darkest colors on the right-hand side of the image represent open seawater. The white parts are icebergs, the remaining parts of the ice shelf, and glaciers on land. The bright blue area is a mixture of seawater and highly fractured ice. The area of the image is about 150 km by 185 km. [NASA/GSFC/LaRC/JPL, MISR Team.]

of the Antarctic Peninsula (Figure 14.33). The fracturing of this piece of the ice sheet produced thousands of icebergs.

Sea Level Rise

If the ice shelves around Antarctica continue to collapse, will sea level rise? It turns out that even if all of Earth's ice shelves were to break off into the ocean over the next few years, sea level wouldn't change much at all. Like icebergs, ice shelves float because ice is less dense than seawater, so they displace a mass of seawater equal to their own weight (Figure 14.34). When they melt, the conversion of ice to water does not change the total mass of the oceans, and there is no change in sea level. (Do an experiment: the level of the drink in your glass doesn't change when the ice cubes melt, for the same reason.)

Melting ice can cause sea level to rise only if it is on land, not floating in water. The melting of glaciers sheds water and new icebergs into the sea, which is why we should be concerned by the observed rapid increase in glacial melting. The destruction of ice shelves contributes to sea level rise only if it destabilizes the continental glaciers that feed the ice shelves, causing the glaciers to flow more rapidly into the oceans. Such an acceleration of glacial flow was observed after the Larsen Ice Shelf collapse.

Sea level also rises as the temperature of ocean water increases, increasing its overall volume by a tiny fraction.

> *Although the melting of floating ice will not raise sea level, the melting of glaciers on land and the warming of ocean water have the potential to do so.*

FIGURE 14.34 Ice shelves and icebergs float and thus displace a mass of water equal to their own mass; therefore, their melting by global warming will not change sea level. The melting of ice sheets on land, however, injects new water into the oceans, causing sea level to rise.

FIGURE 14.35 The low-lying country of Bangladesh would be subject to disastrous flooding if sea level rises due to global warming. [James P. Blair/National Geographic.]

The IPCC estimates that sea level rose about 170 mm because of the twentieth-century warming.

Sea level is currently rising at about 3 mm per year, and this rate is expected to accelerate. Climate models indicate that sea level could rise by as much as a meter during the twenty-first century, creating serious problems for low-lying countries such as Bangladesh (**Figure 14.35**), as well as the Eastern Seaboard and Gulf Coast of the United States, where flooding during coastal storm surges could become much worse.

> *Climate models indicate that sea level could rise by as much as a meter during the twenty-first century.*

Species and Ecosystem Migration

The IPCC has documented effects of global warming on agriculture and forestry at higher latitudes in the Northern Hemisphere, such as earlier spring planting of crops and forest disturbances due to fires and pests. Droughts in the Sahel region of Africa, described in Chapter 1, have led to a reduced growing season there (**Figure 14.36**).

Many plant and animal species will have difficulty adjusting to rapid anthropogenic warming or migrating to more suitable climates. Those that cannot cope with the rapid warming could become extinct. Global warming is already being blamed for a variety of adverse ecological effects, such as the disruption of Arctic ecosystems as permafrost begins to melt, and the spread of tropical diseases as more of the world experiences a tropical climate.

The Potential for Catastrophic Changes to the Climate System

The current atmospheric concentrations of carbon dioxide and methane far exceed anything seen in the natural range determined from ice core data for the last 650,000 years. Our climate system is therefore entering unknown territory.

FIGURE 14.36 Members of the Mali Gao tribe digging for edible roots during the Sahel drought of 1984 and 1985. Global warming is expected to increase droughts in this and other subtropical regions. [Frans Lemmens/Alamy.]

Positive feedbacks within the climate system could greatly enhance anthropogenic climate change, with potentially catastrophic effects.

Some observers think that the credibility of climate change projections suffers from "the Chicken Little problem": too many people are running around yelling "The sky is falling!" Yet most scientists believe that those projections are too conservative because they may not properly take into account some of the positive feedbacks within the climate system that could greatly enhance climate change. Ocean acidification, which we discussed earlier in this chapter, is a good example. Here are a few more:

Destabilization of continental glaciers. The surface melting of the Greenland glacier in 2005 was the largest on record, and there are indications that glacial streams within the ice sheet are accelerating much faster than expected. If the Greenland and Antarctic glaciers begin to shed ice faster than snowfall can generate new glacial ice, sea level could begin to rise much faster than the IPCC predictions.

Shutoff of thermohaline circulation. Changes in precipitation and evaporation patterns are decreasing the salinity of seawater at mid- and high latitudes. Some scientists have speculated that this change could substantially reduce global thermohaline circulation (see Figure 10.3), which is driven by differences in temperature and salinity. Major changes in the Gulf Stream and other aspects of the climate system could result.

Methane release from seafloor sediments and permafrost. Recall from Chapter 9 that a massive release of methane from shallow seafloor sediments about 55 million years ago may have caused abrupt global warming and led to the mass extinction at the Paleocene-Eocene boundary. Today, there is far more methane stored in shallow seafloor sediments and in permafrost than was released at the end of the Paleocene. If global warming begins to thaw these methane deposits, another runaway cycle of extreme warming could begin.

Welcome to the Anthropocene: A New Geologic Epoch

In 2003, atmospheric chemist and Nobel laureate Paul Crutzen proposed the recognition of a new geologic epoch—the **Anthropocene,** or Age of Man—beginning about 1780, when James Watt's coal-powered steam engine launched the industrial revolution. The global changes that mark the Holocene-Anthropocene boundary are just now getting under way, so a future scientist with a full record of the next few thousand years may place the geologic boundary at a somewhat later date. Crutzen's main point, however, is that global changes are proceeding so rapidly that such quibbles are likely to be minor. As with many previous geologic boundaries, the main marker will be a mass extinction.

FIGURE 14.37 The Caribbean island country of Haiti is now 98 percent deforested.
[Daniel Morel/AP World Wide.]

Between 1850 and 1880, as much as 15 percent of the land surface was deforested, and rates of deforestation have continued to rise. According to the United Nations, over 150,000 km² of tropical rain forests—about 1 percent of the total resource—are being converted each year to other land uses, mostly agricultural ones. In 1950, forests covered approximately 25 percent of Haiti (a Caribbean island country the size of Maryland); its forested area now stands at less than 2 percent (Figure 14.37). Other developing nations face a similar problem.

Not surprisingly, the number of extant species, the most important measure of *biodiversity,* is declining. Biologists estimate that there are over 10 million different species alive on the planet today, although only 1.5 million have been officially classified. Extinction rates are difficult to quantify, but most knowledgeable scientists believe that up to one-fifth of all species will disappear during the next 30 years, and as many as one-half may go extinct during the twenty-first century. One respected biologist, Peter Raven, has put the problem bluntly:

We are confronting an episode of species extinction greater than anything the world has experienced for the past 65 million years. Of all the global problems that confront us, this is the one that is moving the most rapidly and the one that will have the most serious consequences. And, unlike other global ecological problems, it is completely irreversible.

Some observers, such as sociobiologist E. O. Wilson, have gone so far as to call the current rapid worldwide decline in biodiversity the "Sixth Extinction," placing it in the same rank as the "Big Five" mass extinctions of the Phanerozoic eon (see Figure 2.18 and Chapter 9). Others consider this extrapolation to be premature, however, because even the rapid losses in biodiversity we see today will not necessarily affect the fossil record as profoundly as, say, the impact-related mass extinction at the end of the Cretaceous period, or even the less severe mass extinction associated with global warming at the Paleocene-Eocene boundary.

EARTH SYSTEM ENGINEERING AND MANAGEMENT

By any measure, the problems we face in confronting anthropogenic carbon emissions and finding new energy sources are daunting. The world's population will expand to more than 9 billion in the next 50 years, and per capita use of energy will grow as societies continue to develop their economies. A continuing reliance by these economies on fossil fuels for energy production—the business-as-usual scenario of Figure 14.27—would cause the rate of carbon emissions to almost double during this period, from the current value of 8 gigatons per year to at least 15 gigatons per year. The CO_2 concentration in the atmosphere would reach 650 ppm and continue to increase thereafter, with potentially disastrous consequences for global ecosystems and civilization. Rewriting the operating manual for Spaceship Earth to control carbon emissions—perhaps civilization's most important task—will require an extraordinary collaboration of Earth scientists, policy makers, and the public.

Energy Policy

There is little question that we will need to make changes in the energy sources we use and the ways in which we use them. But it will be difficult to decide how much money will be needed to curb anthropogenic carbon emissions and whether the benefits of doing so will justify the costs. Too much spending could depress the economy and cause job losses; yet preventing the most drastic effects of climate change might be less costly than coping with these disasters after they happen.

A partial solution—and certainly the most economical one—is to improve energy use efficiency. As we have seen, the U.S. energy system wastes about 60 quads of energy for every 100 quads it produces (see Figure 14.3). In a real sense, using energy more efficiently is like discovering a new source of fuel. Some experts believe that the United States could reduce its emissions of greenhouse gases by as much as 50 percent from 1990 levels by implementing efficiency measures that cost relatively little—for example, insulating buildings, replacing incandescent light bulbs with fluorescent bulbs, increasing the fuel efficiency of motor vehicles, and making greater use of natural gas. The savings could amount to hundreds of billions of dollars a year, much of which is now being spent on expensive imported oil. These modest steps would offer substantial fringe benefits as well, including lowered manufacturing costs and improved air quality.

The most economical way to reduce carbon emissions is to improve the efficiency of our energy system.

Many observers would say that fossil fuels are simply too cheap in the United States. Carbon emissions are not taxed as they are in many other developed nations; consequently, there is little incentive for energy conservation or conversion to new energy sources. If the full economic costs of fossil fuels—the costs of cleaning up acid rain, oil spills, and other environmental damage; the costs of trade deficits; the costs of global warming; the military costs of defending oil supplies—were included in their prices, alternative energy sources might become much more competitive with fossil fuels. Such *full-cost accounting* has not been politically popular in the United States, however.

FIGURE 14.38 A coal-fired power plant near Zibo, a city in Shandong Province, China. In 2007, China replaced the United States as the nation with the highest rate of greenhouse gas emissions. The carbon economies of China, India, and other developing countries will have a huge influence on future climates. [© David Lyons/Alamy.]

We also face the issue of fairness in international politics. The United States, Canada, the European Union, and Japan—with only 25 percent of the world's population—are responsible for about 75 percent of the global increase in CO_2 and other greenhouse gases. These rich industrial nations will be able to pay the costs of reducing their greenhouse gas emissions more easily than the developing countries. China, for example, depends on its huge coal deposits for its economic growth, and its turbocharged economy has just recently (in 2007) become the world's leader in greenhouse gas emissions (**Figure 14.38**). Developing nations argue that they will need financial and technological support from the developed countries to help them cope with the demand to reduce emissions.

People have come to agree that our problems cannot be solved on a national scale and will have to be addressed through international cooperation. New types of treaties, such as the Kyoto Protocol, which was forged in Kyoto, Japan, in December of 1997, have initiated a global program to manage anthropogenic carbon emissions. The Kyoto Protocol entered into force on February 16, 2005, after ratification by Russia in November of 2004. However, some nations—most notably the United States—have refused to participate, emphasizing the potential costs of moving too quickly.

Alternative Energy Resources

As we have seen, no one alternative energy source will be able to quickly replace fossil fuels. However, some renewable energy resources, such as solar power, wind power, and biofuels, are becoming more important contributors to our energy system. If these technologies were aggressively implemented during the next 50 years, together they could reduce carbon emissions by gigatons per year.

Another step that could be taken is to increase the use of nuclear energy. The capacity of nuclear power plants, which today is approximately 350 gigawatts, could easily be tripled in the next 50 years, but this option is unattractive to many people, for reasons we have described. The potential exists for cleaner nuclear technologies, such as *fusion power*—the use of small, controlled thermonuclear explosions to generate energy. But scientific progress toward this goal has been slow, and conceptual breakthroughs will be required.

Increased use of alternative energy resources during the next 50 years could reduce carbon emissions by gigatons per year.

Engineering the Carbon Cycle

What about the possibility of engineering the carbon cycle to reduce the accumulation of greenhouse gases in the atmosphere? Several promising technologies aim to reduce greenhouse gas emissions by pumping the carbon dioxide generated by fossil-fuel combustion into reservoirs other than Earth's atmosphere—a procedure known as **carbon sequestration.**

One obvious reservoir for carbon is the biosphere. Earlier in this chapter, we saw that deforestation releases CO_2 into the atmosphere, whereas plants withdraw it, in surprisingly large amounts. Land-use policies that encourage reforestation and other biomass production and discourage deforestation might help to mitigate anthropogenic climate change.

Biotechnology might provide some ways of increasing the capacity of the biosphere to sequester carbon. One possibility is genetically modifying bacteria to make them more efficient at fixing carbon in soils. Another is to engineer bacteria that are capable of metabolizing methane, sequestering the carbon, and giving off hydrogen.

Another controversial possibility is fertilization of the marine biosphere. We know that small photosynthetic marine organisms (phytoplankton) take up CO_2 from the atmosphere by photosynthesis. In most regions of the ocean, phytoplankton productivity is limited by the lack of nutrients, such as iron. Preliminary experiments in the 1990s suggested that the growth of phytoplankton can be stimulated by dumping modest amounts of iron into the ocean. Unfortunately, it appears that fertilizing the ocean in this manner also stimulates the growth of animals that eat the phytoplankton and quickly return the CO_2 to the atmosphere.

Our ability to sequester carbon in the lithosphere and biosphere will depend on better understanding of the Earth system and its carbon cycle.

This last example suggests that using biotechnology to remove CO_2 from the atmosphere on a large scale will require a much better understanding of terrestrial and marine ecosystems to avoid unintended consequences. However, this type of Earth system engineering is not as far-fetched as it once might have seemed.

In fact, one straightforward technology for carbon sequestration—underground storage of CO_2—offers considerable promise. Carbon dioxide captured from oil and gas wells is already being pumped back into the ground as a means of moving oil toward the wells. If capture and storage of the carbon dioxide from coal-fired power plants were economically feasible, the world's abundant coal resources would become much more attractive as a replacement for petroleum.

Stabilizing Carbon Emissions

Now that we have discussed the problem of carbon emissions and some of the technologies that could contribute to its solution, let's take the bull by the horns. Under the business-as-usual scenario, as we have seen, carbon emissions are expected to increase by at least 7 gigatons per year during the next half-century. How can this increase be stopped? In other words, what would it take to *stabilize carbon emissions at current levels?*

Two scientists from Princeton University, Stephen Pacala and Robert Socolow, have provided a simple quantitative framework to answer this question. They begin by admitting that there is no single solution to the problem—no "silver bullet." Instead, they break the problem into what they call **stabilization wedges,** each of which offsets the projected growth of carbon emissions by 1 gigaton per year in the next 50 years (**Figure 14.39**). Therefore, one wedge roughly corresponds to one-seventh of the problem. Table 14.2 lists seven examples of one-wedge solutions using existing technologies, along with some comparisons that measure the magnitude of the task.

Under the business-as-usual scenario, anthropogenic carbon emissions are expected to increase by at least 7 gigatons per year in the next 50 years.

Implementing each stabilization wedge will be a monumental task. To achieve wedge 1, for example, the average gasoline mileage of the world's entire fleet of passenger vehicles, which will grow to 2 billion by mid-century, will have to be steadily increased from 30 miles per gallon (mpg) to 60 mpg. This calculation assumes that a car is driven 10,000 miles per year, the current annual average. An alternative, not shown in the table, would be to maintain gas mileage at 30 mpg but reduce the average amount of driving by half, to 5000 miles per year. Yet another alternative (wedge 2) would be to convert all cars to biofuels. Growing this much biofuel would take up one-sixth of the world's total cropland, so this strategy could adversely affect agricultural productivity and food supplies.

Some of the stabilization wedges involve controversial or expensive technologies, such as expanding nuclear power by a factor of three (wedge 3), increasing the number of large windmills into the millions (wedge 4), or covering large desert areas with solar panels (wedge 5). At least one of the proposed wedges, the capture and storage of carbon emitted from coal power plants (wedge 6), is at the margin of current technological feasibility. The last option, elimination of tropical deforestation and the reforestation of huge additional land areas (wedge 7), is favored by many people in principle, but it would be difficult to achieve without imposing severe restrictions on developing countries such as Brazil.

Implementing at least seven stabilization wedges will be necessary to kekep carbon emissions at their current level.

The stabilization of carbon emissions at current emission rates would reduce, but not eliminate, the threat of significant climate change. The stabilization scenario (given by the blue lines in Figure 14.27) would still allow

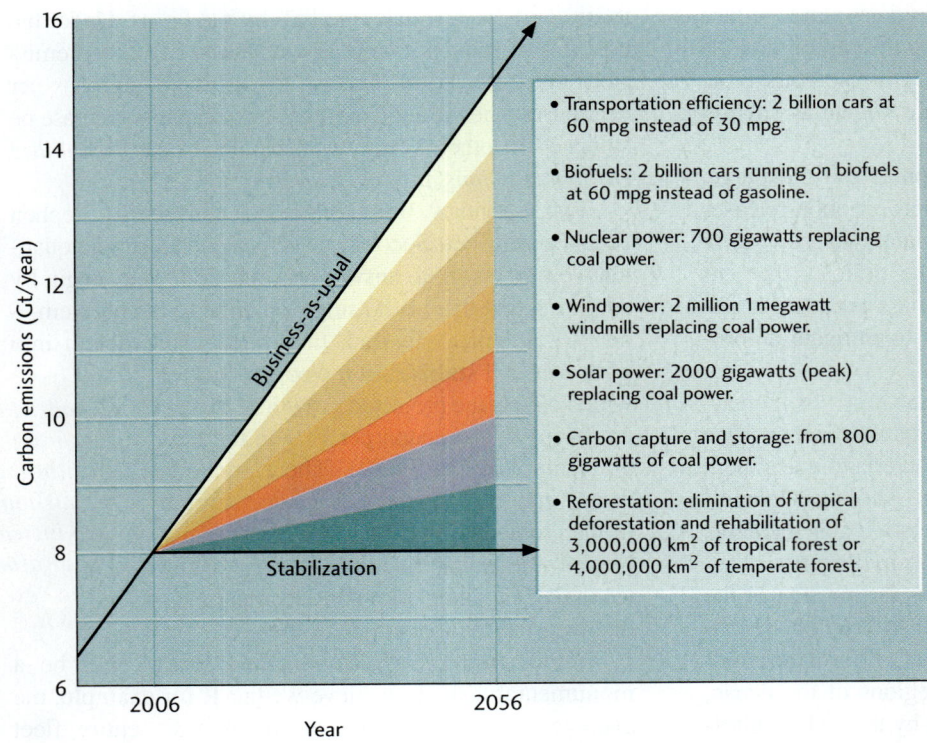

FIGURE 14.39 Under the business-as-usual scenario, carbon emissions are expected to increase by at least 7 gigatons per year during the next 50 years. The problem of stabilizing carbon emissions at their 2006 value can be broken into seven stabilization wedges, each representing a reduction in emissions of 1 gigaton per year by 2056. Possible actions that use existing technologies to achieve one-wedge reductions are listed here and in Table 14.2.

TABLE 14.2	Stabilization Wedges: Seven Solutions That Could Each Reduce Carbon Emissions by 1 Gigaton per Year in 50 Years
Wedge 1	*Transportation efficiency:* Build 2 billion passenger cars with a fuel efficiency of 60 miles per gallon of gasoline instead of 30 miles per gallon. There are currently about 600 million passenger cars worldwide.
Wedge 2	*Biofuels:* Convert 2 billion cars with a fuel efficiency of 60 miles per gallon from gasoline to biofuels. Biofuel production at this scale would require 2.5 million square kilometers (964,000 square miles) of high-yield crops, about one-sixth of world's total cropland.
Wedge 3	*Nuclear power:* Build nuclear power plants with a total electricity-generating capacity of 700 gigawatts to replace coal-fired power plants. This strategy would triple the world's current supply of nuclear energy.
Wedge 4	*Wind power:* Build 2 million windmills, each with an electricity-generating capacity of 1 megawatt, to replace coal-fired power plants. Global wind-power capacity now stands at about 80 gigawatts, or 4 percent of this amount.
Wedge 5	*Solar power:* Build solar electric power plants with a peak ("sunny day") generating capacity of 2000 gigawatts. With today's technology, these power plants would need to cover at least 20,000 square kilometers (7719 square miles) of arid land surface, about the area of California's Riverside County.
Wedge 6	*Carbon sequestration:* Capture and store underground the carbon dioxide produced by coal-fired power plants generating 800 gigawatts of electricity. In 2006, the total electricity-generating capacity of all U.S. power plants was about 1000 gigawatts.
Wedge 7	*Reforestation:* Eliminate tropical deforestation and rehabilitate 3 million square kilometers (1.16 million square miles) of tropical forests or 4 million square kilometers (1.54 million square miles) of temperate forests. The latter area is about half the size of Australia.

Source: S. Pacala and R. Socolow, *Science* 305: 968–972 (2004).

the atmospheric concentration of carbon dioxide to grow to 500 ppm, almost twice the preindustrial value. Further reductions in carbon emissions during the second half of the twenty-first century would be necessary to maintain the concentration below this value. Climate models indicate that such a scenario would still increase the average global temperature by about 2°C, more than three times the total twentieth-century warming (see Figure 14.27b).

Nevertheless, the continued rise of atmospheric CO_2 concentrations is not inevitable. The available inventory of stabilization wedges constitutes a technological framework for concerted action by governments. It will be difficult to develop the broad public consensus necessary to take on the stabilization problem in a serious way, not to mention the international agreements that will be needed to implement the requisite technologies. Yet, as the Pacala-Socolow analysis demonstrates, there is still time for actions that can substantially reduce anthropogenic global change. Decisions to use this opportunity will depend on our understanding of the problem, its potential solutions, and the consequences of inaction.

As French novelist Marcel Proust once wrote, "The real voyage of discovery consists not in seeking new lands, but in seeing with new eyes." We hope this textbook has given you new eyes to see the critical issues of global change and the other problems of Earth science that confront your generation.

SUMMARY >>

Why is the carbon economy an important component of the Earth system? Most of the energy used by human civilization comes from carbon-based resources. The rise of this carbon economy has altered the natural carbon cycle by creating a huge new flow of carbon from the lithosphere to the atmosphere. If that flow continues unabated, carbon dioxide concentrations in the atmosphere will double by the mid-twenty-first century.

What is the origin of oil and natural gas? Oil and natural gas form from organic matter deposited in oxygen-poor sedimentary basins, typically in coastal waters. Under elevated temperatures and pressures, the buried organic matter is transformed into hydrocarbons. Oil and gas accumulate where geologic structures called oil traps create impermeable barriers to their upward migration.

Why is there concern about the world's oil supply? Oil is a nonrenewable resource: at current rates of use, it will be depleted far faster than nature can replenish it. Therefore, as oil is withdrawn from the reservoirs of the world, its availability will diminish and its price will rise. The key issue is not when oil will run out, but when global oil production will reach Hubbert's peak—when it will stop rising and begin to decline. Oil optimists argue that oil resources will meet demand for decades to come; oil pessimists think we are within a few years of Hubbert's peak, if not already there.

What is the origin of coal and what are the consequences of burning it? Coal is formed by the burial, compression, and metamorphism of wetland vegetation. There are huge resources of coal in sedimentary rocks. Coal combustion is a major source of atmospheric carbon dioxide as well as sulfur oxides that cause acid rain. Furthermore, coal mining and toxic substances produced by coal burning present risks to human life and to the environment. Because of its abundance and low cost, however, the use of coal is likely to increase.

What are the prospects for alternative energy sources? Alternative energy sources include nuclear, solar, hydroelectric, wind, and geothermal energy and biofuels. Taken together, these energy sources currently supply only a small percentage of world energy demand. Nuclear energy produced by the fission of uranium, the world's most abundant minable resource, could be a major energy source, but only if the public can be assured of its safety and security. With advances in technology and reductions in cost, renewable sources such as solar energy, wind energy, and biomass could become major contributors in the twenty-first century.

What is the carbon cycle? The carbon cycle describes the continual movement of carbon between the atmosphere and its other principal reservoirs: the lithosphere, hydrosphere, and biosphere. Major fluxes of the carbon cycle include the movement of carbon dioxide between the biosphere and the atmosphere through photosynthesis, respiration, and decay; gas exchange at the interface between the atmosphere and the sea surface; and the weathering and precipitation of calcium carbonate.

How much global warming will there be in the twenty-first century, and what will be its consequences? Atmospheric concentrations of greenhouse gases will continue to rise throughout the twenty-first century, primarily because of human activities. The magnitude of the increase will depend on whether human society takes active steps to limit its greenhouse gas emissions. Projections of climate warming during the twenty-first century are highly uncertain, but the range of likely warming is 1° to 6°C. This warming will disrupt ecosystems and increase the rate of species extinctions. The oceans will warm and expand, raising sea level as much as a meter. The north polar ice cap will continue to shrink rapidly, and much of the Arctic Ocean is expected to become ice-free during the summer.

How might we stabilize carbon emissions at their current levels? If human civilization continues to rely on fossil fuels, anthropogenic carbon emissions will increase by at least 7 gigatons per year during the next 50 years. This problem could be addressed by implementing seven stabilization wedges, each offsetting the projected growth of carbon emissions by 1 gigaton per year.

KEY TERMS AND CONCEPTS

Anthropocene (p. 408)
biofuel (p. 394)
carbon cycle (p. 395)
carbon economy (p. 381)
carbon sequestration (p. 411)
climate model (p. 402)
coal bed (p. 387)
crude oil (p. 383)
geochemical reservoir (p. 395)
geothermal energy (p. 392)
Hubbert's peak (p. 385)
natural gas (p. 383)
nuclear energy (p. 390)
oil shale (p. 390)
oil trap (p. 383)
oil window (p. 383)
residence time (p. 396)
solar energy (p. 392)
stabilization wedge (p. 411)
tar sands (p. 390)

EXERCISES

1. Describe some of the ways in which human civilization is fundamentally different from the natural geosystems we have studied in this textbook.

2. Which fossil fuel produces the least amount of carbon dioxide per unit of energy: oil, natural gas, or coal?

3. What are the prerequisites required for oil traps to contain oil?

4. Explain which of the following factors are important in estimating the future supply of oil and gas: (a) the rate of oil and gas accumulation, (b) the rate of depletion of known reserves, (c) the rate of discovery of new reserves, (d) the total amount of oil and gas now present on Earth.

5. Are you an oil optimist or an oil pessimist? Explain why.

6. An aggressive drilling program in the Arctic National Wildlife Refuge could produce as much as 12 billion barrels of oil. At current consumption rates, for how many years would this resource supply U.S. oil demand?

7. Which three countries have the largest coal reserves?

8. What issues related to the use of nuclear energy can be addressed by geologists?

9. Contrast the risks and benefits of nuclear fission and coal combustion as energy sources.

10. How would the geochemical cycle of calcium be affected by a global increase in chemical weathering?

11. From the information given in Figure 14.22, estimate the residence time of carbon dioxide (a) in the ocean and (b) in the atmosphere.

12. If we keep pumping carbon dioxide into the atmosphere and Earth's climate warms significantly in the next 100 years, how might the global carbon cycle be affected?

13. An economist once wrote: "The predicted change in global temperature due to human activity is less than the difference in winter temperature between New York and Florida, so why worry?" Should he worry? Why or why not?

14. Do you think we should act now to reduce carbon emissions or delay until the science of climate change is better understood?

15. Will the melting of ice shelves by global warming increase sea level? Explain your answer.

16. What do you think will be the major sources of the world's energy in the year 2030? In the year 2100?

17. Is the United States justified in insisting that developing countries that now use much less fossil fuel than developed countries agree to limit their future carbon emissions?

18. Do you think that future scientists and engineers will be able to intervene and prevent catastrophic changes in the Earth system?

19. Why might a geologist several thousand years in the future consider the industrial revolution to be the beginning of a new geologic epoch?

APPENDIXES

APPENDIX 1

Conversion Factors

LENGTH

1 centimeter	0.3937 inch
1 inch	2.5400 centimeters
1 meter	3.2808 feet; 1.0936 yards
1 foot	0.3048 meter
1 yard	0.9144 meter
1 kilometer	0.6214 mile (statute); 3281 feet

LENGTH

1 mile (statute)	1.6093 kilometers
1 mile (nautical)	1.8531 kilometers
1 fathom	6 feet; 1.8288 meters
1 angstrom	10^{-8} centimeter
1 micrometer	0.0001 centimeter

VELOCITY

1 kilometer/hour	27.78 centimeters/second
1 mile/hour	17.60 inches/second

AREA

1 square centimeter	0.1550 square inch
1 square inch	6.452 square centimeters
1 square meter	10.764 square feet; 1.1960 square yards
1 square foot	0.0929 square meter
1 square kilometer	0.3861 square mile
1 square mile	2.590 square kilometers
1 acre (U.S.)	4840 square yards

VOLUME

1 cubic centimeter	0.0610 cubic inch
1 cubic inch	16.3872 cubic centimeters
1 cubic meter	35.314 cubic feet
1 cubic foot	0.02832 cubic meter
1 cubic meter	1.3079 cubic yards
1 cubic yard	0.7646 cubic meter
1 liter	1000 cubic centimeters; 1.0567 quarts (U.S. liquid)
1 gallon (U.S. liquid)	3.7853 liters

MASS

1 gram	0.03527 ounce
1 ounce	28.3495 grams
1 kilogram	2.20462 pounds
1 pound	0.45359 kilogram

PRESSURE

1 kilogram/square centimeter	0.96784 atmosphere; 0.98067 bar; 14.2233 pounds/square inch
1 bar	0.98692 atmosphere; 10^5 pascals

ENERGY

1 joule	0.239 calorie; 9.479×10^{-4} Btu
1 British thermal unit (Btu)	251.9 calories; 1054 joules
1 quad	10^{15} Btu

POWER

1 watt	0.001341 horsepower (U.S.); 3.413 Btu/hour

APPENDIX 2

Numerical Data Pertaining to Earth

Equatorial radius	6378 kilometers
Polar radius	6357 kilometers
Radius of sphere with Earth's volume	6371 kilometers
Volume	1.083×10^{27} cubic centimeters
Surface area	5.1×10^{18} square centimeters
Percent surface area of oceans	71
Percent surface area of land	29
Average elevation of land	623 meters
Average depth of oceans	3.8 kilometers
Mass	5.976×10^{27} grams
Density	5.517 grams/cubic centimeters
Gravity at equator	978.032 centimeters/second/second
Mass of atmosphere	5.1×10^{21} grams
Mass of ice	$25-30 \times 10^{21}$ grams
Mass of oceans	1.4×10^{24} grams
Mass of crust	2.5×10^{25} grams
Mass of mantle	4.05×10^{27} grams
Mass of core	1.90×10^{27} grams
Mean distance to Sun	1.496×10^{8} kilometers
Mean distance to Moon	3.844×10^{5} kilometers
Ratio: Mass of Sun/mass of Earth	3.329×10^{5}
Ratio: Mass of Earth/mass of Moon	81.303
Total geothermal energy reaching Earth's surface each year	10^{21} joule; 2.39×10^{20} calories; 949 quads
Earth's daily receipt of solar energy	14,137 quads; 1.49×10^{22} joules
U.S. energy consumption, 2004	99.7 quads; 1.05×10^{20} joules

APPENDIX 3

Chemical Reactions

Electron Shells and Ion Stability

Electrons surround the nucleus of an atom in a unique set of concentric spheres called electron shells. Each shell can hold a certain maximum number of electrons. In the chemical reactions of most elements, only the electrons in the outermost shells interact. In the reaction between sodium (Na) and chlorine (Cl) that forms sodium chloride (NaCl), the sodium atom loses an electron from its outer shell of electrons, and the chlorine atom gains an electron in its outer shell (see Figure 3.4, page 48).

Before reacting with chlorine, the sodium atom has one electron in its outer shell. When it loses that electron, its outer shell is eliminated and the next shell inward, which has eight electrons (the maximum that this shell can hold), becomes the outer shell. The original chlorine atom had seven electrons in its outer shell, with room for a total of eight. By gaining an electron, its outer shell is filled. Many elements have a strong tendency to acquire a full outer electron shell, some by gaining electrons and some by losing them in the course of a chemical reaction. The stability of ions with fully occupied outer shells is related to the interactions of electrons in various orbitals around the nucleus.

Many chemical reactions entail gains and losses of several electrons as two or more elements combine. The element calcium (Ca), for example, becomes a doubly charged cation, Ca^{2+}, as it reacts with two chlorine atoms to form calcium chloride. (In the chemical formula for calcium chloride, $CaCl_2$, the presence of two chloride ions is symbolized by the subscript 2.) Chemical formulas thus show the relative proportion of atoms or ions in a compound.

The periodic table.

Common practice is to omit the subscript 1 next to single ions in a formula.

The periodic table organizes the elements (from left to right along a row) in order of atomic number (the number of protons), which also means increasing the numbers of electrons in the outer shell. The third row from the top, for example, starts at the left with sodium (atomic number 11), which has one electron in its outer shell. The next is magnesium (atomic number 12), which has two electrons in its outer shell, followed by aluminum (atomic number 13), with three, and silicon (atomic number 14), with four. Then come phosphorus (atomic number 15), with five; sulfur (atomic number 16), with six; and chlorine (atomic number 17), with seven. The last element in this row is argon (atomic number 18), with eight electrons, the maximum possible, in its outer shell. Each column in the table forms a vertical grouping of elements with similar electron-shell patterns.

Elements That Tend to Lose Electrons

The elements in the leftmost column of the table all have a single electron in their outer shells and have a strong tendency to lose that electron in chemical reactions. Of this group, hydrogen (H), sodium (Na), and potassium (K) are found in major abundance at Earth's surface and in its crust.

The second column from the left includes two more elements of major abundance, magnesium (Mg) and calcium (Ca). Elements in this column have two electrons in their outer shells and a strong tendency to lose both of them in chemical reactions.

Elements That Tend to Gain Electrons

Toward the right side of the table, the two columns headed by oxygen (O), the most abundant element in the Earth, and fluorine (F), a highly reactive toxic gas, group the elements that tend to gain electrons in their outer shells. The elements in the column headed by oxygen have six of the possible eight electrons in their outer shells and tend to gain two electrons. Those in the column headed by fluorine have seven electrons in their outer shells and tend to gain one.

Other Elements

The columns between those farthest on the left and those farthest on the right have varying tendencies to gain, lose, or share electrons. The column toward the right side of the table headed by carbon (C) includes silicon (Si), of major abundance in the Earth. Both silicon and carbon tend to share electrons.

The elements in the last column on the right, headed by helium (He), have full outer shells and thus no tendency either to gain or to lose electrons. As a result, these elements, in contrast with those in other columns, do not react chemically with other elements, except under very special conditions.

APPENDIX 4

Properties of the Most Common Minerals of Earth's Crust

Mineral or Group Name	Structure or Composition	Varieties and Chemical Composition	Form, Diagnostic Characteristics	Cleavage, Fracture	Color	Hardness
LIGHT-COLORED MINERALS, VERY ABUNDANT IN EARTH'S CRUST IN ALL MAJOR ROCK TYPES FELDSPAR	FRAMEWORK SILICATES	*POTASSIUM FELDSPARS* $KAlSi_3O_8$ Sanidine Orthoclase Microcline	Cleavable coarsely crystalline or finely granular masses; isolated crystals or grains in rocks, most commonly not showing crystal faces	Two at right angles, one perfect and one good; pearly luster on perfect cleavage	White to gray, frequently pink or yellowish; some green	6
		PLAGIOCLASE FELDSPARS $NaAlSi_3O_8$ Albite $CaAl_2Si_2O_8$ Anorthite		Two at nearly right angles, one perfect and one good; fine parallel striations on perfect cleavage	White to gray, less commonly greenish or yellowish	
QUARTZ		SiO_2	Single crystals or masses of 6-sided prismatic crystals; also formless crystals and grains or finely granular or massive	Very poor or nondetectable; conchoidal fracture	Colorless, usually transparent; also slightly colored smoky gray, pink, yellow	7
MICA	SHEET SILICATES	*MUSCOVITE* $KAl_3Si_3O_{10}(OH)_2$	Thin, disc-shaped crystals, some with hexagonal outlines; dispersed or aggregates	One perfect; splittable into very thin, flexible, transparent sheets	Colorless; slight gray or green to brown in thick pieces	$2-2\frac{1}{2}$
		BIOTITE $K(Mg, Fe)_3AlSi_3O_{10}(OH)_2$	Irregular, foliated masses; scaly aggregates	One perfect; splittable into thin, flexible sheets	Black to dark brown; translucent to opaque	$2\frac{1}{2}-3$
		CHLORITE $(Mg,Fe)_5(Al,Fe)_2Si_3O_{10}(OH)_8$	Foliated masses or aggregates of small scales	One perfect; thin sheets flexible but not elastic	Various shades of green	$2-2\frac{1}{2}$
DARK-COLORED MINERALS, ABUNDANT IN MANY KINDS OF IGNEOUS AND METAMORPHIC ROCKS AMPHIBOLE	DOUBLE CHAINS	*TREMOLITE-ACTINOLITE* $Ca_2(Mg,Fe)_5Si_8O_{22}(OH)_2$	Long, prismatic crystals, usually 6-sided; commonly in fibrous masses or irregular aggregates	Two perfect cleavage directions at 56° and 124° angles	Pale to deep green Pure tremolite white	5–6
		HORNBLENDE Complex Ca, Na, Mg, Fe, Al silicate				
PYROXENE	SINGLE CHAINS	*ENSTATITE-HYPERSTHENE* $(Mg,Fe)_2Si_2O_6$	Prismatic crystals, either 4- or 8-sided; granular masses and scattered grains	Two good cleavage directions at about 90°	Green and brown to grayish or greenish white	5–6
		DIOPSIDE $(Ca,Mg)_2Si_2O_6$			Light to dark green	
		AUGITE Complex Ca, Na, Mg, Fe, Al silicate			Very dark green to black	

(Continued)

APPENDIX 4 Properties of the Most Common Minerals of Earth's Crust

	Mineral or Group Name	Structure or Composition	Varieties and Chemical Composition	Form, Diagnostic Characteristics	Cleavage, Fracture	Color	Hardness
LIGHT-COLORED MINERALS, TYPICALLY AS ABUNDANT CONSTITUENTS OF SEDIMENTS AND SEDIMENTARY ROCKS	OLIVINE	ISOLATED TETRAHEDRA	$(Mg,Fe)_2SiO_4$	Granular masses and disseminated small grains	Conchoidal fracture	Olive to grayish green and brown	$6\frac{1}{2}–7$
	GARNET		Ca, Mg, Fe, Al silicate	Isometric crystals, well formed or rounded; high specific gravity, 3.5–4.3	Conchoidal and irregular fracture	Red and brown, less commonly pale colors	$6\frac{1}{2}–7$
	CALCITE	CARBONATES	$CaCO_3$	Coarsely to finely crystalline in beds, veins, and other aggregates; cleavage faces may show in coarser masses; calcite effervesces rapidly in acid, but dolomite effervesces slowly and only if crushed into powder	Three perfect cleavages, at oblique angles; splits to rhombohedral cleavage pieces	Colorless, transparent to translucent; variously colored by impurities	3
	DOLOMITE		$CaMg(CO_3)_2$				$3\frac{1}{2}–4$
	CLAY MINERALS	HYDROUS ALUMINO-SILICATES	*KAOLINITE* $Al_2Si_2O_5(OH)_4$ *ILLITE* Similar to muscovite $+Mg,Fe$ *SMECTITE* Complex Ca, Na, Mg, Fe, Al silicate $+ H_2O$	Earthy masses in soils; bedded; in association with other clays, iron oxides, or carbonates; plastic when wet; montmorillonite swells when wet	Earthy, irregular	White to light gray and buff; also gray to dark gray, greenish gray, and brownish depending on impurities and associated minerals	$1\frac{1}{2}–2\frac{1}{2}$
	GYPSUM	SULFATES	$CaSO_4 \cdot 2H_2O$	Granular, earthy, or finely crystalline masses; tabular crystals	One perfect, splitting to fairly thin slabs or sheets; two other good cleavages	Colorless to white; transparent to translucent	2
	ANHYDRITE		$CaSO_4$	Massive or crystalline aggregates in beds and veins	One perfect, one nearly perfect, one good; at right angles	Colorless, some tinged with blue	$3–3\frac{1}{2}$
	HALITE	HALIDES	NaCl	Granular masses in beds; some cubic crystals; salty taste	Three perfect cleavages at right angles	Colorless, transparent to translucent	$2\frac{1}{2}$
	OPAL-CHALCEDONY	SILICA	SiO_2 [Opal is an amorphous variety; chalcedony is a formless microcrystalline quartz.]	Beds in siliceous sediments and chert; in veins or banded aggregates	Conchoidal fracture	Colorless or white when pure, but tinged with various colors by impurities in bands, especially in agates	$5–6\frac{1}{2}$
DARK-COLORED MINERALS, COMMON IN MANY ROCK TYPES	MAGNETITE	IRON OXIDES	Fe_3O_4	Magnetic; disseminated grains, granular masses; occasional octahedral isometric crystals; high specific gravity, 5.2	Conchoidal or irregular fracture	Black, metallic luster	6

Properties of the Most Common Minerals of Earth's Crust APPENDIX 4 AP-7

Mineral or Group Name	Structure or Composition	Varieties and Chemical Composition	Form, Diagnostic Characteristics	Cleavage, Fracture	Color	Hardness
HEMATITE	IRON OXIDES	Fe_2O_3	Earthy to dense masses, some with rounded forms, some granular or foliated; high specific gravity, 4.9–5.3	None; uneven, sometimes splintery fracture	Reddish brown to black	5–6
"LIMONITE"		GOETHITE [the major mineral of the mixture called "limonite," a field term] $FeO(OH)$	Earthy masses, massive bodies or encrustations, irregular layers; high specific gravity, 3.3–4.3	One excellent in the rare crystals; usually an early fracture	Yellowish brown to dark brown and black	5–5½
KYANITE	ALUMINO-SILICATES	Al_2SiO_5	Long, bladed or tabular crystals or aggregates	One perfect and one poor, parallel to length of crystals	White to light-colored or pale blue	5 parallel to crystal length; 7 across crystals
SILLIMANITE		Al_2SiO_5	Long, slender crystals or fibrous, felted masses	One perfect parallel to length, not usually seen	Colorless, gray to white	6–7
ANDALUSITE		Al_2SiO_5	Coarse, nearly square prismatic crystals, some with symmetrically arranged impurities	One distinct; irregular fracture	Red, reddish brown, olive-green	7½
FELDSPATHOIDS		NEPHELINE $(Na,K)AlSiO_4$	Compact masses or as embedded grains, rarely as small prismatic crystals	One distinct; irregular fracture	Colorless, white, light gray; gray-greenish in masses, with greasy luster	5½–6
		LEUCITE $KAlSi_2O_6$	Trapezohedral crystals embedded in volcanic rocks	One very imperfect	White to gray	5½–6
SERPENTINE		$Mg_6Si_4O_{10}(OH)_8$	Fibrous (asbestos) or platy masses	Splintery fracture	Green; some yellowish brownish or gray; waxy or greasy luster in massive habit; silky luster in fibrous habit	4–6
TALC		$Mg_3Si_4O_{10}(OH)_2$ masses or aggregates	Foliated or compact masses or aggregates	One perfect, making thin flakes or scales; soapy feel	White to pale green; pearly or greasy luster	1
CORUNDUM		Al_2O_3	Some rounded, barrel-shaped crystals; most often as disseminated grains or granular masses (emery)	Irregular fracture	Usually brown, pink, or blue; emery black. Gemstone varieties: ruby, sapphire	9

LIGHT-COLORED MINERALS, MAINLY IN METAMORPHIC AND IGNEOUS ROCKS AS COMMON OR MINOR CONSTITUENTS

(*Continued*)

APPENDIX 4 Properties of the Most Common Minerals of Earth's Crust

	Mineral or Group Name	Structure or Composition	Varieties and Chemical Composition	Form, Diagnostic Characteristics	Cleavage, Fracture	Color	Hardness
DARK-COLORED MINERALS, COMMON IN METAMORPHIC ROCKS	EPIDOTE	SILICATES	$Ca_2(Al,Fe)Al_2Si_3O_{12}(OH)$	Aggregates of long prismatic crystals, granular or compact masses, embedded grains	One good, one poor at greater than right angles; conchoidal and irregular fracture	Green, yellow-green, gray, some varieties dark brown to black	6–7
	STAUROLITE		$Fe_2Al_9Si_4O_{22}(O,OH)_2$	Short prismatic crystals, some cross-shaped, usually coarser than matrix of rock	One poor	Brown, reddish, or dark brown to black	$7–7\frac{1}{2}$
METALLIC LUSTER, COMMON IN MANY ROCK TYPES, ABUNDANT IN VEINS	PYRITE	SULFIDES	FeS_2	Granular masses or well-formed cubic crystals in veins and beds or disseminated; high specific gravity, 4.9–5.2	Uneven fracture	Pale brass-yellow	$6–6\frac{1}{2}$
	GALENA		PbS	Granular masses in veins and disseminated; some cubic crystals; very high specific gravity, 7.3–7.6	Three perfect cleavages at mutual right angles, giving cubic cleavage fragments	Silver-gray	$2\frac{1}{2}$
	SPHALERITE		ZnS	Granular masses or compact crystalline aggregates; high specific gravity, 3.9–4.1	Six perfect cleavages at 60° to one another	White to green, brown, and black; resinous to submetallic luster	$3\frac{1}{2}–4$
	CHALCOPYRITE		$CuFeS_2$	Granular or compact masses; disseminated crystals; high specific gravity, 4.1–4.3	Uneven fracture	Brassy to golden-yellow	$3\frac{1}{2}–4$
	CHALCOCITE		Cu_2S	Fine-grained masses; high specific gravity, 5.5–5.8	Conchoidal fracture	Lead-gray to black; may tarnish green or blue	$2\frac{1}{2}–3$
MINERALS FOUND IN MINOR AMOUNTS IN A VARIETY OF ROCK TYPES AND IN VEINS OR PLACERS	RUTILE	TITANIUM OXIDES	TiO_2	Slender to prismatic crystals; granular masses; high specific gravity, 4.25	One distinct, one less distinct; conchoidal fracture	Reddish brown, some yellowish, violet, or black	$6–6\frac{1}{2}$
	ILMENITE		$FeTiO_3$	Compact masses, embedded grains, detrital grains in sand; high specific gravity, 4.79	Conchoidal fracture	Iron-black; metallic to submetallic luster	5–6
	ZEOLITES	SILICATES	Complex hydrous silicates; many varieties of minerals, including analcime, natrolite, phillipsite, heulandite, and chabazite	Well-formed radiating crystals in cavities in volcanics, veins, and hot springs; also as fine-grained and earthy bedded deposits	One perfect for most	Colorless, white, some pinkish	4–5

GLOSSARY

Words in *italic* have separate entries in the Glossary. Specific minerals are defined in Appendix 4.

ablation The total amount of ice that a glacier loses each year. (Compare *accumulation*.)

absolute age The actual number of years elapsed from a geologic event until now. (Compare *relative age*.)

accretion A process of continental growth in which buoyant fragments of *crust* are attached (accreted) to existing continental masses by horizontal transport during plate movements. (Compare *magmatic addition*.)

accumulation The amount of snow added to a glacier annually. (Compare *ablation*.)

acid rain Acid *precipitation* caused by *anthropogenic* emissions of sulfur dioxide into the atmosphere, where it reacts with water vapor to form sulfuric acid.

aftershock An *earthquake* that occurs as a consequence of a previous earthquake of larger magnitude. (Compare *foreshock*.)

albedo The fraction of *solar energy* reflected by the surface of a planet or moon.

alluvial fan A cone- or fan-shaped accumulation of *sediment* deposited where a *stream* leaves a mountain front and enters a broad, relatively flat valley.

andesite An *extrusive igneous rock* intermediate in composition between *rhyolite* and *basalt*. The extrusive equivalent of *diorite*.

Anthropocene The Age of Man; a new geologic epoch beginning about A.D. 1780, proposed by Nobel laureate Paul Crutzen.

anthropogenic Arising from human activity.

anticline A large upfolding of layered *rocks* into an arch. (Compare *syncline*.)

aquiclude A relatively impermeable *rock* or *soil* that bounds an *aquifer* above or below and acts as a barrier to the flow of *groundwater*.

aquifer A porous *formation* that stores and transmits *groundwater* in sufficient quantity to supply wells.

artesian flow A spontaneous flow of *groundwater* through a confined *aquifer* to a point where the elevation of the ground surface is lower than that of the *groundwater table*.

asteroid One of the more than 10,000 small celestial bodies orbiting the Sun, most of them between Mars and Jupiter.

asthenosphere The weak, *ductile* layer of *rock* in the upper *mantle* below the *lithosphere* on which the tectonic plates slide. (From the Greek *asthenes,* meaning "weak.")

atom The smallest unit of an element that retains the physical and chemical properties of that element.

barrier island A long offshore sandbar that builds up to form a barricade between open ocean waves and the main *shoreline*.

basalt A fine-grained, dark, *extrusive igneous rock* composed largely of mafic *minerals,* particularly plagioclase feldspar and pyroxene. The extrusive equivalent of *gabbro*.

batholith An irregular, coarse-grained *pluton* that covers at least 100 km^2.

beach A *shoreline* made up of *sand* and pebbles.

bed load The *sediment* carried by a current along Earth's surface by sliding and rolling. (Compare *suspended load*.)

bedding Parallel layers of *sediment* formed as particles settle to the bottom of a water body or a land surface.

bedding sequence A pattern of interbedded and vertically stacked layers of different *sedimentary rock* types.

bioclastic sediment A shallow-water *sediment* made up of the fragmented shells of marine organisms and consisting primarily of two calcium carbonate *minerals*—calcite and aragonite—in variable proportions.

biofuel Liquid hydrogen fuel derived from plant biomass.

biological sediment A *sediment* that forms near its place of deposition as a result of mineral precipitation within organisms as they grow. (Compare *chemical sediment; clastic sediment*.)

biosphere The sum total of all living organisms; a major domain of the *Earth system*.

bioturbation The process by which organisms rework *muds* and *sands* by burrowing through them.

blueschist A *metamorphic rock* formed under conditions of high pressure and moderate temperature and containing glaucophane, a blue amphibole.

bolide A chunk of *rock* or other debris from interplanetary space.

bomb A large *pyroclast* ejected by a *volcanic eruption* as a blob of *lava* that cools in flight or as a chunk torn loose from previously solidified volcanic *rock*.

braided stream A *stream* whose *channel* divides into an interlacing network of channels, which then rejoin in a pattern resembling braids of hair.

breccia A volcanic *rock* formed by the *lithification* of large *pyroclasts*. (Compare *tuff*.)

brittle Pertaining to the behavior of a material that undergoes little deformation under increasing force until it breaks suddenly. (Compare *ductile*.)

building code A set of standards for the design and construction of new buildings that specifies the intensity of shaking a structure must be able to withstand in an *earthquake*.

caldera A large, steep-walled, basin-shaped depression formed by a violent *volcanic eruption* in which large volumes of *magma* are discharged rapidly from a large *magma chamber*, causing the overlying volcanic structure to collapse catastrophically through the roof of the emptied chamber.

Cambrian explosion A rapid evolutionary radiation during the early Cambrian period that gave rise to every major animal group that exists on Earth today, as well as others that have since become extinct.

carbon cycle The *geosystem* that describes the continual movement of carbon among the atmosphere and its other principal *geochemical reservoirs*: the *lithosphere*, hydrosphere, and *biosphere*.

carbon economy A civilization fed by an energy system that depends on carbon-based energy sources, particularly *fossil fuels*.

carbon sequestration The practice of pumping carbon dioxide generated by fossil-fuel combustion into *geochemical reservoirs* other than the atmosphere.

carbonate rock A *sedimentary rock* formed from the *lithification* of *carbonate sediment*.

carbonate sediment An accumulation of carbonate *minerals* *precipitated* organically or inorganically.

cementation A chemical *diagenetic* process in which *minerals* are *precipitated* in the pores between *sediment* particles and bind them together.

channel A well-defined trough through which the water in a *stream* flows.

chemical bond An electrostatic attraction between negatively charged electrons and positively charged protons that forms in a *chemical reaction*.

chemical reaction An interaction between the *atoms* of two or more chemical elements in certain fixed proportions that produces a new chemical compound.

chemical sediment A *sediment* formed by the *precipitation* of dissolved products of *chemical weathering*, usually at or near its place of deposition. (Compare *biological sediment; clastic sediment*.)

chemical weathering The *weathering* that occurs when the *minerals* in a *rock* are chemically altered or dissolved. (Compare *physical weathering*.)

chert A *sedimentary rock* made up of chemically or biologically *precipitated* silica.

clastic sediment A *sediment* formed by the transportation and accumulation of particles produced by the *weathering* of preexisting *rock*. (Compare *biological sediment, chemical sediment*.)

clay A fine-grained *clastic sediment* made up of particles less than 0.0039 mm in diameter and consisting largely of clay minerals.

claystone A *sedimentary rock* made up exclusively of *clay*.

cleavage (1) The tendency of a *crystal* to split along planar (flat) surfaces. (2) The geometric pattern produced by such breakage.

climate The average surface conditions produced by weather cycles and their variation.

climate model A representation of the *climate system*, such as a computer program, that can reproduce one or more aspects of its behavior.

climate system The *geosystem* that includes all the components of the *Earth system* and all the interactions among those components, needed to describe how *climate* behaves in space and time.

coal A biological *sedimentary rock* composed almost entirely of organic carbon formed by the *diagenesis* of swamp vegetation.

coal bed A layer or stratum of *coal*.

coastline The narrow zones where land meets the sea.

color A property of a *mineral* imparted by transmitted or reflected light.

compaction A decrease in the volume and *porosity* of a *sediment* that occurs when its particles are squeezed closer together by the weight of the overlying sediment.

compressive force A force that squeezes or shortens a body. (Compare *shearing force; tensional force*).

conglomerate A *sedimentary rock* composed of pebbles, cobbles, and boulders. The *lithified* equivalent of *gravel*.

contact metamorphism Changes in the *mineralogy* and *texture* of *rock* resulting from high temperatures, in a small area, such as the area immediately surrounding an igneous intrusion. (Compare *regional metamorphism*.)

continental drift Large-scale movements of continents across Earth's surface driven by the *plate tectonic system*.

continental glacier A thick, slow-moving sheet of ice that covers a large part of a continent or other large landmass. (Compare *valley glacier.*)

continental shelf A broad, flat, submerged platform consisting of a thick layer of sediments that extends from the shoreline to the edge of the continental slope.

convergent boundary A plate boundary where two plates move toward each other. (Compare *divergent boundary.*)

core The dense central part of Earth below a depth of 2900 km, composed principally of iron and nickel, comprising a liquid *outer core* and a solid *inner core.*

core-mantle boundary The boundary between Earth's *mantle* and *core,* about 2890 km below Earth's surface.

country rock The rock surrounding *intrusive igneous rock.*

crater (1) A bowl-shaped pit found at the summit of most *volcanoes,* centered on the vent. (2) A depression caused by the impact of a *meteorite.*

craton The stable nucleus of a continent comprising ancient shields and platforms, usually of Archean age.

cross-bedding A *sedimentary structure* consisting of bedded material deposited by currents of wind or water and inclined at angles as much as 35° from the horizontal.

crude oil Raw petroleum; a diverse class of liquids composed of complex hydrocarbons.

crust The thin outer layer of Earth, averaging from about 8 km thick under the oceans to about 40 km thick under the continents, consisting of relatively low-density silicates that melt at low temperatures.

crystal An ordered three-dimensional array of *atoms* in which the basic arrangement is repeated in all directions.

crystal habit The shape in which a *mineral*'s individual *crystals* or aggregates of crystals grow.

crystallization The formation of a solid mineral from a gas or liquid whose constituent *atoms* come together in the proper chemical proportions and ordered three-dimensional arrangement.

dacite A light-colored, fine-grained *extrusive igneous rock* with the same general composition as *andesite.* The extrusive equivalent of *granodiorite.*

decompression melting Melting of *mantle* material that occurs spontaneously, without the introduction of any additional heat, as the material rises toward Earth's surface and pressure decreases below a critical point. (Compare *fluid-induced melting.*)

delta A platform of *sediments* deposited where a *river* enters an ocean or lake and its current slows.

density The mass per unit volume of a substance, commonly expressed in grams per cubic centimeter (g/cm^3). (Compare *specific gravity.*)

desertification Changes in a region's *climate* that transform semiarid lands into deserts.

diagenesis The physical and chemical changes in *sediments* that result from their exposure to increasing temperatures and pressures as they are buried ever deeper in Earth's crust.

differentiation The transformation of an accretion of matter into a body whose interior is divided into concentric layers that differ both physically and chemically.

dike A sheetlike igneous intrusion that cuts across layers of bedded *country rock.* (Compare *sill.*)

diorite A coarse-grained *intrusive igneous rock* intermediate in composition between *granite* and *gabbro.* The intrusive equivalent of *andesite.*

dip The amount of tilting of a *rock* layer; the angle at which the layer inclines from the horizontal. (Compare *strike.*)

dip-slip fault A fault on which there has been relative movement of the blocks up or down the *dip* of the fault plane.

discharge (1) The volume of *groundwater* leaving an *aquifer* in a given time. (Compare *recharge.*) (2) The volume of water that passes a given point in a given time as it flows through a *channel* of a certain width and depth.

distributary A *stream* that receives water and *sediment* from the main *channel,* branches off downstream, and thus distributes the water and sediment into many channels; typically found on a *delta.*

divergent boundary A plate boundary where two plates move apart and new lithosphere is created. (Compare *convergent boundary.*)

divide A topographic rise between two *streams* along which all rainfall runs off down one side or the other.

dolostone An abundant *carbonate rock* composed primarily of dolomite and formed by the replacement of *carbonate sediments* and *limestone.*

drainage basin An area of land, bounded by *divides,* that funnels all its water into the network of *streams* draining the area.

drought A period of months or years when precipitation is much lower than normal.

ductile Pertaining to the behavior of a material that undergoes smooth and continuous deformation under increasing stress without fracturing and does not spring back to its original shape when the stress is released. (Compare *brittle.*)

dune An elongated mound or ridge of *sand* formed by a current of wind or water.

Earth system All the parts of our planet and all their interactions; the collection of all of Earth's open, interacting (and often overlapping) *geosystems*.

earthquake The violent motion of the ground that occurs when brittle *rock* under stress suddenly breaks along a new or preexisting fault.

El Niño An anomalous warming of the eastern tropical Pacific Ocean that occurs every 3 to 7 years and lasts for a year or so.

elastic rebound theory A theory of *faulting* and *earthquake* generation holding that as the crustal blocks on either side of a fault are deformed by tectonic forces, they remain locked in place by friction, accumulating elastic strain energy, until they fracture and rebound to their undeformed state.

environment The complex of physical, chemical, and biological factors (such as *climate, soil,* and living organisms) that act on an organism and ultimately determine its form and survival.

eolian Pertaining to wind. Most commonly refers to wind-related sediment transport or *erosion* of bedrock.

eon The largest division of geologic time, embracing multiple *eras*.

epicenter The geographic point on Earth's surface directly above the *focus* of an *earthquake*.

epoch A division of geologic time representing one subdivision of a geologic *period*.

era A division of geologic time representing one subdivision of an *eon* and including multiple *periods*.

erosion The set of processes that loosen *soil* and *rock* and move them downhill or downstream.

evaporite rock A *sedimentary rock* formed from *evaporite sediments*.

evaporite sediment *Chemical sediment precipitated* from evaporating seawater or from water in arid-region lakes that lack *river* outlets.

evolution Systematic change in organisms over time, driven by the process of *natural selection*.

exhumation The transportation of *metamorphic rocks* back to Earth's surface.

extrusive igneous rock A fine-grained or glassy *igneous rock* formed from *magma* that erupts at Earth's surface and cools rapidly. (Compare *intrusive igneous rock*.)

fault mechanism The type of fault rupture (*normal, thrust,* or *strike-slip*) that produces an *earthquake;* determined by the orientation of the fault rupture and the direction of slipping.

fault slip The displacement of the two blocks on either side of a fault that occurs during an *earthquake*.

faulting The process by which tectonic forces cause *rock* to break and slip on both sides of a fracture.

faunal succession, principle of See *principle of faunal succession*.

felsic rock Light-colored *igneous rock* containing felsic *minerals,* such as quartz, orthoclase feldspar, and plagioclase feldspar, that are poor in iron and magnesium and rich in silica. (Compare *mafic rock; ultramafic rock*.)

fjord A *U-shaped valley* carved by a glacier and occupied by the sea.

flexural basin A *sedimentary basin* that develops at a *convergent boundary* where one plate pushes up over the other and the weight of the overriding plate causes the underlying plate to bend or flex downward.

flood Inundation that results when water flows into a region faster than it can flow out.

floodplain A flat area about level with the top of a *channel* that lies on either side of the channel; the part of a valley that is flooded when a *stream* spills over its banks.

fluid-induced melting Melting of *rock* induced by the presence of water, which lowers the melting temperature of the rock. (Compare *decompression melting*.)

focus The point along a fault at which fault slipping initiates an *earthquake*.

folding The process by which tectonic forces deform an area of *crust* so that layers of *rock* are pushed into folds.

foliated rock A *metamorphic rock* that displays *foliation*. Foliated rocks include *slate, phyllite, schist,* and *gneiss*. (Compare *granoblastic rock*.)

foliation A set of flat or wavy parallel cleavage planes, produced by deformation under directed pressure, that is characteristic of regionally metamorphosed rocks.

foreshock A small *earthquake* that occurs in the vicinity of, but before, a main shock. (Compare *aftershock*.)

formation A distinct group of *rock* layers that can be identified throughout a region by its physical properties and possibly by the assemblage of *fossils* it contains.

fossil A trace of an organism that has been preserved in the *geologic record*.

fossil fuel An energy *resource* created by the burial and heating of dead organic matter, such as *coal, crude oil,* or *natural gas*.

fractional crystallization The process by which the *crystals* formed in a cooling *magma* are segregated from the remaining liquid.

fracture The tendency of a *crystal* to break along irregular surfaces other than *cleavage* planes.

gabbro A dark gray, coarse-grained *intrusive igneous rock* with an abundance of mafic *minerals,* particularly pyroxene. The intrusive equivalent of *basalt.*

geochemical reservoir A repository for the temporary storage of materials transported through the *Earth system* in geochemical cycles.

geodynamo The *geosystem* that produces and maintains Earth's *magnetic field,* driven by convection in the *outer core.*

geologic cross section A diagram showing the geologic features that would be visible by exposing a vertical slice through a part of Earth.

geologic map A map representing the rock formations exposed at Earth's surface.

geologic record The information that has been preserved in rocks as they have formed at various times throughout Earth's history.

geologic time scale A worldwide history of geologic events that divides Earth's history into intervals, many of which are marked by distinctive sets of fossils and bounded by times when those sets of fossils changed abruptly.

geology The branch of Earth science that studies Earth's history, its composition and internal structure, and its surface features.

geosystem One of the open, interacting, and often overlapping subsystems of the *Earth system* that produces specific types of behavior. Geosystems include the *plate tectonic system,* the *climate system,* and the *geodynamo.*

geotherm The curve that describes how Earth's temperatures increase with depth.

geothermal energy Energy derived from water heated during passage through a subsurface region of hot *rock.*

glacial cycle A climate cycle that involves the alternation between a cold glacial period, or *ice age,* and a warm *interglacial period.*

global change A change in one or more components of the *Earth system* that is worldwide in scale, such as global warming of Earth's *climate.*

gneiss A light-colored, poorly *foliated* rock with coarse bands of segregated light and dark *minerals* throughout, formed by high-grade metamorphism.

graded bedding A *sedimentary structure* in which each bed progresses from the coarsest particles at the bottom to the finest particles at the top.

granite A felsic, coarse-grained *intrusive igneous rock* composed of quartz, orthoclase feldspar, sodium-rich plagioclase feldspar, and micas. The intrusive equivalent of *rhyolite.*

granoblastic rock A nonfoliated *metamorphic rock* composed mainly of *crystals* that grow in equant shapes, such as cubes and spheres, rather than in platy or elongate shapes. Granoblastic rocks include hornfels, quartzite, marble, greenstone, amphibolite, and granulite. (Compare *foliated rock.*)

granodiorite A light-colored, coarse-grained *intrusive igneous rock* that is similar to *granite* in containing abundant quartz but whose predominant feldspar is plagioclase, not orthoclase. The intrusive equivalent of *dacite.*

gravel The coarsest *clastic sediment,* consisting of particles larger than 2 mm in diameter and including pebbles, cobbles, and boulders.

greenhouse effect A global warming effect that results when a planet with an atmosphere containing *greenhouse gases* radiates *solar energy* back into space less efficiently than it would without such an atmosphere.

greenhouse gas An atmospheric gas that absorbs and re-radiates solar energy, contributing to the *greenhouse effect.* Greenhouse gases include water vapor, carbon dioxide, ozone, and methane.

greenschist A chlorite-containing *schist* formed by low-grade metamorphism of *mafic rocks.*

groundwater Water stored in *rocks* beneath Earth's surface.

groundwater table The boundary between the *unsaturated zone* and the *saturated zone* in subsurface *soil* or *rock.*

half-life The time required for one-half of the original number of *atoms* of a radioactive *isotope* to decay.

hardness A measure of the ease with which the surface of a *mineral* can be scratched.

Heavy Bombardment A time early in the history of the solar system when crater-forming bolide impacts were very frequent.

Hubbert's peak The point in time when global oil production will peak and then begin to decline. Named after the geologist M. King Hubbert, who in 1956 correctly predicted that U.S. oil production would peak in the early 1970s.

hurricane A great storm that forms over tropical seas, between 8° and 20° latitude, in areas of high humidity, light winds, and warm sea surface temperatures, producing winds of at least 119 km/hour (74 miles/hour) and large amounts of rainfall.

hydrologic cycle The cyclical movement of water from the ocean to the atmosphere by evaporation, to Earth's surface by *precipitation,* to *streams* through *runoff* and *groundwater,* and back to the ocean.

hydrology The science that studies the movements and characteristics of water on and under Earth's surface.

hydrothermal solution A hot water solution formed when circulating *groundwater* or seawater comes into contact with a hot magmatic intrusion, reacts with it, and carries off significant quantities of elements and ions released by the reaction. These elements and their ions may then interact with one another to deposit ore minerals, usually as the solution cools.

hydrothermal water Hot water deep in the *crust*.

hypothesis A tentative explanation of a phenomenon, based on observational data and experiments, that serves as a basis for further investigation.

ice age The cold interval of a *glacial cycle*, during which Earth cools, water is transferred from the hydrosphere to glaciers of the cryosphere, and sea level drops. (Compare *interglacial period*.)

igneous rock A *rock* formed by the solidification of *magma*.

infiltration The movement of water into *rock* or *soil* through cracks or small pores between particles.

inner core The central part of Earth; a solid metallic sphere 1250 km in radius, composed of iron and nickel suspended within the liquid *outer core*.

intensity scale A scale for estimating the intensity of *earthquake* shaking directly from an earthquake's effects on humans, built structures, and the ground surface.

interglacial period The warm interval of a glacial cycle during which ice sheets melt, glaciers retreat to higher latitudes and altitudes, and water is transferred from the cryosphere to the hydrosphere. (Compare *ice age*.)

intermediate igneous rock An *igneous rock* midway in composition between mafic and felsic, neither as rich in silica as *felsic rock* nor as poor in it as *mafic rock*.

intrusive igneous rock A coarse-grained *igneous rock* formed from *magma* that intrudes into *country rock* deep in Earth's *crust* and cools slowly.

ion An *atom* or group of atoms that has an electrical charge, either positive or negative, resulting from the loss or gain of one or more electrons.

island arc A chain of volcanic islands formed on the overriding plate at a *convergent boundary* by *magma* that rises from the *mantle* as water released from the subducting lithospheric slab causes *fluid-induced melting*.

isochron A contour that connects rocks of equal age on the seafloor.

isotope One of two or more species of *atoms* of the same element that have different numbers of neutrons.

isotopic dating The use of naturally occurring radioactive elements to determine the ages of *rocks*.

joint A crack in a *rock* along which there has been no appreciable movement.

karst topography An irregular hilly terrain characterized by *sinkholes*, caverns, and a lack of surface *streams*, formed in regions with extensive limestone formations.

kettle A hollow or undrained depression that often has steep sides and may be occupied by a pond or lake, formed in glacial terrains when outwash is deposited around a residual block of ice that later melts.

lahar A torrential mudflow of wet volcanic debris produced by a *volcanic eruption*.

laminar flow A flow in which straight or gently curved streamlines run parallel to one another without mixing or crossing between layers. (Compare *turbulent flow*.)

landslide A downhill flow of rock, mud, or soil that occurs when the friction that keeps the material from slipping downslope is overcome by the downward force of gravity, often due to water saturation.

lava *Magma* that flows out onto Earth's surface.

limestone A *carbonate rock* composed mainly of calcium carbonate in the form of the *mineral* calcite.

lithification The transformation of *sediment* into solid *rock* by *compaction* and *cementation*.

lithosphere The strong, rigid outer shell of Earth that forms the tectonic plates. The lithosphere contains the *crust* and the uppermost part of the *mantle* down to an average depth of about 100 km.

longshore current A shallow-water current that runs parallel to the shore.

lower mantle A relatively homogeneous region of the *mantle* that extends from the base of the transition zone at a depth of 660 km to the *core-mantle boundary* at a depth of about 2890 km.

low-velocity zone A layer of Earth beginning at a depth of about 100 km where *S-wave* speed abruptly decreases, corresponding to the top part of the *asthenosphere*.

luster The way the surface of a *mineral* reflects light. (See Table 4.3.)

mafic rock Dark-colored *igneous rock* containing mafic *minerals*, such as pyroxenes and olivines, that are rich in iron and magnesium and relatively poor in silica. (Compare *felsic rock*; *ultramafic rock*.)

magma Hot, molten *rock*.

magma chamber A *magma*-filled cavity in the *lithosphere* that forms as rising pools of magma push aside surrounding solid rock.

magmatic addition A process of continental growth in which low-density, silica-rich *rock* differentiates in Earth's *mantle* and is transported vertically to the *crust*. (Compare *accretion*.)

magmatic differentiation A process by which *rocks* of varying composition arise from a uniform parent *magma* as various *minerals* are withdrawn from it by *fractional crystallization* as it cools, changing its composition.

magnetic anomaly One of a pattern of long, narrow bands on the seafloor, almost perfectly symmetrical with respect to the crest of a mid-ocean ridge, in which the intensity of the *magnetic field* alternates between high and low values.

magnetic field The contours of magnetic force that surround a magnetized body or electric current.

magnetic time scale The detailed history of Earth's *magnetic field* reversals as determined by measuring the thermoremanent magnetization of *rock* samples.

magnitude scale A scale for estimating the size of an *earthquake* using the logarithm of the largest ground motion registered by a *seismograph* (Richter magnitude) or the logarithm of the area of fault rupture times fault slip (moment magnitude).

mantle The region between the *crust* and the *core* composed primarily of *ultramafic rocks*, constituting almost two-thirds of Earth's mass.

mass extinction A short interval during which an unusually large proportion of a particular group of organisms (such as species, genus, or family) living at the time disappear from the *geologic record*.

meander A curve or bend in a *stream* that develops as the stream erodes the outer bank of a bend and deposits *sediment* against the inner bank; typical of streams flowing on low slopes in plains or lowlands, where *channels* typically cut through unconsolidated sediments or easily eroded bedrock.

metamorphic facies Groupings of *metamorphic rocks* of various *mineral* compositions formed under different grades of metamorphism from different parent rocks.

metamorphic rock A *rock* formed by the transformation of pre-existing solid rock under the influence of high pressures and temperatures.

metasomatism Change in a *rock's* composition by fluid transport of chemical substances into or out of the rock.

meteoric waters Rain, snow, or other forms of water derived from the atmosphere.

meteorite A *bolide* from outer space that strikes Earth; most commonly a tiny piece of an *asteroid* ejected from its orbit during a collision with another asteroid.

microbe A single-celled microorganism.

microfossil A trace of an individual *microbe* preserved in the *geologic record*.

Milankovitch cycle A periodic variation in Earth's movement through the solar system that affects the amount of solar energy received at Earth's surface. These cycles include the eccentricity of Earth's orbit; the tilt of Earth's axis of rotation; and the precession of the rotation axis.

mineral A naturally occurring, solid crystalline substance, usually inorganic, with a specific chemical composition.

mineralogy (1) The branch of geology that studies the composition, structure, appearance, stability, occurrence, and associations of *minerals*. (2) The relative proportions of a *rock's* constituent minerals.

Mohs scale of hardness An ascending scale of mineral *hardness* based on the ability of one *mineral* to scratch another. (See Table 4.2.)

moraine An accumulation of rocky, sandy, and clayey material carried by glacial ice and deposited as *till*.

mud A fine-grained *clastic sediment*, mixed with water, in which most of the particles are less than 0.062 mm in diameter.

mudstone A blocky, poorly bedded, fine-grained *sedimentary rock* produced by the *lithification* of *mud*.

natural gas Methane gas (CH_4), the simplest hydrocarbon.

natural hazard An event produced by natural processes that has the potential to kill people and damage buildings and other man-made structures.

natural levee A ridge of coarse material that confines a *stream* within its banks between floods, even when water levels are high.

natural resource (1) A supply of energy, water, and/or materials that is available from the natural environment. (2) The entire amount of a given material, including the amount that may become available for use in the future, including *reserves*, plus known but currently unrecoverable supplies, plus undiscovered supplies that geologists think may eventually be found. (Compare *reserve*.)

natural selection The process by which populations of organisms adapt to new environments.

nebular hypothesis The idea that the solar system originated from a diffuse, slowly rotating cloud of gas and fine dust (a "nebula") that contracted under the force of gravity and eventually evolved into the Sun and planets.

negative feedback A process within a system in which one action produces another (the feedback) that reduces the original action and tends to stabilize the system. (Compare *positive feedback*.)

nonrenewable resource A *natural resource* that is produced by natural processes at a rate much slower than the rate at which it is being used up; for example, *fossil fuels.* (Compare *renewable resource.*)

normal fault A *dip-slip fault* on which the *rocks* above the fault plane have moved downward relative to the rocks below the fault plane, extending the structure horizontally.

nuclear energy Energy produced by radioactive fission, which can be used to heat steam and drive turbines to create electricity.

obsidian A dense, dark, glassy volcanic *rock,* usually of felsic composition.

oil shale A fine-grained, clay-rich *sedimentary rock* containing relatively large amounts of organic matter.

oil trap A geologic structure capped by an impermeable barrier that blocks the upward migration of *crude oil* or *natural gas,* allowing them to collect beneath the barrier.

oil window The limited range of pressures and temperatures, usually found at depths between about 2 and 5 km, at which *crude oil* forms from buried organic matter.

ore A rich *mineral* deposit from which valuable metals can be recovered profitably.

original horizontality, principle of See *principle of original horizontality.*

orogeny Mountain building by tectonic forces, particularly through the *folding* and *faulting* of *rock* layers, often with accompanying volcanism.

outer core A dense liquid zone, composed principally of iron and nickel, below Earth's *mantle* that surrounds its solid *inner core.*

P wave The first type of *seismic wave* to arrive at a *seismograph* from the *focus* of an *earthquake;* a type of compressional wave.

Pangaea A supercontinent that coalesced in the late Paleozoic era and comprised all the present continents, then began to break up in the Mesozoic era.

partial melting Incomplete melting of a *rock* that occurs because the *minerals* that compose it melt at different temperatures.

pegmatite An extremely coarse-grained *granite* formed from the last material in a large igneous intrusion to crystallize, and often containing rich concentrations of rare *minerals.*

peridotite A coarse-grained *ultramafic intrusive igneous rock* composed primarily of olivine with smaller amounts of pyroxene; the dominant rock in Earth's *mantle* and the source rock of *basaltic magmas.*

period A division of geologic time representing one subdivision of an *era.*

permeability The ability of a solid to allow fluids to pass through it.

phase change A change in a material's state (e.g., from solid to liquid, or from one mineral structure to another) without a change in composition, usually due to a change in pressure and/or temperature.

phyllite A *foliated rock* that is intermediate in metamorphic grade between *slate* and *schist,* containing small *crystals* of mica and chlorite that give it a more or less glossy sheen.

physical weathering *Weathering* in which solid *rock* is fragmented by mechanical processes that do not change its chemical composition. (Compare *chemical weathering.*)

plate tectonic system The *geosystem* that includes all the components of the *Earth system,* and all the interactions among those components, needed to describe the creation, destruction, and movements of Earth's lithospheric plates.

plate tectonics The theory that describes and explains the creation and destruction of lithospheric plates and their movement over Earth's surface.

playa lake A permanent or temporary lake in an arid mountain valley or basin, where dissolved minerals are often concentrated and precipitated as the water evaporates.

pluton A large igneous intrusion, ranging in size from a cubic kilometer to hundreds of cubic kilometers, formed deep in the *crust.*

point bar A curved sandbar deposited along the inner bank of a *stream,* where the current is slower.

polymorph One of two or more alternative possible *crystal* structures for a single chemical compound; for example, the *minerals* quartz and cristobalite are polymorphs of silica (SiO_2).

porosity The percentage of a *rock's* volume consisting of open pores between particles.

porphyry An *igneous rock* of mixed *texture* in which large *crystals* (phenocrysts) "float" in a predominantly fine-grained matrix.

positive feedback A process within a system in which one action produces another (the feedback) that enhances the original action and tends to destabilize the system. (Compare *negative feedback.*)

potable Pertaining to water that tastes agreeable and is not dangerous to human health.

precipitation (1) Water that falls to Earth's surface by condensation of atmospheric water vapor in the form of rain, snow, sleet, hail, or mist. (2) The condensation of a solid from a solution during a chemical reaction.

principle of faunal succession A stratigraphic principle stating that the *sedimentary rock* strata contain *fossils* that vary in a definite sequence so that strata of the same age from different outcrops will contain the same types of fossils.

principle of original horizontality A stratigraphic principle stating that *sediments* are deposited as essentially horizontal beds.

principle of superposition A stratigraphic principle stating that each *sedimentary rock* stratum in a tectonically undisturbed sequence is younger than the one beneath it and older than the one above it.

principle of uniformitarianism A principle stating that the geologic processes we see in action today have worked in much the same way throughout geologic time; that is, "the present is key to the past."

pumice A volcanic *rock*, usually *rhyolitic* in composition, containing numerous cavities (vesicles) that remain after trapped gas has escaped from solidifying *lava*.

pyroclast A *rock* fragment ejected into the air by a *volcanic eruption*.

pyroclastic flow Hot cloud of hot ash, dust, and gases ejected by a *volcanic eruption* that rolls downhill at high speeds.

rain shadow An area of low rainfall on the leeward slope of a mountain range.

recharge The *infiltration* of water into any subsurface rock formation.

recurrence interval The average time between large *earthquakes* at a particular location; according to the elastic rebound theory, the time required to accumulate the strain that will be released by fault slipping in a future earthquake.

reef A moundlike or ridgelike organic structure constructed of the carbonate skeletons and shells of millions of organisms.

regional metamorphism Metamorphism caused by high pressures and temperatures that extend over large regions, typical of plate boundaries where two continents converge. (Compare *contact metamorphism*.)

relative age The age of one geologic event in relation to another. (Compare *absolute age*.)

relative humidity The amount of water vapor in the air, expressed as a percentage of the total amount of water the air could hold at that temperature if it were saturated.

renewable resource A *natural resource* that is produced by natural processes at a rate rapid enough to compensate for the rate at which it is being used up; for example, wood. (Compare *nonrenewable resource*.)

reserve The supply of a *natural resource* that has already been discovered and can be exploited economically and legally at the present time. (Compare *natural resource*.)

reservoir (1) A place where water is stored. (2) A *geochemical reservoir*.

residence time The average time that an *atom* of a particular element spends in a *geochemical reservoir*.

resource See *natural resource*.

rhyolite A light brown to gray, fine-grained *extrusive igneous rock* with a felsic composition. The extrusive equivalent of *granite*.

rift basin A *sedimentary basin* that develops at a *divergent boundary* at an early stage of plate separation as the stretching, thinning, and heating of the underlying continental lithosphere occur. (Compare *thermal subsidence basin*.)

ripple A *sedimentary structure* consisting of a very small *dune* whose long dimension is at right angles to a current.

risk The damage that is likely to be caused when a *natural hazard* strikes a particular location, usually measured in lost lives and dollars.

river A major branch of a *stream* system.

rock A naturally occurring solid aggregate of *minerals* or, in some cases, nonmineral solid matter.

rock cycle The set of geologic processes that convert *rocks* of each of the three major types—*igneous, sedimentary,* and *metamorphic*—into the other two types.

Rodinia A supercontinent older than *Pangaea* that formed about 1.1 billion years ago and began to break up about 750 million years ago.

runoff The sum of all *precipitation* that flows over Earth's surface, gradually collecting into *streams* and *rivers*.

S wave The second type of *seismic wave* to arrive at a *seismograph* from the *focus* of an *earthquake*. S waves are shear waves that cannot travel through liquids or gases.

salinity The total amount of dissolved substances in a given volume of water.

saltation The transportation of *sand* or smaller *sediment* particles by currents in such a manner that they move along in a series of short intermittent jumps.

sand A medium-grained *clastic sediment*, consisting of particles ranging from 0.062 to 2 mm in diameter.

sandstone A *sedimentary rock* that is the *lithified* equivalent of *sand*.

saturated zone The level below the *groundwater table*, in which the pores of the *soil* or *rock* are completely filled with water. (Compare *unsaturated zone*.)

schist An intermediate-grade *metamorphic rock* characterized by coarse, wavy *foliation* known as schistosity.

scientific method A general procedure, based on systematic observations and experiments, by which scientists propose and test hypotheses that explain some aspect of how the physical universe works.

seafloor spreading The mechanism by which new oceanic *crust* is formed at a *spreading center* at the crest of a mid-ocean ridge. As two plates move apart, hot *magma* wells up into the rift between them to form new seafloor, which spreads laterally away from the rift as it is replaced by ever newer crust in a continuing process of plate creation.

sediment Material deposited on Earth's surface by physical processes (wind, water, and ice), chemical processes (*precipitation* from oceans, lakes, and rivers), or biological processes (organisms, living and dead).

sedimentary basin A region of considerable extent (at least 10,000 km^2) where a combination of subsidence and sediment deposition has formed thick accumulations of *sediment* and *sedimentary rock*.

sedimentary environment A geographic location characterized by a particular combination of *climate* conditions and physical, chemical, and biological processes.

sedimentary rock A *rock* formed by the burial and *diagenesis* of layers of *sediment*.

sedimentary structure Any kind of *bedding* or other feature (such as *cross-bedding, graded bedding,* or *ripples*) formed at the time of *sediment* deposition.

seismic hazard The intensity of shaking and ground disruption by *earthquakes* that can be expected over the long term at some specified location.

seismic risk The *earthquake* damage that can be expected over the long term at some specified location.

seismic tomography A technique that uses differences in the velocities of *seismic waves* produced by *earthquakes* and recorded on *seismographs* around the world to construct three-dimensional images of Earth's interior.

seismic wave A ground vibration produced by an *earthquake* or explosion. (See also *P wave; S wave; surface wave.*)

seismicity map A map that shows the *epicenters* of *earthquakes* recorded around the world or in a particular region over a given period of time.

seismograph An instrument that records the *seismic waves* generated by *earthquakes*.

shadow zone (1) A zone beyond 105° from the *focus* of an *earthquake* where *S waves* are not recorded because they are not transmitted through Earth's liquid *outer core*. (2) A zone at angular distances between 105° and 142° from the focus of an earthquake where *P waves* are not recorded because they are refracted downward into the core and emerge at greater distances after the delay caused by their detour through the core.

shale A fine-grained *sedimentary rock* composed of *silt* plus a significant component of *clay*, which causes it to break readily along *bedding* planes.

shearing force A force that pushes two sides of a body in opposite horizontal directions. (Compare *compressive force; tensional force.*)

shield volcano A broad, shield-shaped *volcano* many tens of kilometers in circumference and more than 2 km high built by successive flows of fluid *basaltic lava* from a central vent.

shoreline The line where the ocean surface intersects the land.

siliciclastic sediment *Clastic sediment* produced by the *weathering* of *rocks* composed largely of silicate *minerals*.

siliciclastic sedimentary environment A *sedimentary environment* dominated by *siliciclastic sediments*.

sill A sheetlike igneous intrusion formed by the injection of *magma* between parallel layers of bedded *country rock*. (Compare *dike.*)

silt A fine-grained *clastic sediment* consisting of particles between 0.0039 and 0.062 mm in diameter.

siltstone A *sedimentary rock* that is the *lithified* equivalent of *silt*.

sinkhole A small, steep depression in the land surface formed when the roof of a *limestone* cave collapses suddenly.

slate A fine-grained *foliated rock* that is easily split into thin sheets, formed primarily by low-grade metamorphism of *shale*.

soil An unconsolidated or loose mixture of materials at the surface of the Earth, including fragments of *rocks* and *minerals*, organic matter, and water.

solar energy Energy derived from the Sun.

solar nebula A disk of gas and dust that surrounded the proto-Sun and from which, according to the *nebular hypothesis*, the planets of the solar system formed.

sorting The tendency for variations in current velocity to segregate *sediments* according to their size.

specific gravity The weight of a substance divided by the weight of an equal volume of pure water at 4°C. (Compare *density.*)

spit A narrow extension of a *beach* formed by *longshore currents* that carry *sand* to the downcurrent end of the beach.

spreading center A *divergent boundary* marked by a rift at the crest of a mid-ocean ridge where new oceanic *crust* is formed by *seafloor spreading*.

stabilization wedge A change in human energy use that can offset the projected growth of carbon emissions by 1 gigaton per year in the next 50 years. Approximately 7 stabilization wedges will have to be implemented to keep global carbon emissions from growing beyond the current value of 8 gigatons per year.

stock A *pluton* less than 100 km^2 in area.

storm surge A dome of seawater formed by a *hurricane* that rises above the level of the surrounding ocean surface.

stratigraphic succession A chronologically ordered set of *rock* strata.

stratigraphy The description, correlation, and classification of strata in *sedimentary rock*.

stratosphere The cold, dry layer of the atmosphere above the *troposphere* that extends from about 11 to 50 km in altitude. (Compare *troposphere*.)

stratospheric ozone depletion The thinning of the layer of ozone in the *stratosphere*, which shields Earth from cell-damaging ultraviolet radiation, caused by chemical reactions involving *anthropogenic* compounds such as chlorofluorocarbons (CFCs).

stratovolcano A concave-shaped composite *volcano* formed from alternating layers of *lava* flows and beds of *pyroclasts*.

streak The *color* of the fine deposit of mineral powder left on an abrasive surface when a *mineral* is scraped across it.

stream Any body of water, large or small, that flows over the land surface.

stream valley The entire area between the tops of the slopes on both sides of a *stream*.

strike The compass direction of a *rock* layer, fault, or other planar structure where it intersects with a horizontal surface. (Compare *dip*.)

strike-slip fault A fault on which movement of the blocks has been horizontal, parallel to the *strike* of the fault plane.

subduction The sinking of oceanic *lithosphere* beneath overriding oceanic or continental lithosphere at a *convergent boundary*.

superposition, principle of See *principle of superposition*.

surface wave A type of *seismic wave* that travels around Earth's surface from the *focus* of an *earthquake* and arrives at a *seismograph* later than *S* waves.

suspended load The material temporarily or permanently suspended in the flow of a current. (Compare *bed load*.)

sustainable development Development that meets the needs of the present without compromising the ability of future generations to meet their own needs.

syncline A large downfolding of layered *rocks* into a trough. (Compare *anticline*.)

tar sand A deposit of *sand* or *sandstone* containing a tarlike hydrocarbon called natural bitumen.

tensional force A force that stretches a body and tends to pull it apart. (Compare *compressive force; shearing force*.)

texture The sizes and shapes of a *rock's mineral crystals* or grains and the way they are put together.

thermal subsidence basin A *sedimentary basin* that develops at a *divergent boundary* in the later stages of plate separation as *lithosphere* that was thinned and heated during the earlier rifting stage cools, leading to an increase in *density*, which in turn leads to subsidence below sea level. (Compare *rift basin*.)

thermohaline circulation A global three-dimensional pattern of oceanic circulation driven by differences in the temperature and salinity—and therefore in the density—of ocean waters.

thrust fault A low-angled reverse fault—one with a *dip* of less than 45°—on which the overlying rocks have moved upward relative to rocks below the fault, compressing the structure horizontally.

tides The twice-daily rise and fall of ocean water caused by gravitational attraction between Earth and the Moon.

till An unstratified and poorly sorted *sediment*, containing a wide range of particle sizes, that is deposited directly by melting glaciers.

tillite The *lithified* equivalent of *till*.

topography The general configuration of varying heights that gives shape to Earth's surface, which is measured with respect to sea level.

transform fault A plate boundary at which the plates slide horizontally past each other, forming a *strike-slip fault*, and *lithosphere* is neither created nor destroyed.

troposphere The lowest layer of the atmosphere, which has an average thickness of about 11 km and contains most of the atmosphere's mass. (Compare *stratosphere*.)

tsunami A sea wave, generated by an *earthquake* that lifts the seafloor, that propagates across the ocean and increases in size when it reaches the shore.

tuff A volcanic *rock* formed by the *lithification* of small *pyroclasts*. (Compare *breccia*.)

turbidity current A turbulent submarine avalanche of *sediment* and water that moves downslope; typical of continental slopes.

turbulent flow A flow in which streamlines mix, cross, and form swirls and eddies. (Compare *laminar flow*.)

twentieth-century warming The rise in Earth's mean surface temperature by about 0.6°C between the end of the nineteenth century and the beginning of the twenty-first century.

ultramafic rock *Igneous rock* consisting primarily of mafic *minerals* and containing less than 10 percent feldspar. (Compare *felsic rock, mafic rock.*)

unconformity A surface between two *rock* layers in a *stratigraphic succession* that were laid down with a time gap between them.

uniformitarianism, principle of See *principle of uniformitarianism.*

unsaturated zone The level above the *groundwater table*, in which the pores of the *soil* or *rock* are not completely filled with water. (Compare *saturated zone.*)

U-shaped valley A deep valley with steep upper walls that grade into a flat floor; the typical shape of a valley eroded by a glacier.

valley glacier A river of ice that forms in the cold heights of a mountain range and flows down a valley. (Compare *continental glacier.*)

vein A sheetlike deposit of *ore precipitated* from a *hydrothermal solution* in fractures or *joints* of *country rock*.

viscosity A measure of a fluid's resistance to flow.

volcanic ash Extremely small fragments, usually of glass, that form when escaping gases force a fine spray of *magma* from a *volcano;* the finest *pyroclasts.*

volcanic eruption A geologic event that deposits molten rock from a planet's interior onto its surface and spews gases and solid materials into its atmosphere.

volcano A structure consisting of a *magma chamber,* a plumbing system that transports *magma* to Earth's surface; one or more vents at the surface, and often a hill or mountain constructed from the accumulation of *lava* and *pyroclasts* erupted from the vents.

wave-cut terrace A level surface, formed at a rocky *shoreline* by wave *erosion* beneath the surf zone, that may be visible at low *tides.*

weathering The chemical and physical processes that break up and decompose *rocks* into *sediment* particles.

Wilson cycle The sequence of tectonic events in the formation and breakup of supercontinents that comprises (1) rifting during the breakup of a supercontinent, (2) passive margin cooling and *sediment* accumulation during *seafloor spreading* and ocean opening, (3) *magmatic addition* and *accretion* during *subduction* and ocean closure, and (4) *orogeny* during the continent-continent collision that forms the next supercontinent.

INDEX

Boldface indicates a glossary term.   *Italics* indicates a figure.   *t* indicates a table.

Aa, *129,* 129–130
Ablation, 268, **268,** *270*
Abrasion, 149–150, *150*
Absolute ages, 200
   and geologic time scale, *211,* 211–212, 223
Abyssal plain, *335,* 335–336
Acadian orogeny, 237, *237*
Acasta gneiss, 238, 239, *239*
Accretion, 234, *234*
Accumulation, 267–268, *268, 270*
Acid rain, 25, **25**–26
   from coal, 10, 387, 389
   from volcanoes, 247
Acid tests, 87, *87*
Active margins, 233
Adamantine luster, 90, 90*t*
Adelaide River, *316*
Adelman, Morris, 386
Aerosols, 259
Aftershocks, 348, **348**–349
Agassiz, Louis, 272–273
Ages, absolute vs. relative, 200
Aggregates, 93
Agriculture. *See also* Irrigation
   and deforestation, 409
   and desertification, 337–338
   rise of, 267, 400
   and water contamination, 299, *299*
Air. *See* Atmosphere
Alaska earthquake, *343,* 343–344, 365, 368
Albedo, **260**–261
   and climate feedback, 262, 266, 278
Aleutian Islands, 61
Allegheny Plateau, *230,* 231
Allegheny River, *307*
Alluvial environments, *153,* 154
Alluvial fans, *319,* **319,** 339
Alpine fault, 365
Alps Mountains
   and compressive tectonics, 182
   orogeny of, 235–236
Aluminum, recycling of, 10
Amazon River, 289, 289*t,* 323
Amethyst, *81*
Ammonites, *200*
Amorphous substances, 77
Amphiboles
   cleavage of, 89, 89–90
   as double-chain silicates, 84, *85*
   in igneous rocks, 95, 95*t,* 112*t,* 113
Amphibolite, as granoblastic rock, 191, 192*t*
Amphibolite facies, **194,** 194*t*
Andes Mountains, *59,* 62, 234
Andesite, *113,* **114,** 114*t*
Andesitic lavas, 130
Angular unconformities, *199,* 203–204, *204, 205,* 207
   as structural traps, 383, *383*
Anhydrite
   as evaporite mineral, 166
   in sedimentary rocks, 163*t*
   as sulfate, 87
Anions, *79,* 80
   sizes of, 81, *81*
Antarctica
   collapse of ice shelves on, 405–406, *406*
   continental glacier of, 256, *257*

   deserts on, 336
   ice cores from, 265, *265, 266*
Anthracite, 388, *388*
Anthropocene epoch, *408*–409
Anthropogenic changes, **24**–25. *See also* Human activities
Anticlines, **179,** *179*
   as structural traps, 383, *383*
Apatite
   hardness of, 88*t*
   and isotopic dating, 210*t*
   in sedimentary rocks, 163*t*
Apollo [spacecraft] missions, 215, 222
Appalachian Mountains, 62, 69
   and compressive tectonics, 182
   and lead-zinc deposits, 102
   orogeny of, *236,* 237
   and passive margins, 233
   as tectonic province, *230,* 230–231
Aquicludes, *293,* **293**–294, *294*
Aquifers, **14**–15, *15, 293,* **293**–294
Aragonite
   in biological sediments, 148
   as carbonate sediment, 163–166, 163*t*
   as hydrothermal deposit, 301, *301*
   luster of, 90
Archaea domain, 240, *240*
Archean eon, *211,* 212, 223, 224, 239
Arctic National Wildlife Refuge, 387
Arenal volcano, *131*
Arêtes, **269,** *269*
Argillite, 192*t*
Argon, and isotopic dating, 210*t*
Artesian flows, *293,* **293**–294, *294*
Asama volcano, *132*
Asbestos
   crystal habit of, *92,* 93
   and disease, 93
Ash, volcanic, *111,* **132,** *132*
Ash-flow deposits, 136
Asymmetrical folds, *180, 180*
Asteroid belt, 218, *218*
Asteroids, 218. *See also* Bolides
   and Late Heavy Bombardment, 224
Asthenosphere, **45,** *359,* 360
   and Earth system, *43*
   and plate tectonics, 52
Atlantis myth, 136
Atmosphere. *See also* Climate change; Global warming; Wind belts, prevailing
   and calcium cycle, 397
   and carbon cycle, 397, *398*
   carbon dioxide in, 381–382, *400,* 400–403, *402*
   and climate system, *253,* 253–255, *396*
   composition of, 254–255
   and Earth system, 42, *43,* 44
   evolution of, 46–47, 212, 241–242
   formation of, 220–221, *221*
   layers of, 253, *253*
   viscosity of, 309
   as water reservoir, 283, 283*t*
Atomic mass, 78
Atomic number, 78
Atoms, 77–78, *78*
Atwood, Robert, 343
Australian Desert, 336, *336*
Axial planes, *179, 179*

Backshore, *331,* 332
Backwash, 327, *327*
Bacteria domain, 240, *240*
   genetic engineering of, 411
Bahamas, as carbonate platform, *164, 164*
Baltica, *236,* 237
Bam earthquake, 368
Banded iron formations, 242, *243*
Barrier islands, **333,** *334*
Basalt, *113,* **114,** 114*t*
   as extrusive igneous rock, 95, *95*
   flood, 136, *136*
   as igneous rock, 94
   and magmatic differentiation, 118–119
   metamorphism of, 192*t,* 195
   texture of, 109, *109, 111,* 131, *131*
Basaltic lavas, 128–130, *129*
Basin and Range province, 182, *231,* 232, 363, 365–366
Batholiths, *119,* **120**
Beaches, *153,* 154, *325, 331,* **331**–332, *332*
   and currents, *307*
Beagle [research vessel], 203
Becquerel, Henri, 209
Bed load, **310,** *310*
Bedding, **96**–97, *156,* **156**–157
Bedding sequences, *158,* **158**–159
   on Mars, 227, *227*
Bedrock, 173
Beds, 156
Beryl
   cleavage of, 90
   gem varieties of, 91, *91*
Bhuj earthquake, 365, 368
Big Bang, 216
Bioclastic sediments, *147*–148, *148*
Biodiversity, 409
   and wetlands, 290
Biofuels, 7, 9, **394**–395, *395*
   and Earth system management, 410, 411, 412*t*
Biogeochemical cycles, 396
Biological sedimentary environments, **155**–156, 155*t*
Biological sediments, 96, *96,* **147**–148, *148*
   and rocks, 163–166, 163*t,* 167–168
Biomass, 5–6, 299
   and carbon economy, *381, 381*
   as energy resource, 394–395
Biosphere, **23,** 23–24
   and carbon cycle, 397, *398*
   and climate system, *253,* 259–260, *396*
   and Earth system, 42, *43,* 44
   as water reservoir, 283, 283*t*
Biotite. *See also* Micas
   as index mineral, *193*
   and isotopic dating, 210*t*
   as rock-forming mineral, 93
   as sheet silicate, *85*
Bioturbation, 157–**158,** *158*
Bitumen, natural, 390
Bituminous coal, 388, *388*
Blackfoot River, *318*
Blue Ridge province, *230,* 231, 237
Blueschist, **195**
   as metamorphic rock, *97*

Blueschist facies, *194*
Bolides, *16,* 16–17, **21**–22, *22,* 47. *See also* Asteroids; Comets; Meteorites
   and mass extinctions, *245,* 245–246, *246*
Bombs, volcanic, *111,* **132,** *132*
Borates, 155*t,* 156, 167
Borax, 340
Bottomset beds, 323, *323*
Boulders, 161, 161*t*
Bowen's reaction series, 117, *117*
Brahmaputra River, 289*t*
Braided streams, *271, 314,* **319,** *319*
Breakers, 326, *327*
Breccias, **132,** *132*
British thermal units (Btu), 7
Brittle behavior, **175**–176
Bryce Canyon National Park, stratigraphy of, 207, *208*
Btu. *See* British thermal units (Btu)
Building codes, 367, **373**
Buoyancy, 40, *40*
Burial, *145,* 146, 159, *159*
Burial metamorphism, **187,** 188
Bushveldt Complex, *103*

Calcite
   in biological sediments, 96, 148
   and calcium cycle, 397
   as carbonate, 84–85, *85*
   as carbonate sediment, 163–166, 163*t*
   as cement, *160*
   cleavage of, 89, *89*
   hardness of, 88*t*
   as hydrothermal deposit, 301, *301*
   in metamorphic rocks, 98
   as mineral, 77, *77*
   and oceanic chemical balance, 150
   in sedimentary rocks, 95*t,* 97
Calcium cycle, 397, *397*
Calderas, **133,** *135, 135*
   collapses of, *137,* 138–139
Caledonian orogeny, 237, *237*
Cambrian explosion, *243,* 243–245, *244,* **244,** *245*
Cambrian period, *206, 207, 208, 209, 211,* 212, 244
Campbell, Colin, 386
Canadian Shield, *228,* 228–229
   as craton, 233, *233*
   orogenies in, 237
Cape Ann earthquake, 365
Cape Cod National Seashore, 273, *333, 334, 334*
Carbon
   isotopes of, 78
   and isotopic dating, 210*t,* 211, 273
Carbon cycle, 260, 379, 381–382, **395**–400, *398*
   engineering of, 410–411
   and human activities, 398–400, *399*
Carbon dioxide
   atmospheric, 254
   and biosphere, 260
   and carbon economy, *381, 381*
   and climate history, 265, *266*
   and erosion of limestone, 297

I-1

I-2 INDEX

Carbon dioxide (*continued*)
   and global warming, 26, 247, *264*, 264–265, *400*, 400–401, *402*
   as greenhouse gas, 261, *261*
   and ice ages, 278
   stabilizing, 411, 412*t*, 413
   from volcanoes, 221, *221*, 247
Carbon economy, 380–382, **381**, *381*
Carbon-neutral fuels, 394–395
Carbon sequestration, **411**, 412*t*
Carbonate environments, 155–156, 155*t*
Carbonate ions, 86, *86*
Carbonate platforms, 163, *164*
Carbonate rocks, **163**–166, 163*t*, *164*, *165*
Carbonate sediments, **163**–166, 163*t*
Carbonates, 83*t*, **84**–86, *86*
   as rock-forming minerals, 97
   in sedimentary environments, 155*t*
Carboniferous period, 206, *206*, *208*, *211*
   ice ages during, 276
Carlsbad Caverns, 297, *297*
Cascade Mountains, 62, 234, 286
Casita volcano, 21
Casler, Michael, *395*
Cation substitution, 81
Cations, 79, 80
   sizes of, 81, *81*
Cavendish, Henry, 39
Caves, 297, *297*, *298*, 298
Cementation, 96, *159*, **160**, *160*
   in glaciers, 267
Cenozoic era, 206, *206*, *211*, 212
Central eruptions, 133–135, *134*, *135*
Central vents, 133, *134*
Ceres [asteroid], 218
Cernan, Gene, 215
Cerro Negro volcano, *134*
CFCs. *See* Chlorofluorocarbons (CFCs)
Chalcedony, 163*t*
Chalcocite
   as ore mineral, 101, *102*
Chalcopyrite, as ore mineral, 102, *102*
Challenger Deep, 37
Challenger [research vessel], 51
Channeled Scablands, *272*, 272
Channels, **306**, *314–315*
   patterns of, *318*, 318–319, *319*
Charleston earthquake, 365
Chemical bonds, **80**
   and cleavage, 89
   and hardness, 88
   and luster, 90
Chemical contaminants, 299, *299*
Chemical reactions, **78**–80
Chemical sedimentary environments, **155**–156, 155*t*
Chemical sediments, 96, *96*, *147*, 147
   and rocks, 163*t*, 166–168
Chemical weathering, **146**, *146*
Chernobyl accident, 391
Chert, 163*t*, *165*, **167**
Chicxulub crater, *245*, 246
Chile earthquake, 365
Chitina River, *319*
Chlorite
   and foliation, 189
   as index mineral, 193, *193*
Chlorofluorocarbons (CFCs), 26
Chromite, concentrations of, 104
Chromium, ores of, 102, *103*
Chrysotile, crystal habit of, *92*
Cinder cones, 133, *134*
Cinnabar, as ore mineral, 101, *101*
Cirques, *269*, 269–270
Clark, William, 313, 322
Clastic sediments, **147**
Clay [minerals]
   in sedimentary rocks, 95*t*, 97
   as sheet silicates, 84

Clay [particle size], 161*t*, **163**
   varved, *271*, 272
Claystone, 161*t*, **163**
Clean Air Act, 25
Cleavage, 88, *89*, **89**–90, 92*t*
   and foliation, *189*
Climate, **44**, 252
   and exhumation, 196
   history of, 13–14
   and hydrology, 285–290
   and rock cycle, 98
   and streams, 306
Climate change, 400–403
   and desertification, 337
   effects of, 403–409, 403*t*
   and mass extinctions, 207
   and water resources, 14, 288
Climate models, 402, **402**–403, *403*
Climate system, 43, **44, 252**, *253*, *259*
   components of, 252–260, 396
Climate variation, 262–267
   and glacial cycles, 266–267
   history of, 265–266
   regional, 263–264
   and twentieth-century warming, *264*, 264–265
Coal, 6, 7, 10, 143, 145, 387–390
   and carbon economy, 381, *381*
   classification of, 77, 93, 163*t*
   formation of, *159*, 167–168
   mining of, *7*, 389, 389–390
   resources of, 382, *382*, 388, *389*
   and water contamination, 299
Coal beds, **387**, *388*
Coarse-grained sediments and rocks, 94, 160–162, 161*t*
   and sediment transport, *314*–315
Coastal plain, 231, *232*
Coastal processes, 325–331
Coastlines, *307*, **307**, 325. *See also* Shorelines
Cobbles, 161–162, 161*t*
Coesite, 188
Color, mineral, **90**–91, *91*, 92*t*
Colorado National Monument, *35*
Colorado Plateau, *231*, 232
   stratigraphy of, 207, *208*
Colorado River, 282, 289, *322*
Columbia River basalts, 125, 136, *136*
Columbia supercontinent, 238
Columbus, Christopher, 36
Comets, 218, 221
Compaction, 96, *159*, **160**
Compressional waves. *See* P waves
Compressive forces, *175*, **176**
   and fault mechanisms, *355*, 355
Compressive tectonics, *181*, 182–183, *183*
Computer simulations
   of climate system, 262, *402*, 402–403, *403*
   of droughts, 282
   of Heavy Bombardment, 224
   of Moon formation, *219*
   of natural systems, 33, 46, 159
Concentration factors, 100
Conchoidal fracture, 90
Cone of depression, 295, *295*
Confined aquifers, 293, *293*
Confining pressure, 175, 186
Conglomerates, *161*, **161**, 161*t*
   formation of, *159*
Congo River, 289*t*
Conservation, 8
Contact metamorphism, 97, **98**, *187*, **188**
   and granoblastic rocks, 191
   and metamorphic facies, *194*
Contamination. *See* Pollution
Continental drift, 52, **52**–**53**, *53*. *See also* Plate tectonics
Continental environments, 152, *153*, 154

Continental glaciers, **256**, *257*
   and climate change, 408
   and ice ages, 266, *266*, 273–274
Continental-margin/slope environments, *153*, 154
Continental rises, *335*, 335–336
Continental-shelf environments, *153*, 154
Continental shelves, 151, **152**, *335*, 335–336
   and ice ages, 266
Continental slopes, *335*, 335–336
Continents, evolution of, 228–239, *239*
   around the world, 232–233
   and craton formation, 238–239
   growth of, 233–234
   modification of, 235–238
   North America, 228–232
Convection
   and geodynamo, 46
   and heat flow, 220
   in mantle, 361, *362*
   and plate tectonics, **44**, 45, 53, 72–74, *73*, *362*, 363
Convergent plate boundaries, *58–59*, **59**, 60–62. *See also* Subduction zones
   earthquakes along, *364*, 365
Copper
   and metasomatism, 186
   mining for, *103*
   ores of, 100, 101, 102, *102*, 103–104, 229
   recycling of, 10–11
Coral, 164, *164*. *See also* Carbonate platforms; Reefs
   and ocean acidification, *399*, 399–400
   as sedimentary rock, 96, *96*
Cordillera, 230, *231*, 231–232
   and accretion, *234*
   as orogenic belt, 233, *233*
Core, 38, **39**–40
   discovery of, *357*, 357–358
   formation of, 219–220, *220*
   layering and composition of, *359*, 361
Core-mantle boundary, 40, **360**–361, *362*
Coronae, *225*, 225–226
Corundum
   gem varieties of, 91, *91*
   hardness of, 88*t*
Country rock, **111**
   intrusions into, *119*, 119–122, *120*, *121*
Covalent bonds, 80
   and cleavage, 89
   and hardness, 88
   and luster, 90
Covelite, as ore mineral, 100, 101
Crater Lake, *135*
Craters, volcanic, **133**, *135*
Cratons, **232**–233, *233*
   formation of, 238–239
Cretaceous period, 206, 207, *208*, *211*, 212
Cretaceous-Tertiary boundary, *245*, 245
Crevasses, *270*
Cristobalite, as polymorph, 83
Cross-bedding, **156**, *156*
Cross-cutting relationships, 204, *205*, 206
   and cratering, 222
Cross-sections, geologic, **174**, *174*
Crude oil, **383**. *See also* Petroleum
Crust, 38, **40**
   age of continental, 68, 228
   age of oceanic, 68, *68*
   deformation of, 175–176
   elemental abundance of, 84

initial formation of, 219–220, *220*
   layering and composition of, 358–359, *359*
Crutzen, Paul, 26, 408
Cryosphere
   and climate change, *404*, 404–406, *405*, *406*
   and climate system, 253, *255*, 255–258, *256*, *257*, *258*
   and Earth system, 42, *43*, 44
   and ice ages, 266, *266*, 274
Crystal faces, 81, *81*, *82*
Crystal habit, 92, **92**–93, 92*t*
Crystal size, 109
Crystal structures, 80–81
Crystalline substances, 77
Crystallization, **80**, 81–82
   fractional, *117*, 117–118, *118*
   and isotopic dating, 210
   laboratory studies of, 109–110
Crystals, *81*, **81**–82, *82*
Cubic structures, 80
Currents, 148–150
   air, 307–308, *309*, 309–310
   created by tides, 328–329
   created by waves, *326*, 326–328, *327*
   longshore, *327*, 327–328, *332*, 332
   oceanic, *254*, 255, 307
   physics of, *308*, 308–310
   rip, *327*
   rivers and streams, 306–307, *307*
   turbidity, 154, 156, *335*, 335
Cyanobacteria, 242
Cyclones, 17, *17*. *See also* Hurricanes

Dacite, *113*, **114**, 114*t*
Dalton, John, 77
Dams
   and flood control, 321
   and hydroelectric energy, 282, 392, *393*
   and retreat of beaches, 333
   and water reservoirs, 289, *290*
Darwin, Charles, 203
Dating techniques. *See* Isotopic dating; Stratigraphy
Daughter nuclides, 209
David [statue], *171*
De Beers Consolidated Mines, Ltd., 75–76
Decompression melting, **116**, *123*, 124
Deep-sea drilling, *66*, 66–67
Deep-sea environments, *153*, 154
Deep-sea trenches, as convergent plate boundaries, 61
Deffeyes, Ken, 386
Deforestation, 24, *24*, 408, *409*
   and global warming, 264, 400, 411
Deformation, 175–176
   styles of, 181–184
Deltas, *153*, 154, *315*, **323**, *323*, *324*
Density, **83**
   of Earth, 39, *41*, 359, *359*, 361
   of minerals, 91–92, 92*t*, 255–256, *359*
Deposition, 145, *146*, 148–150, 310–313
   in deserts, 338–340, *339*
   at shorelines, *332*, 332–334, *333*, *334*
Desalination, 285
Desert environments, *153*, 154, 336–340, *337*
   and wind, 307–308
Desert varnish, 338, *338*
Desertification, 308, **337**
Development. *See also* Economics
   on barrier islands, 334
   on floodplains, 320–321, *321*
Devonian period, *206*, *208*, *211*
Diagenesis, 145, *146*, *159*, **159**–160, *160*

and metamorphism, 188
Diamond, 76
    atomic structure of, 79, *79*, 80
    classification of, 77
    cleavage of, 89
    concentrations of, 104, 229
    hardness of, 88*t*
    marketing of, 75–76
    and metamorphism, 188
    as polymorph, 82–83, *83*
Diesel, Rudolf, 394
Dietz, Robert, 55
Differential stress, 186
Differentiation
    of Earth, **218**, 219–220, *220*
    magmatic, **117**–119, *118*
    of planets, 221–222
Dikes, *119*, 120–**121**, *121*
    and cross-cutting relationships, 204, *205*
    in oceanic crust, *122, 123,* 124
    sheeted, *122*, 124
Dinosaurs, 203
    mass extinction of, *245,* 245–246, *246*
Diorite, **113**, *113*, 114*t*
Dip, *173,* 173–174
Dip-slip faults, **176***, 177*
Directed pressure, 186, *189*
Discharge, **293**, **320**
    balancing of, with recharge, 294–295, *295*
    and flooding, 320, *320*
Disconformities, 203, *204*
Diseases
    from mineral dust, 93, 389
    and stratospheric ozone depletion, 26
    from water contamination, 299
Disseminated deposits, 102
Distributaries, *315,* **323**, *323*
Divergent plate boundaries, *58,* **59**, 60, *60, 61*
    earthquakes along, 363, *364*
Divides, *321,* **321**–322
Dolomite
    as carbonate, 86
    cleavage of, 89
    in sedimentary rocks, 95*t,* 97, 163*t,* 165–166
Dolostone, 163*t,* 165
Domains, biological, 240, *240*
Double-chain silicates, 84, *85. See also* Amphiboles
Drainage basins, *321,* **321**–322, *322*
Drainage networks, 322, *322*
Drake, Edwin L., *9,* 143
Drake Passage, 276
Droughts, 13–14, **286**, *286*
    and climate change, 407, *407*
    and groundwater table, 295, *295*
Drumlins, 271, *271*
Ductile behavior, **175**–176
Dunes, *311,* **311**–313, *312, 313*
    at beaches, *331,* 332
    cross-bedding in, 156, *156, 157, 157*
    in deserts, 339
Dust storms, 311, *311*

Earth
    age of, 68
    chemical composition of, *41,* 41–42
    density of, 39, *41,* *359,* *359,* 361
    early history of, 218–221, *220*
    history of life on, 239–247, *242*
    layers of, *38,* 38–42, 358–361, *359*
    orbital variations of, 274–276, *275*
    origin of, 46
    shape and surface of, 36–38
    as spaceship, 3–4
    as terrestrial planet, 217, *218,* 221*t*

Earth system, 4, 32, *42,* **42**–46, *43*
    engineering and management of, 409–413
Earthquakes, **18**–20, *19,* 343–378, **344**
    destructiveness of, *343,* 366–375, *367*
    elastic rebound theory of, 344–347, *345*
    fault rupture during, *347,* 347–349, *348*
    and interior of Earth, 356–363, *357, 359, 362*
    intraplate, 365
    measuring, using seismic waves, 349–356
    and plate tectonics, 363, *364,* 365
    predicting, 375–376
    reducing risks of, 371–375
    and seafloor spreading, *54,* 55, *55*
    size of, 351–355, 354*t*
    warning systems for, 373, 375–376
East African rift, 152, 182, *182*, 363
Ebb tide, 328
Eccentricity, 275, *275*
Eclogite, *97,* 188, 195
Eclogite facies, *194,* 194*t*
Ecology, 3
    and climate change, 407
Economic basement, 188
Economics. *See also* Development; Ores
    and exclusive economic zones, 12
    and mineral concentrations, 100–104
    and natural resources, 6–7, 10–11, *11,* 382
    and oxides, 86
    and water resources, 14
Efficiency of energy production, 381, *381,* 390
    and Earth system management, 409, 411, 412*t*
El Capitan Mountain, *155*
El Chichón volcano, 259
El Niño, *263,* **263**–264
El Salvador earthquake, 368
Elastic rebound theory, **344**–347, *345*
    and earthquake prediction, 375–376
Electron sharing, 79, 80
Electron transfer, 79, 80
Electrons, 78, *78*
Emerald, 91, *91*
Emergency responses, 373–374
End moraines, *270,* 271, 273
Endeavor [space shuttle], *31, 31*
Energy balance, *260,* **260**–261, *261*
Energy resources, 5–10. *See also* Coal; Fossil fuels; Hydroelectric energy; Natural Gas; Nuclear energy; Petroleum; Solar energy
    biofuels, 7, 9, 394–395, *395*
    consumption of, *6,* 7–8, *8,* 9–10, *10,* 385
    and Earth system management, 409–410
    geothermal, 303, 393–394, *394*
    production of, *385,* 385–386, *386*
    wind, 393, *393*
Enhanced greenhouse effect, 264–265
Enstatite. *See also* Pyroxenes
    as single-chain silicate, 84, *85*
Environment, **3**–4
    built *vs.* natural, 3
    and global change, 23–28
    human impact on, 379–413
Environmental damage
    by dams, 392
    by deforestation, 24, *24,* 264
    by mining, *12,* 12–13, 388–390, *389*
    by petroleum production, 386–387, *387*
    by wetlands destruction, 290

Eocene period, 206, 211
Eolian processes, **308**
Epicenters, *347,* **347**–348
    locating, *352–353*
Epidote, *81*
Epochs, **206**, *206*
Eras, **206**, *206*
Eratosthenes, 36, 38, 39
Ergs, 339
Erosion, **95**–96, *96, 145,* 146–148
    by glaciers, *268,* 268–271, *269, 270*
    by groundwater, *297,* 297–298, *298*
    at shorelines, *332,* 332–334, *333, 334*
    and unconformities, 203
    by wind, 338
Erratics, 271
Eruption clouds, *137,* 139
Ethanol, 7, 394–395
Ethics in science, 33
Eukaryota domain, 240, *240*
Euphrates River, 320
Evaporation, *284,* 284–285
Evaporite environments, 155*t,* 156
Evaporite minerals, 154, 166–167
    and playa lakes, 340
Evaporite rocks, 163*t,* **166**–167
Evaporite sediments, **166**–167
Everest, George, 31
Evolution of life, 46–48, *47,* **244**
    by natural selection, 244
    and plate tectonics, 53
    theory of, 203
Evolutionary radiation, 243
Ewing, John, 51
Ewing, Maurice "Doc," 51, 53–54, *54*
Exclusive economic zones, 12
Exhumation, **195**
    and climate, 196
Exploration
    for groundwater, 358
    for minerals, 11–13, *12*
    for natural gas, 144, 335–336, 358
    for petroleum, *8,* 8–9, *9,* 143–144, 335–336, 358
    seismic, 358
    of space, 215–216
Extinctions, 245–247. *See also* Mass extinctions
    and climate change, 28, 407
Extreme events, 16, 400
    and mass extinctions, 207
Extrusive igneous rocks, **95**, *95*
    texture of, 111, *111,* 131, *131*
Exxon Valdez [ship], 387

Fahrenheit, Daniel, 264
Failure of rocks, 344, *345*
Fatalities. *See* Natural hazards
Fault mechanisms, *355,* **355**–356, *356*
Fault slip, *345,* **346**
Fault systems, 365–366, *366*
Faults, 18, *19,* *172,* **172**, 176–178, *177*
    and cross-cutting relationships, 204, *205*
    rupture of, during earthquakes, *347, 347*–349, *348*
    as structural traps, 383, *383*
Faunal succession, principle of, **202**–203
Feedbacks, 262, 266, 276, 278, 400, 408
Feldspars
    and calcium cycle, 397
    as framework silicates, 84, *85*
    in igneous rocks, 95, 95*t,* 112*t,* 113
    in metamorphic rocks, 95*t,* 98
    in sedimentary rocks, 95*t,* 97
Felsic minerals, 112, 112*t*
Felsic rocks, *113,* **113**, 114*t*
    where found, 359

Fibrous crystal habit, 92, *92*
Fibrous fracture, 90
Fine-grained sediments and rocks, 94, 160–161, 161*t,* 162–163
    and sediment transport, *314–315*
Finger Mountain, 121, *121*
Fission, 390
Fissure eruptions, *135,* 135–136, *136*
Fjords, *269,* **271**
Flake tectonics
    on Earth, 239
    on Venus, 226
Flank collapses, *137,* 138
Flash flooding, 338, *339*
Flexural basins, **152**
Flint, 167
Flood basalts, 136, *136*
    and mass extinctions, 207, 247
Flood tide, 328
Flooding, *18,* **18**, 313
    flash, 338, *339*
    glacial, 272, *272*
    and global warming, 27, 406–407, *407*
    and human activities, 321, *321*
    and natural levees, 320, *320*
Floodplains, *314–315,* **316**, *316*
    human activities on, 320–321, *321*
Fluid-induced melting, **124**, *125*
Fluids. *See* Hydrothermal solutions; Lavas; Magmas; Solutions; Water
Fluorite, hardness of, 88*t*
Fluxes, 395–396, *396*
Foci of earthquakes, *347,* **347**–348, *348*
    locating, 351, *352–353*
Fold and thrust belts, *181,* 182
Fold axes, 179, *179*
Folds, *172,* **172**, *178,* 178–180, *179, 180*
Foliated rocks, **189**–191, *190,* 192*t*
Foliation, *97,* **98**, *189,* *189*
"Fool's gold." *See* Pyrite
Foraminifers, 163–164
    as temperature proxies, 274
Ford, Henry, 394
Foreset beds, 323, *323*
Foreshocks, *348,* **348**–349
Foreshore, 331, *331*
Formations, **202**
Fossil-fuel economy, 6–7
    and carbon cycle, 398–400, *399*
    and global warming, 264
    and industry, 5–10
    and public policy, 409–410
Fossil fuels, **6**
    resources of, 382–390
Fossils, **46**, 69
    diversity of, *245*
    and geologic time, *200,* 201, *202,* 202–203
    micro-, *240,* **240**–241
    multicellular, *244*
Fractional crystallization, *117,* **117**–118, *118*
Fracture, **90**, 92*t*
Framework silicates, 84, *85. See also* Quartz; Silica
Full-cost accounting, 409
Fuller, Buckminster, 4
Fusion, 410

Gabbro, *113,* **114**, 114*t*
    in oceanic crust, *122, 123,* 124
    texture of, *111*
    where found, 359
Galena
    cleavage of, 89
    luster of, 90
    as ore mineral, 101, *101*
Ganges River, 289*t*

Garnet
  cleavage of, 90
  as index mineral, *193*
  in metamorphic rocks, 95*t*, 98
  as porphyroblasts, 192, *192*
Garnierite, as ore mineral, 100
Gasoline, 6, 394
Gems
  colors of, 91, *91*
  oxides as, 86
  as raw materials, 10
Genes, 239
Geobiology, 3
Geochemical cycles, 395–397, *396*
  bio-, 396
  calcium, 397, *397*
  carbon, 260, 379, 381–382, 395–400, *398*
Geochemical reservoirs, **395**, *396*
Geochemistry, 3
Geodynamo, **46**
  and magnetic reversals, 63
Geodynamo system, *43*, 45–46
Geologic cross-sections, **174**, *174*
Geologic maps, **174**, *174*
Geologic record, **34**, *35*
Geologic time scale, 46–48, *47*, **206**
  absolute, *211*, 211–212
  biological, *242*
  continental, *239*
  planetary, 222–224, *223*
  relative, *206*, 206–207, *208*
Geology, 2–3
  as science, 33–36
Geophysics, 3
Geosystems, **43**–44. *See also* Earth system
Geothermal energy, 303, **393**–394, *394*
Geothermal gradient, 185–186, **360**, *360*
  and geothermal energy, 393
  and metamorphism, 188
Geothermometers, 186
Geysers, *302*, 302–303, 394, *394*
  and hydrothermal solutions, 102
Ghawar oil field, *384*, 385
Glacial budgets, 267–268, *268*
Glacial cycles, **265**–267, *266*
Glacial drift, 271
Glacial environments, *153*, 154
Glacial lakes, 272
Glacial maxima, 266
Glacial periods. *See* Ice ages
Glacial striations, *268*, 268–269
Glacial till, *270*, **271**, *273*
Glacier Bay National Park, *268*
Glacier National Park, *316*, 404
Glaciers, **255**, 256, *256*, *257*
  formation of, 267–268
  retreat of, 251, 404, *405*, 406
  and stream transport, *314*, 319
  as water reservoirs, 283, 283*t*
Glassy substances, 77, 82
Glaucophane, 195
Global change, 23–28, **24**
  and acid rain, 25–26
  and global warming, 26–28, *27*
  and human activities, 24–25, 260
  and stratospheric ozone depletion, *26*, 26
Global Positioning System (GPS), *34*, *67*, 67, 200, 349
Global warming, 26–28, *27*
  and human activities, *264*, 264–265, 382
  and industrialization, 401, *401*
  and mass extinctions, 246–247
  predictions of future, 401–403, *402*, *403*
Gneiss, *190*, **190**–191, 192*t*
Gobi Desert, 336, *336*
Gold

concentrations of, 100, *100*, 101, *101*, 104, 229
  luster of, 90
  panning for, 104, *104*
Gold Rush, 104, 122
Gondwana, 69, *70*, *71*, 236, *236*
  continental glaciers in, 276, *277*
GPS. *See* Global Positioning System (GPS)
Graded bedding, **156**–157
Grains, 82
Grand Canyon National Park, *36*
  angular unconformities in, 204, *204*, 207, *208*
  dikes in, *121*
  springs in, *291*
  stratigraphy of, *201*, 207, *208*
Granite, **113**, *113*, 114*t*
  as igneous rock, 94, *94*
  as intrusive igneous rock, 95, *95*
  and magmatic differentiation, 118–119
  as rock, *93*
  texture of, 109, *109*, 110, *110*, *111*
  where found, 359
Granoblastic rocks, 191, **191**–192, *192*, 192*t*
Granodiorite, **113**, *113*, 114*t*
Granofels, as granoblastic rock, 191–192
Granular texture, 98
Granulite, as granoblastic rock, 191–192, 192*t*
Granulite facies, *194*, 194*t*
Graphite, as polymorph, 82–83, *83*
Gravel, **161**–162, 161*t*
  diagenesis of, *159*
  porosity and permeability of, 291*t*
Gray, Peter, *34*
Greasy luster, 90*t*
Great Barrier Reef, *399*
Great Basin
  geothermal gradient beneath, 117, 186
  and tensional tectonics, 182
Great Basin Desert, 336, *336*
Great Rift Valley, *58*, 60
Great Salt Lake, 154, 167
Great Trigonometrical Survey of India, 31, 37
Greenhouse effect, 260–262, **261**, *261*
  enhanced, 264–265
Greenhouse gases, **26**
  atmospheric, 254–255, 261–262
  and biosphere, 260
  and global warming, 26–27, 400
Greenland
  continental glacier of, 256, *257*
  ice cores from, 265, *265*
Greenschist, **195**
Greenschist facies, *194*, 194*t*
Greenstone
  as granoblastic rock, 191, 192*t*
Grenville province, *236*
Grotzinger, John, 1, *28*, 227
Ground moraines, *270*
Groundwater, 14–15, *15*, **283**, 290–290
  and aquifers, 293–294
  deep crustal, 301–303, *302*
  erosion by, *297*, 297–298, *298*
  exploration for, 358
  and human activities, 296–297
  and hydrologic cycle, 284, *284*
  hydrothermal, 301–302, *302*, 393–394, *394*
  speed of flow, 295–296, *296*
  as water reservoir, 283, 283*t*
Groundwater table, 292, **292**–293
  perched, 294, *294*
  seasonal changes in, 294–295, *295*
Gulf of Aqaba, *61*
Gulf of California, 60, *61*

Gulf of Mexico, 385
  petroleum production in, 385, 386
Gulf Stream, 255, 408
Gullies, 317, *317*
Gutenberg, Beno, 40, 358
Gypania, 243
Gypsum
  and calcium cycle, 397
  crystal habit of, *92*
  as evaporite mineral, 166, 167
  gem varieties of, *92*
  hardness of, 88*t*
  precipitation of, 147
  in sedimentary rocks, 95*t*, 97, 155*t*, 163*t*, 165
  as sulfate, 87, *87*

Hadean eon, *211*, 212, *213*, **223**, 224
Haicheng earthquake, 376
Half-life, **209**, *210*, 210*t*
Halides, 83*t*
Halite, *79*, 82, *82*
  in chemical sediments, 96
  cleavage of, 89
  as evaporite mineral, 154, 155*t*, 156, 166, 167
  precipitation of, 147, *147*
  in sedimentary rocks, 95*t*, 97, 163*t*, 165
Hall, James, 199
Hallucigenia, 243
Hanging valleys, *269*, 270
Hardness, **88**, 88*t*, 92*t*
Hawaii
  basaltic lava on, *129*, 129–130
  as hot spot, 73, 126, *126*
  sedimentary environment of, 152
  volcanic eruptions on, 20, *20*, 133, *134*
Hazards. *See* Natural hazards
Heat reservoirs, 393
Heat transport
  in the atmosphere, 254, 263
  in the hydrosphere, 254
  in the mantle, 220, 361, *362*
Heavy Bombardment, *222*, **222**, 224
  and continental crust, 228
Heavy oil, 390
Hector Mine earthquake, 366
Heezen, Bruce, 51, 54, *55*
Heimaey Island eruption, 140
Helium, in space, 216
Hematite
  density of, 91
  in desert varnish, 338, *338*
  formation of, 242
  as ore mineral, 100
  as oxide, 86, *86*
  in sedimentary rocks, 163*t*
  streak of, 90, *91*
Hercynian orogeny, *236*
Herodotus, 323
Hess, Harry, 55
High-grade metamorphic rocks, 185, *185*, 190
High-pressure metamorphism, *187*, 188
Himalaya Mountains, *59*, 62
  and carbon cycle, 398
  and compressive tectonics, 182
  orogeny of, *235*, 235–236
  and regional metamorphism, 187
Holmes, Arthur, 53, *72*
Holocene epoch, 206, *206*, *211*, 252, 400–401
  as interglacial period, 266, 267
Hominids, 48
Homo sapiens, 48, 212, 400
Horizontal folds, *179*
Horizontal transport of mantle material, 234
Hornblende. *See also* Amphiboles

as double-chain silicate, 84, *85*
  and isotopic dating, 210*t*
Hornfels
  as granoblastic rock, 191, 192*t*
  as metamorphic rock, 97
Hornfels facies, *194*
Hot spots, 73, *73*, **126**, 361
  magmatic processes at, 125–126, *126*
Hot springs, *302*, 302–303
  and hydrothermal solutions, 102, 301*t*, 301–302
Hot Springs (AR), 301, 302
Huang Ho [river], 320
Hubbert, M. King, 385
Hubbert's peak, *385*, **385**, 386, *386*
Hubble Space Telescope, 216
Human activities
  and biosphere, 260
  and carbon cycle, 398–400, *399*
  and desertification, 337–338
  and the environment, 3–4, 379–413
  and floodplains, 320–321, *321*
  and global change, *24*, 24–25, 252
  and global warming, *264*, 264–265
  and groundwater, 296–297
  and mass extinctions, 246
  and streams, 307
  and water quality, 299, 299–301
Human health. *See* Diseases
Humidity, relative, **285**
Hurricane Katrina, *17*, 17–18, 321, 330, *330*
Hurricane Mitch, 21
Hurricanes, *17*, **17**–18, *329*, 329–331, *330*
  and global warming, 27
Hutton, James, 35, 110, 199–200, 204
Hydrocarbons, 380, 383
  reservoirs of, 383, *383*
Hydroelectric energy, 282, 392, 393, *393*
  and carbon economy, 381, *381*
Hydrogen
  in space, 216
  from volcanoes, 221, *221*
Hydrologic cycle, 14, *284*, **284**–285
Hydrology, 14, **282**
  and climate, 285–290
Hydrosphere
  and calcium cycle, *397*
  and carbon cycle, 397, *398*
  and climate system, *253*, 254, 255, *396*
  and Earth system, 42, *43*, 44
  and hydrologic cycle, 282
  and ice ages, 266, 274
Hydrothermal solutions, **101**–102
  metasomatism by, 186–187
Hydrothermal waters, **301**–302, *302*
Hydroxides, 83*t*
Hypothesis, **32**

Iapetus Ocean, *236*, 237
Ice, as mineral, 255–256
Ice ages, **266**, 266–267
  ancient, 276–278, *277*
  discovery of, 272–273
  Pleistocene, 272–276
Ice cores, climate history from, 265, *266*, *267*, 400
Ice sheets. *See* Continental glaciers
Ice shelves, 256–257, *257*, 405–406, *406*
Iceberg calving, 256, *256*, 268
Iceland, volcanic eruptions on, *135*, 135–136, 140
Igneous intrusions, *119*, 119–122, *121*
Igneous ore deposits, 102–103, *103*
Igneous processes, 107–141
  and plate tectonics, 108, *108*, *123*, 123–126, *125*, *126*
Igneous rocks, **94**–95, *95*

intrusive vs. extrusive, 95, *95*, 110–111, *111*
mineralogy of, 95, 95*t*, 112–115, 112*t*, *113*
texture of, 94, *94*, *109*, 109–111, 131, *131*
Iguaçú Falls, *317*
Immiscible magmas, 119
Index minerals, 193, *193*
Indus River, 320
Industrial revolution
    and Anthropocene epoch, 408
    and energy use, 5, 6
    and environment, 380
    and global warming, 401, *401*
    and water use, 13
Infiltration, *284*, **284**–285
Inner core, *38*, **40**
    discovery of, *357*, 358, 361
    and Earth system, *43*
    formation of, 219–220, *220*
Inner planets, 217, *218*
Inorganic substances, 77
Instruments. *See* Laboratory instruments
Intensity scale, **354**–355, 354*t*, *355*
Interglacial periods, **266**, *266*
    history of, 274, *274*
Intergovernmental Panel on Climate Change (IPCC), 26, 402, 404
Intermediate igneous rocks, *113*, **113**–114, 114*t*
Intermediate-term earthquake prediction, 376
Intraplate earthquakes, 365
Intrusions, igneous, *119*, 119–122, *121*
Intrusive igneous rocks, **95**, *95*
    texture of, 110–111, *111*
Ionic bonds, 80
    and cleavage, 89
    and hardness, 88
    and luster, 90
Ions, *79*, **79**–80
    carbonate, 86, *86*
    and mineral color, 90–91, *91*
    silicate, 84, *85*
    sizes of, 81, *81*
    sulfate, 87
    sulfide, 86
IPCC. *See* Intergovernmental Panel on Climate Change (IPCC)
Iridium, *245*, 245–246
Iron
    and composition of Earth, *41*, 41–42, 220, 361
    on Mars, 226
    in meteorites, 39, *39*
    and ocean fertilization, 411
    ores of, 100, *101*, **101**, 103, 145, 229, 242, *243*
    and remediation, 300
    residence time of, 396
Iron formations, 163*t*, 242, *243*
Irrigation, 13, 15, 282, 286, *287*, 288
    and groundwater, 296–297
Island arcs, **61**
    magmatic processes under, 124–125, *125*
Isochrons, **68**, *68*
    and history of plate movements, *69*
Isograds, 192–194, *193*
Isolated-tetrahedra silicates, 84, *85*. *See also* Olivine
Isotopes, **78**, 209
    radioactive decay of, *209*, 209–211, *210*
    of uranium, 390–391
Isotopic dating, **209**–211, *210*
    of solar system, 217, 223
Isthmus of Panama, 259, 276
Izmit earthquake, 368

Jack Hills, 212, *213*
Johnston, David A., 107
Joints, *180*, **181**
    erosion along, 297
Jordan, Tom, 1, *28*
Jordan Valley, 152
Jupiter, 218, *218*
Jurassic period, 206, *206*, *208*, 211

Kalahari Desert, 336, *336*
Kames, 272, *273*
Kansu earthquake, 368
Kant, Immanuel, 216
Kanto earthquake, 371
Kaolinite. *See also* Clay [minerals]
    as sheet silicate, 84
Karst topography, **298**, *298*
Kennecott Copper Mine, *103*
Kettle lakes, *271*, **272**, *273*
Keystone thrust fault, *183*
Kilauea volcano, *129*, 138
Kilobar, 186
Kimberley diamond mines, 104
Kimberlites, 76
Kluane National Park, *255*
Kobe earthquake, 367, *367*, 368
Krakatau volcano, 20, 130, 259, 370
Kupferschiefer ores, 104
Kyanite
    as index mineral, *193*
    in metamorphic rocks, 95*t*, 98
Kyoto Protocol, 410

La Niña, *263*, 263–264
La Plata River, 289*t*
Laboratory instruments
    Global Positioning System (GPS), *34*, 67, *67*, 200, 349
    magnetometers, 63
    mass spectrometers, *35*, 210
    seismographs, 38, 349, *349*, 351
Laboratory studies
    and composition of core, 361
    of crystallization, 109–110
    of deformation, 175, *175*, 187
    of phase changes, 360
    of seismic waves, 361
Lahars, **137**, *137*
Lake environments, 152, *153*, 154
Lake Missoula, 272, *272*
Lake Powell, *281*, 281–282, 283
Lakes
    glacial, 272
    kettle, *271*, **272**, *273*
    oxbow, *315*, *318*, 319
    playa, *339*, 339–340
    saline, 154
    and stream transport, *314*
    as water reservoirs, 283, 283*t*, 289
Laki fissure, 135
Lambton, William, 31
Laminar flow, **308**, *308*
Landers earthquake, 366
Landforms
    glacial, *268*, 268–272, *269*, *270*, *271*, *272*
    volcanic, 133–136, *134*, *135*, *136*
Landslides, 20, **20**–21, *21*
    and earthquakes, 343, 368
    on Mars, 226, *226*
    submarine, 335, *335*
Larochelle, A., 64
Larsen Ice Shelf, 405–406, *406*
Late Heavy Bombardment, *222*, 224, 226
    and continental crust, 228
Lateral moraines, *270*
Lateral variations, 361
Laurasia, 69, *71*
Laurentia, *236*, 236–237
Laurussia, *236*, 237, *237*

Lava streams, 128, *129*
Lavas, **109**, 111
    types of, 128–131
    and volcanoes, 128
Law of the Sea, 12
Layered intrusions, 118–119
Lead
    and isotopic dating, 210–211, 210*t*, 212
    and metasomatism, 186
    ores of, 101, *101*, 102
Left-lateral faults, *177*, 178
    and fault mechanisms, 355, *355*
Lehmann, Inge, 40, 358, *358*, 361
Levees, *314*–*315*, **320**, *320*
    artificial, 321
Lewis, Meriwether, 313, 322
Life
    and climate, 259–260
    and climate feedback, 262
    early history of, *240*, 240–241, *241*
    evolution of, 46–48, *47*, 53, 212, 239–247, *242*
    extinctions of, *245*, 245–247, *246*
    extraterrestrial, 226, 227
    radiation of, *243*, 243–245, *244*, *245*
    universal tree of, 240, *240*, *244*
Lignite, 388, *388*
Limbs, 179, *179*
Limestone, 163*t*, **165**, *165*
    erosion of, by groundwater, *297*, 297–298, *298*
    as sedimentary rock, 97
Limonite, 338, *338*
Liquefaction of soils, 368
Lithification, **96**, *159*, **160**
Lithosphere, **45**, *359*
    and calcium cycle, 397
    and carbon cycle, 397, *398*
    and carbon economy, 380, 381–382
    and climate system, *253*, 258–259, *396*
    and Earth system, *43*, 44
    fate of, *362*, 363
    tectonic plates of, 52, 56–57
Loess, 272
Loma Prieta earthquake, 366–367, 368
Long Island, 273
Long-term earthquake forecasting, 375
Long Valley caldera, 135, 138
Longshore currents, *327*, **327**–**328**, *332*, *332*
Longshore drift, *327*, 328, *332*, *332*
Low-grade metamorphic rocks, *185*, **185**, *190*
Low-velocity zone, *359*, **360**
Lower mantle, 41, *359*, **360**
    and Earth system, *43*
Lunar highlands, 222, *222*
Lunar maria, 222, *222*
Luster, **90**, 90*t*, 92*t*
Lyell, Charles, 265

Maat Mons, *225*
Mafic minerals, 112, 112*t*
Mafic rocks, *113*, **114**, 114*t*
    where found, 359
Magellan [spacecraft], 224–225
Magma chambers, **116**
Magma ocean, 219, *219*
Magmas, **82**
    differentiation of, 117–119, *118*
    formation of, 115–117
Magmatic addition, **233**–234
Magmatic differentiation, **117**–119, *118*
Magmatic processes, 122–126, *123*, *124*, *126*
Magmatic stoping, 120, *120*
Magnesium, and composition of Earth, *41*, 41
Magnetic anomalies, **63**

seafloor, 63–64, *64*, *65*
Magnetic chrons, 63, *65*
Magnetic field, *45*, **45**–46
Magnetic reversals, 46, 63
    and magnetic anomalies, 63
    and seafloor spreading, 63–65, *64*, *65*
Magnetic time scale, **63**, *65*
    and plate velocities, 65–66
Magnetite
    concentrations of, 104
    density of, 91
    formation of, 242
Magnetization, thermoremanent, 63, *64*
Magnetometers, 63
Magnitude scale, **353**, *354*
Mainshocks, 348, *348*
Malachite, as ore mineral, *102*
Mammoth Cave, 297
Mammoth Mountain, 138
Mantle, *38*, **39**–40
    composition of, 114, *359*, 359–360, *360*
    formation of, 219–220, *220*
    structure of, *359*, 359–360, *360*, 361, *362*, 363
    upper vs. lower, 41
    vertical vs. horizontal transport of material in, 233–234
Mantle convection. *See* Convection
Mantle plumes, 73, *73*, **125**, 361
    magmatic processes at, 125–126, *126*
    super, *362*, 363
Mantle wedges, 124, *125*
Mapping
    of geologic structure, *173*, 173–174, *174*
    of India, 31
    by satellites, 31, *31*, 34, 42, 405
Marble
    in art and architecture, *171*, 171–172, 191
    deformation of, *175*
    as granoblastic rock, 191, *191*, 192*t*
    as metamorphic rock, 98
    and recrystallization, 186
    as rock, 93
Mariana Islands, *58*
Marianas Trench, 37, *58*, 61
Marine environments, *153*, 154
    evaporite, *166*, 166–167
Mariner [spacecraft], 224
Marquette, Craig, *34*
Mars, 217, *218*, 221, 221*t*, *226*, 226–227, *227*
    sulfates on, 87
    topography of, *225*
Mars Express [spacecraft], 227
Mars Global Surveyor [spacecraft], 225
Mass extinctions, 16–17, 47–48, 74, **206**
    and bolides, *245*, 245–246, *246*
    and flood basalts, 247
    and geologic time scale, *206*, 206–207
    and human activities, 408–409
    and methane, 246–247
Mass spectrometers, *35*, 210
Matter, structure of, 77–80
Matterhorn, 272, *273*
Matthews, Drummond H., 64
Mauna Loa volcano, 133, *134*
Meanders, *314*, *316*, *318*, **318**–319
    and bedding sequences, 158, *158*
Mediterranean Sea, as marine evaporite environment, *166*
Medium-grained sediments and rocks, 160–161, 161*t*, *162*

Megathrusts, 182, 365, 368, *368*
Melting. *See also* Magmas
    decompression, 116, *123*, 124
    of Earth, 218–219, *219*
    fluid-induced, 124, *125*
    and geothermal gradient, 360
    of ice, 255
    partial, 114, 115
Mercalli, Giuseppe, 354
Mercalli intensity scale, modified, 354–355, 354*t*, *355*
Mercury [element]
    ores of, 101, *101*
    pollution by, 390
Mercury [planet], 217, *218*, 221*t*, 222, 224, *224*
Mesopotamian basin, 152
Mesosaurus, 53, *53*
Mesozoic era, 206, *206*, *211*, 212
Messenger [spacecraft], 224
Metallic bonds, 80
Metallic luster, 90, 90*t*
Metamorphic facies, *194*, **194**–195, 194*t*
Metamorphic grade, 185, *185*, *190*
    and parent rock, 194*t*, 195
    and regional metamorphism, 192–195, *193*, *194*
Metamorphic processes
    causes of, 185–187
    and climate, 196
    and plate tectonics, 186, *187*, 187–188, 195–196, *196*
    types of, 187–188
Metamorphic rocks, **94**, *97*, 97–98, **185**
    mineralogy of, 95*t*, 98
    regional vs. contact, *97*, 97–98
    textures of, 188–192, *189*, *190*, *192*
Metasomatism, **186**–187
Meteor Crater, *36*, *223*
Meteoric waters, **302**
Meteorites, *39*, **218**
    and age of Earth, 209, 217
    and density of Earth, 39
Meteorology, 3
Methane
    and biosphere, 260
    and carbon-cycle engineering, 411
    and climate history, 265, *266*, 400, *400*
    formation of, 383
    as greenhouse gas, 261, *261*
    and mass extinctions, 247, 408
Micas
    cleavage of, *88*, 89
    and foliation, 189, *189*
    in igneous rocks, 95, 95*t*, 112*t*, *113*
    in metamorphic rocks, 95*t*, 98, 186
    as sheet silicates, 84, *85*
Micaschist, as metamorphic rock, 97
Michelangelo, 171
Michigan Basin, *229*, 230
Microbes, **239**–241, *240*
    as contaminants, 299, *299*
Microfossils, *240*, 240–241
Mid-Atlantic Ridge, *54*, *58*, 60, *60*
Mid-ocean ridges
    as divergent plate boundaries, 60, *60*
    magmatic processes at, 123–124, *124*
    and partial melting of peridotite, 114, 115
Migliaccio, Ralph, 343
Migmatite, *190*, 191
Milankovitch, Milutin, 275
Milankovitch cycles, 274–276, *275*, **275**, 400
Mineralogy, 75–104, **76**. *See also* Minerals; Rocks
    of rocks, 94, 95, 95*t*, 97, 98
Minerals, **76**–77, *77*
    classification of, 83*t*

crystal structure of, 80–81
crystallization of, 81–82
evaporite, 154
exploration for, 11–13, *12*
formation of, 80–83
physical properties of, 87–93, 92*t*
as raw materials, 10–13
rock-forming, 83–87, 83*t*
and weathering, 147*t*
Mining
    for coal, *7*, *389*, 389–390
    environmental damage by, *12*, 12–13, 389–390
    for minerals, *100*, 102, *103*
    for raw materials, 10
    and water use, 287
Miocene period, 206, *206*, *211*
    desert varnish from, 338
    global cooling during, 276
Miscible magmas, 119
Mississippi River, 289, *289*, 289*t*
    delta of, 323, *324*
    flooding by, 18, *18*
    human activities along, 320–321, *321*
    natural levees of, 320, *320*
    and sedimentary processes, 145, *145*, 148, 154
Mississippian period. *See* Carboniferous period
Models, scientific, 33
Modified Mercalli intensity scale, 354–355, 354*t*, *355*
Mohorovičić, Andrija, 40
Mohorovičić discontinuity (Moho), 40, *123*, 359
Mohs, Friedrich, 88
Mohs scale of hardness, **88**, 88*t*
Mojave Desert, 336, *336*
Molina, Mario, 26
Moment magnitude, 353, *354*
Monongahela River, *307*
Monroe, Marilyn, 75, *75*
Monsoons, 285
Mont Pelée, 132–133
Montreal Protocol, 26
Moon, *215*, 221*t*, *222*
    albedo of, 260
    formation of, 219, *219*
    missions to, 215
    and planetary time scale, 222–224, *223*
    temperature of, 261
    and tides, 328, *328*
Moraines, *270*, **271**
Morley, Lawrence, 64
Mother Lode, 104, 122
Mount Etna, 133, *135*
Mount Everest, 31, 37, 144
Mount Fuji, 20, 133, *134*
Mount Huascarán, *21*
Mount Kilimanjaro, 251
Mount Pinatubo, 132, 139, 259
Mount Rainier, 20, 133, 139, *139*
Mount St. Helens, 20
    changing shape of, *34*
    and convergent plate boundaries, 62
    eruption of, 107, *107*, 130, *130*
    hazards of, 138
    predicting eruption of, 139
    as stratovolcano, 133
    volcanic dome of, 133, *134*
Mount Tambora, 259
Mount Unzen, *132*
Mount Vesuvius, 133
Mountains. *See* Orogeny; *specific mountains*
Mud, 161*t*, **162**
    diagenesis of, *159*
Mudflows, *20*, 21, *21*
Mudstone, 161*t*, **162**

Muscovite. *See also* Micas
    cleavage of, *88*, 89, 90
    and isotopic dating, 210*t*
    as sheet silicate, 84, *85*

Namacalathus, 243
Namib Desert, 306, *307*, *311*, 336, *336*
Native elements, 83, 83*t*, 100, *100*
Natural bitumen, 390
Natural gas, 6, 7, **383**, 387
    and carbon economy, *381*, 381
    classification of, 163*t*, 168
    exploration for, 144, 335–336, 358
    formation of, *159*, 167–168, 382–384, *383*
    and geothermal gradient, 188
    resources of, 382, *382*
Natural hazards, **16**–23
    earthquakes, 343–344, 353–355, 366–375, *367*, *370*, *371*, *372*
    examples of, 17–22
    floods, 320–321, *321*
    hurricanes, *17*, 17–18, *329*, 329–331, *330*
    primary vs. secondary, 368
    reducing risks of, 22–23, 139–140, 371–375
    tsunamis, 343, 344, 368–370, *369*
    volcanoes, 132–133, 136–140, *137*, *138*
Natural resources, **4**–16. *See also* Reserves; Resources
    energy, 5–10
    minerals and raw materials, 10–13
    resources vs. reserves, 4–5, *5*
    soils, 15–16, *16*
    water, 13–15
Natural selection, **244**
Nebulae, 216
Nebular hypothesis, **216**, *217*
Negative feedbacks, **262**
Neogene period, 206, *206*, *211*
Neptune, 218, *218*
Neutrons, 78, *78*
Nevado del Ruiz volcano, 20, *20*, 62, 137
New Madrid earthquake, 355, *355*, 365
Newton, Isaac, 39
Nickel
    and composition of Earth, 220, 361
    in meteorites, 39, *39*
    ores of, 100, 103, 229
Nile River, 307, 320, 323
Nisino-sima volcano, *130*
Nitrates, 155*t*, 156, 167
Nitrogen
    atmospheric, 254
    and isotopic dating, 210*t*, 211
    residence time of, 396
    from volcanoes, 221, *221*
Nitrous oxide, *400*, 401
Nonconformities, 203, *204*, 207
Nonrenewable resources, **6**, 15, 382
Normal faults, **177**, *177*
    and fault mechanisms, 355, *355*
    and tensional tectonics, 182
North America, *228*
    Appalachian fold belt, *230*, 230–231
    Canadian Shield, *228*, 228–229
    coastal plain, 232
    Cordillera, *231*, 231–232
    interior platform, *229*, 229–230
North Atlantic Oscillation, 264
Northridge earthquake, 183, 365, 367, *367*
Northwest Passage, *404*
Nuclear energy, 6, *6*, 9, *381*, 382, **390**–391
    and Earth system management, 410, 411, 412*t*

radioactive wastes from, 299, *299*, 391, *391*
Nuclei [atomic], 78, *78*
    radioactive decay of, *209*, 209–211, *210*
Nuclei [cellular], 240

Oases, 338
Oblique-slip faults, 176, *177*
Obsidian, 93, **131**
Oceanography, 3
Oceans
    acidification of, 396, 399–400, 408
    chemical mixing in, 150–151
    formation of, 220–221, *221*
    as water reservoirs, 283, 283*t*
Oceanus [research vessel], *34*
Offsets, 176
Offshore, 331, *331*, *335*, 335–336
Ogallala aquifer, 15, *15*, 296
Ohio River, *307*, 320
Oil. *See* Petroleum
Oil shales, 10, 162–163, **390**
Oil traps, **383**, *383*
Oil wells, *5*, 6, 7, *9*, 335–336
Oil window, **383**
Oldham, Robert D., 40, 358
Oligocene period, 206, *211*
Olivine
    cation substitution in, 81
    density of, 91
    in igneous rocks, 95, 95*t*, 112*t*, *113*
    as isolated-tetrahedra silicate, 84, *85*
    as mineral, 77
    phase changes in, 360
Olympus Mons, *226*, 227
Onyx, 172
Opal, 163*t*, 167
An Operating Manual for Spaceship Earth [Fuller], 4
Ophiolite suites, *122*, 123, 235
Opportunity [spacecraft], 227
Ordovician period, 206, *211*
    ice ages during, 276, *277*
Ores, 12, **100**–102, *101*, *102*
    hydrothermal, 301, 303
    and metasomatism, 186
    sulfides as, 86, 101–102, *102*
Organic materials, and isotopic dating, 210*t*, 211
Organic sedimentary rocks, 163*t*, 167–168
Orientation
    of bedding, *173*, 173–174
    of foliation, 189, *189*
Original horizontality, principle of, **200**
Orogeny, *172*, *235*, 235–238, *236–237*
Orthoclase. *See also* Feldspars
    as framework silicate, *85*
    hardness of, 88*t*
    as rock-forming mineral, 93
Ostrom, Will, *34*
Otto, Nikolaus, 394
Outcrops, *172*, 173
Outer core, *38*, **40**
    discovery of, *357*, 358, 361
    and Earth system, 43
    formation of, 219–220, *220*
Outer planets, 218, *218*
Outwash, *270*, 271
Outwash plains, 271, *271*
Overthrusts, 182, *183*
Overturned folds, 180, *180*
Owens Valley earthquake, 366
Oxbow lakes, *315*, *318*, 319
Oxides, 83*t*, 84, 86, *86*
    as ore minerals, 100
Oxygen
    atmospheric, 254
    and composition of Earth, *41*, 41–42, 361

and evolution of the atmosphere, 46–47, 212, 241–242, 245
isotopes of, as temperature proxies, 265, 274, 274
and methane in oceans, 247
residence time of, 396
Ozone
depletion of stratospheric, **26,** *26*
as greenhouse gas, 261, *261*
stratospheric, 255

P-T paths, 195–196
P waves, *350,* **351**
and fault mechanisms, 356, *356*
and interior structure of Earth, 39, 356–363, *357, 359, 362*
shadow zone of, *357,* 357–358
Pahoehoe, 129, *129,* 130
Pakistan earthquake, 368
Paleocene-Eocene mass extinction, 246–247, 408
Paleocene epoch, *206, 211*
Paleoclimatology, 265, *266*
Paleocontinents, 69
Paleogene period, *206, 206, 208,* 211
Paleontology, 203
Paleozoic era, *206, 206, 211,* 212, 244
Palisades intrusion, 117–118, *118,* 120
Pangaea, *52,* **53,** 67
and Appalachian Mountains, 230
assembly of, 69, *70,* 236–237, *236–237*
breakup of, 69, *71,* 152
Panthalassa Ocean, *70*
Parent nuclides, 209
Partial melting, 114, **115**
Passive margins, 233
Patagonian Desert, *336*
Paterson, Clare, 209
Pearly luster, 90, *90t*
Peat, *155t,* 167, *375,* 388, *388*
Pebbles, 161–162, *161t*
Pegmatites, 103, **122,** *122*
Pennsylvanian period. See Carboniferous period
Perched water table, 294, *294*
Peridotite, *113,* **114**
in mantle, 359–360
and oceanic crust, 124, 359
partial melting of, 114, 115, *359, 360*
Periods, **206,** *206*
Permafrost, 258, *258*
and climate change, 404–405, *405*
Permeability, *291,* **291,** *291t, 300*
Permian mass extinction, 247
Permian period, *206,* 207, *208, 211*
ice ages during, 276, *277*
Perovskite, density of, 91
Peru earthquake, 368
Petrified Forest, *200*
Petroleum, 6, *7*
and carbon economy, 381, *381*
classification of, *163t*
exploration for, *8,* 8–9, *9,* 143–144, 335–336, 358
in flexural basins, 152
formation of, *159,* 167–168, *382–384, 383*
and geothermal gradient, 188
refining of, *379*
reserves of, *384,* 384–386
resources of, 382, *382*
Phanerozoic eon, *211,* 212, *223,* 241, 244
Phase changes, **360**
Phenocrysts, 111, *111*
Phosphorite, *163t*
Photomicrographs, 109, *109*
Photosynthesis, **241,** 260, 397
Phreatic explosions, 130, *130*

Phreatic zone, 292, *292*
Phyla, 244
Phyllite, *190,* **190,** *192t*
Physical laws, 32–33
Physical weathering, **146,** *146*
Piedmont, *230,* 231
Pillow lavas, *122, 123,* 124, 130
Pitchblende, 391
Placers, 104
Plagioclase. See also Feldspars
and calcium cycle, 397
as framework silicate, *85*
and isotopic dating, *210t*
as rock-forming mineral, *93*
Planetesimals, 217, 220
Planets, *218*
formation of, 216–218, *217*
Plate boundaries, 56–63
convergent, *58–59, 59,* 60–62
divergent, *58, 59,* 60, *60, 61*
transform-fault, *59, 59, 62,* 62–63
and volcanoes, 128, *128*
Plate tectonic system, *43,* 44–45
Plate tectonics, 45, 51–74, **52,** *56–57*
and climate, 259, 276
and continental deformation, *181,* 181–184, *182, 183, 184*
and currents, 306
discovery of, 52–55
early history of, 220, 239
and earthquakes, 344, 363, *364,* 365
and igneous processes, 108, *108, 123,* 123–126, *125, 126*
and mantle convection, *44,* 45, 53, 72–74, *73, 362,* 363
and metamorphic processes, 186, *187,* 187–188, 195–196, *196*
and plate movements, 63–67
and reconstruction of plate history, 67–69, *70–71*
and rock cycle, 98–100
and sedimentary processes, 144, *144, 151,* 151–152, 162
types of boundaries in, 56–63
Platinum, ores of, 102
Platy crystal habit, 92
and foliation, 189, *189*
Playa lakes, *339,* **339**–340
Playas, 340
Playfair, John, 199, *200*
Pleistocene period, *206, 211,* 400
ice ages during, 259, 272–276, 398
Pliocene period, *206, 206, 211*
ice ages during, 276
Plunging folds, 179, *179*
Pluto, *218*
Plutons, *119,* **120**
Point bars, *314, 318,* **319**
Pollution
from coal, 10, 299, 388–390
of groundwater, 15, 299–301
remediation of, 300
at shorelines, 332
of water, 299, *299*
Polymorphs, **82**–83, *83*
Population growth, 380, *380,* 401, 409
Porosity, **160**
and groundwater, 290–291, *291, 291t, 300*
Porphyroblasts, 192, *192, 192t*
Porphyry, **111,** *111*
Positive feedbacks, **262,** 400, 408
Potability, **300**–301
Potassium, and isotopic dating, *210t*
Potholes, *316,* 316–317
Precambrian time, 207, *208,* 209, *211, 211,* 212, 243
ice ages during, 276, *277*
Precession, 275, *275*
Precipitation [meteorology], *284,* **284**–285, *288*

and groundwater recharge, 293
and streamflow, 313
Precipitation [mineral formation], 82, 147, *147,* 150–151
Precursors, 376
Prediction
of earthquakes, 375–376
of future global warming, 401–403, *402, 403*
of volcanic eruptions, 139–140
Preferred orientation, 189, *189*
Press, Frank, 54
Pressure
confining, 175, 186
and density, 91
and depth, 185, *185*
and diagenesis, 160
directed, 186
and magma formation, 115–116
and metamorphism, 186, 193, *193, 194*
and P-T paths, 195–196
and polymorphs, 82–83
Primary hazards, 368
Principle of faunal succession, **202**–203
Principle of original horizontality, **200**
Principle of superposition, **200**–201
Principle of uniformitarianism, **35,** 200
Professional meetings, 33, *33*
Proterozoic eon, *211,* 212, *223*
ice ages during, 276, *277*
Protons, 78, *78*
Public policy
and acid rain, 25–26
and building codes, 367, 373
and deep-sea resources, 12
and Earth system engineering, 409–413, *412t*
and energy use, 409–410
and global warming, 28, 403–404
and groundwater management, 296–297
and hurricanes, 334
and natural hazards, 17, 18, 22–23, 139–140
and seismic hazards, 367, *372,* 372–373, 375–376
and soil management, 16
and stratospheric ozone depletion, 26
and strip mining, *389,* 390
and water quality, 300–301
and water use, 281–282, *282t,* 285, 286–288, *287*
Pumice, 93, *111,* **131**
Pyrite
as ore mineral, 101, *101*
as sulfide, 87, *87*
Pyroclastic deposits, 131–133, *132*
Pyroclastic flows, *132,* **132**–133
Pyroclasts, *111,* **111,** *132*
Pyroxenes
cleavage of, *89,* 89–90
in igneous rocks, 95, *95t, 112t, 113*
in metamorphic rocks, *95t,* 98
as single-chain silicates, 84, *85*

Quad, 6–7
Quartz
cleavage of, 90
crystal habit of, 92
crystals of, *81, 82*
fracture of, 90
as framework silicate, 84, *85*
hardness of, *88t*
in igneous rocks, 95, *95t, 112t, 113*
as mineral, 77
polymorphs of, 83, 188
as rock-forming mineral, *93*
in sedimentary rocks, *95t,* 97, 160, *163t*

Quartzite, as granoblastic rock, 191, *191, 192t*
Quaternary period, *211*

Radiation [electromagnetic]
balance of, 260, *260, 261*
and climate feedback, 262, 276
Radiation [evolutionary], *243,* 243–245, *244, 245*
Radioactive wastes, 299, *299,* 391, *391*
Radioactivity
as dating technique, 209–211, *210*
discovery of, 209
hazards of, 391
as interior heat source, 219
Radium, 209
Rain shadows, 259, **285**–286, *286*
Rainfall, 285–286, *286.* See also Precipitation [meteorology]
and groundwater recharge, 293, *293,* 297
Rapids, *314,* 317
Raven, Peter, 409
Raw materials, 10–13
Recharge, **293**
artificial, 296
balancing of, with discharge, 294–295, *295*
Recharge area, 293, *293*
Recurrence intervals, **375**
Recycling, 10–11
Red beds, 242, *243*
Red Sea, 60, *61,* 182, *182*
Reefs, *153,* 154, **164,** *164,* 325
and ocean acidification, *399,* 399–400
as sedimentary rock, 96, *96,* 165
Reflection of seismic waves, 356–357, *357*
Reforestation, 411, *412t*
Refraction of seismic waves, 356–357, *357*
Regional metamorphism, *97,* **97**–98, *187,* **187**–188
and metamorphic facies, *194*
and metamorphic grade, 192–195, *193*
Reid, Henry Fielding, 344
Relative ages, **200**
and geologic time scale, *206, 206–207, 208,* 222
Relative humidity, **285**
Remediation, 300, *301*
Remote sensing
of desertification, 337, *337*
of magnetic anomalies, 63, *66*
of planets, 224
and satellite mapping, 31, *31,* 34, *42,* 405
Renewable resources, **5**–6, 9–10, 14, 382
Research, scientific, 32, *32*
Reserves, **4**–5, *5,* 382, *384,* 384–385. See also Natural resources; Resources
Reservoirs, 283
geochemical, 395, *396*
Residence times, 284, **396**–397
Resinous luster, *90t*
Resources, 4–5, *5.* See also Natural resources; Reserves
of fossil fuels, 382–390
Resurgent calderas, 133, 135
Reverse faults, **177,** *177*
and fault mechanisms, 355, *355*
Rhine Valley, 60, 182
Rhyolite, **113,** *113, 114t*
texture of, *111*
Rhyolitic lavas, 130–131
Richter, Charles, 351, 354
Richter magnitude, 351, 353, *354*
Rift basins, *151,* 151–**152**

Rift valleys
    as divergent boundaries, 54, 60, *60*, *61*, *151*
    and tensional tectonics, 182, *182*
Right-lateral faults, *177*, 178
    and fault mechanisms, 355, *355*
Ring of Fire, 20, 55, *55*, 363
Rio Grande Valley, 152
Rip currents, *327*
Ripples, **157**, *157*, *312*, **312**–313, *313*
Risk, **22**–23
    reducing seismic, 371–375
    reducing volcanic, 139–140
Rivers, **306**–307, *307*
    as water reservoirs, 283, 283*t*
Rock cycle, *98*–*99*, **98**–100
    metamorphic stages of, 172
    sedimentary stages of, *145*, 145–151
Rock-forming minerals, 83–87, 83*t*
Rocks, 93, **93**–94. *See also* Igneous rocks; Metamorphic rocks; Sedimentary rocks
Rocky Mountains, 230
Rodinia, **69**, *70*, 236, *236*, 237
Ross Ice Shelf, 256–257, *257*, 405
Rowland, Sherwood, 26
Rubidium, 209, *209*, 210, 210*t*
Ruby, 91, *91*
Ruddiman, William, 400
Runoff, *284*, **284**–285, *288*, 289–290, *290*
Rutherford, Ernest, 209

S waves, *350*, **351**
    and interior structure of Earth, 39, 356–363, *357*, *359*, *362*
    shadow zone of, *357*, 357–358
Saguaro National Park, *339*
Sahara Desert, 286, 306, *336*, 337, *337*, *339*
Sahel drought, 407, *407*
St. Elias Mountains, *255*
St. Lawrence River, 320
Saline lakes, 154
Salinity, **150**
Salt. *See* Halite
Salt dome traps, *383*, 384
Saltation, *310*, **310**–311
Salton Trough, *181*, 184
San Andreas fault
    earthquake prediction along, 375
    and elastic rebound theory, 344, *345*, *346*
    as fault, 176, *176*, 178
    and regional fault systems, 365–366, *366*
    and shearing tectonics, 183–184
    as transform-fault boundary, 59, 62, 63, 232, *232*, 363
San Fernando earthquake, 365, 373
San Francisco earthquake, 344, *345*, *346*, 348, 371
San Gabriel Mountains, *1*, 365, *366*
San Juan River, *318*
Sand, 161*t*, **162**
    diagenesis of, *159*
    porosity and permeability of, 291*t*
Sand budgets, 332, *332*
Sandstone, *161*, 161*t*, **162**
    formation of, *159*
    porosity and permeability of, 291*t*
    as sedimentary rock, 94, *94*, *96*
    and water quality, 301
Santorini volcano, 136
Sapphire, 91, *91*
Satellite mapping, 31, *31*, 34, 42, 405
Saturated solutions, 82
Saturated zone, *292*, **292**
Saturn, 218, *218*
Saudi Arabian Desert, *336*
Scarps, 178, *178*

Schist, 190, **190**, *192*, 192*t*
    as metamorphic rock, 94, *94*, 97
Schistosity, 190, *190*
Schmidt, Harrison "Jack," 215, *215*
Scientific instruments. *See also* Global Positioning System (GPS)
    magnetometers, 63
    mass spectrometers, *35*
    seismographs, 38
Scientific method, **32**–33, 52
Scientific models, 33
Scientific research, 32, *32*
Sea ice, 257, *258*
    and climate change, 404, *404*
Sea level, 37, *38*
    and global warming, 27, 406–407, *407*
    and ice ages, 150, 203, 273–274
    and unconformities, 203
Seafloor spreading, **53**–55
    as magnetic tape recorder, 63–66, *64*, *65*
Secondary hazards, 368–371
Sedimentary basins, *151*, **151**–152
    and fossil-fuel resources, 382
Sedimentary environments, **152**–156, *153*
Sedimentary landforms, glacial, *270*, *271*, 271–272, *272*
Sedimentary mineral deposits, 103–104, *104*
Sedimentary processes, 143–169
    and plate tectonics, *144*, 144, *151*, 151–152, 162
Sedimentary rocks, *94*, 95–97, *96*
    classification of, *160*, 160–161, 161*t*
    mineralogy of, 95*t*, 97
Sedimentary sequences, on Mars, 227, *227*
Sedimentary structures, *156*, **156**–159, *157*, *158*
Sedimentation. *See* Deposition
Sediments, 95–96, *96*
Seine River, 320
Seismic exploration, 358
Seismic hazard, *370*, **371**, *372*
Seismic imaging, *8*, *9*
Seismic risk, **371**, *371*
    reducing, 371–375
Seismic tomography, 361, *362*, **363**
Seismic waves, 18, **38**, 344
    and interior structure of Earth, 220, 356–363, *357*, *359*, *362*
    types of, *350*, 351
Seismicity maps, 363, *364*
Seismograms, 349, *350*
Seismographs, 38, *349*, **349**, 351
Seitz, Gordon, 375
Selenite, crystal habit of, *92*
Settling velocity, 149
Shadow zones, *357*, 357–358
Shale, *161*, 161*t*, **162**–163
    formation of, *159*
    metamorphism of, 192*t*, 195
    porosity and permeability of, 291*t*
Shear waves. *See* S waves
Shearing forces, **176**
    and fault mechanisms, 355, *355*
Shearing tectonics, *181*, 183–184
Sheet silicates, 84, *85*. *See also* Clay [minerals]; Micas
Sheeted dikes, *122*, 124
Shield volcanoes, **133**, *134*
Shipwrecks, 305, *305*
Shock metamorphism, *187*, 188
Shoemaker, Eugene, 215, 222, *223*
Shoreline environments, *153*, 154
    and currents, 307, *307*
Shorelines, **325**, *325*
    erosion and deposition at, *332*, 332–334, *333*, *334*

shaping of, *331*, 331–334, *332*, *333*, *334*
    wave motion at, *326*, 326–328, *327*
Short-term earthquake prediction, 375–376
Shuttle Radar Topography Mission (SRTM), 31, *31*
Siccar Point, 199, *199*, 204
    angular unconformity at, 199, *199*, 204
Siderite, in sedimentary rocks, 163*t*
Sierra Nevada Mountains, 234
Silica
    in igneous rocks, 112–114, *113*, 114*t*
    precipitation of, 148
    in sedimentary rocks, 155*t*, 156, 167
Silicate ions, 84, *85*
Silicates, 83*t*, 84, *85*
    and density of Earth, 39
    as ore minerals, 100
    as rock-forming minerals, 95, 97, 98
Siliceous environments, 155*t*, 156, 167
Silicic rocks, 112
Siliciclastic rocks, classification of, 160–163, *161*, 161*t*
Siliciclastic sedimentary environments, **154**–155, *155*
Siliciclastic sediments, 96, *96*, **147**
    classification of, 160–163, 161*t*
Silicon, and composition of Earth, *41*, 41–42
Silicosis, 93
Silky luster, 90*t*
Sillimanite, as index mineral, *193*
Sills, **119**, **120**–121, *121*
Silt, 161*t*, **162**
    porosity and permeability of, 291*t*
Siltstone, 161*t*, **162**
Silurian period, *206*, 211
Silver, concentrations of, *101*
Single-chain silicates, 84, *85*. *See also* Pyroxenes
Sink areas, 145, 151–152
Sinkholes, *297*, **297**–298, *298*
Skeleton Coast, *305*, 306
Sklodowska-Curie, Marie, 209
Slate, *190*, **190**, 192*t*
Slaty cleavage, *189*, 190
Slip directions, 176
Slip faces, 312, *312*
Smith, William, 202, 203
Smog
    from coal, 10
    and ozone, 26
Snake River basalts, 125
Snow lines, 267
Snowball Earth, 276–278, *277*
Sodium, residence time of, 396
Sodium carbonate, 340
Sodium chloride
    atomic structure of, 79, *79*, 80
    and oceanic chemical balance, 150
Soils, **15**–16, *16*
    liquefaction of, 368
Sojourner [spacecraft], *227*
Solar constant, 260, *261*
Solar energy, 5, *9*, 9–10, **392**, *392*. *See also* Sun
    and Earth system, *42*, 42–43
    and Earth system management, 410, 411, 412*t*
    and Milankovitch cycles, 274–276, *275*
    and prevailing wind belts, 310
    and radiation balance, *260*, 260–261, *261*
    source of, 216
Solar forcing, 252
    and Milankovitch cycles, 276
Solar nebula, 216–**217**
Solar system, *218*

origin of, 216–218, *217*
Solutions, 82. *See also* Hydrothermal solutions
Sonoran Desert, *336*
Sorting, **149**, *149*, 162
    in glacial sediments, 271
Source areas, 145, 146–148
Source beds, 383, 390
Southern Oscillation, *263*, 263–264
Space exploration, 215–216
Spaceship Earth, 3–4, 379–380
Speciation, 244
Specific gravity, 91
Sphalerite, as ore mineral, 101, *101*
Spinel, as oxide, 86, *86*
Spirit [spacecraft], 227, *227*
Spitak earthquake, 368
Spits, **333**, *334*
Splintery fracture, 90
Spreading centers, *58*, **60**. *See also* Divergent plate boundaries; Mid-ocean ridges
    magmatic processes at, *123*, 123–124
Springs, 290, *291*, 293, *294*. *See also* Groundwater
SRTM. *See* Shuttle Radar Topography Mission (SRTM)
Stabilization wedges, **411**, *412*, 412*t*, 413
Stacks, *325*, 333
Stalactites, 297, *297*
Stalagmites, 297, *297*
Staurolite
    and foliation, *189*
    as index mineral, *193*
    in metamorphic rocks, 95*t*, 98
    as porphyroblasts, 192
Steno, Nicolaus, 201
Stishovite, 188
Stocks, *119*, **120**
Storm surges, **330**, *330*
Strain, and earthquakes, 344–346, *345*
Stratification. *See* Bedding
Stratigraphic succession, **202**
    on Mars, 227, *227*
Stratigraphic traps, *383*, 384
Stratigraphy, **201**–206
    and cross-cutting relationships, 204, *205*, 206
    and fossils, 202–203
    principles of, *201*, 201–202
    and unconformities, 203–204, *204*, *205*
Stratosphere, 26, **253**, *253*
Stratospheric ozone depletion, **26**, *26*
Stratovolcanoes, *133*, **134**
Streak, **90**, *91*, 92*t*
Streak plates, 90, *91*
Stream valleys, *314*–315, **316**–317
Streamlines, 308, *308*
Streams, **306**–307, *307*, *314*–315
    braided, 271
    desert, 338, *339*
    disappearing, 298, *298*
    and streamflow, 313
Strength of rocks, 344, *345*
Stress, 186. *See also* Pressure
    differential, *186*
    and earthquakes, 344–346, *345*
Striations, glacial, *268*, 268–269
Strike, *173*, **173**–174
Strike-slip faults, **176**, *177*
    and fault mechanisms, 355, *355*
Strip mines. *See* Mining
Stromatolites, *240*, 241, *241*
Strontium, 209, *209*, 210*t*
Structural traps, 383, *383*
    and compressive tectonics, 182
Sub-bituminous coal, 388, *388*
Subduction, *58*–59, **61**

Subduction zones
 earthquakes in, 348
 magmatic processes at, 124–125, *125, 233*
 metamorphic processes at, 195–196, *196*
Submarine fans, 335, *335*
Subsidence, 151
Sudbury ore body, 103
Suess, Eduard, 53, 259
Sulfate ions, 87
Sulfates, 83*t*, 84, 87, *87*
Sulfide ions, 86
Sulfides, 83*t*, 84, 86–87, *87*
 as ore minerals, 100, *101*
Sulfur, and composition of Earth, 361
Sulfur dioxide, 247
 residence time of, 396
Sumatra earthquake, 182, 369, *369*
Sun. *See also* Solar energy
 formation of, 216, *217*
Superposition, principle of, **200**–201
Surf zones, 326, *327*, 331, *331*
Surface tension, 292
Surface waves, *350,* **351**
Suspended load, **310,** *310*
Suspension, *310,* 310–311
Sustainable development, **4,** 382
Swash, 327, *327*
Swash zones, 331, *331*
Swells, 326, *327*
Switchgrass, 395, *395*
Symmetrical folds, *180*
Synclines, **179,** *179*

Tabular deposits, 101
Taconic orogeny, *236,* 237
Talc, hardness of, 88, 88*t*
Tangshan earthquake, 368, 376
Tar sands, 10, **390**
Tectonic provinces, 228, *228,* 233, *233*
Tectonics, 52. *See also* Plate tectonics
 and deformation, 176, 181–184
 and streamflow, 313
Tektites, 188
Temperature. *See also* Global warming
 atmospheric, 252*t*, 261
 and deformation, 175
 and density, 91–92
 and depth, 116–117, 185, 359–360, *360*
 and diagenesis, 160
 and magma formation, 115
 and metamorphism, 185–186, 193, *193, 194*
 and P-T paths, 195–196
 and polymorphs, 82–83
Tensional forces, **176**
 and fault mechanisms, 355, *355*
Tensional tectonics, *181,* 182, *182*
Terraces
 tide, 331, *331*
 wave-cut, **333,** *333*
Terrestrial planets, 217, *218*
Tertiary period, *211*
Tethys Ocean, 69, *70, 71,* 235
 and oil reserves, 384
Tetrahedra, 80, 84, *85*
Textures, **94,** *94*
 of igneous rocks, *109,* 109–111, 131, *131*
 of metamorphic rocks, 188–192, *189, 190, 192,* 192*t*
Thames River, 320
Tharp, Marie, 54, *55*
Theories, **32**
Theory of the Earth [Hutton], 200
Thera volcano, 136
Thermal subsidence basins, *151,* **152**

Thermoelectric energy, 287
Thermohaline circulation, *254,* **255**
 and climate change, 408
 and ice ages, 267
Thermometers, 264
Thermoremanent magnetization, 63, *64*
Thin sections. *See* Photomicrographs
Thompson, Lonnie, 251, *251,* 404
Thoreau, Henry David, 333
Three Gorges Dam, 392, *393*
Three Mile Island accident, 391
Thrust faults, **177,** *177*–178
 and compressive tectonics, 182–183
 and continental collisions, 235, *235*
 on Mercury, 224, *224*
Tiber River, 320
Tibetan Plateau, *59,* 62
 and carbon cycle, 398
 permafrost in, *258*
Tidal flats, *153,* 154, 328, *329,* 331
Tide terraces, 331, *331*
Tides, **328,** *328*–329
 and landforms, 307
Tigris River, 320
Till, *270,* **271,** 273
Tillites, **276,** *277*
Tilt angle, *275,* 275
Time. *See* Ages; Geologic time scale; Isotopic dating; Stratigraphy
Tomography, seismic, 361, *362,* **363**
Topaz, hardness of, 88, 88*t*
Topography, **37,** *38*
 and buoyancy of continental crust, 40, *40*
 and climate, 259
 karst, 298, *298*
 of terrestrial planets, 225, *225, 226*
Topset beds, 323, *323*
Trace elements, and mineral color, 91, *91*
Trade winds, 263, 310
Transform-fault plate boundaries, **59,** *59, 62,* 62–63
 earthquakes along, 363, *364,* 365
 and history of plate movements, 69
 and shearing tectonics, 183–184
Transition zone, 41
Transportation of sediments, *145,* 146, *148,* 148–150, *310,* 310–313
 by streams, 313, *314*–*315,* **316**–324
Travertine, 172, 301, *301*
Tree rings, climate history from, 265
Triassic period, *206, 208, 211*
Trilobites, 203, *203,* 243
Tropical depressions, 330
Troposphere, **253,** *253*
Tsunamis, **19,** 368, *368*
 and compressive tectonics, 182
 and convergent plate boundaries, 61
 destructiveness of, 343, 344, *369,* 369–370
 and megathrusts, 365
 and volcanic eruptions, 130, 136, 138
 warning systems for, 373–374
Tuffs, **132**
Tunguska bolide, 22, *22*
Turbidites, 157, 335, *335*
Turbidity currents, 154, 156, **335,** *335*
Turbulent flow, **308,** *308*
Twentieth-century warming, *264,* **264**–265, *400,* **401,** 407
Typhoons, 17. *See also* Hurricanes

U-shaped valleys, *269,* 270
Ultra-high-pressure metamorphism, 188

Ultramafic rocks, *113,* **114**
 where found, 359
Unconfined aquifers, 293
Unconformities, **203**–204, *204*
Uniformitarianism, principle of, **35,** 200
Universal tree of life, 240, *240,* 244
Unsaturated zone, **292,** *292*
Upper mantle, 41
Ural Mountains, 69
 orogeny of, *236*
Uraninite, 391
Uranium
 and isotopic dating, 209, 210–211, 210*t,* 212
 radioactivity of, 209
 reserves of, 390–391
 resources of, 382, *382,* 390
 in sedimentary rocks, 145
 as trace element, 91
Uranus, 218, *218*
Urbanization
 on floodplains, 320–321, *321*
 and natural hazards, 22–23

V-shaped valleys, *269,* 270, 316, *318*
Vadose zone, 292, *292*
Valles caldera, 135
Valley and Ridge province, 183, *230,* 231
Valley glaciers, *255,* **256,** *256*
Valleys
 hanging, *269,* 270
 rift, 54, 60, *60, 61,* 151, 182, *182*
 stream, *314*–*315,* **316**–317
 U-shaped, *269,* 270
 V-shaped, *269,* 270, 316, *318*
Vallis Marineris, *226*
Variscan orogeny, *236,* 237
Varves, *271,* 272
Veins, *101,* **101**–102, **122,** *122*
Venus, 217, *218,* 221, 221*t,* 224–226, *225*
 topography of, *225*
Vertical transport of mantle material, 233–234
Vesicles, 131, *131*
Vine, Frederick J., 64
Viscosity, **114,** 308–309
Vitreous luster, 90, 90*t*
Volatiles
 and formation of oceans and atmosphere, 220–221, *221*
 in solar system, 217, 218
Volcanic ash, *111,* **132,** *132*
Volcanic bombs, *111,* **132,** *132*
Volcanic domes, 133, *134*
 on Venus, 225, *225*
Volcanic eruptions, **20,** *20*
 central, 133–135, *134, 135*
 control of, 140
 fissure, *135,* 135–136, *136*
 and formation of oceans and atmosphere, 221, *221*
 prediction of, 139–140
Volcanic hazards, 132–133, 136–139, *137, 138*
 reducing risks of, 139–140
Volcanic processes, 126–128
Volcanic rocks. *See* Extrusive igneous rocks
Volcanism, 126
 and carbon cycle, 398, *398*
 and climate, 259
 on Mars, 226, *226*–227
 on Venus, 224–226, *225*
Volcanoes, *127,* **127,** *128.* See also *specific volcano*
Vostok Station, 265, *265*

Water
 distribution of, 283–284, 283*t*

 fresh, 256, 257, 285
 hydrothermal, 301–303
 and magma formation, 116
 on Mars, 226, 227
 and metamorphism, 187
 potable, 300–301
 total supply of, 284
 viscosity of, 309
 from volcanoes, 221, *221*
Water quality, *299,* 299–301
Water resources, 13–15
 and climate change, 288
Water table. *See* Groundwater table
Water use, 281–282, 282*t,* 285, 286–288, *287*
Water vapor
 atmospheric, 254
 and climate feedback, 262
 as greenhouse gas, 261, *261*
Waterfalls, *269,* 270, *314,* 317, *317*
 and hydroelectric energy, 392, *393*
Watt, James, 6, 408
Wave-cut terraces, **333,** *333*
Waves, **326,** 326–327, *327*
 currents created by, 326–328, *327*
 and landforms, 307
Weather, 44. *See also* Climate
 and climate change, 404
Weathering, **95,** *96, 145,* 145–148, **146,** *146*
 and calcium cycle, 397
 in deserts, 338, *338*
 and joints, *180,* 181
Wegener, Alfred, 53, 67
Wells, 290, 291, 293, 295. *See also* Groundwater
 artesian, 294, *294*
 deep-sea, 336
 and groundwater table, 295, *295*
Westerlies, *309,* 309–310
Wetlands, 289–290, *290*
 and formation of coal, 388
Wiechert, Emil, 39, 40
Wilson, Edward O., 409
Wilson, J. Tuzo, 55, 237
Wilson cycles, **237**–238, *238,* 239
Wind belts, prevailing, 253, *309,* 309–310
Windmills, 393, *393*
 and Earth system management, 410, 412*t*
Wisconsinan glaciation, *266,* 266–267, 273
Wrangell-St. Elias National Park, *256*
Wrangellia, *234*
Würm glaciation, 273

Xenoliths, 120, *120*

Yangtze River, 289*t,* 320, 392, *393*
Yellowstone National Park
 ash-flow deposits in, 136
 as hot spot, 126, *126*
 hydrothermal waters in, 301, *301*
 volcanoes in, 108–109, 131, 133, 135, 138
Yucca Mountain, 391, *391*
Yule, George, 171

Zeolite, 195
Zeolite facies, *194*
Zinc
 and metasomatism, 186
 ores of, 101, *101,* 102
Zion Canyon National Park
 cross-bedding in, *156*
 stratigraphy of, 207, *208*
Zircon, and isotopic dating, 210–211, 210*t,* 212, *213,* 220